Modern Well Design

Modern Well Design is a comprehensive guide that explores the complexities of well design and drilling operations. It delves into drilling fluids, geomechanical evaluation, casing design, and the latest developments in digitalisation and emerging technologies.

This is an essential read for professionals looking for practical solutions to common drilling challenges and to advance their knowledge of drilling fluids, hydraulic optimisation, and geomechanical evaluation. With a special emphasis on digitalisation, this book offers useful insights into the latest trends in Artificial Intelligence (AI) applications, real-time monitoring, and predictive modelling for efficient and safe drilling operations. This edition also seeks to update readers on emerging technologies like downhole sensors and autonomous drilling.

It is designed for petroleum engineers, drilling professionals, researchers, and students in the field of energy and petroleum engineering. Whether you are an industry veteran or a newcomer, *Modern Well Design* will equip you with the necessary knowledge and tools to excel in the dynamic world of drilling and well operations.

Modern Well Design

Third Edition

Bernt S. Aadnøy, Rasool Khosravanian, and
Mesfin Belayneh Agonafir

CRC Press
Taylor & Francis Group
Boca Raton London New York

CRC Press is an imprint of the
Taylor & Francis Group, an **informa** business

Designed cover image: a schematic illustration of offshore well drilling operations with a semi-submersible rig in a marine environment. The design includes annotated elements such as the rig, riser, BOP stack, casing, cementing, and formation—all developed in-house.

Third edition published 2026
by CRC Press
4 Park Square, Milton Park, Abingdon, Oxon, OX14 4RN

and by CRC Press
2385 NW Executive Center Drive, Suite 320, Boca Raton FL 33431

First edition published by CRC Press 1996
Second edition published by CRC Press 2011

CRC Press is an imprint of Informa UK Limited

British Library Cataloguing-in-Publication Data
A catalogue record for this book is available from the British Library

Library of Congress Cataloging-in-Publication Data
Names: Aadnøy, Bernt Sigve author | Khosravanian, Rasool author |
Agonafir, Mesfin Belayneh, 1968– author
Title: Modern well design / Bernt S. Aadnøy, Rasool Khosravanian, and Mesfin Belayneh Agonafir.
Description: Third edition. | Abingdon, Oxon ; Boca Raton, FL : CRC Press, 2026. |
First edition published by CRC Press 1996.
Second edition published by CRC Press 2011. |
Includes bibliographical references and index.
Identifiers: LCCN 2025035981 (print) | LCCN 2025035982 (ebook) |
ISBN 9781041022084 hardback | ISBN 9781041022114 paperback | ISBN 9781003618171 ebook
Subjects: LCSH: Oil well drilling
Classification: LCC TN871.2 .A13 2026 (print) | LCC TN871.2 (ebook)
LC record available at https://lccn.loc.gov/2025035981
LC ebook record available at https://lccn.loc.gov/2025035982

ISBN: 9781041022084 (hbk)
ISBN: 9781041022114 (pbk)
ISBN: 9781003618171 (ebk)

DOI: 10.1201/9781003618171

Typeset in Times
by Newgen Publishing UK

Contents

Preface

The current trend in the oil industry is to drill more difficult wells in a more cost-effective way. To be able to reach these goals, both the planning and the drilling operation need continuous improvements. Although modern computer systems give us access to more data than ever, both cost and well failure statistics show that there is a considerable potential for improvements. It is our belief that the underlying understanding of the geology and the wells are important elements to make progress. The purpose of this book is therefore to present a unified picture of the well process.

The main idea behind this book is to provide a systematic approach to improve the planning and the design of wells. To be able to improve, each new well should be designed individually, and should be based on experiences from earlier wells.

The book will treat the subject as a design process, attempting to bring forward some of the improvements seen in later years addressing the latest trends and technologies in the industry. It offers comprehensive insights into advancements in drilling fluids, well integrity, and plug and abandonment (P&A) of wells, including well barriers, planning, and P&A operations. Additionally, it explores the growing significance of digitalisation in drilling operations, the integration of cutting-edge technologies, and drilling industries systems approach to enhance the rate of penetration (ROP).

In particular, practical borehole stability analysis and methods to derive geomechanical prognosis are key subjects. Also, the book suggests ways to present the well design, for easy verification and modification. Therefore, in addition to being used as a textbook, it is also intended to be used as a guide by well designers.

Many people have provided stimulation in the writing of this book. First of all, the many students that we have trained in well design over the years but also the many friends and former colleagues in the oil industry have made significant contributions, as many of the topics of this book have been implemented in the field.

Bernt S. Aadnøy
Rasool Khosravanian
Mesfin A. Belayneh
Stavanger, August 2025

Preface

Author Biography

Bernt S. Aadnøy is Professor Emeritus at the University of Stavanger and the Norwegian University of Science and Technology (NTNU). He worked for several oil companies earlier in his career such as Phillips Petroleum, Saga Petroleum, and Statoil. He also founded Ullrigg, the full offshore size drilling research rig in Stavanger. He has published ten books in four languages and nearly 400 scientific publications. He holds a BS degree in Mechanical Engineering from the University of Wyoming, an MS degree in Control Engineering from the University of Texas, and a Doctorate (Dr.Ing) in rock mechanics from NTNU. He has also studied petroleum and geophysics at the University of Texas. Professor Aadnøy has received numerous awards for his work, one among them being the 1999 SPE International Drilling Engineering Award. He is also the only Norwegian national to be elected Honorary Member of the Society of Petroleum Engineers and the American Institute of Mining, Metallurgical, and Petroleum Engineers (AIME). He was elected the 2018 SPE Oil Professional of the Year.

Rasool Khosravanian holds a BS in Drilling from Kerman University and an MS and PhD in Industrial Engineering from the University of IUST. He served as a faculty member and assistant professor from 2011 to 2018 at AUT, accumulating 15 years of experience in academia and industry. From 2019 to 2022, Rasool held a postdoctoral position in the Department of Energy and Petroleum Engineering (IEP) at the University of Stavanger, Norway. Following this, he worked for Aker BP in the DIG D&W as a Halliburton employee, focusing on digitalisation in drilling and wells organisations. Rasool has six years of professional experience with EPD companies and has been a strategic planner for large organisations. His research interests include large-scale optimisation, data mining, and AI for technology, digitalisation, and automation. He published a book in February 2022 and has authored approximately 40 journal papers in international journals. Rasool is a member of the Society of Petroleum Engineers (SPE) and Tekna in Norway.

Mesfin Belayneh Agonafir holds a BSc in Physics/Mathematics from Addis Ababa University (AAU), Ethiopia, an MSc in Applied Geophysics from the Norwegian University of Science and Technology (NTNU), Norway, and a PhD and postdoctoral qualification in Geomechanics/drill string mechanics, with a focus on both experimental and analytical modelling, from the University of Stavanger, Norway. He has worked as a geophysicist, seismic data processor, and research engineer at NORCE (Norwegian Research Centre). Since 2010, he has served as Associate Professor at the University of Stavanger, teaching and conducting research in drilling and well engineering through 2025.

Author Biography

Symbols and Units

SYMBOLS

The general symbols used in this book are listed below. In addition, specific nomenclature and subscripts are listed in each chapter to quickly identify the variables used.

A	Area (m^2)
c	Fluid compressibility (1/bar)
d	Pressure gradient as specific gravity to water
D	Depth (m)
Diam.	Diameter (m)
E	Modulus of elasticity (N/m^2)
f	Fanning friction factor
F	Force (N)
h	Distance or height (m)
L	Length (m)
N	Rotational speed (revolutions/min)
P	Pressure (bar) or pressure gradient (s.g.)
q	Flow rate (litres/min)
t	time (hrs), or thickness (m)
T	Temperature (°C)
V	Volume (m^3)
α	Temperature expansion coefficient (1/°C)
δ	Difference between two readings of a parameter
ϕ	Wellbore azimuth (degrees)
γ	Wellbore inclination (degrees)
σ	Stress (N/m^2)
μ	Viscosity (cP)
ν	Poisson's ratio
ρ	Density (kg/m^3)
τ	Shear stress (pa)
σr	Radial stress (pa)
$\sigma\theta$	Hoop stress (pa)
σz	Axial stress (pa)
Pw	Well pressure (pa)
σv	Overburden stress (pa)
σh	Minimum horizontal stress (pa)
σH	Maximum horizontal stress (pa)

ABBREVIATIONS

AARs	After-action reviews
AI	Artificial Intelligence
APB	annular pressure buildup
API	American Petroleum Institute
ASM	Along-string measurement
ASV	Annular safety valve
BHA	Bottom hole assembly
BOP	Blowout preventer
CBL	Cement bond logging tool
CWP	Collaborative well planning
DAE	Drilling Application Engineer
DAS	Distributed acoustic sensing
DVS	Distributed vibration sensing
DDR	Daily drilling reports
DHSV	Downhole safety valve
DIFs	Drill-in fluids
DLS	Dogleg severity
DOS	Drilling optimisation specialist
DSS	Distributed strain sensing
DST	Drill stem tests
DTS	Distributed temperature sensing
DWOS	Drilling a well on a simulator
DWP	Digital well planning
ECD	Equivalent circulating density
ELOT	Extended leak-off-test (s.g.)
ERD	Extended reach drilling
FIT	Formation integrity test (s.g.)
G&G	Geology & Geophysics
HMI	Human machine interface
HP	Hydraulic horsepower at drill bit
HPHT	High-pressure high-temperature
HSE	Health, Safety, and Environment
ICVs	Interval control valves
ID	Inner diameter
ISO	International Organisation for Standardisation
KPIs	Key performance indicators
LCM	lost circulation material
LOT	Leak off pressure test (s.g.)
LSYS	Lower shear yield stress
LWD	Logging-while-drilling
ML	Machine learning
MPD	Managed pressure drilling
MPT	Mud pulse telemetry
MSL	Mean sea level
MWD	Measurement-while-drilling
NCS	Norwegian continental shelf
NCT	Normal compaction trend
NPT	Non productive time

OBM	Oil-based mud
OD	Outer diameter
OPC	Ordinary Portland cement
OWR	Oil-to-water ratio
P&A	Plug and Abandonment
PT	Pressure and temperature gauges
PV	Plastic viscosity
PVT	Pressure-Volume-Temperature
PWC	Perforate wash and cement
QA	Quality Assurance
QC	Quality Control
R&D	Research and Development
RAP	Rig Action Plan
RFT	Repeat formation tests
RKB	Drill floor depth reference point
ROP	Rate of penetration
ROT	Real-time operation
RPM	Revolutions per minutes
RSS	Rotary steerable systems
RTM	Real-time monitoring
RTWE	Real-time well engineering
s.g.	Specific gravity relative to water
SF	Sea floor depth reference
SF	Static sag factor
TOB	Torque on bit
TRL	Technology readiness levels
TVD	True vertical depth
UCS	Uniaxial compressive strength
USIT	Ultrasonic imaging tool
WBE	Well barrier elements
WBM	Water-based mud
WDT	Wired pipe telemetry
WOB	Weight on drill bit
WOC	Wait on cement
XMT	Christmas tree
YS	Yield strength

UNITS

In this book we have adapted the units commonly used in drilling operations. Because pressure often is related to the density of the drilling fluid in the well, we are usually referring to equivalent mud density or specific gravity instead of a pressure or a pressure gradient. The following equation is used for these calculations throughout this book:

$$P(\text{bar}) = 0.098 \times d(\text{s.g.}) \times D(\text{m})$$

Introduction to the Well Design Process 1

Petroleum wells have changed character in the last decades as compared to earlier times. We have had a considerable improvement in equipment and technology, but we are also facing wells that are more difficult to drill, and we are required to make the wells more cost-effective. The result of these requirements is to put more emphasis on the well design process. The wells should preferably be designed for easy implementation. The design should also provide flexibility if changes are introduced during the drilling operation. One of the key elements in any design is cost-effectiveness. This is, of course, an element that should be considered in all parts of the design process.

There are many computer programs available for well design purposes. However, the quality of designs from software depends on the knowledge of the well designer. The objective of this book is to provide basic knowledge and design examples and to approach the construction of the well as a systematic design process.

First, the objectives and the design premises have to be established before the actual design is carried out. In the numerous examples given in the book, certain assumptions are used. It is the intention that these should always be re-evaluated and changed when new conditions arise. This book also uses simple physics principles. Usually, this is adequate. However, in certain instances, more detailed studies are required. It is fully in line with the intention of the book to start the design process in a simple way, but to increase the complexity when needed.

Secondly, the book also aims to cover technologies related to digitalisation and automation, along with data transfer and big data analytics for modelling and monitoring drilling processes. Additionally, the chapter highlights the processes, technologies, and materials used in plug and abandonment operations.

The brief descriptions of the chapters are highlighted as follows:

- The **second chapter** deals with two important elements when drilling a well: mud weight and hydraulic design. The mud weight schedule is designed from a simple principle called the median line principle. By keeping the mud density close to the virgin in-situ stress, it has been demonstrated that borehole stability problems have been minimised. The main intention of the hydraulic design is to provide a sufficient flow rate to obtain good hole cleaning. The two elements covered in Chapter Two are very critical for a successful and problem-free drilling operation. The hydraulic and hole-cleaning performance of drilling fluids depends on their density and rheological properties. This chapter presents the formulation of oil- and water-based drilling fluids and the determination of rheological parameters using various models. It also covers the characterisation of viscoelasticity and sagging properties. To address formation damage, the formulation of drill-in fluids is discussed. Additionally, the chapter highlights the application of nanotechnology in drilling fluids, including their formulation and examples.
- **Chapter Three** deals primarily with borehole stability-related issues. Firstly, methods to normalise field data to the same reference level are derived. The second section defines methods to derive field stresses and fracture gradients. The basic idea is to normalise the data and obtain correlations rather than conventional modelling. A fracture model for shallow depth is also given. If a borehole caliper log is available, simple correlations are defined to obtain the critical mud

DOI: 10.1201/9781003618171-1

density required to minimise mechanical hole collapse. The drillability log is discussed as an important tool for field interpretation. Finally, a generalised fracture model is presented, which is applicable at any water depth. All these elements serve to provide a simple basis for well design and to give simple tools to analyse drilling problems.

- The design premises are treated in **Chapter Four**. Of particular interest is the connection to the geomechanical evaluation. In addition, to define minimum fracture pressure and kick margins required to drill a section of the hole, the maximum leak-off value is also used. Part of this chapter is devoted to demonstrating various methodologies to select depths for casing strings, and finally, the long-term perspective of the well is considered by defining completion and production requirements.
- In the casing design, **Chapter Five**, the design criteria and the failure mechanisms are first defined. Design of the casing test pressure is also covered, with an example of ways to test critical wells. A complete, well-designed example was presented. The last section of Chapter Five deals with three-dimensional tubular design.
- **Chapter Six** is a continuation of the previous chapter. Here, the particular considerations for critical wells are discussed. A high-pressure, high-temperature (HPHT) well is used as a design example. It is handled in three parts: the shallow casing strings, the intermediate, and the deep casing strings. Part of the chapter is devoted to establishing fracture prognosis for use in design and operation. This chapter is intended to identify many of the problems encountered in the design of HPHT wells.
- There are certain operational issues also to consider. Therefore, **Chapter Seven** is intended to define issues related to platform types, the execution of drilling operations, and aspects of well friction. The emerging issue of well integrity is also presented.
- At a late stage in production, the well no longer produces sufficiently, and it has to be plugged and abandoned (P&A) permanently in due course. Also, exploration wells are usually plugged. **Chapter Eight** is based on the NORSOK D10 standard and defines the P&A operation, including alternative materials to ensure long-term well integrity.
- **Chapter Nine** covers the application of digital technologies in the drilling process, focusing on concepts like digital twins, automation, and digitalisation. It highlights the role of big data analytics and the use of real-time data in monitoring and modelling downhole drilling operations. In addition, the chapter briefly highlights the role of the Drilling Well on Simulator (DWOS) in training drilling engineers. This simulator enables engineers to gain a deeper understanding of the drilling process by replicating real-world conditions, providing hands-on experience in managing drilling operations, and troubleshooting potential issues before they occur in the field.
- **Chapter Ten** highlights emerging new technologies in drilling operations, including downhole sensors, 4D calipers, and 4D wellbore geometry, which enhance real-time monitoring and precision. It also covers automated mud systems, wired pipe, and along-string measurement systems to improve efficiency and data collection. Additionally, the chapter explores the development of autonomous drilling, marking a shift toward fully automated, data-driven processes that enhance accuracy, safety, and operational efficiency.
- In recent years, drilling companies have managed drilling at a higher rate of penetration (ROP). In addition to drill bit designs, several factors control the ROP, such as operational and drilling fluid properties. However, **Chapter Eleven** introduces a high-level approach to improving the Rate of Penetration (ROP), rather than focusing solely on mathematical models that often highlight non-optimal or low ROP values. It begins with Time vs. Depth Analysis, followed by the introduction of the Rig Action Plan (RAP)—a structured set of operational guidelines for drilling operations.

  Attachments to the RAP may include engineering simulations, performance assessments, and lessons learned from offset well operations, among others. The chapter also presents strategies to overcome ROP limitations, such as optimizing hydraulics and enhancing hole cleaning. A Field Case example is presented to demonstrate this.

- In order to improve well planning, post-analysis of earlier failures is required. **Appendix A** shows an evaluation of six wells. This is used to establish a drilling time curve based on realistic data. Appendix A also shows a summary of borehole stability-related problems and gives a simple system for experience transfer.
- **Appendix B** discusses another aspect of these wells, namely the volumetric behaviour of the drilling fluid, which may disturb kick control and also change the effective bottom-hole pressure of the well.

1.1 WELL INTEGRITY, DESIGN CRITERIA, HSE, QUALITY, AND OPERATIONAL RISKS FOCUS FOR WELL DESIGN

High risks and potentially high rewards characterise the petroleum industry business. When a company discovers a profitable oil or gas reserve, it can recover its operational costs and generate substantial profits. In contrast, the entire investment may be lost. Due to these inherent risks, companies must focus on minimising costs, operational optimisation, safety, and environmentally responsible operations at every stage, from exploration to decommissioning. Throughout the petroleum project lifecycle phases, it's crucial to maintain technical integrity, operational quality, and safety.

Following standards and best practices is vital in petroleum operations. International standards, such as those from the American Petroleum Institute (API) and the International Organization for Standardization (ISO), provide critical specifications for well construction.

Additionally, national guidelines apply; for example, Norway's Petroleum Safety Authority (PSA) has developed NORSOK D-010 for the Norwegian Continental Shelf (NCS). This guideline covers well integrity from design to abandonment and defines the criteria for safe well construction and operation.

1.1.1 NORSOK D-010: Well Integrity

Wellbore structural integrity is crucial for the longevity of the well and for preventing leakage to the surface. Over time, wells may experience leakage due to various operational factors. The NORSOK D-010 Well integrity guidelines recommend minimising the risk of uncontrolled fluid release throughout the well's life cycle. The NORSOK D-010 guideline outlines three primary solutions to mitigate the risk level of leakage (NORSOK D10, 2013).

1. *Technical Solution*: Focuses on material selection and well design to ensure robustness and prevent failure.
2. *Operational Solution*: Involves regular pressure testing and monitoring to verify the integrity of the wellbore.
3. *Organisational Solution*: Emphasises training and competence development for personnel, along with proper documentation and knowledge transfer to ensure the well's integrity is maintained throughout its lifecycle.

When planning the permanent abandonment of a well, it's essential to consider future scenarios, including potential chemical and geological reactions and pressure developments over time. This ensures the well has a reduced risk of future leakage.

1.1.2 NORSOK D-010: Design Criteria

The NORSOK D-010 Design criteria provide detailed guidelines for ensuring the integrity and reliability of wellbore casings. These include the following requirements (NORSOK D10, 2013).

- *Material Quality*: Casings must be made from high-quality materials that withstand aggressive downhole environments, including exposure to corrosive elements like H_2S and CO_2.
- *Load Conditions*: Casings must be designed to handle all possible realistic load conditions throughout the well's life. These conditions must account for the following factors:
    - *Biaxial loading stress*
    - *Casing wear*
    - *Bending in the deviated sections of the hole*
    - *Temperature*
    - *Corrosion*
    - *Plastic deformation of formations and reservoir compaction*
    - *Pressure during completion, workover, and kill operations.*
- *Safety Factors for Tubular Design*: The NORSOK D-010 specifies the following safety factors (SF) for tubular components:
    - *Collapse SF = 1.1*
    - *Burst SF = 1.1*
    - *Axial SF = 1.25*
    - *Tri-axial SF = 1.25*

These safety factors are intended to ensure the integrity of the casing under various downhole stresses. It's important to note that API standards may specify different operational safety factors, which should be considered depending on the specific well conditions. Figures 1.1 and 1.2 show the design limit and

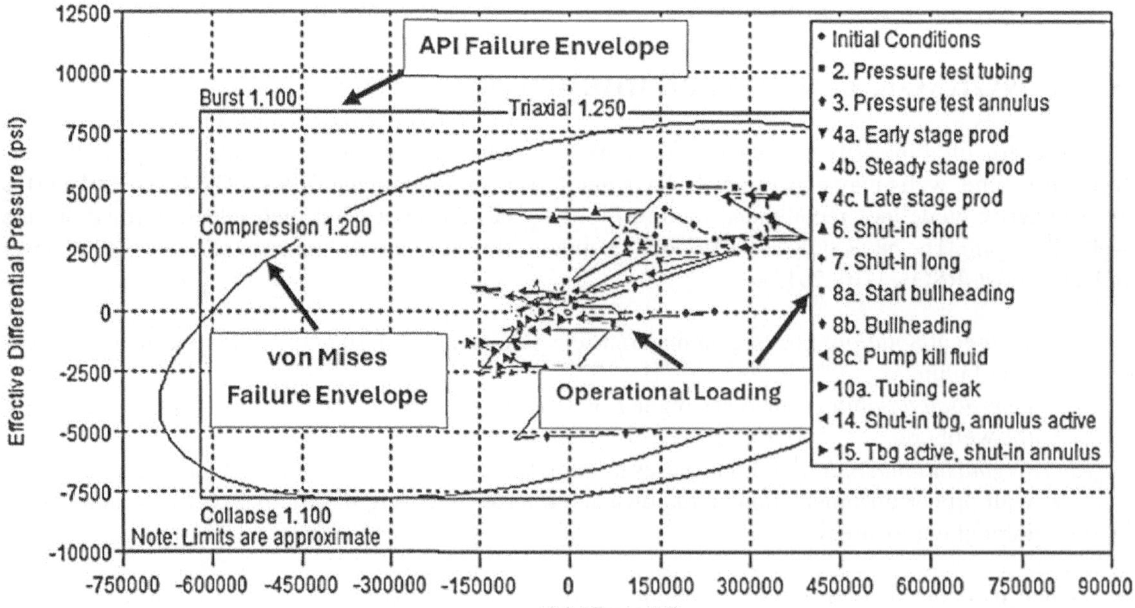

FIGURE 1.1 Illustration of design limit plot for production tube exposed to 14 loading.

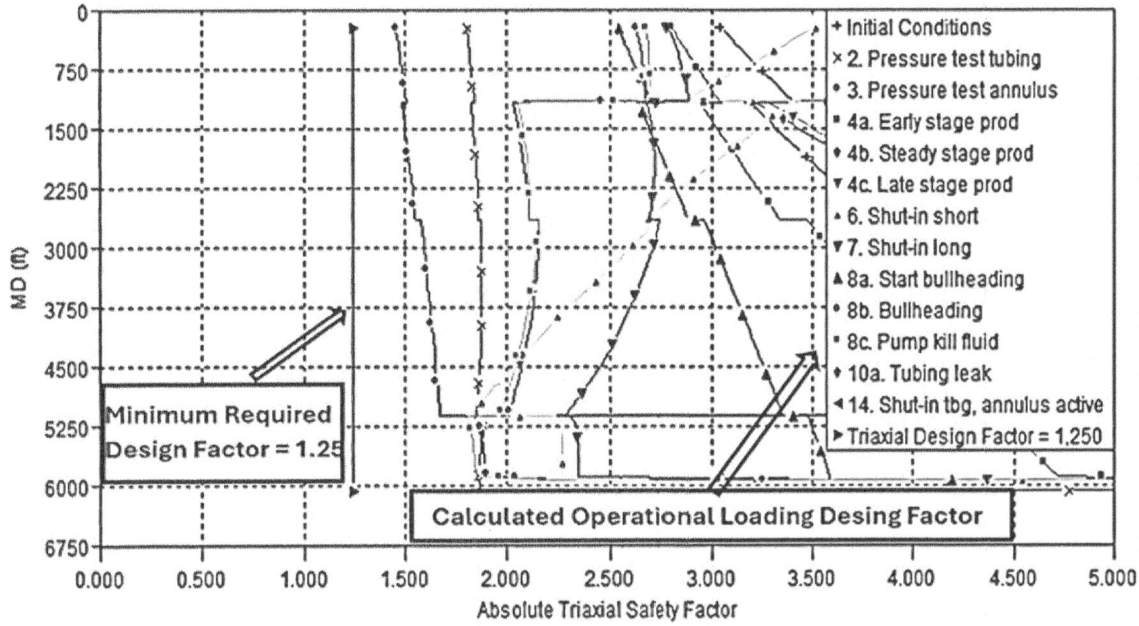

FIGURE 1.2 Illustration of triaxial safety factor for production tube exposed to 14 loading.

triaxial safety factor plots simulated for a production tube loaded under 14 different loading scenarios. Results show that all the loads are within the failure envelope, and the calculated triaxial safety factors are higher than the NORSOK D10 minimum requirement, i.e., 1.25. In this case, the tubing carries all the load and can be qualified for the operation.

1.1.3 HSE, Quality, and Operational Risks

The International Labor Organization (ILO) reports significant work-related accidents, fatalities, and diseases annually, underscoring the importance of effective safety management systems in the petroleum industry. Due to the nature of operations in remote, high-risk environments, the upstream petroleum sector is particularly vulnerable to accidents. Factors such as human error, operational flaws, and design shortcomings are key contributors to these incidents. Notable historical examples, like the Piper Alpha disaster (1988) (Paté-Cornell, 1993) and the Macondo spill (Baram, 2011), highlight the need for continuous improvement in safety practices. National safety authorities, including Norway, constantly refine regulations to mitigate these risks.

1.1.3.1 Risk Assessment and Management

Risk assessments are crucial for identifying and mitigating hazards in every well operation. The basic formula for risk is:

$$\text{Risk} = \text{Impact} \times \text{Likelihood}$$

This involves evaluating risks based on their potential consequences and the likelihood of occurrence. Once identified, appropriate measures are taken to minimise these risks. Continuous monitoring during operations and rapid remedial action are essential in reducing the chances of catastrophic failure.

1.1.3.2 HSE Focus on Drilling Operations

Safety is a central concern in drilling activities. Though specific approaches may vary between companies, the following are the typical HSE practices that help minimise risk in drilling operations:

- Pre-Job Meetings (Toolbox Talks)
    - The objective is to discuss safety issues before the task begins.
    - The meeting process is that the team leader reviews the job, outlines risks, and specifies preventive measures. Everyone is encouraged to share concerns and ask questions to ensure mutual understanding and safety.

- Quality Handover
    - NORSOK D10 emphasises the importance of personnel competency and documentation transfer in ensuring well integrity.
    - The goal is smooth transitions between shifts, with all relevant safety information and operational statuses communicated to ensure continuity.

- Safety Management Procedures
    - The objective is to prevent injuries, environmental damage, and equipment failure.
    - Safety must be a culture, so it must be a core value for everyone involved. It is not just a management responsibility; all personnel must adhere to safety protocols.

- Discharge Permits and Environmental Limits
    - All personnel *must follow* local regulations to discharge oil-based cuttings, produced water, and other waste.
    - All personnel must know *discharge permits* and *limits* to minimise environmental impacts and prevent illegal discharges.

- Chemical Use and Disposal
    - Where possible, use environmentally friendly chemicals, such as *water-based mud*, especially in sensitive areas like fishing zones.
    - Ensure chemicals used are within *discharge limits*.
    - Properly dispose of chemicals to minimise environmental impact.

- Operational Safety Requirements
  To reduce operational and environmental risks, companies must implement and adhere to the following safety measures:
    - *Preventing Spills*:
        - Ensure systems are in place to minimise and contain any spills.
    - *Preventing Falling Objects*:
        - Use *barricades, toe boards, warning signs*, and other protective measures to prevent falling objects from injuring personnel.
    - *Defining Safe Work Areas*:
        - Clearly define hazardous areas, restricting access to authorised personnel only.

- Operational Procedures and Risk Awareness
    - *Loyalty to Procedures*:
        - All personnel must know and follow established safety protocols.
    - *Understanding Risks*:
        - Clear communication of risks related to specific operations is essential for ensuring everyone is on the same page.

- *Communication*:
  - Maintain effective *radio communication* to monitor the operational status and safety in real time.

- *Training*:
  - Regular training on equipment handling and safety procedures is necessary for all involved personnel.

- *Environmental Considerations*:
  - Be aware of environmental constraints, such as weather conditions (e.g., wind or rig heave), and adjust operations accordingly.

Sticking to these main HSE and quality control practices significantly reduces risks and guarantees the safety of personnel, equipment, and the environment. Through effective risk assessments, practical safety measures, and clear communication, the petroleum industry can continue improving its safety standards, minimising human and environmental impact.

1.1.3.3 Quality Focus for Drilling Activity

Drilling operations, particularly during exploration activities, demand substantial investment. The quality of well construction directly impacts the well's longevity. Poorly constructed wells can incur significant maintenance costs and result in non-productive time (NPT), further increasing costs and delays. Therefore, implementing rigorous Quality Assurance (QA) and Quality Control (QC) measures in drilling and completion equipment is critical to minimising these undesired costs and optimising operational efficiency.

1.1.3.3.1 Example of Quality Control Importance: North Sea Well Incident

An incident in a North Sea well highlighted the critical importance of quality control in wellbore design. In this case, a 9-5/8" N80 53.5 lbs/ft casing was mistakenly joined with a 47 lbs/ft casing. The collapse rating for the 53.5 lbs/ft casing was 456 bar, while the collapse rating for the 47 lbs/ft casing was only 328 bar, representing a 28% reduction in strength.

During a later production phase, it was discovered that the production casing had collapsed at a depth of approximately 700 meters, as illustrated in Figure 1.3.

Root Cause Analysis:

- *Weak Casing Identified as the Cause*: Investigators determined that the casing failure occurred at the point where the weaker casing (47 lbs/ft) was used.
- *Lack of Documentation*: No records were found explaining why the weaker casing was incorporated into the production string, raising concerns about the adequacy of quality control and documentation processes.
- *Potential Failure Mode*: Investigators hypothesised that the failure could have been caused by a leak in the production casing or at the packer/bottom hole (PBR). This leak could have led to pressure building up behind the casing, ultimately causing the collapse.

From the case study, we understand that there was a lack of operational QA during the casing design and installation phase, where no checks were conducted to verify the casing's strength and placement. The failure resulted in replacing casing and tubing, increasing *operational costs,* and *NPT*.

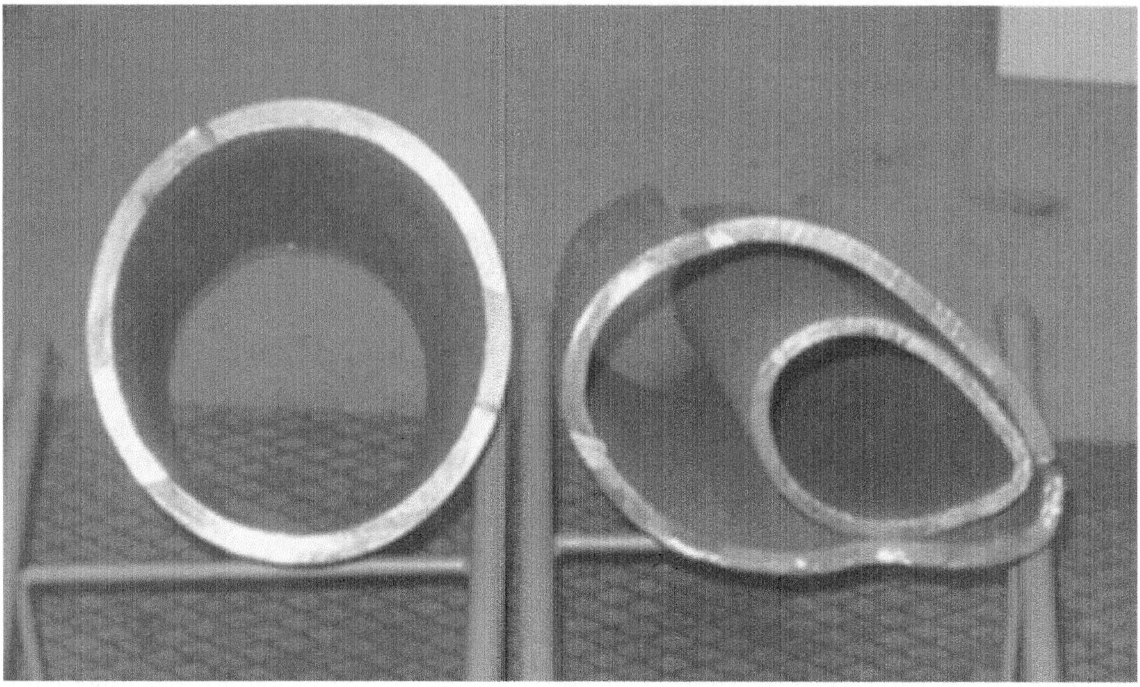

FIGURE 1.3 Production casing/tubing collapse in NCS.

Source: Torbergsen et al. (2012).

This example demonstrates the necessity of *QA/QC procedures* for every aspect of well construction, from design to equipment installation and beyond.

1.1.3.3.2 Main Aspects of Quality Focus in Drilling Operations

- Pre-Deployment Equipment Inspection
  - *Logistics Costs*:
    - The petroleum industry spends significant costs associated with logistics, including transporting drilling equipment offshore. If the wrong equipment is sent, it leads to delays and NPT, further driving up costs.
  - *QA/QC Measures*:
    - All equipment must undergo rigorous checks before being shipped offshore. Technical drawings and specifications should be reviewed to ensure compatibility and compliance with operational requirements.

- Quality Control During Rig Operations
  - *Equipment Checks*:
    - Before running any equipment into the wellbore, thorough inspections and quality checks must be conducted to prevent *structural failures* like the one illustrated in the case study above (Figure 1.3).
  - *Operational Readiness*:
    - Ensure all equipment meets the required specifications and operational standards before deployment. This includes confirming that *pressure ratings, strength tolerances, and compatibility with formation and environmental conditions are met.*

- Effective Communication
  - *Offshore and Onshore Coordination*:
    - It is vital to maintain open, efficient communication between *offshore teams, onshore personnel*, and the *logistics base*. This allows for smooth equipment mobilisation and demobilisation.
  - *Real-time Updates*:
    - This communication should include status updates on equipment readiness, potential issues, and timeline adjustments, ensuring no delays in operations.

- Documentation and Knowledge Sharing
  - *Experience Reviews*:
    - It's important to document operational experiences during drilling activities to avoid repeating mistakes and to enhance future performance.
  - *After-Action Reviews (AAR)*:
    - AARs provide an opportunity to reflect on the operation by addressing:
    - *What was planned?*
    - *What happened?*
    - *What lessons were learned?*
    - *What can be improved for next time?*

By systematically reviewing past operations and applying the lessons learned, future drilling activities can be executed with higher efficiency and reduced risk.

1.2 SUMMARY

Drilling operations' success depends heavily on equipment quality, design, and execution. Implementing a robust QA/QC system, conducting comprehensive pre-deployment inspections, maintaining effective communication, and regularly performing After-Action Reviews (AARs) are essential to ensuring the construction's long-term success. By focusing on quality at every phase of drilling, operators can avoid costly failures, reduce non-productive time, and maximise the value of each well.

Drilling Design 2

2.1 DRILLING FLUID

2.1.1 Introduction

Drilling fluids are typically formulated by mixing weight materials, filtrate loss control agents, pH stabilisers, viscosity modifiers, and corrosion inhibitors in water or oil. Understanding the properties of these chemical additives is crucial for achieving the desired performance of the drilling fluids. Additionally, a deep understanding of the geological formation and the chemical additives' performance *mechanisms* enables the design of custom-tailored drilling fluids that ensure safe and efficient operations under specific drilling conditions.

The Main Functions of Drilling Fluid:
Drilling fluids play a crucial role in the success of rotary drilling operations. Drilling fluids serve several critical functions, including:

- Maintaining well pressure.
- Cooling and lubricating the drill string and bit.
- Suspending solids and preventing cuttings or weight material from settling when circulation stops.
- Transporting cuttings from the bottom to the surface.
- Controlling filtrate loss by forming a filter cake across permeable formations.

To achieve these functions, drilling fluids must be carefully formulated to meet specific performance, providing sufficient density, heat transfer, lubrication, filtrate loss control, and viscosity.

The Primary Considerations in Drilling Fluid Selection (Caenn et al., 2011; Bleler, 1990):

When selecting a drilling fluid, three main factors need to be evaluated:

- *Performance*: The ability of the fluid to meet operational requirements.
- *Environmental impact*: Ensuring the fluid's environmental impact is minimised.
- *Cost*: Balancing performance and environmental factors with budget.

An appropriately designed drilling fluid is essential for efficient and safe drilling operations. In contrast, poorly designed drilling fluids can lead to various problems, such as:

- Well pressure fluctuations.
- Increased torque and drag.
- Drill bit wear.
- Stuck drill strings.

DOI: 10.1201/9781003618171-2

This chapter covers:

- Conventional drilling fluid formulation, characterisation, and performance examples.
- Drill in drilling fluid.
- Emerging advanced nano-based drilling fluids.

2.1.2 Drilling Fluids Characterisation Methods

Rheology is the study of fluid flow and deformation. The rheological properties of drilling fluids and cement slurries are critical in determining well pressure, fluid flow, hole-cleaning efficiency, and cement placement. Therefore, accurately characterizing these rheological parameters is essential for formulating effective drilling fluids and ensuring the success of the well construction process.

The right rheological modelling is essential for estimating the rheological parameters of drilling fluids and cement slurries. Optimizing the rheological parameters contributes to areas such as:

- *Hydraulics and Flow Control*:
  - Predict pressure drops and flow rates and ensure cutting transport efficiency, optimal pump rates, and wellbore cleaning.
- *Suspension and Hole Cleaning*:
  - Ability to suspend and transport solid particles (such as drilled cuttings or cement).
- *Cement Placement*:
  - Ensure smooth cement flow into the annular space, preventing issues like channelling or poor bonding.

Fluid rheology's shear stress-shear rate is categorised into Newtonian and non-Newtonian fluids.

2.1.2.1 Newtonian Fluids

Fluids that do not contain particles and polymers, such as water and oil. According to Newtonian fluid, shear stress is proportional to shear rate, with *constant viscosity*.

$$\tau = \mu\dot{\gamma} \tag{2.1}$$

where μ is a constant viscosity.

2.1.2.2 Non-Newtonian Fluids

Non-Newtonian rheological models are essential for describing the behaviour of most drilling and cement fluids, as these fluids exhibit variable viscosity that changes with shear rate. Unlike Newtonian fluids, their flow behaviour is non-linear. Several non-Newtonian models are used to characterise viscosity and flow behaviour. Table 2.1 summarises widely used models, including Bingham Plastic, Power Law, Herschel-Bulkley (H-B), Robertson and Stiff, Casson, and Collins Graves, among others.

Example 1: Compute the Rheological Parameters

The fluid's rheological parameters are determined using VG meter viscosity data and the non-Newtonian rheological models outlined in Table 2.2. For model evaluation, the relative mean absolute error (ARME) is calculated as the percentage difference between predicted and observed values, using the formula:

TABLE 2.1 Rheological models

MODELS	*EQUATION*		*PARAMETERS*	*REFERENCE*
Bingham Plastic	$\tau = YS + PV\dot{\gamma}$	(2.2)	$PV[cP] = \theta_{600} - \theta_{300} YS\left[\frac{lbf}{100sqft}\right] = 2\theta_{300} - \theta_{600}$	(API RP 13D, 2009)
Power law	$\tau = k\dot{\gamma}^n$	(2.3)	$n = 3.32\log\left(\frac{\theta_{600}}{\theta_{300}}\right)$ $k\left[\frac{lbf-s}{100sqft}\right] = \frac{\theta_{300}}{511^n} = \frac{\theta_{600}}{1022^n}$ where θ_{600} and θ_{300} are the shear stress reading values at 600 and 300 RPM	(API RP 13D, 2009)
Herschel-Bulkley	$\tau = \tau_y + K\dot{\gamma}^n$	(2.4)	$\tau_y = (2\theta_3 - \theta_6) * 1.067$ (Power and Zamora, 2001) where θ_3 and θ_6 are the shear stress reading values at 3 and 6 RPM. n and k are determined from curve fitting	(Zamora et al., 2002); API RP 13D, 2009)
Robertson-Stiff	$\tau = A(\dot{\gamma} + C)^B$	(2.5)	$C = \frac{\gamma_{min} \times \gamma_{max} \times \gamma^{*2}}{2\gamma^* - \gamma_{min} - \gamma_{max}}$ where γ* is determined by interpolation after the τ* value is estimated from the maximum and minimum shear stress. $\tau^* = \sqrt{\tau_{min} \times \tau_{max}}$ - A and B parameters will be determined by curve fitting - The yield stress is estimated by setting the shear rate to zero and is given by AC^B.	(Robertson et al., 1976)
Casson	$\tau^{0.5} = \tau_c^{0.5} + \mu_c^{0.5}\dot{\gamma}^{0.5}$	(2.6)	Where τ_c and μ_c are Casson Yield stress and viscosity, respectively. these values are determined by curve fitting.	(Casson 1959)
Collins-Graves	$\tau = (A + B\dot{\gamma})(1 - e^{-c\dot{\gamma}})$	(2.7)	A, B, and C Determined by curve fitting.	(Collins et al., 1978)

$$ARME = \left[\frac{1}{N}\sum_{i=1}^{N}\left|\frac{\tau_{measured} - \tau_{calculated}}{\tau_{measured}}\right|\right] * 100 \tag{2.8}$$

The laboratory VG meter measures the viscosity responses as dial reading and RPM. Before modelling, the measured dataset will be converted to the shear rate and the shear stress as:

$$\dot{\gamma} = 1.703 \times \text{RPM}$$
$$\tau = 1.067 \text{ Reading}$$

where, the shear rate $\dot{\gamma}$ is in 1/s and the unit of the shear stress is in $lbf/100ft^2$

Table 2.3 presents the computed rheological parameters and the corresponding relative mean absolute error (ARME) values. The model predictions alongside the measured data are shown in Figures 2.1–2.6, displaying a visual representation of the model performance. The traditional Bingham

TABLE 2.2 Measured VG meter and its conversion to shear rate and shear stress

RPM	*DIAL READING*	*CONVERT*	*SHEAR RATE, 1/S*	*SHEAR STRESS, LBF/100FT²*
3	7.50		5.11	8.00
6	8.50		10.22	9.07
30	14.5		51.09	15.47
60	21.0		102.18	22.41
100	29.0		170.30	30.94
200	45.0		340.60	48.02
300	59.0		510.90	62.95
600	99.0		1021.80	105.63

TABLE 2.3 Rheological parameters and AME values

MODEL	*PARAMETERS*	*ARME%*
Bingham Plastic	PV=40 cP, YS =19 lbf/100ft²	43.34
Power law	k =0.56184 lbf-s/100ft², n =0.74628	24.17
Collins-Graves	A = 14.042 lbf/100ft², B = 0.0915 lbf-s/100ft², C = 0.01134 1/s	9.24
Herschel-Bulkley	τ_y = 6.936 lbf/100ft², k= 0.2889 lbf-s/100ft², n= 0.85	2.74
Robertson & Stiff	A = 0.6454 lbf/100ft², B = 0.731, C = 26.427 1/s^B	0.84
Casson	τ_c = 4.879 lbf/100ft², m_c = 0,064 lbf-s/100ft²	1.70

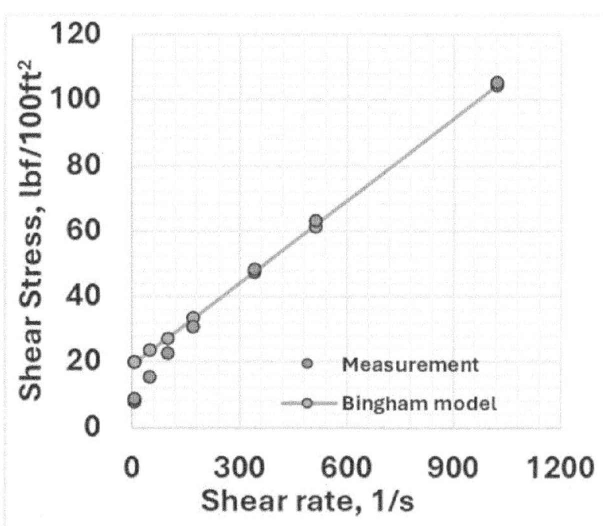

FIGURE 2.1 Measured data vs Bingham Plastic model.

Plastic and Power Law models deviate from the measurements at lower shear rates, resulting in higher ARME values. In contrast, the Collin-Graves, Herschel-Bulkley, Robertson and Stiff, and Casson models demonstrate minimal ARME values, effectively capturing the entire VG meter dataset. It is important to note that the accurately predicted rheological parameters-based hydraulic model calculation is not necessarily always valid for all fluids. It is always essential to compare the hydraulic model predictions with the measurements.

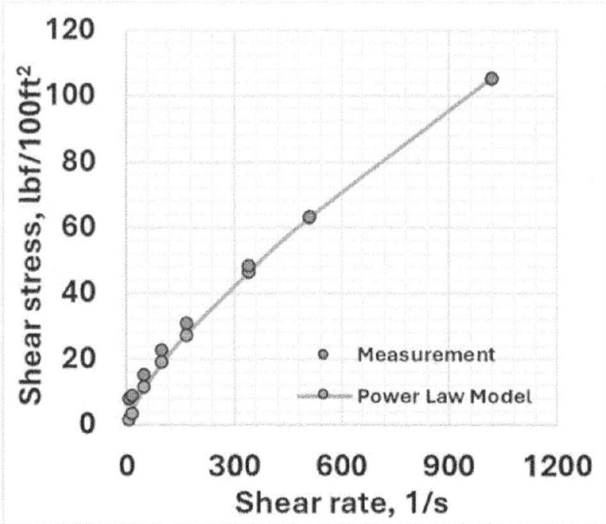

FIGURE 2.2 Measured data vs Power Law model.

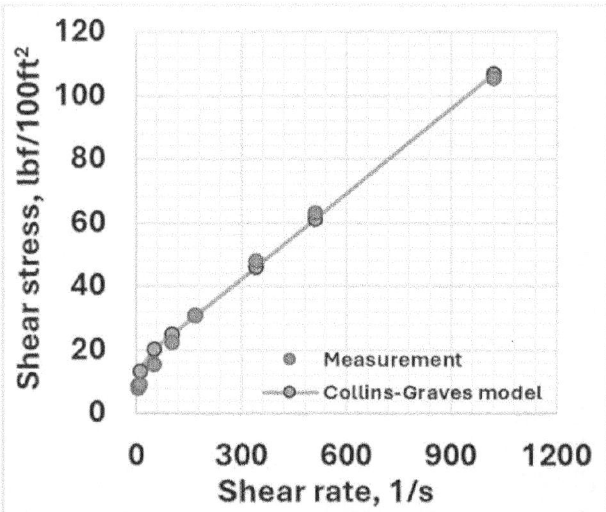

FIGURE 2.3 Measured data vs Collin-Graves model.

2.1.2.3 Barite Sag Evaluation

2.1.2.3.1 Static Sag Measurement

The static sag factor quantifies a drilling fluid's density stability under static conditions. During the standard static sag evaluation, fluid samples are aged at 50°C for 16 hours while positioned vertically. After aging, the sag factor is determined by measuring the densities of samples taken from the top and bottom of the fluid column. Free oil must first be removed from the top sample to ensure accurate readings, as it does not contain solid particles and could skew the results. The densities of the samples are then estimated based on their measured mass and volume. The sag factor is calculated using Equation (2.9):

$$SF = \frac{\rho_{bottom}}{\rho_{top} + \rho_{bottom}} \tag{2.9}$$

Where SF is the static sag factor and ρ_{top} and ρ_{bottom} are the densities of the drilling fluids at the top and the bottom, respectively.

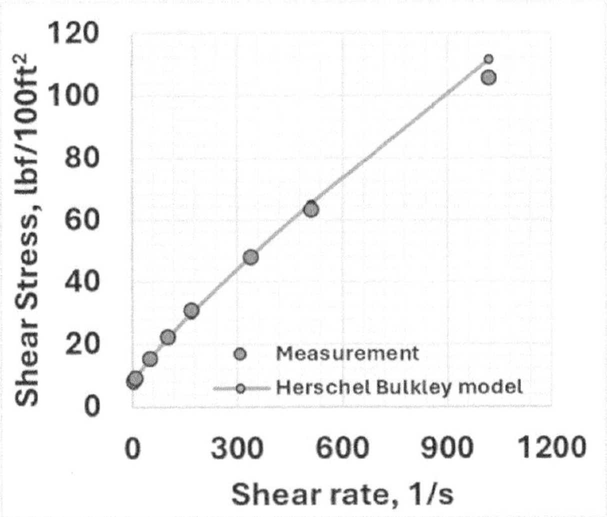

FIGURE 2.4 Measured data vs Herschel Bulkley model.

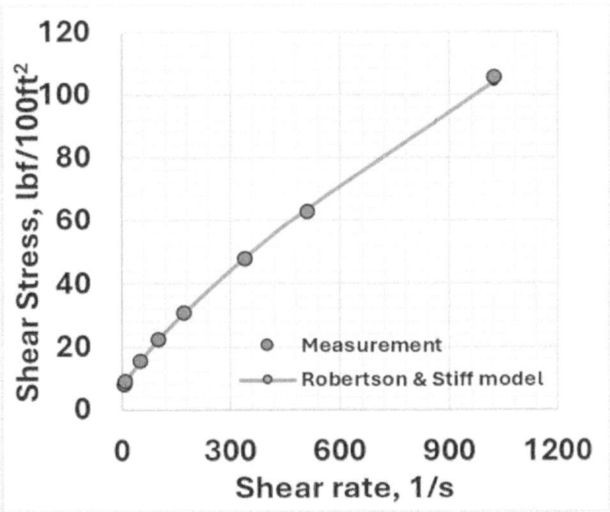

FIGURE 2.5 Measured data vs Robertson & Stiff model.

According to Maxey (2007), if the sag factor exceeds 0.53, the internal gel structure of the drilling fluid is too weak to maintain solid particles in suspension. This leads to a higher potential for barite sagging, causing more barite material to precipitate out of the fluid.

2.1.2.3.2 Dynamic Sag Measurement

The dynamic sag index is measured using a sagging shoe and VG meter. The procedure begins by filling the VG viscometer heating cup with drilling fluid, which is then heated and maintained at 50°C. The bob is

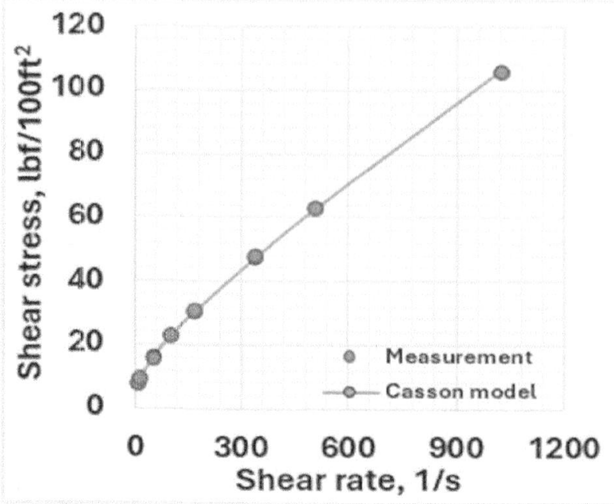

FIGURE 2.6 Measured data vs Casson model.

rotated to shear the fluid at 100 RPM for 30 minutes. After this period, 20 ml of sagged fluid is taken from the bottom of the sagging shoe. This procedure is based on one of the drilling fluid companies.

The sag factor is calculated by comparing the initial mud weight with the sagged mud weight, as shown in the equation below:

$$Sag\ factor = \frac{Total\ mud\ weight\ after\ 40\text{min}\,(gm)}{2{*}Initial\ mud\ weight,\ gm} \tag{2.10}$$

A sag factor of 0.5 indicates a homogeneous fluid system with no significant change in density. A sag factor greater than 0.5 suggests a change in density, indicating the potential for solids to settle out, resulting in a heterogeneous fluid. The change in density (or mass) is calculated using the following formula:

$$\Delta\rho_{mud} = \frac{m_{final} - m_{initial}}{Volume} \tag{2.11}$$

where m_{final} is the mass taken after testing, m_{initial} is the mass taken before testing, and volume is the sample volume.

This calculation helps quantify the degree of density variation and provides insight into the fluid's stability under static conditions.

2.1.2.4 Viscoelasticity

The viscoelasticity of a fluid describes the fluid as possessing both viscous and elastic behaviours. The viscoelastic fluid responds to stress by flowing (viscous behaviour) and deforming elastically (returning to its original shape).

Drilling fluids with viscoelastic properties provide these two behaviours, allowing them to adapt to changing conditions. For example, during drilling, the drilling fluid must flow to remove drill cuttings (viscous behaviour), and it must also maintain enough elasticity, reflecting the gel strength of drilling fluids, to suspend those particles when circulation is stopped.

The elastic and viscous components of drilling fluids can be quantified through oscillatory tests *such as amplitude sweep and frequency sweep*, which measure parameters like storage modulus (G') and loss modulus (G").

1. *Storage Modulus (G')*:
   - Represents the **elastic** component fluids, i.e., the fluids' ability to store energy and return to their original shape after deformation.
   - **G'** is associated with the solid-like behaviour of the fluid.

2. *Loss Modulus (G")*:
   - Represents the **viscous** component of the material, i.e., the energy lost as heat during deformation.
   - **G"** is associated with the liquid-like behaviour of the fluid.

2.1.2.4.1 Oscillatory Tests for Drilling Fluids

A rheometer is used to measure the viscoelastic properties of the drilling fluids. The sample is placed between parallel plates and a dynamic oscillatory loading is applied. Figure 2.7 illustrates the loading and the deformation. The bottom of the plate is fixed while the top cycles with the shear forces (±F), the deflection path (s), and the deflection angle φ.

The oscillatory shear stress and shear rate are described mathematically as (Bui et al., 2012):

$$\gamma(t) = \gamma_o \sin(\omega t) \tag{2.12}$$

$$\tau(t) = \tau_o \sin(\omega t + \delta) \tag{2.13}$$

Expanding and collecting terms, we get:

$$\tau(t) = \tau_o \left[\sin(\omega t)\cos\delta + \cos(\omega t)\sin(\delta)\right] \tag{2.14}$$

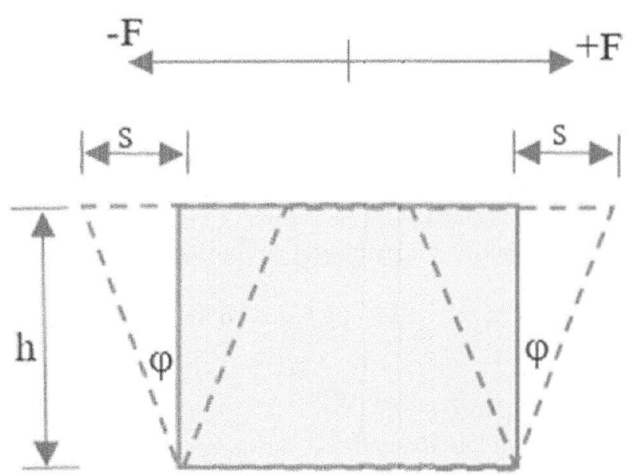

FIGURE 2.7 Oscillatory tests: Shear forces, deflection path, deflection angle, and shear gap (h).

TABLE 2.4 Viscoelastic behaviour with changing angles

IDEALLY, VISCOUS FLOW BEHAVIOUR	*THE BEHAVIOUR OF VISCOELASTIC LIQUID*	*50/50 EQUAL VISCOUS AND ELASTIC PORTIONS*	*THE BEHAVIOUR OF VISCOELASTIC GEL OR SOLID*	*IDEALLY ELASTIC BEHAVIOUR*
δ = 90° tan δ → ∞ G' → 0	90° > δ > 45° tan δ > 1 G" > G'	δ = 45° tan δ = 1 G' = G"	45° > δ > 0° tan δ < 1 G' > G"	δ = 0° tan δ → 0 G" → 0

Source: Mezger (2011).

$$\tau(t)=\gamma_o\left[\left(\frac{\tau_o}{\gamma_o}\cos\delta\right)\sin(\omega t)+\left(\frac{\tau_o}{\gamma_o}\sin\delta\right)\cos(\omega t)\right] \tag{2.15}$$

$$\tau(t)=\left[G'\sin(\omega t)+G''\cos(\omega t)\right] \tag{2.16}$$

From equations 2.15 and 2.16, the storage (G') and loss (G") moduli can be given as:

$$G'=\frac{\tau_o}{\gamma_o}\cos\delta \tag{2.17}$$

$$G''=\frac{\tau_o}{\gamma_o}\sin\delta \tag{2.18}$$

$$\tan\delta=\frac{G''}{G'} \tag{2.19}$$

The ratio G"/ G' is the damping factor (or loss factor).

Figure 2.8 shows the phase shift angle or loss angle (δ):

- The phase angle is the time lag between the applied stress and the resulting strain.
- It indicates the elastic (solid-like) and viscous (liquid-like) behaviours.
    - **δ = 0°**: Purely elastic (solid-like behaviour).
    - **δ = 90°**: Purely viscous (liquid-like behaviour).
    - **δ = 45°**: Equal contribution of elastic and viscous components.

Table 2.4 shows the behaviour of viscoelasticity based on the phase angle and the elastic and viscous moduli.

2.1.2.4.2 Amplitude Sweep Test and Results:

In the amplitude sweep test, the oscillation amplitude varies, keeping frequency (e.g., 10 rad/s) and temperature constant. Figure 2.8 shows typical results. This test helps identify the Linear Viscoelastic Region (LVER), where both the storage modulus (G') and loss modulus (G'') remain constant. The gel characteristic is in the LVE range (G'> G"). When the amplitude exceeds the LVER, the fluid transitions to

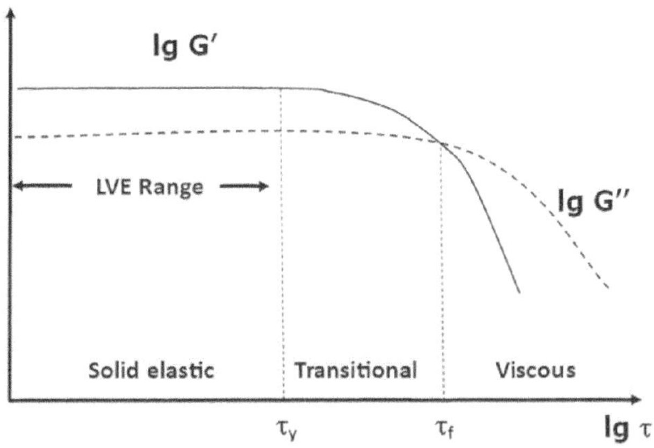

FIGURE 2.8 Stress amplitude sweep of a fluid sample showing the yield stress, τ_y, and flow point at G′ = G″ and the existence of a gel structure in the LVE range (G′ > G″).

non-linear behaviour, causing G' and G'' to deviate. The fluid possesses equal viscous and elastic behaviour at the flow point. Beyond this point, the fluid becomes more viscous than elastic. At the flow point, the phase angle reaches 45°. The yield stress is typically estimated from when the G' begins to deviate, with the corresponding shear yield stress values (τ_y). Figure 2.8 illustrates this.

2.1.3 Water-Based Drilling Fluids

Compared with oil-based drilling fluid (OBM), water-based drilling fluid (WBM) is more environmentally friendly, easier to dispose of, and more cost-effective. Unless formulated with shale-inhibitive additives, WBM can interact with shale, leading to swelling. WBMs are typically formulated with various chemicals to achieve desired properties, such as weight, viscosity, filter loss, pH control, and corrosion inhibition. Table 2.5 shows the typical chemical additives and their functions.

Example 2: Water-Based Drilling Fluid-Well Strengthening Performance Assessment
Instead of formulating laboratory-based drilling fluids, we evaluated three industry-formulated fluids in the North Sea. The exact chemical ingredients and fluid properties are not disclosed.
Drilling Fluids Descriptions:

1. *Drilling Fluid A*: Formulated with standard drilling-grade barite.
2. *Drilling Fluid B*: Uses engineered weight particles approximately 10 times smaller than traditional drilling-grade barite, designed to prevent barite sagging and maintain fluid stability.
3. *Drilling Fluid C*: Incorporates a specialised bubble system.

2.1.3.1 Evaluation Methodology
The mechanical strength of the drilling fluids was assessed by evaluating their ability to form a barrier on a test core plug (1 cm well size, 10 cm diameter) covered with a PVC pipe. The evaluation involved hydraulic fracturing, specifically through fracture initiation and reopening pressure tests, to determine the fluid's effectiveness in strengthening the wellbore.

During the test, the core plug was exposed to an external confining pressure of 4 MPa, while well pressure was gradually increased until fracturing occurred. Before the test, the drilling fluids were circulated through the well to establish a mud cake barrier.

TABLE 2.5 Typical water-based drilling fluid chemical additives

ADDITIVES	*FUNCTION*
Water	Used to mix all the chemical additives and make a solution. • Water as part of the drilling fluid provides functions such as cooling, lubricating, maintaining pressure, cutting transport, etc.
Bentonite	Used to provide drilling fluid properties such as • Increase viscosity (Yield stress PV) and provide gel strength. - *(Control well pressure and cutting suspension/transport)* • to form a filter cake - *Control formation damage and fluid loss and provide well strength.* • to provide thixotropic properties - *(suspend cuttings during static condition)*
Soda ash ((Na_2CO_3), Lime (CaO)	Used to adjust the proper pH value for the drilling fluid's optimum performance, such as: • *Control of Viscosity and Rheology* • *Prevent Clay Flocculation and maintain Clay dispersion.* • *Control of Filtration Properties / Filter cake formation* • *Prevent Corrosion of the tubular* • *Controls shale swelling or disintegration*
Polymers (PAC, CMC, Xanthan Gum)	Used to control the viscosity of the drilling fluid. • *Viscosity* • *Reduced filtrate*
Starch	Used to control the drilling fluid. • *Viscosity* • *Reduced filtrate*
Salt (KCl, NaCl, $CaCl_2$, Mg Cl_2, Cs-formate, K-Formate)	Used to control. • *Minimise sensitive shale hydration/swelling.* • *Increase the density,* • *Adjust the rheological properties,* • *Stabilise the wellbore in salt-sensitive formations.*
Deflocculants (Thinner/Dispersants)	Used to control. • *Flocculation (It de-floccules disperse clay in suspension)*
Corrosion inhibitors	Used to control. • *Corrosion*
Biocides	Used to control. • *Bacteria that degrade the Drilling Fluid additives*
Defoamers	Used to • *To prevent or break down the foam formation*

Figure 2.9 shows the core plug after testing. Two fracture re-opening tests were conducted to evaluate the sealing performance of the drilling fluids:

- The fluids were circulated and then stopped.
- Then, wait for 10 and 60 minutes to allow the fracture to close under the applied confining pressure.

The results from the fracture initiation and re-opening tests (after 10 and 60 minutes) are summarised in Table 2.6. Since the core specimens were made with identical concentrations and ingredients, their tensile strength is assumed to be consistent. However, due to natural inhomogeneity in the samples, fracture responses varied between specimens. Multiple repeat tests were conducted for each drilling fluid, and the values shown in the table are the averages.

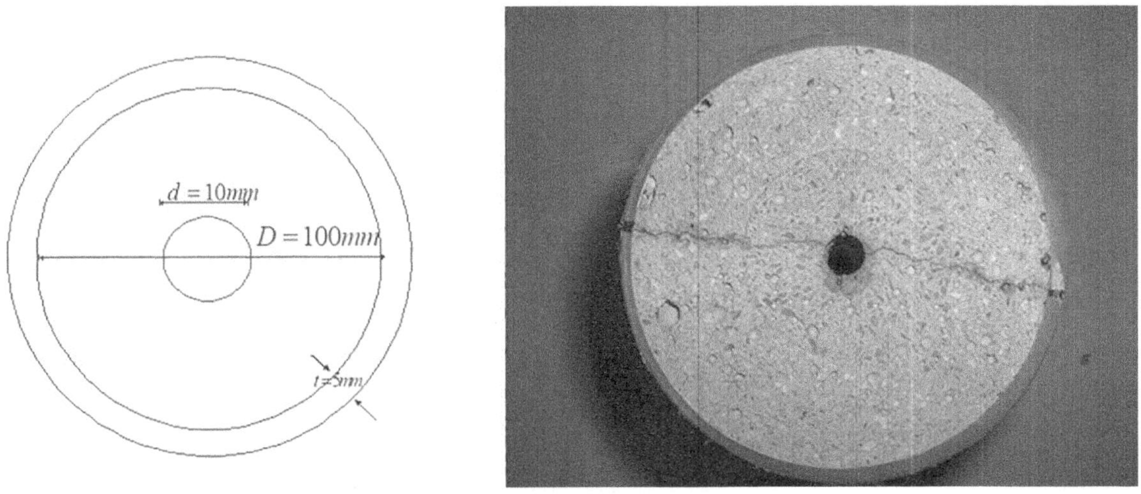

FIGURE 2.9 Concrete core plug geometry and the core plug specimen after fracturing.

TABLE 2.6 Fracturing and reopening test results using commercial drilling fluids

DRILLING FLUID	*FRACTURING [MPA]*	*REOPENING AFTER 10 MIN [MPA]*	*REOPENING AFTER 60 MIN [MPA]*
Fluid A	53.35	12.5	12.0
Fluid B	45.35	11.2	10.2
Fluid C	18.20	5.2	5.2

The results from conventional drilling fluids indicate that the fluid's mechanical strength directly impacts wellbore strength. The fluid barrier's effectiveness depends on the mud cake formation, which is influenced by its mechanical properties, internal packing, porosity, and permeability. The observed low reopening pressure after 10- and 60-minute testing suggests a need to enhance the sealing capability of drilling fluids in fractured formations.

2.1.4 Oil-Based Drilling Fluids

Compared to WBMs, OBMs offer superior performance in HPHT wells. The main advantages include:

- No shale swelling issues.
- Enhanced lubricity, which minimises torque, drag, and wear.
- Reduced risk of tubular corrosion.

However, OBMs are expensive and environmentally unfriendly. In addition, disposal of OBM requires careful management, and recycling is costly. Table 2.7 outlines the chemical ingredients used in a typical OBM and their respective functions.

Example 3: Characterisation of Typical Oil-Based Drilling Fluids – Issues and Solutions
Table 2.8 shows the typical chemical composition of an emulsion OBM. OBMs are often classified by their Oil-Water Ratio (OWR), with common ratios including 60/40, 70/30, 80/20, and 90/10. All OBMs contain the same viscosity control additive (Organoclay). Fluids with a lower OWR absorb more water, leading to higher viscosity. As the OWR increases, viscosity and gel strength tend to decrease. To characterise the properties of these four OBM formulations, the chemical additives listed in Table 2.8 were mixed to achieve a density of 1.75 s.g.

TABLE 2.7 Typical oil-based drilling fluid chemical additives

CHEMICAL INGREDIENT	*FUNCTION*
Based Oil (mineral oil, diesel, or synthetic oil)	- Base oil is the continuous phase in the drilling fluid. The solution provides several functions, such as lubricity, shale inhibition, and corrosion control.
Brine water (Fresh Water + $CaCl_2$ Salt)	- Brine water is formulated by mixing water with Salt
Emulsifiers	- Emulsify the base oil and brine (#1 and 2) - Control the viscosity. - Control the electrical stability
Filtrate control	- Control filtrate loss
Weight material (Barite),	- Control density
Lime	- Control pH - Control the electrical stability with an emulsifier
Organoclays	- Organic compound surface-modified clay minerals used to control viscosity

TABLE 2.8 Drilling fluid chemical additives

CHEMICAL	*60/40*	*70/30*	*80/20*	*90/10*
Base oil [g]	331	378	443	498
Emulsifier [g]	25	25	24	24
pH control [g]	25	25	25	25
Viscosity control [g]	7	7	7	7
Filtrate loss control [g]	10	10	10	10
Water [ml]	275	206	136	68
Salt [g]	85	63	42	21
Weight material [g]	991	1027	1061	1095

Source: Belayneh et al. (2016).

Drilling fluids were characterised using the Fann VG meter and the dynamic sagging shoe test.

- *Sagging Behaviour:*

Dynamic sag was measured at 50°C with a 100 RPM rotation for 30 minutes, following the procedure outlined in Section 2.1.2.3. The computed sag factor is shown in Figure 2.10. According to Maxey (2007), the sag potential limit is 0.53. Therefore, all OBMs except the 60/40 OWR exceed this sag limit. The severity of sagging increases as the OWR decreases.

- *Fann Viscometer (VG-meter) Measured Data*:

Figures 2.11–2.14 show the viscometer readings at room temperature, 50°C, and 80°C. The VG-meter data indicate that the viscosity of the drilling fluids decreases as the OWR increases. Similarly, gel strength, measured at 3 RPM, decreases with higher OWR values. These properties directly correlate with the sagging phenomenon. Comparing the sag factor data, it is evident that drilling fluids with a stronger internal gel structure (e.g., 60/40 OWR) exhibit a lower tendency for sagging. As gel strength decreases (e.g., 70/30, 80/20, and 90/10 OWR), the sag factor increases, indicating a higher risk of sagging.

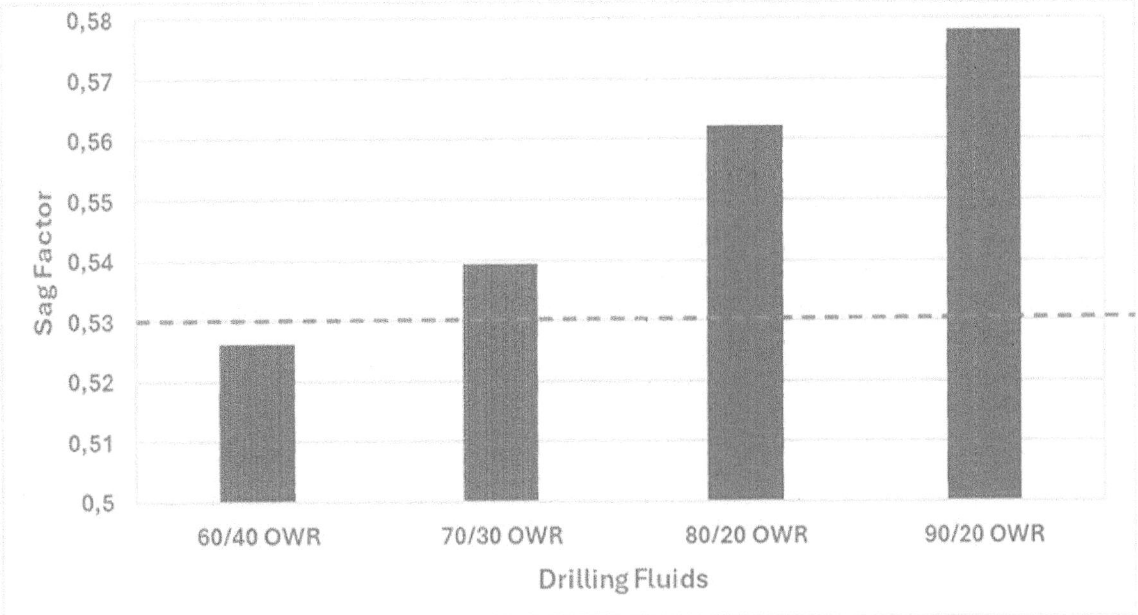

FIGURE 2.10 Dynamic sagging of the drilling fluids.

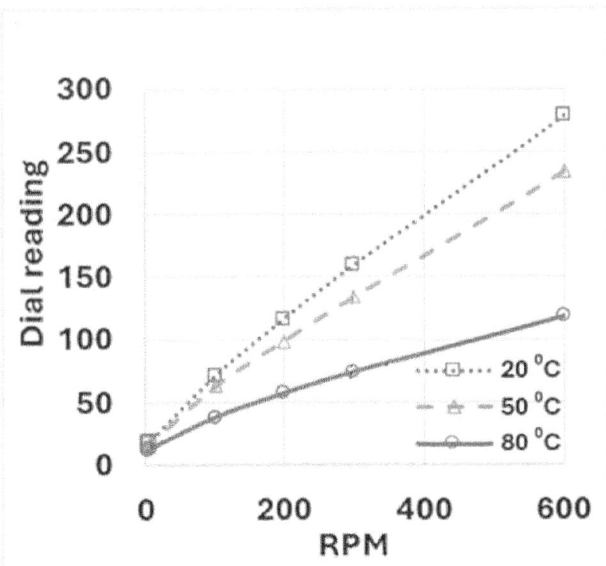

FIGURE 2.11 Fann VG-meter data of 60/40 OBM.

Lower Shear Yield Stress (LSYS) and Sagging Potential

Field studies indicate that drilling fluids with insufficient Lower Shear Yield Stress (LSYS) are more prone to sagging. To mitigate barite sagging, it is recommended to treat drilling fluids with appropriate chemicals to achieve an LSYS of 7–15 lbf/100ft^2 (Scott et al., 2004).

At 50°C, the LSYS values of the drilling fluids were compared with their sagging factor. Figure 2.15 shows that fluids with an LSYS above the recommended lower limit (7 lbf/100ft^2) showed a sagging factor below the sagging limit. In contrast, fluids with an LSYS below this threshold exhibited sagging factors above the limit. However, additional testing is needed to confirm the repeatability of these results and validate the LSYS range recommended value.

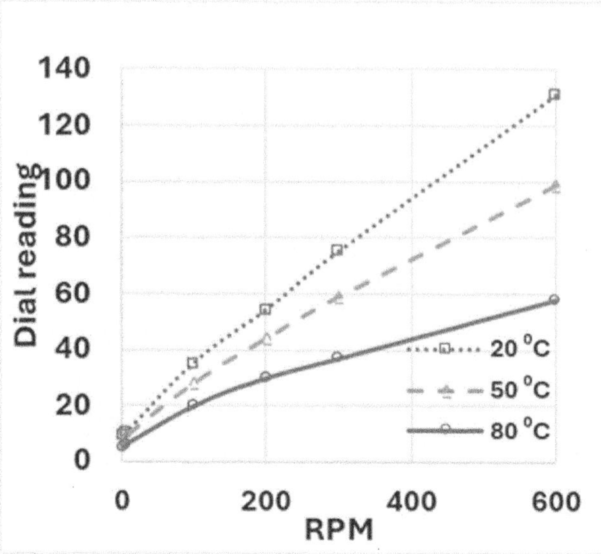

FIGURE 2.12 Fann VG-meter data of 70/30 OBM.

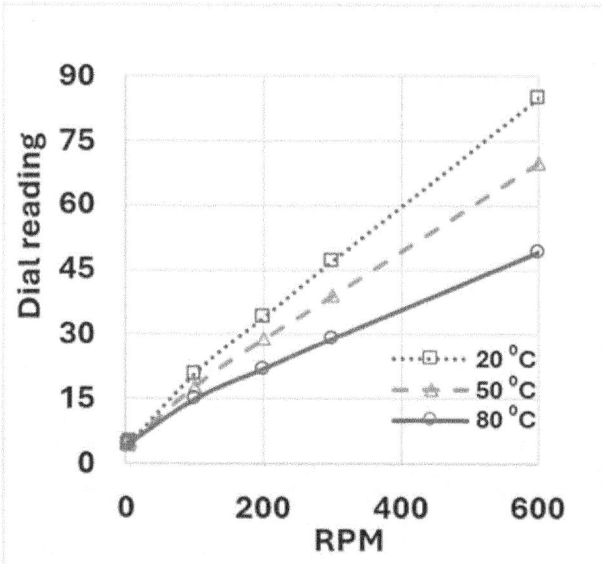

FIGURE 2.13 Fann VG-meter data of 80/20 OBM.

- *Viscoelasticity Behaviour*:

Figure 2.16 shows the amplitude sweeps test results of the four oil-based drilling fluids at 20°C and ω = 10 rad/s, revealing important insights into their viscoelastic properties. Except for the 90:10 oil-water ratio (OWR), all fluids show a flow point where the loss modulus (G'') and storage modulus (G') intersect, indicating a transition from solid-like to liquid-like behaviour. The modulus values increase as the oil-water ratio decreases, suggesting that higher water content leads to a stronger internal gel structure. Figure 2.17 also displays the shear stress against the phase angle for better visualisation. The phase angle of the 90:10 OWR is higher than 45°, indicating that it behaves more like a viscous fluid with no elastic/solid character. In contrast, the other fluids exhibit more balanced elastic and viscous properties. Compared with the barite sagging, the viscoelasticity value correlates with the barite sagging test results in Figure 2.10.

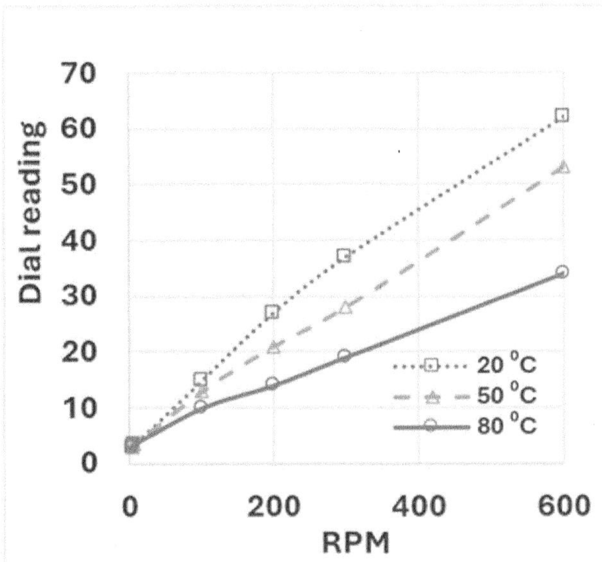

FIGURE 2.14 Fann VG-meter data of 90/10 OBM.

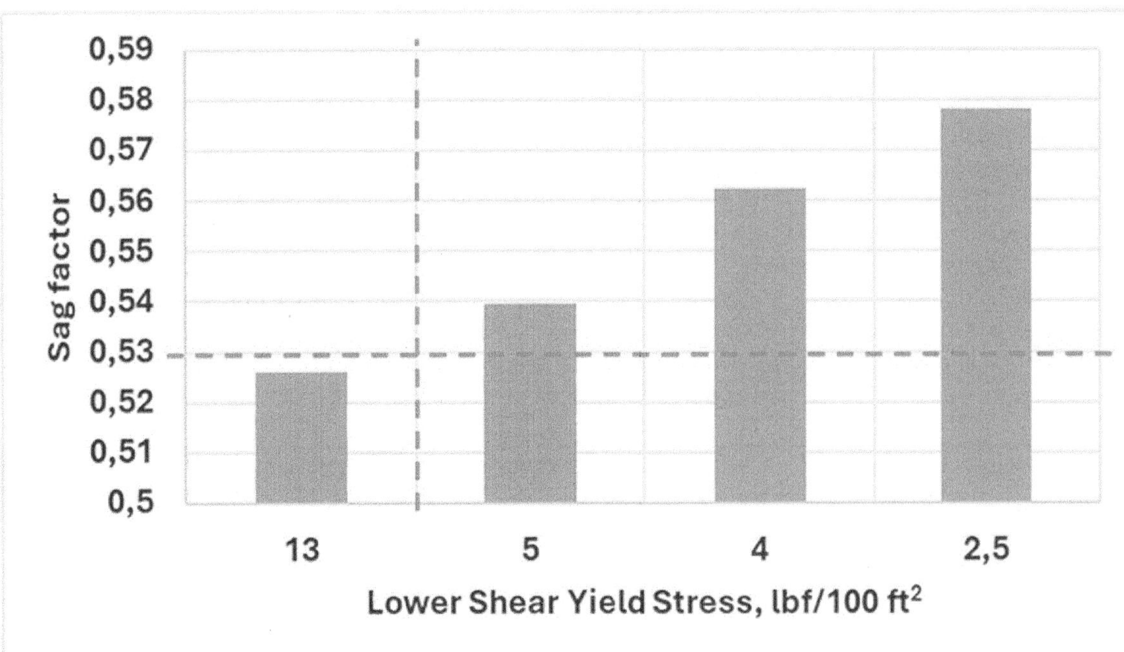

FIGURE 2.15 Lower shear yield stress vs barite sag factor.

Example 4: Lubricity of OBM vs WBM

Accurate torque and drag predictions are essential for designing the drill string and achieving the target depth. Most drilling software models assume a constant friction coefficient between the drill string and wellbore/casing, typically back-calculated from hook load measurements in cased and open-hole sections (Kårstad and Aadnøy, 2009). However, this approach can sometimes yield unrealistic friction values if the model does not adequately capture the dynamics of torque and drag.

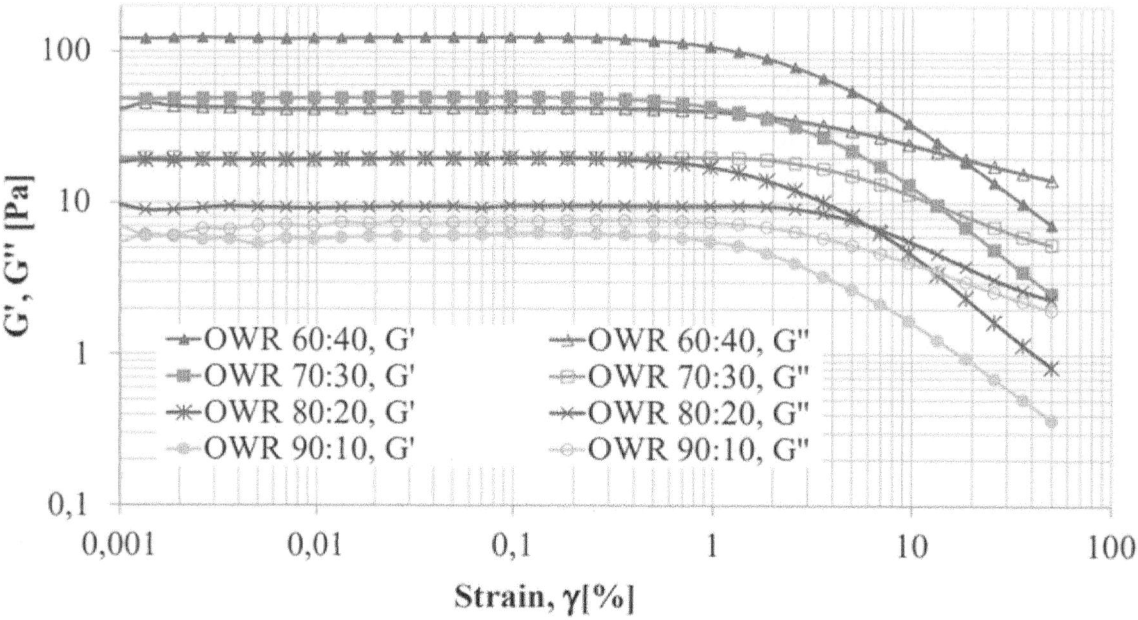

FIGURE 2.16 Amplitude sweeps of four OBMs.

Higher friction increases torque and drag, potentially causing wear on the tubing/casing and reducing the ability to reach the target depth. It can also compromise tubular integrity. For example, in the Gullfaks A-42 well, a 35% casing wall thickness loss was observed due to mechanical interactions with tool joints. Similarly, 47% of the production tubing wall thickness was reduced in the bend section in the North Sea well.

Tribology, which focuses on friction, wear, and lubrication between surfaces, plays a crucial role in drilling. The lubricating properties of drilling fluids reduce friction between the drill string and the wellbore, minimizing wear. Effective lubricants form a protective film that reduces friction and wear, providing long-term stability in terms of thermal resistance and mechanical degradation. As shown in the measured data trend (Figure 2.18), a linear regression line can be assumed for friction as a function of temperature as $\mu = a + bT$, resulting in the following models for the different drilling fluids. Please note that this model is valid for drilling fluids and testing conditions. Results show the need to improve the lubricity of WBM. This can be done by mixing the fluid with friction-reducing lubricants and nanoparticles.

FLUID	*COEFFICIENT OF FRICTION MODEL*
Water	μ= 0.4831 + 0.0015×T
WBM (with air bubble)	μ = 0.3359 + 0,0016×T
WBM (without air bubble)	μ = 0.2634 + 0.0009×T
OBM	μ = 0.1276 + 0.0002×T

2.1.5 Drill-In Fluids

Conventional drilling fluid systems (WBM and OBM) typically use barite as a weight control additive. However, during drilling in the reservoir section, fluids can be flushed with solids and chemicals, potentially impairing formation permeability, which is known as the skin effect, which reduces well productivity.

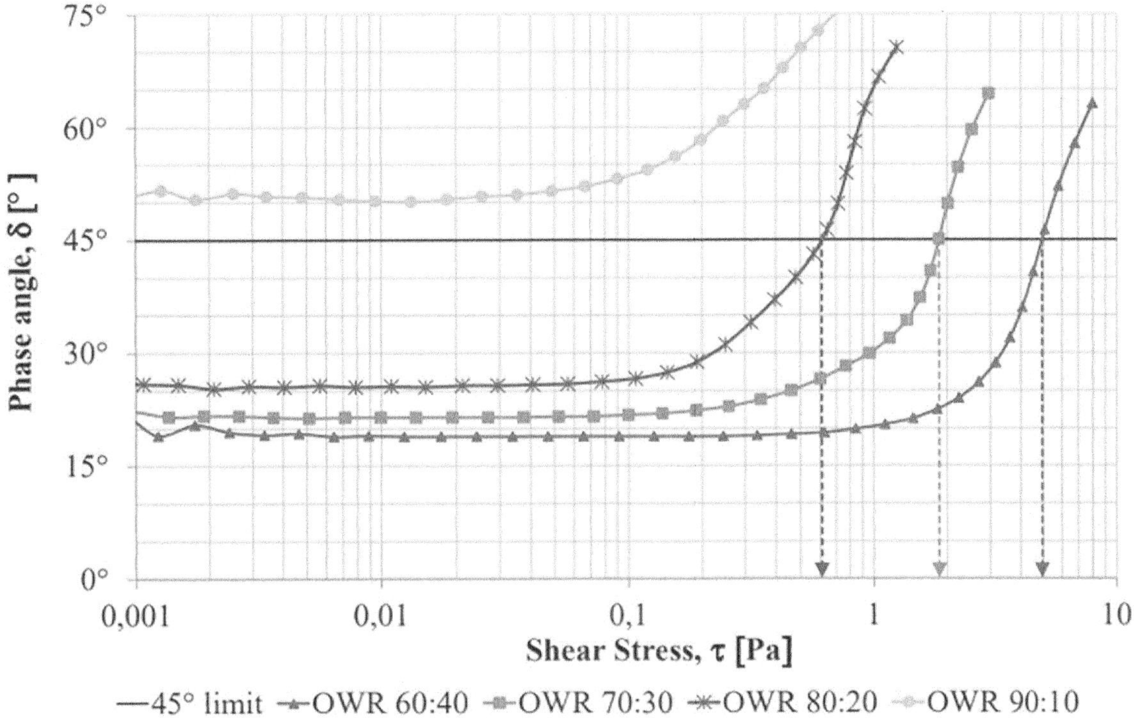

FIGURE 2.17 Shear stress vs phase angle.

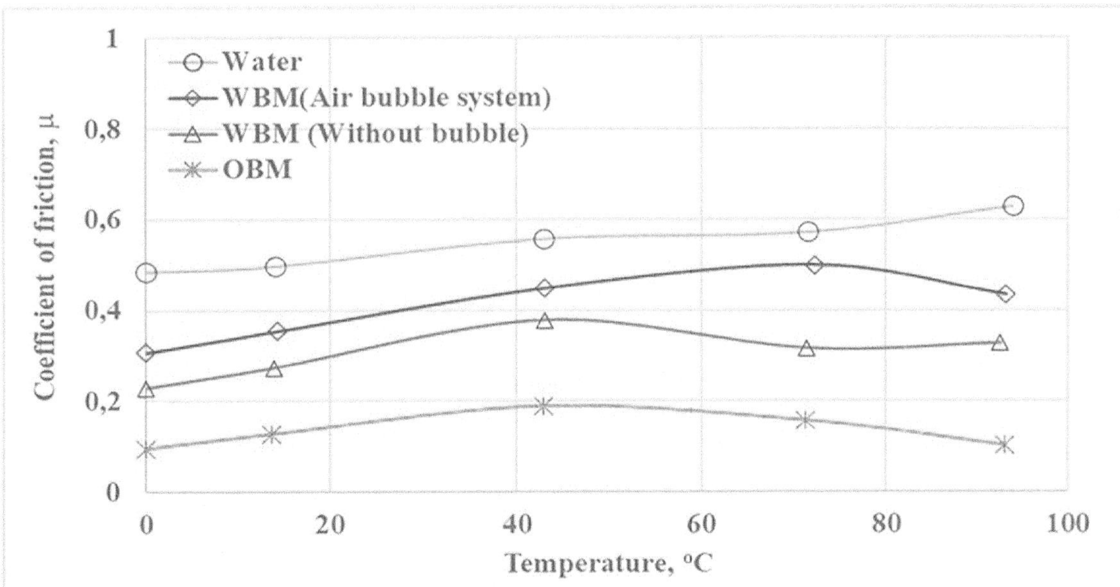

FIGURE 2.18 Temperature effect on the lubricity of drilling fluids.

Source: Data digitised from Kårstad and Aadnøy (2009).

TABLE 2.9 Cesium formate fluid formulation

CHEMICAL ADDITIVES	*QUANTITY (KG/M³)*	*FUNCTIONS*
Cesium formate	1617	Density control
Potassium formate	336	
Purified Xanthan Gum	1.57	Viscosity &
High-temperature Modified Starch	15.7	Fluid loss control
pH buffer: $K_2CO_3/KHCO_3$	11.4	pH Stability control
Calcium Carbonate medium (D_{50} = 24 mm)	50	Fluid loss & Filter cake
Calcium Carbonate fine (D_{50} =11 mm)	50	

Source: Jøntvedt (2018).

To minimise formation damage and improve productivity, specially designed drill-in fluids (DIFs) are used, particularly in horizontal reservoir sections.

As shown in Table 2.9, the drill-in fluids, often based on brines like cesium formate (2.2–2.4 s.g.) or potassium formate (1.57 s.g.), replace barite for weight control. They contain Xanthan gum for viscosity, cutting suspension and transport, and calcium carbonate ($CaCO_3$) as an acid-soluble lost circulation material. Unlike barite, $CaCO_3$ can dissolve and help restore formation permeability (Jøntvedt, 2018).

Performance Comparison: Cesium Formate vs Non-Aqueous-Based Mud (OBM)
Two reservoir sections were drilled in the HPHT Martin Linge wells: one with non-aqueous-based mud (OBM) using barite and another with cesium/potassium formate-based fluid.

Results showed that (Jøntvedt, 2018):

- OBM with barite caused a stuck-pipe incident, requiring a sidetrack.
- Cesium/potassium formate-based fluid (2.07 s.g.) exhibited:
  - Improved wellbore stability.
  - No stuck-pipe incidents.
  - Effective hole cleaning.

Despite the benefits, including reduced formation damage and improved wellbore stability, the cesium/potassium formate fluid resulted in a 25% increase in drilling torque, indicating higher friction. To address this, lubricity must be improved. Lab testing is crucial to select the right type and concentration of lubricant, balancing friction reduction with formation permeability preservation.

2.1.6 Advanced Nanoparticle-Based Drilling Fluids

2.1.6.1 Introduction

Nanotechnology (particle size 1–100 nm) is gaining attention for its impressive results in drilling fluids, cementing, and enhanced oil recovery. Nanoparticles, with their higher surface area-to-volume ratio, enable better packing in drilling fluids and cement, improving overall performance and efficiency.

This section presents a literature review based on commercial and in-house nanoparticle synthesis and their performance in WBMs and OBMs.

The literature review research results have shown that incorporating nanoparticles into *conventional oil-based and water-based* drilling fluids can significantly enhance their performance across a range of properties, such as:

In Water-Based Drilling Fluids (KCl and Bentonite based) (Sharma et al. 2012; Abdo et al. 2013; Sadeghalvaad et al. 2015; Taha et al. 2015; Vryzas et al. 2015; Aftab et al. 2016; Aftab et al. 2017; Al-saba et al. 2018; Dehghani et al. 2019; Ahmed et al. 2020; Dong et al. 2022; Dehghani et al. 2023; Murtaza et al. 2023; Ibrahim et al. 2024):

- Increase/decrease rheological properties.
- Decrease filtrate loss.
- Decrease filter cake thickness.
- Decrease the permeability of shale by plugging the pore spaces.
- Increase the fluid's lubricity.
- Increase electrical and thermal conductivities.
- Reduce shale swelling.
- Reduced cake permeability.
- Increase thermal stability of the fluid.

In Oil-Based Drilling Fluids (Zakaria et al. 2012; Nwaoji et al. 2013; Alvi et al. 2021):

- Increase fluid lubricity.
- Decrease filtrate loss.
- Improve wellbore strengthening.
- Reduced barite sagging factor.
- Decrease mud cake thickness.

Nanoparticles' *size*, *shape*, *concentration*, and *surface chemistry* are fundamental in controlling the performance of drilling fluids. Tailoring nanoparticle characteristics to the specific fluid type and drilling conditions enables more effective drilling operations and greater adaptability to complex, high-pressure, high-temperature environments. With ongoing research into nanoparticle synthesis, testing, and environmental impact evaluation, nanotechnology is set to play a crucial role in the future of drilling fluid optimisation and cementing systems.

2.1.6.2 Nanoparticles Synthesis and Effect on Drilling Fluids

This section highlights the in-house synthesised iron oxide nanoparticles (from the University of Calgary and the University of Stavanger) and the experimental test results obtained from commercially available nanoparticles.

Example 1: University of Calgary's Fe(OH)$_3$ Nanoparticle Synthesis and Application
Nassar and Husein (2007) synthesised iron oxide nanoparticles ($Fe(OH)_3$) using the AOT microemulsion (w/o) method. (Nassar et al., 2007):
Materials:

- Aerosol-OT (AOT): Surfactant
- Isooctane: Oil phase
- Iron chloride ($FeCl_2$): Precursor salt
- Sodium hydroxide (NaOH): Precipitating agent

Synthesis:
Two synthesis schemes were used. Using Scheme I,

- A microemulsion system was prepared by mixing AOT, water, and isooctane.
- Iron chloride ($FeCl_2$) was then added as the precursor.

- After thorough mixing, a reducing agent was introduced to separate the nanoparticles from the solution.
- The system was stirred at 300 RPM and 25°C to produce $Fe(OH)_3$ nanoparticles, suspended with some settling at the bottom.

Example 2: Impact of Iron Oxide Nanoparticles in 90:10 OBM (Zakaria et al., 2012)
The research group evaluated the impact of the synthesised iron oxide nanoparticles in a 90/10 oil-based drilling fluid (OBM). The nanoparticles were blended both in-situ and ex-situ, and the study also explored the effect of micro-sized lost circulation material (LCM). The focus was on filtrate loss, measured using the low-temperature, low-pressure API filter press, and on the characterisation of the filter cake. Results after 30 minutes of testing showed the following:

- LCM-blended system: Reduced filtrate loss by 9%.
- Ex-situ blended system: Reduced filtrate loss by 68%.
- In-situ blended nanoparticles: Reduced filtrate loss by 77% and decreased mud cake thickness by 22.6%.

The nanoparticle-based OBM exhibited a thinner filter cake compared to conventional LCM-based fluids. The reduced filtrate loss was attributed to the nanoparticles' ability to enhance the packing of the mud cake by blocking small pores and interacting with clay particles. The main findings are:

- Particle size and in-situ blending were critical for maintaining particle stability and facilitating effective chemical interactions with the drilling fluid components.
- Proper nanoparticle concentration is essential for achieving the desired fluid properties.

These results highlight the potential of nanoparticles in improving filtrate control and mud cake quality.

Example 3: Impact of Iron Oxide Nanoparticles on Wellbore Strengthening (Nwaoji et al., 2013)
To assess the performance of nanoparticle-based drilling fluids in enhancing wellbore stability, the University of Calgary research team conducted hydraulic fracturing tests on sandstone and concrete core specimens. This study blended iron (III) hydroxide nanoparticles with graphite-based lost circulation materials (LCM) in water-based and oil-based mud systems.

The main findings showed that:

- *Iron (III) hydroxide nanoparticle + graphite blend*: Increased fracture pressure by 70% (1,668 psi) compared to the standard water-based system.
- *Calcium carbonate nanoparticles + graphite blend*: Increased fracture pressure by 36% over the unblended invert emulsion mud.
- The nano-blended mud system demonstrated superior fracture healing, as a successful reopening test indicated.

Examples 2 and 3 underscore the significant potential of nanoparticles in improving drilling fluid properties. The nanoparticle's ability to reduce filtrate loss, form a compact mud cake, and strengthen the wellbore highlights its importance in drilling operations. However, achieving optimal performance relies on selecting the right *type*, *concentration*, and *mixing method* to tailor the nano-based drilling fluid to the specific requirements of the wellbore environment.

Example 4: University of Stavanger Nano-Based Drilling Fluids Test Results
This section presents the synthesis of iron oxide (Fe_2O_4) nanoparticles and evaluates performance in a 90:10 OWR OBM.

Nanoparticle synthesis:
Using the methods outlined in Bandyopadhyay (2014), iron oxide nanoparticles (NPs) were synthesised via thermal decomposition. (Alvi et al., 2021).
Materials:

- Octadecene
- Oleylamine
- Iron pentacarbonyl ($Fe(CO)_5$) precursor

Synthesis:

- Mix octadecene and oleylamine in a round-bottom flask under an argon atmosphere, stirring vigorously for 30 minutes at 120°C.
- Increase the temperature to 180°C, then inject iron pentacarbonyl ($Fe(CO)_5$) into the mixture using a syringe.
- Maintain the reaction at 180°C for 20 minutes.
- Cool the solution to 20°C and decant the supernatant.
- Magnetically separate the iron oxide nanoparticles (NPs) and wash twice with acetone.

The produced 14-nm-sized nanoparticles were spherical. The nanoparticles were dispersed in hexane for testing in the drilling fluid.

Drilling Fluid Formulation
A 90/10 oil-to-water ratio oil-based drilling fluid was prepared following a drilling fluid company recipe. The additives and their percentages are listed in Table 2.10. All chemical additives were mixed for 5 minutes, while barite was mixed for 25 minutes. To formulate the NP-based drilling fluid, 0.5 wt.% and 1.0 wt.% of NPs dispersed in hexane were added to the base fluid (ex-situ) and mixed for 15 minutes to ensure uniform distribution.

Results
The base and the nanoparticle-based drilling fluids were characterised. Results showed that 0.5 wt.% nanoparticle additives:

- *reduced the filtrate losses of the base fluid by 70%,*
- *reduced friction coefficient by 39%,*
- *reduced the thickness of the filter cake by 55%,*
- *reduced the dynamic sag factor, bringing it below the sag indicator limit, and*
- *reduced the cake permeability by 88%.*

TABLE 2.10 Oil-based drilling fluid formulation (base and nanoparticle blended)

ADDITIVES	*BASE FLUID (WT.% %)*	*BASE FLUID (WT. %) + NPS (WT.%)*
EDC 95/11	28,48	28,48
Emulsifier	1,43	1,43
Lime	1,43	1,43
Water	3,87	3,87
$CaCl_2$	1,19	1,19
Filtrate loss control	0,57	0,57
Viscosifer	0,4	0,4
Barite	62,62	62,62
F_2O_4 NP in Hexane		0,5 and 1,0 wt.%

Source: Alvi et al. (2021).

2.1.6.3 Environmental and Health Impacts Assessment

The nanoparticle-based drilling fluid test results presented above demonstrated the potential of nanoparticles in conventional water and oil-based drilling fluids. However, it is important to emphasise that the nanoparticles' potential effects on ecosystems, human health, and regulatory compliance must be carefully assessed before their application.

2.2 SELECTION OF OPTIMAL MUD WEIGHT

2.2.1 Introduction

The specific nomenclature used in Section 2.2 is as follows:

ECD = equivalent circulating density
$\sigma_r = P_w$ = borehole pressure
σ = tangential (hoop) stress
σ_v = vertical stress
σ_a = average horizontal in-situ stress
P_{wf} = fracturing pressure
P_o = pore pressure

In this section, borehole problems such as fracturing, collapse, lost circulation, differential sticking, and others are discussed in a rock mechanical context. It is shown that by maintaining the mud weight close to the level of the in-situ stresses, most of the borehole problems will be minimised. A design methodology called "the median line principle" is derived. The enclosed field case also demonstrates a reduction in drilling problems by using this methodology. In addition to problems during drilling, zonal isolation in the reservoir is identified as a crucial consequence of hole problems.

This section will first give a discussion of some borehole problems. A simple rock mechanical model will be presented and seen in the context of the borehole problems. The median line principle will be defined, and finally, a field study will demonstrate the improvements obtained so far. Six production wells from a field offshore Norway constitute the field study.

Figure 2.19 specifically shows the aim of this section. The low mud weight schedule has traditionally been used mainly because of pore pressure estimation purposes, but also because one believed that a low mud weight increased the drilling rate. The high mud weight schedule has been used in problem wells and in highly deviated wells, but to a limited extent because of fear of losing circulation and of differential sticking.

In this section we will demonstrate that neither of the two approaches discussed above are preferred from a borehole stability point of view. In fact, the "median line" mud weight also shown in Figure 2.19 is beneficial and will provide a common optimum for many of the parameters of the drilling process.

2.2.2 Borehole Problems

Many elements affect the success of a drilling operation. Since the main function of a drilling rig is to penetrate and to seal off formations, any single technical failure may halt this progress, thereby causing additional expenditures. The cost of an offshore drilling operation is dictated by the rig rate. Therefore, the success of a drilling operation is strongly dependent on avoiding problems that cause downtime.

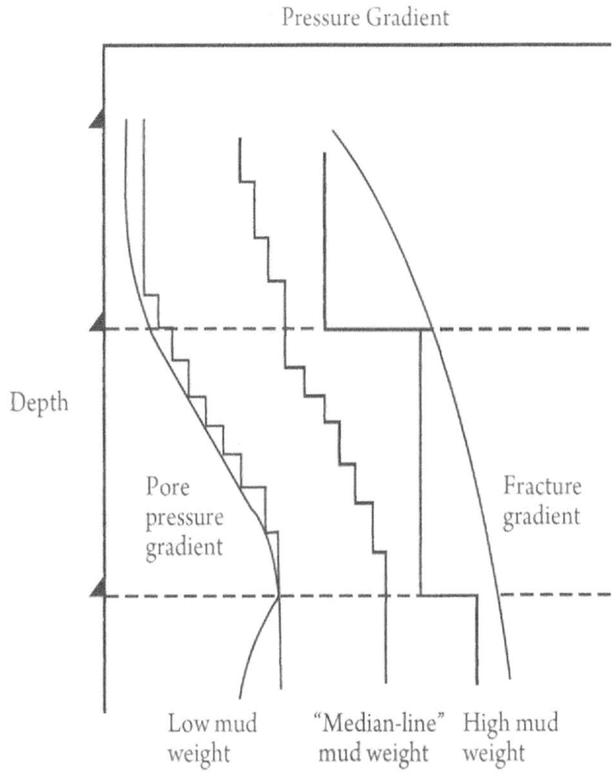

FIGURE 2.19 Typical mud weights used.

Bradley et al. (1990) brought borehole problems into a wider perspective by identifying the human element as a key factor in avoiding stuck pipe situations. In addition to sound engineering practices, the operational culture may, therefore, also strongly affect the outcome of a potential borehole problem. Furthermore, we will point out technical aspects of borehole problems not covered in this chapter. Although the mud weight selection is a key factor, related elements require good planning as well. Examples are torque and drag considerations in well path planning as discussed by Sheppard et al. (1987), and evaluation of stuck pipe experience as discussed by Hemkins (1987). Of course, hole cleaning and reaming practices must also be adequate. We will not give a detailed discussion of all the other elements, only point at the fact that no single element will replace a good overall well planning.

With the above view in mind, we will proceed to the main topic, optimal mud weight selection.

2.2.2.1 Higher Mud Weight, the Whole Truth?

The mud weight is a key factor in a drilling operation. The difference between success or failure is nearly always tied to the mud weight program. Too low mud weight may result in collapse and fill problems, while too high a mud weight may result in mud losses or pipe sticking. In an attempt to tie effects of high mud weight to drilling problems, Table 2.11 was produced, showing some beneficial effects of mud weight.

The elements of Table 2.11 will briefly be discussed in the following:

Borehole collapse. It is well known that borehole collapse occurs when the mud weight is too low because the hoop stress around the hole wall is very high, often resulting in rock failure (Aadnoy and Chenevert, 1987). The most important remedy is often to increase the mud weight.

TABLE 2.11 Effects of high mud weight

ELEMENT	*ADVANTAGE*	*DEBATABLE*	*DISADVANTAGE*
Reduce borehole collapse	X		
Reduce fill	X	X	
Reduce pressure variations	X		
Reduce washout	X	X	
Reduce tight hole	X	X	
Reduce clay swelling	X	X	
Increase differential sticking		X	X
Increase lost circulation			X
Reduced drilling rate		X	X
Expensive mud			X
Poor pore pressure estimation		X	X

Fill. Fill is the problem of cleaning the well. Cuttings or collapsed fragments may accumulate in the lower part of the well and lead to problems such as inability to reach bottom with the casing. Fill is commonly associated with the flow rate and the carrying capacity of the mud. There is also a strong connection to mud chemistry.

An increased mud weight should therefore reduce the potential for borehole collapse, thereby reducing the potential for fill.

Pressure variations. If the mud weight is kept more constant, the well is subjected to more static pressures. As pressure variations may lead to borehole failures (a fatigue type effect), a higher and more constant mud weight should be preferred. In addition to maintaining a more constant mud weight, also the equivalent circulating density (ECD) and the surge and swabbing pressures should be kept within limits.

Washouts. The theory behind borehole washout is that the jet action through the bit nozzles hydraulically erode the borehole wall away. The result is often believed to be an enlarged borehole of considerable size.

We believe that it is difficult to hydraulically wash out a consolidated rock at several kilometres' depth. What sometimes may happen is that the mud weight is too low, resulting in a failed hole wall. The washout is, therefore, often actually a collapse. The hydraulic action just removes already broken fragments. Field cases have shown that by increasing the mud weight by a small amount, the result is an in-gauge hole despite the same high flow rate.

Tight hole. A high mud weight will balance the rock stresses and keep the borehole more in-gauge. However, it is still likely that the hole will decrease in diameter the first day after it is drilled by swelling, still requiring wiper trips or back-reaming. Therefore, we propose to allow for an increase in mud weight, but not a reduction. Tight holes may also be caused by fill packing around the bottom-hole-assembly, combined with doglegs.

As shown later in this chapter, the tight hole conditions may be reduced or eliminated by increasing the mud weight. However, sound wiping or back reaming practices should still be maintained.

Clay swelling. Changes in fluid chemistry philosophy have been seen (Clark 1976; O'Brien and Chenevert 1973; Simpson et al. 1989; Steiger 1982). A good review of fluid chemistry is given by Santarelli and Carminati (1995). One key problem has been to inhibit reactive clays, as they often contribute to borehole problems such as collapse. However, field experience indicates that a sufficiently high mud weight may in some wells keep the hole stable even with a reduced degree of chemical inhibition, provided that the open hole exposure time is short. Therefore, the clay swelling problem should be reduced by increasing the mud weight. However, some wells seem to show hole enlargement irrespective of borehole pressure.

Differential sticking. An increased mud weight will lead to a higher-pressure overbalance, and the drilling assembly will be more easily subjected to differential sticking. From this point of view a high mud weight is detrimental.

However, it is also becoming clear that what we sometimes believe is differential sticking is often something else. Collapse and fill may pack around the bottom-hole-assembly resulting in sticking, and tight hole may be another contributor. Also, if we have intermittent layers of shales and sandstones, the shales may often collapse, exposing the sands directly towards the drilling assembly.

Figure 2.20a illustrates a borehole section where there are breakouts in the shale layers but in-gauge sand stringers in between. This situation is highly sensitive to differential sticking due to sand exposure. Figure 2.20b illustrates the same situation with an in-gauge hole. Since all layers now are in-gauge, it is possible that the contact between the hole and the drilling assembly occurs in the shale layers as well, reducing the potential for differential sticking in the sand layers.

A high mud weight is preferred from a collapse point of view. However, a high mud weight may, in general, increase the likelihood for differential sticking. Here is a potential conflict, which can be handled by keeping the mud weight below the critical level for differential sticking.

Lost circulation. Sometimes a weak stringer or a fault is penetrated, resulting in loss of drilling fluids. In general, the mud weights must be kept below this critical limit. Also fractured formations may set restrictions on the mud density, as discussed by Santarelli and Dardeau (1992).

Reduced drilling rate. It is commonly believed that a high overbalance results in a slow drilling. It is our opinion that the drilling rate is mainly a formation characteristic and that the effect of overbalance is of lesser significance. A reduction in drilling rate should also be measured against the cost of borehole problems.

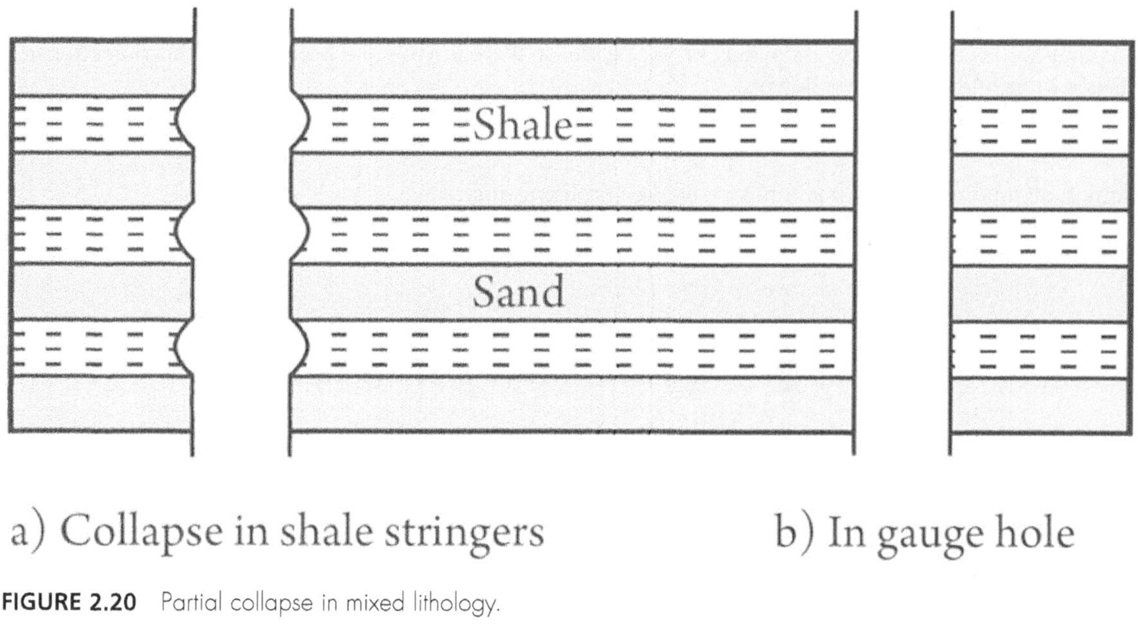

FIGURE 2.20 Partial collapse in mixed lithology.

TABLE 2.12 Likely relations between some borehole problems

PROBLEM	*COLLAPSE*	*FILL*	*WASHOUT*	*TIGHT HOLE*	*DIFF. STICK*	*LOST CIRC.*
Collapse	X					
Fill	X	X	X			
Washout	X		X			
Tight hole	X	X		X		
Diff. stick.	X	X		X	X	
Lost circ.						X

Mud cost. A higher weight mud program is often more expensive. This additional cost is usually negligible if it results in less drilling problems.

Pore pressure estimation. During drilling the geologist estimates the pore pressure using various criteria. One factor of particular concern is the recording of excess gas. This helps to quantify the pore pressure at the particular depth. A high mud weight may suppress high gas readings. A high mud weight may therefore not be preferable during wildcat drilling. During production drilling this requirement is often relaxed.

Mud weight summary. From the above discussion it may be concluded that a relatively high mud weight is acceptable and preferable from many points of view. However, particular concern has to be paid towards:

- lost circulation,
- differential sticking,
- background gas readings in exploration drilling, and
- naturally fractured formations.

Also, the mud chemistry must not be neglected. We have assumed an inhibited mud system in the above discussion. Table 2.12 summarises some likely connections between various borehole problems. Please observe that the mud weight is a common denominator between these.

From the above discussion, it is clear that the mud weight should preferably be on the high side. However, we still have a wide mud weight window. Figure 2.21 shows the allowable mud weight range. In many wells this allowable range may be very wide, so there is a definite need to limit this range further. This will be pursued in the following.

2.2.2.2 Mud Properties

Important mud properties to minimise hole wall problems include:

- chemical inhibition,
- low filtrate loss in permeable zones, and
- coating in impermeable zones.

Another very important property of the drilling fluid can be described as follows:

- Experience shows that new drilling fluid exacerbates fracturing/lost circulation situations. During leak-off testing it is our experience that used mud gives higher leak-off values than new mud. This is believed to be caused by the solids content from drilled cuttings. Therefore, one design criterion applied is to increase mud weight gradually to ensure that there are drilled solids present. In a new hole section one therefore usually start out with a lower mud weight. After drilling out about 100 m below the previous shoe, the mud weight is gradually increased. It is believed that by using this procedure we have avoided potential lost circulation situations. In Section 2.2.3 the practical applications of these observations will be demonstrated.

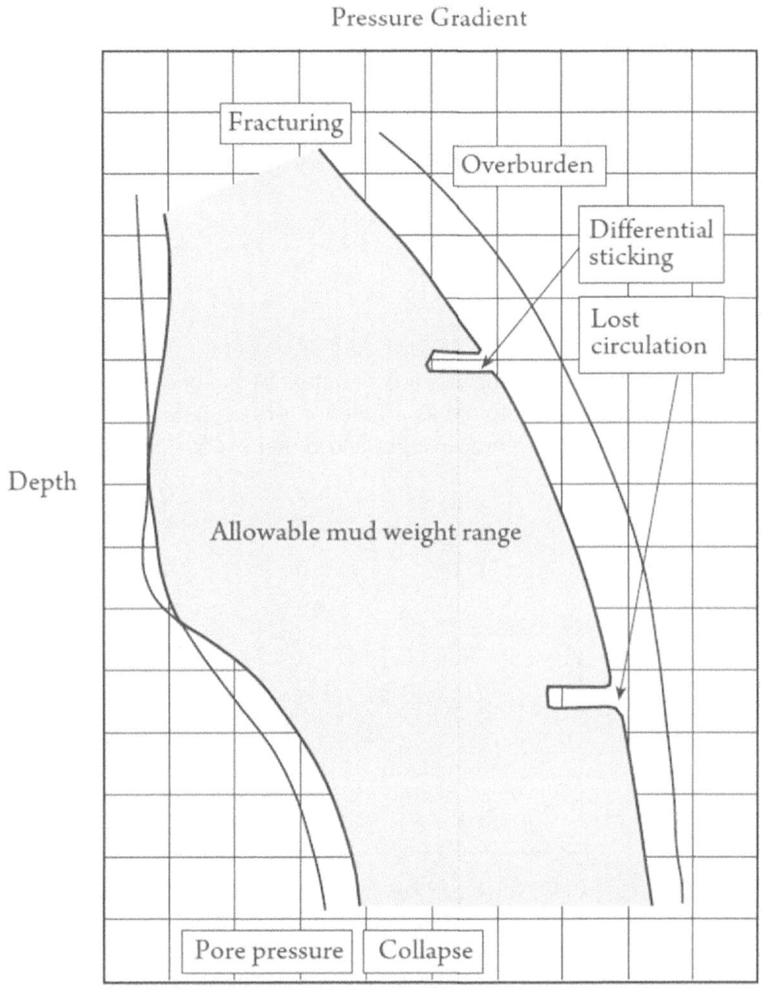

FIGURE 2.21 Allowable mud weight range considering common borehole problems.

2.2.3 Rock Mechanics

2.2.3.1 Stresses Acting on the Borehole Wall

The Kirsch equation is commonly used to calculate the stresses around the borehole. The stress level defines the loading on the borehole wall, and the rock strength the resistance to withstand this load. A number of publications have been written on this subject, and McLean and Addis (1990) and Aadnoy and Chenevert (1987) give a good overview.

It is well established that the stability of a borehole falls into two major groups:

- Borehole fracturing at high borehole pressures. This is actually a tensile failure, where the consequence may be loss of circulation. In a pressure control situation, this is of concern, and further drilling may be halted until circulation is re-established.
- Borehole collapse at low borehole pressures. This is a shear failure caused by high hoop stress around the hole, exceeding the strength of the rock. There are many variations of the collapse phenomenon. In some cases, the rock may yield, resulting in a tight hole. In other cases, a more catastrophic failure occurs, resulting in collapse, which again may lead to hole-cleaning problems.

Figure 2.22 illustrates the stresses acting on the borehole wall when the mud pressure is varied. Figure 2.22a shows the three main stresses acting on the borehole. The radial stress acting on the borehole wall is actually the pressure exerted by the drilling fluid. The axial stress is equal to the overburden load for a vertical well. However, around the circumference of the hole is the tangential stress acting. This is also called hoop stress. This stress depends strongly on the borehole pressure. As equations, these three stresses can in their simplest form be expressed as:

$$\text{Radial stress: } \sigma_r = P_w$$
$$\text{Tangential stress: } \sigma = 2\sigma_a - P_w \tag{2.20}$$
$$\text{Vertical stress: } \sigma_v = constant$$

To understand the borehole failure mechanisms in the context of the borehole stresses, Figure 2.22b was made. The three stress components are plotted as a function of borehole pressure. The vertical stress, or the overburden, is not influenced by the mud weight and remains constant. The radial stress is equal to

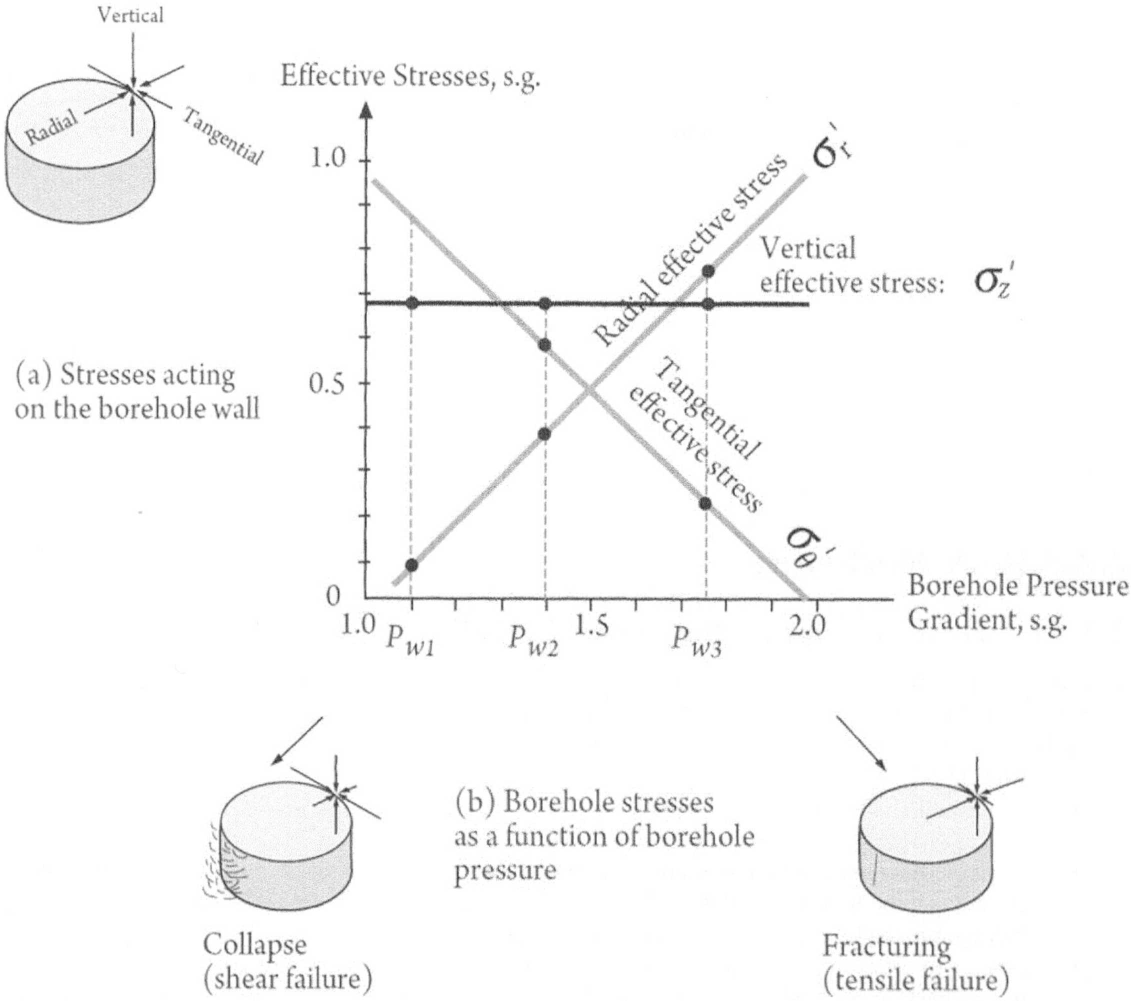

FIGURE 2.22 Borehole stresses with varying mud weight.

the borehole pressure and therefore has a unit slope in the diagram. The tangential stress decreases with increased borehole pressure.

At low borehole pressures, the tangential stress is high. Since there is a significant difference between the radial and tangential stress, a considerable shear stress arises. It is this shear stress that ultimately results in borehole collapse. At high borehole pressure, on the other hand, the tangential stress goes into tension. Since rocks are weak in tension, the borehole will fracture at high borehole pressures, usually resulting in an axial fracture. These two failure types are indicated in Figure 2.22b. More complex failure modes can be evaluated (Maury, 1993), but this will not be pursued here.

From the discussion above, we observe that low and high borehole pressures produce high stress conditions and bring the hole towards a failure state. By further inspection of Figure 2.22b we see that at a given point, the radial and the tangential stresses are equal. Here, the mud weight is equal to the in-situ stress, and there are no longer abnormal stresses. This will be further discussed in the following section.

2.2.3.2 The In-Situ Stress State

We will assume a relaxed depositional basin with a so-called hydrostatic stress state. That is, around a vertical hole, the horizontal stress level is the same in all directions. Having a leak-off pressure and a pore pressure, the fracturing pressure is reached when the effective hoop stress is zero, or $\sigma_\theta - P_o = 0$ from Equation (2.20). The following equation results (Aadnoy and Chenevert, 1987):

$$\sigma_a = \frac{1}{2}\left(P_{wf} + P_o\right) \tag{2.21}$$

The average horizontal stress is equal to the average between the fracturing pressure and the pore pressure. A tectonic stress situation with non-hydrostatic horizontal stresses gives a more complex picture. A short example is given at the end of the chapter. However, the proposed method could also be used in this case, as the fracturing gradient implicitly takes both the actual stress situation and the borehole inclination into account. For example, for a deviated well the design fracture gradient may be corrected for borehole inclination (Aadnoy and Larsen, 1989). This will be further discussed in Section 3.2.

Equation (2.21) can explain several of the borehole problems we have just discussed. Let us first discuss the implications. We assume that the elements of Equation (2.21) are known, and will, in the following, discuss what happens if the actual mud weight is equal to, lower than, or higher than the in-situ stress of Equation (2.21). Figure 2.23 illustrates the responses of varying mud weight.

Figure 2.23a. Using a mud weight *equal to* the horizontal stress σ_a, the immediate surrounding rock is undisturbed by the drilling of the hole. This is the ideal mud weight, and the hole diameter will remain constant.

Figure 2.23b. Using a mud weight *lower than* the horizontal stress σ_a, the stress will locally change. A hoop stress is created, causing the borehole to decrease in diameter. This can result in either:

- borehole collapse, or
- tight hole.

Figure 2.23c. Using a mud weight *higher than* the horizontal stress σ_a, the borehole pressure will attempt to increase the hole diameter, ultimately causing fracturing if the mud weight becomes too high.

As implied from the above discussion, the mud weight/borehole stress relationships can be used to describe common borehole problems. This can be defined as the *median line principle*, which is defined by Equation (2.21).

The mid-point between the fracturing pressure and the pore pressure defines the borehole pressure that is equal to the ideal in-situ stress. Maintaining the mud pressure close to this level causes least disturbance on the borehole wall.

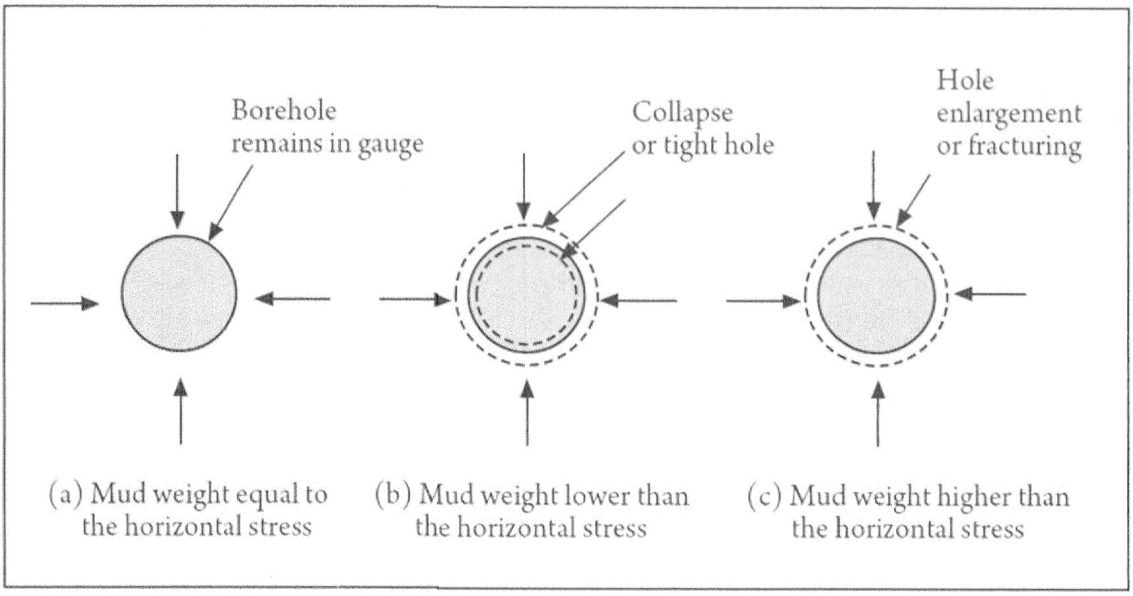

FIGURE 2.23 Effects of varying the borehole pressure.

The median line principle will in the following be used to define the actual mud weights to be used in a drilling operation.

2.2.3.3 The Median Line Principle

Figure 2.24 shows pressure gradient plots for a well. This will first be used to give a general description, then be used in a discussion of drilling problems in Section 2.2.4. Shown are five pressure gradients. The median line is drawn using the previously defined Equation (2.21). The casing seats are selected based on:

- Fracture gradient and pore pressure gradient prognosis.
- Kick scenario.
- Seal off likely lost circulation intervals.
- Minimise effects of borehole stability problems.
- Casing landing considerations.

In the following the mud weight selection for each of the intervals will be described. Details on the geology can be found in Dahl and Solli (1992).

The 26/24 in. hole. The 30 in. conductor casing is set with about 100 m penetration. The fracture gradient below the 30 in. casing is fairly low. Therefore, in the 26/24 in. hole the mud weight is below the median line during most of the interval.

The 16 in. casing interval. Drilling out below the 18-5/8 in. casing, the mud weight of Figure 2.24 is below the median line for two main reasons:

- To give the open hole time before increasing mud weight to minimise the risk of breaking down below the casing shoe.
- It is preferable to have a low mud weight during leak-off testing. The leak-off pressure plot covers a larger pressure range, improving the interpretation.

After drilling below the 18 5/8 in. casing about 100 m, the mud weight is gradually increased to exceed the median line and kept above for the rest of the section. The main reason for staying above the median line is to minimise tight hole conditions.

The 12-1/4 in. hole. Drilling out below the 13-3/8 in. casing, circulation was lost in several wells. Figure 2.24 shows the current approach where the mud weight is initially below the median line. After drilling out 100 m, it is attempted to keep the mud weight above the medium line for the rest of the section. However, at the bottom, the mud weight of Figure 2.24 drops below the medium line for the following reasons:

- To minimise the risk of lost circulation.
- To minimise the risk for differential sticking.

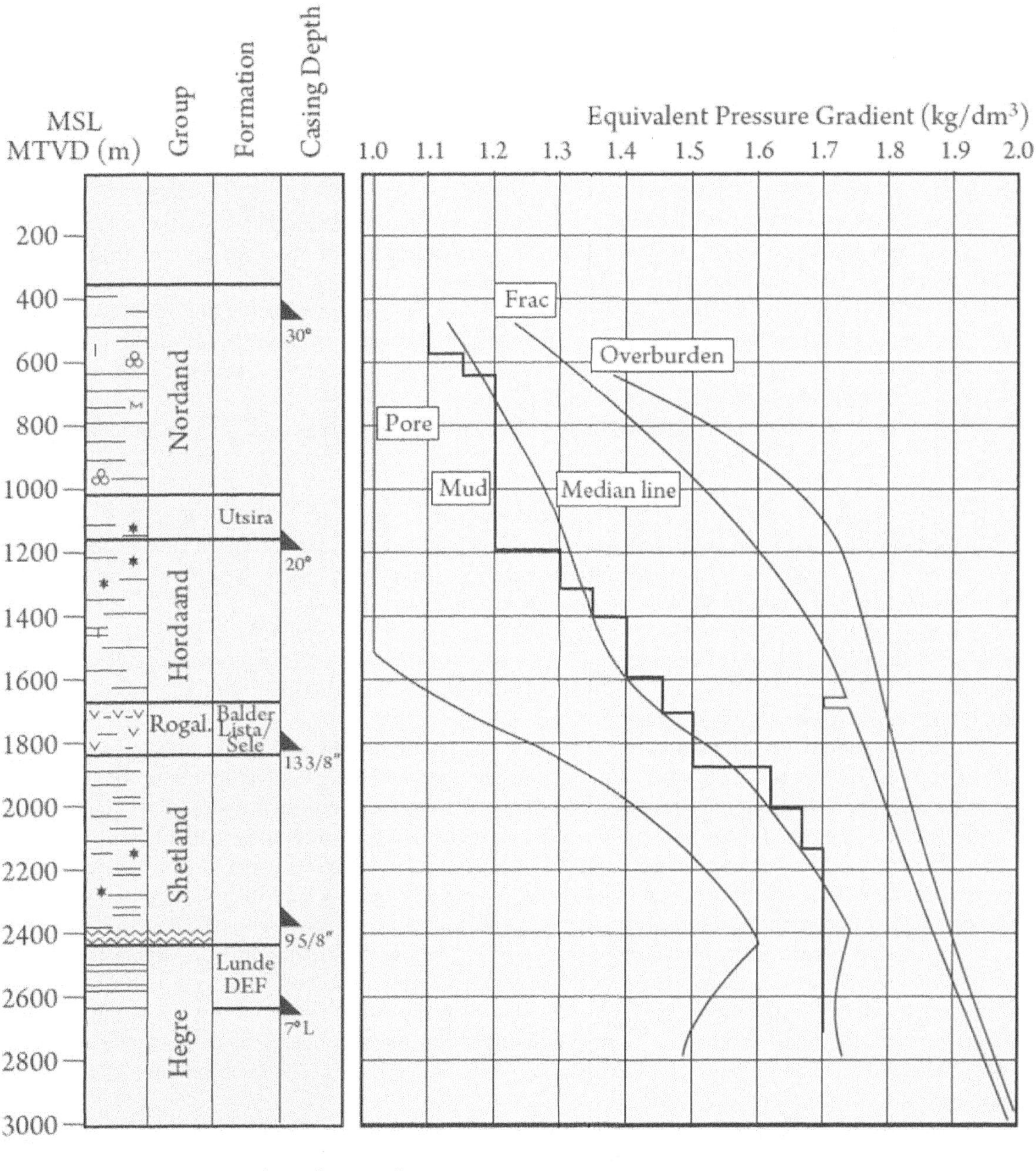

FIGURE 2.24 Pressure gradients for a well.

The 8-1/2 in. hole. The last section of Figure 2.24 penetrates the reservoir. In this case the mud weight is maximum, and it is kept constant throughout the section. Lost circulation and differential sticking experiences have resulted in using a mud weight lower than the median line in the reservoir section.

A final comment to the application: In an open hole section the mud weight should only be increased, and not decreased, as tight hole may result. Furthermore, we have chosen to increase the mud weight in steps of 0.05 g/cm^3, for the convenience of the mud engineer.

2.2.4 Field Cases

Of the six predrilled wells, three were drilled according to the high mud weight profile shown in Figure 2.19, and the three last wells were drilled according to the median line principle. In the following, the three bottom sections of one of each group will briefly be discussed from a drilling problem point of view.

Well 3 (high mud weight profile). In the 16-inch section, the mud weight was initially 1.2 s.g., but increased towards 1.45 s.g. at about 1300 m. Tight hole was not observed during drilling, but at about 1500 m a wiper trip showed 50-ton overpull. After drilling to final depth of the section, a wiper trip to 1400 m showed 30-ton overpull. A final wiper trip after logging resulted in severe tight hole problems, and the hole had to be reamed. After increasing the mud weight to 1.51 s.g., the hole was not tight, except for the bottom 100 meters. Because of these problems, the casing was installed 79 m above planned shoe depth.

It was believed that a more gradual mud weight increase would successively push the hole open, resulting in less tight hole. Actually, this strategy was used on wells 4, 5, and 6, the latter being discussed in the following.

Well 6 (median line principle). The pressure gradients for this well are shown in Figure 2.24. This was the last of the six wells predrilled. Therefore, many parameters are optimised such as the drilling mud composition, chemistry, operational practices, and many other factors. The mud weight schedule is also optimised based on previous experiences.

Figure 2.24 shows the resulting pressure gradients on well 6. Just before finishing this well the casing program was altered to eliminate the 7 in. liner, which resulted in setting the 9-5/8 in. casing to TD. In addition, the coring program was dropped.

The 16 in. section was drilled and cased off with no reported problems. The mud weight was gradually increased, contrary to well 3 where a more constant high mud weight gave tight hole conditions.

In the 12-1/4 in. section, only minor tight hole conditions were reported, but a slight mud weight increase and reaming cured the problems. The mud weight was kept below the median line during most of the 12-1/4 in. section because of the fear of differential sticking. No lost circulation incidents were reported. The tight hole conditions identified in well 3 are much more severe than those reported in well 6. In well 3, the casing point had to be changed, while similar effects were not observed in well 6.

In the reservoir section, the mud weight was also kept below the median line. Significant tight hole was not reported. However, there were several signs of possible differential sticking, which indicates that the mud weight is possibly on the high side. However, the riser margin (at a water depth of about 300 m) restricts the mud weight reduction possibilities considerably, because it limits the operating window between the pore pressure and the fracturing pressure gradients. The riser margin is discussed in Section 4.2.4.

The mud weight schedule has been varied during drilling of the six production wells. The last three wells have been drilled using the median line mud weight design. Figure 2.25 shows the specific reaming time for each of the six wells. A considerable time was spent reaming the open hole sections of the first three wells, while the three last wells only needed a little reaming. A gradual reduction is apparent, with

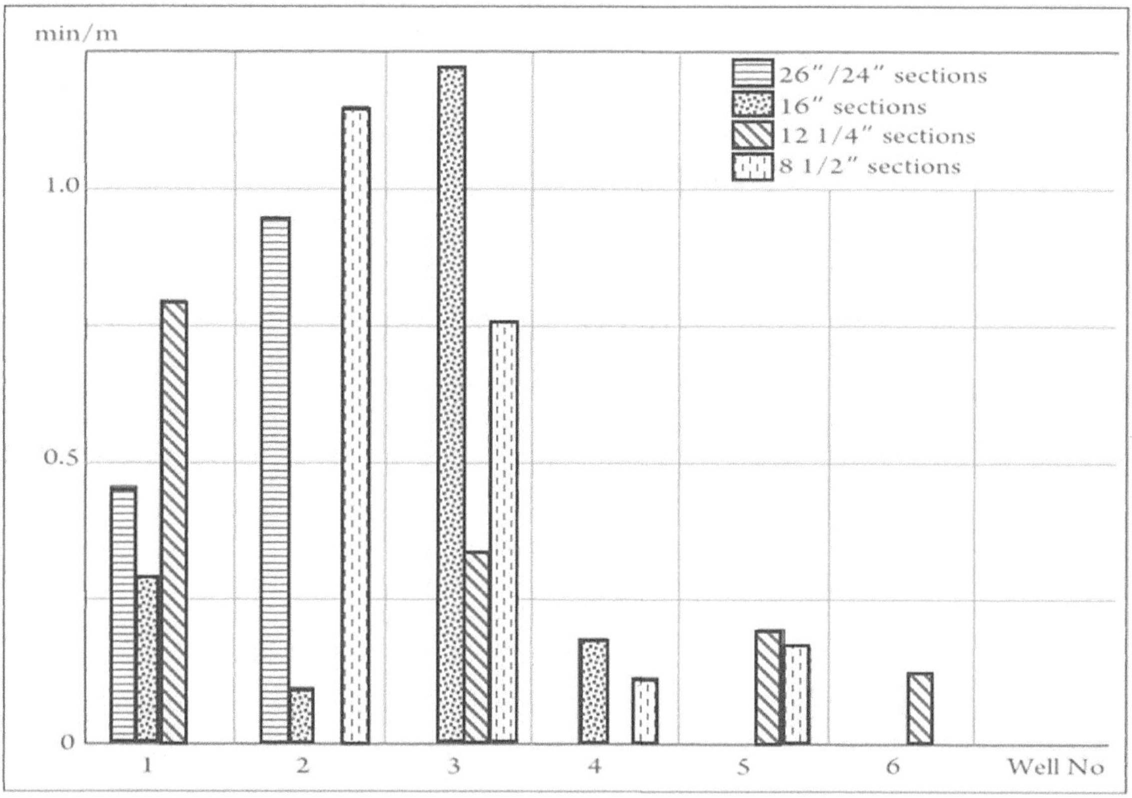

FIGURE 2.25 Specific reaming time for each well.

the last well having only minor reaming. We believe that the mud weight program is a significant contributor here. The amount of reaming necessary is considered a measure of the general condition of the borehole.

2.2.5 Application of the Median Line Principle

Common borehole problems are discussed and evaluated in a rock mechanics context. The result is the "median line principle", which simply says that the mud weight should be kept close to the in-situ stress field in the surrounding rock mass. In this way, the borehole problems are minimised since a minimum of disturbance is introduced on the borehole wall.

The mud weight methodology was applied in the last three wells in a field study of six wells. The enclosed field study shows a considerable reduction in tight hole conditions, which is considered a good indicator of the general condition of the hole.

The median line mud weight design methodology can be summarised as follows:

1. Establish a pore pressure gradient curve and a fracturing gradient curve for the well. The fracture gradient curve should be corrected for known effects like wellbore inclination and tectonic stresses.
2. Draw the median line between the pore and the fracture gradient curve.
3. Design the mud weight gradient to start below the median line immediately below the previous casing shoe.

4. Mark out depth intervals prone to lost circulation and differential sticking and their acceptable mud weight limits, if known.
5. Design a stepwise mud weight schedule around the median line that also takes into account limitations from 3 and 4 above.
6. Avoid reducing the mud weight with depth. If a median line reversal occurs, keep the mud weight constant.

2.2.6 Tectonic Stresses

This section is intended for those who are more interested in the rock mechanics aspect, and who wants to work in more detail.

In this chapter the mud weight is designed based on an assumption of equal horizontal stresses in the formation. This should always be a starting point and will for most applications provide a reasonable mud weight schedule.

However, the readers will observe that in Section 3.2, methods are given to determine anisotropic stresses. For these cases, the median line principle can be modified. Assuming that the two horizontal stresses are of different magnitude and given by σ_H and σ_h, the fracturing pressure is given by (Bradley, 1979; Aadnoy and Chenevert, 1987):

$$P_{wf} = 3\sigma_h - \sigma_H - P_o \tag{2.22}$$

An example will demonstrate the effect of stresses. The first case assumes equal horizontal stresses and the optimal mud weight is defined by Equation (2.21), which is:

$$\sigma_a = 0.5(P_{wf} + P_o)$$

This will be compared to the second case. Now assuming anisotropic horizontal stresses, for example $\sigma_h = 0.8\sigma_H$, Equation (2.22) can be solved for the smallest horizontal stress as follows:

$$\sigma_h = 0.571(P_{wf} + P_o)$$

Assuming all factors are equal, but for the horizontal stresses, the two cases illustrate that for an anisotropic stress state the ideal mud weight should be higher. However, for this case, the difference between the fracturing pressure and the minimum horizontal stress is smaller than for the first case.

For anisotropic or unequal horizontal in-situ stresses the mud weight should actually be higher than for equal horizontal stresses. However, the example above also demonstrates that this situation may be easier subjected to circulation losses. In general, high in-situ stress anisotropy usually leads to a smaller mud weight window.

2.3 MUD LOSSES DURING DRILLING

2.3.1 Introduction

The two most costly drilling problems are stuck pipe and circulation losses. Statistics show that these dominate unplanned events that may take 10–20 % of the total time spent on a well. Very high cost is therefore associated with these problems.

We will in this section address the problem of circulation losses. They can occur at any time during a drilling operation and are very common in depleted reservoirs. Usually, the loss problem must be cured before drilling can resume. Using water-based drilling fluids, the problem is often reduced by pumping lost-circulation-materials (LCM) into the wellbore. In some cases, cementing is required. Using oil-based drilling fluids is much worse. If circulation losses occur with oil mud, it can be difficult to control the losses, and large amounts of mud may be lost before control is regained. This is believed to be related to the wettability contrast between the rock and the mud. A capillary barrier prevents filtrate losses to the rock, maintaining the low viscosity of the mud, thereby allowing for further fracture propagation.

Mud companies have many recipes to stop mud losses. Basically, all of these use particles in various combinations as bridging materials. These are often proprietary and will not be addressed further here. Instead, we will explain the mechanisms believed to cause circulation losses. A research program has been carried out at the University of Stavanger over many years. Some of this work is given by Aadnoy et al. (2008). This section will mainly report results from this work. A new mechanistic model for fracturing called "the elastoplastic barrier model" evolved from this work.

Nomenclature

σ_y yield stress of bridging particles
σ_a horizontal in-situ stress
t barrier thickness
a borehole radius
P_{wf} fracturing pressure
P_o pore pressure
LCM lost circulation material

2.3.2 Experimental Work

Figure 2.26 shows a fracturing cell where specially prepared hollow concrete cores are fractured. The setup also allows for mud circulation to ensure that mud particles are well distributed inside the hole. The cell is rated to 69 Mpa, and the axial load, the confining pressure and the borehole pressure can be varied independently. Many oil and water-based drilling fluids have been tested, as well as novel ideas like changing rock wettability or creating other chemical barriers. Cores with circular, oval, and triangular holes have also been tested to study effects of hole geometry.

Figure 2.27 shows typical results from the fracturing experiments. The commonly used Kirsch equation is used as a reference. The Kirsch equation defines the theoretical fracture pressure with a non-penetrating situation, such as when using drilling muds. From Figure 2.27, it is seen that only one of the measured fracture pressures agrees with the theoretical model; the two others are much larger. Several conclusions have come out of this research:

- the theoretical Kirsch model underestimates the fracture pressure in general, and
- there is significant variation in fracture pressure depending on the quality of the mud.

This shows that the fracture pressure can potentially be increased by designing a better mud. Actually, the results of Figure 2.27 explain the variability we observe in the field; sometimes, a higher leak-off is observed. For some reason, the mud is more optimal in these cases. Aside from standard mud measurements like filter cake thickness, today's mud measurements do not adequately show the fracturing resistance of a drilling mud.

Figure 2.28 shows a mud cell provided with six outlets containing artificial fractures of various dimensions. The mud is circulated with a low-pressure pump to develop a filter cake across the slots. At this stage, a high-pressure pump increases the pressure until the mud cake breaks down. In this way,

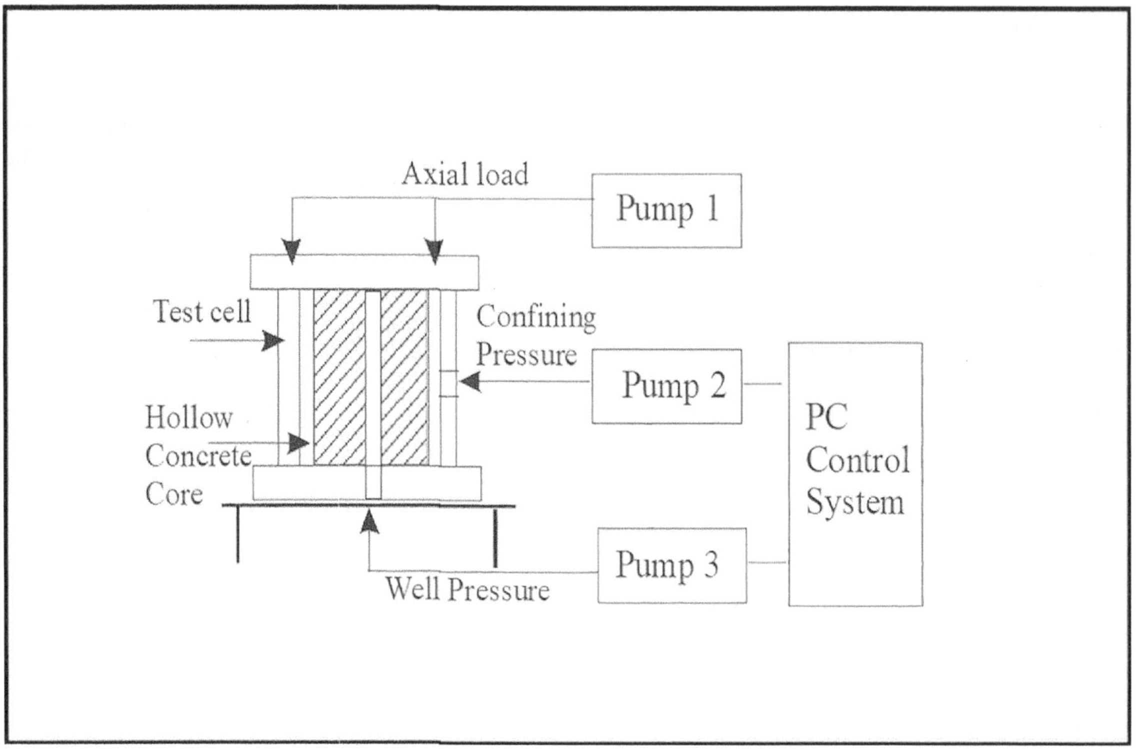

FIGURE 2.26 Fracturing cell for testing of concrete cores.

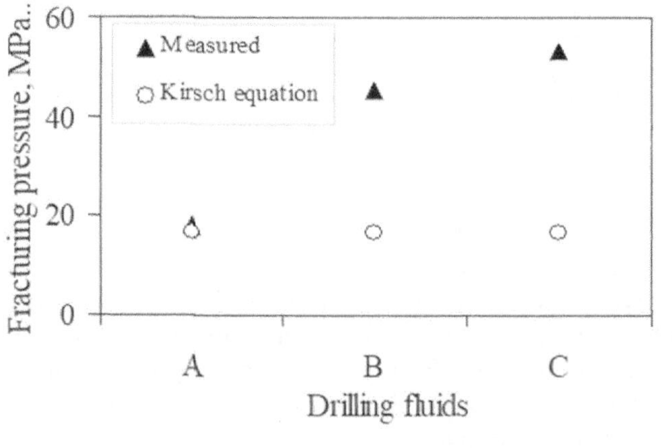

FIGURE 2.27 Examples of theoretical and measured fracture pressures.

we can study the stability and the strength of the mud cake. We have used common muds and additives and observed that reducing the number of additives often gives a better mud. We have also studied non-petroleum products to look for improvements. Some of this will be discussed later.

2.3.3 Fracturing Models

The so-called Kirsch equation is almost exclusively used to model fracture initiation in the oil industry. It is a linear elastic model that assumes that the borehole is penetrating, that is, fluid is pumped into the formation, or it is non-penetrating which means that a mud cake prevents filtrate losses. The latter gives a higher fracture

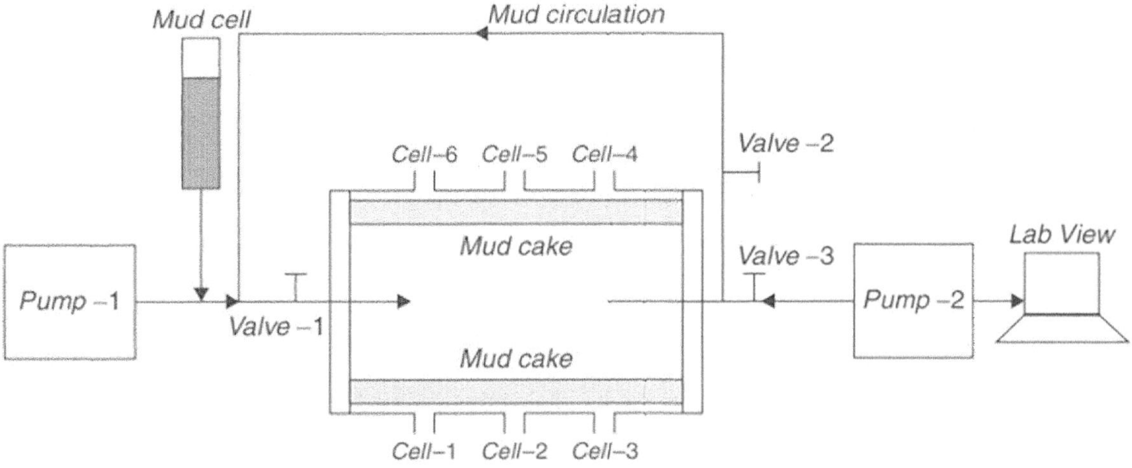

FIGURE 2.28 Apparatus to determine the fracture strength of mud cake.

pressure. In the following, we will only present the simplest versions of the fracturing equations, applicable for vertical holes with equal horizontal stresses, typically for relaxed depositional basin environments.

2.3.3.1 Penetrating Model

This is the simplest fracture model, which is defined as:

$$P_{wf} = \sigma_a \tag{2.23}$$

For well operations like hydraulic fracturing and stimulation, the penetrating model applies. It requires a clean fluid with no filtrate control, such as water, acids, and diesel oil. It simply states that the borehole will fracture when the minimum in-situ stress is exceeded.

All of our fracturing experiments confirm that this theoretical model works well using pure fluids. It should, therefore, be used in well operations involving clean fluids, such as stimulation and acidizing. Please note that Equation (2.23) is valid for fracture initiation. Fracture propagation requires other models.

2.3.3.2 Non-Penetrating Model

In a drilling operation, the fluids build a filter cake barrier. For this case, the Kirsch equation becomes:

$$P_{wf} = 2\sigma_a - P_o \tag{2.24}$$

This equation, in general, underestimates the fracture pressure, as demonstrated in Figure 2.27. The problem rest with the assumptions of a perfect (zero filtrate loss) mudcake.

We found that the mud cake behaves plastically. The new model, therefore, assumes a thin plastic layer, which is the mud cake, followed by a linearly elastic rock. This is called an elastoplastic fracture model. The explanation for the higher fracture pressure is that when a fracture opens, the mud cake does not split up but deforms plastically, maintaining the barrier. This model can be described as (Aadnoy and Belayneh, 2004):

$$P_{wf} = 2\sigma_a - P_o + \frac{2\sigma_y}{\sqrt{3}} \ln\left(1 + \frac{t}{a}\right) \tag{2.25}$$

The additional strength obtained with the elasto-plastic model is directly proportional to the yield strength of the particles forming the barrier. This model accurately describes the measured data shown in Figure 2.27.

2.3.4 Description of the Fracturing Process

In Figure 2.29, we have shown the various steps in the fracturing process.

Event 1: Filter cake formation. A small filtrate loss ensures the formation of a filter cake. During mud flow, a thin filter cake builds up. The thickness of the cake depends on the equilibrium between the filtrate attraction and the erosion due to the flow.

Event 2: Fracture initiation. Increasing the borehole pressure, the hoop stress in the rock goes from compression towards tension. The filtrate loss ensures that the filter cake is in place. The in-situ stresses, which control the borehole hoop stress, resist the pressure. At a critical pressure, the borehole starts to fracture.

Event 3: Fracture growth. A further increase in borehole pressure results in an increase in fracture width. In-situ stress is opposing this fracture growth. The filter cake will remain in place because a stress bridge is formed across the fracture. This is the plastic part of the elastoplastic model. This bridge acts as a natural rock road bridge, the higher the top load, the more compressive forces inside the curvature. The factor that prevents this bridge from collapsing is the mechanical strength of the particles of the filter cake. In this phase, both the rock stress and the filter cake strength resist failure.

Event 4: Further fracture growth. Further pressure increases lead to further fracture opening. The stress bridge expands and becomes thinner with a small thickness. Due to the geometry increase, it becomes weaker.

Event 5: Filter cake collapse. At a critical pressure, the filter cake is no longer strong enough, and the "rock bridge" collapses. This occurs when the yield strength of the particles is exceeded. At this point, communication is established, and we have mud losses towards the formation.

2.3.5 Some Research Findings

2.3.5.1 Properties of the Mud Cake

Our research has concluded that two main characteristics of a filter cake can give a high fracture pressure. These are related to the filtrate properties required to form a filter cake and also the strength of the particles in the mud. The bridge model presented in the previous section depends on the mechanical strength of the particles. For this reason, the choice of lost circulation material determines the maximum fracture pressure that can be obtained.

Figure 2.30 shows some results of fracturing under similar conditions as a function of Mohs number (or hardness or compressive strength). Clearly, calcium carbonate is the weakest particle. In addition to the particle strength, shape and size distribution are important factors. In general, we find that a steep particle size distribution curve works best. Although the data presented shows that calcium carbonate is the weakest, it is still a good LCM material for wells with low pressure overbalance. For higher overbalance situations, stronger particles are recommended.

2.3.5.2 Synergy Between Various LCM Additives

Usually, several additives are used in a drilling fluid with the function to resist fluid losses. In some cases, up to ten different additives are included in a mud. Our research has shown that in many cases, too many additives give a poor mud. We have carried out test programs varying both the number of additives and the concentration of each. After carrying out many tests, we concluded that some were useless or actually detrimental, some were good, and some others were good or bad depending on the combination

Event	Fig.	Main controlling parameters
Filter cake formation	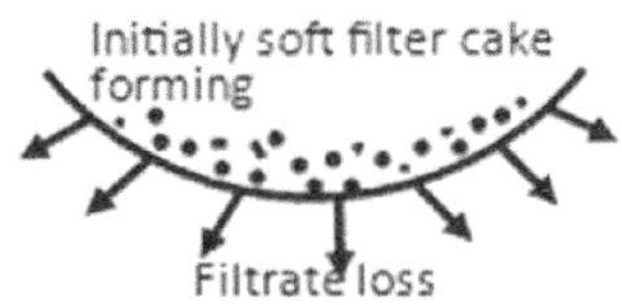	Filtrate loss
Fracture initiation	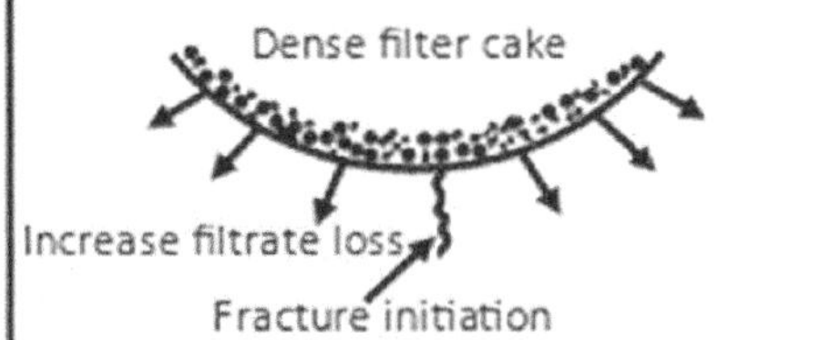	Filtrate loss, Stress
Fracture growth	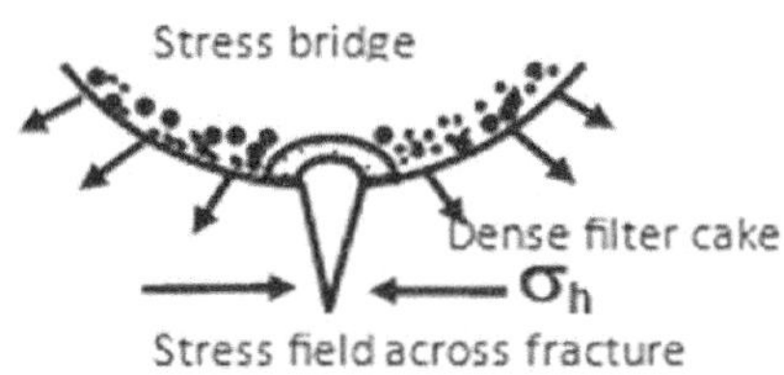	Bridge stress Rock stress
Further fracture growth	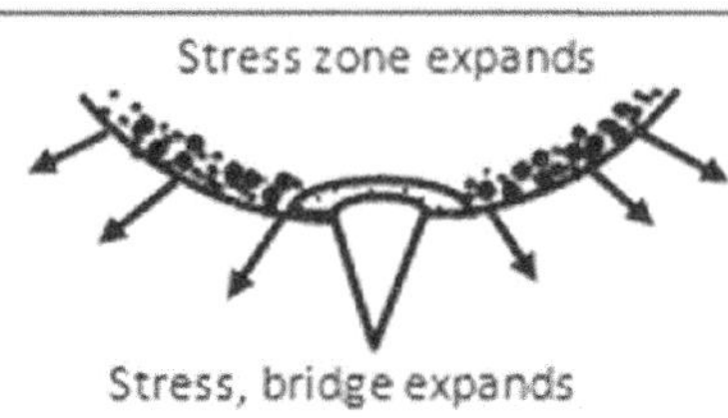	Bridge / Rock stress Particle strength
Filter case collapse	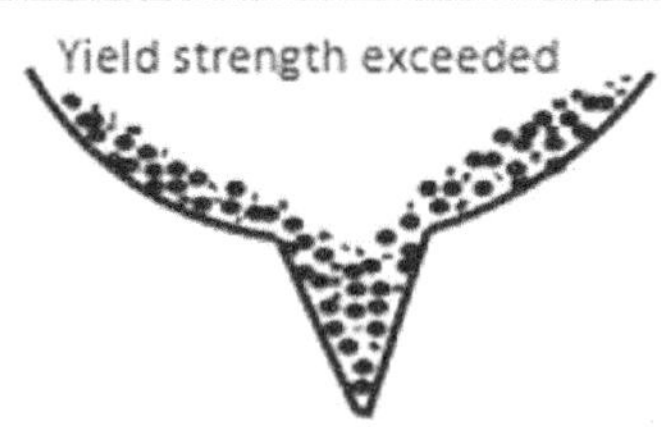	Particle strength

FIGURE 2.29 Qualitative description of the fracturing process.

and concentration of additives. We also observed that there is synergy between the additives; two poor additives can be good when combined.

2.3.5.3 Effects of Carbon Fibres

In addition to the commercial mud additives, a number of non-petroleum additives have been tested to search for additives that improve the drilling operation. One of the issues is to build a bridge in a fractured

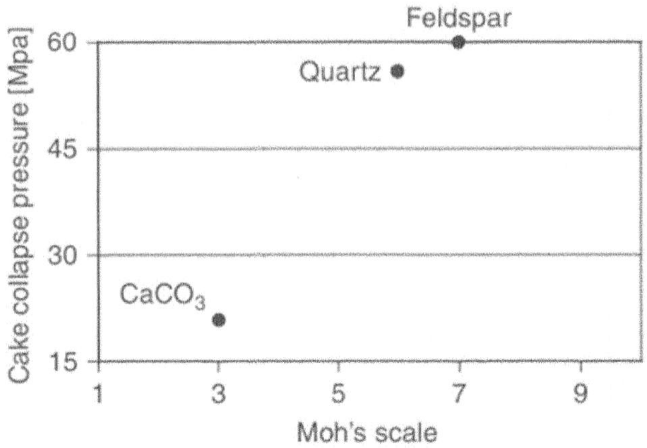

FIGURE 2.30 Barrier strength versus Mohs scale.

rock. This may require relatively large particle sizes in some instances, affecting both mud weight and rheology. Searching for alternatives, we tested out various carbon fibres. Polymer-type particles are not sufficiently strong, so we searched for additives with a high mechanical strength. Carbon fibres worked quite well as they have high mechanical strength, are relatively non-abrasive, and have relatively low density.

General Observations

- To create a stable bridge to prevent losses, the largest particle diameter should be equal to or exceed the fracture width. However, at present, we do not have reliable methods to determine fracture widths.
- A minimum particle concentration is required to provide sufficient bridging material.
- If a high differential pressure is expected in the well, particles with high compressive strength (high Mohs numbers) should be used.
- There is strong synergy between various additives. Two poor additives may work well in a mixture. The only way to determine this synergy effect is by laboratory testing.
- The number of additives to the mud should be kept at a minimum.
- There is large discrepancy between new and used mud with a large potential for improvement.
- Particle placement is important.
- A stronger fracture healing is seen with water-based mud than with oil-based mud. It is believed due to water wet rock, allowing filtrate losses. Water-based muds are preferred from this perspective.
- The fracture propagation pressures for water- and oil-based muds are similar. However, due to the healing effect, propagation pressure increases during losses for water-based fluids, as opposed to oil-based fluids, where the fracture propagation pressure is nearly constant.

2.3.6 Shallow Well Field Case

A shallow well was drilled and the operator expected loss and well integrity problems during the operation. The mud was designed and tested at the fracturing lab at the University of Stavanger. Mud samples were sent onshore, tested, and recommendations were implemented during the operation. The project was quite successful, as the mud quality was ensured during the entire operation. Figure 2.31 shows the LOT data obtained from the well as compared to reference wells, showing a clear improvement. This demonstrates that it is possible to improve the fracture strength of the borehole.

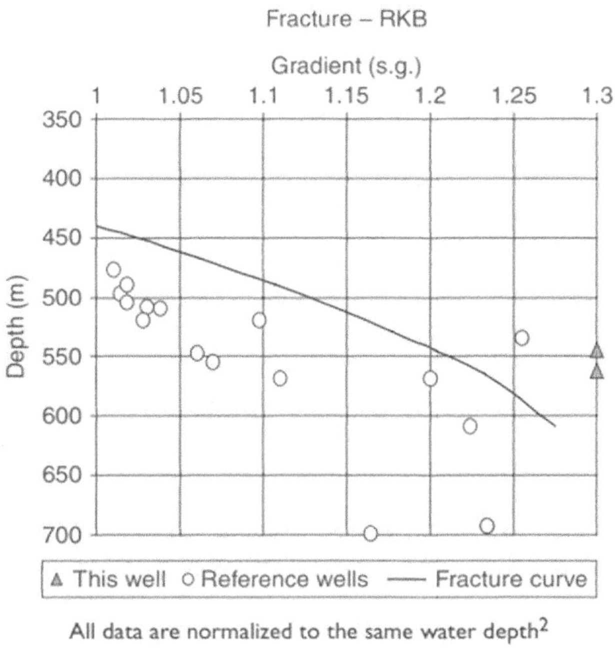

FIGURE 2.31 LOT data from reference wells and new well.

Note: All data are normalised to the same water depth[2].

TABLE 2.13 Proposed additives for drilling mud to minimise losses

ADDITIVE	*CACO3 COARSE*	*GRAPHITE*	*MICA FINE*	*CELLULOSE*	*CARBON FIBRE*
Our proposal	3	-	3	-	-
(6 ppb)	-	3	3		
	2	2	-	-	2
	2	2	-	2	-

2.3.7 Examples of Recommended Mud Recipes

In the following, some mud recipes will be proposed. Because conditions vary in drilling operations, these should be used as guidelines only, the correct recipe is only obtained after laboratory testing.

The application of LCM material is typically a reactive event; a cure is required after the loss event has taken place. A proactive approach requires a mud of such quality that a loss may not occur. The following exemplifies mud designs for both cases. Table 2.13 shows that a good filter cake is obtained with only a few additives. In particular, adding carbon fibre has a positive effect.

Table 2.14 shows a pill designed to stop circulation losses. The operators' recipe was tested in our lab and was not found to behave in an optimal way. Our proposed pill does not contain calcium carbonate and graphite. Adding a small amount of carbon fibres has a significant effect on stopping mud losses.

TABLE 2.14 Design of a lost circulation pill

ADDITIVE:	*CONSISTS OF:*	*OPERATORS RECIPIE (PPB)*	*OUR PROPOSAL (PPB)*
A	$CaCO_3$ coarse	15	
B	$CaCO_3$ fine	15	
C	Fine polymer	20	30
D	Medium polymer	20	20
E	Graphite	40	
F	Mica fine	20	20
G	Mica medium		20
H	Cellulose	30	45
Carbon fibre			Some

2.4 HOLE CLEANING

2.4.1 Introduction

It is important to bring the drilled cuttings out of the wellbore. If they accumulate, the drill string may get stuck. Also, excess cuttings in the annulus may lead to an increase in bottomhole pressure, that again may lead to circulation losses.

When a wellbore increases in size due to wellbore collapse, sometimes larger rock pieces must be transported out of the wellbore for the same reasons as given above.

The flow rate and the drilling rate must stay within certain limits to ensure good hole cleaning. Usually, hydraulic simulators are used to determine the minimum flow rate. In this chapter, we will present a simpler approach using a few charts and equations.

2.4.2 Hole Cleaning

Luo, Bern, and Chambers (1992) wrote an excellent paper on hole cleaning, where they presented the physics of hole cleaning but also a practical way to apply the models. They refined and expanded this work in Luo, Bern, Chambers, and Kellingray (1994). This section is based on the information given in these publications. These papers still capture the state-of-the-art to a large extent, as seen in API (2006). The papers strictly address hole cleaning for wells exceeding 30 degrees inclination. For lower angle inclinations, we have extrapolated the angle factor of Table 2.15 to vertical.

2.4.2.1 Cuttings Transport Mechanisms

The cuttings will sink due to gravity. The fluid velocity and the viscosity will attempt to carry them up the well. In addition to these opposite effects on particles, there are depositional/erosional forces acting on cutting beds. The complete model is complex and requires numerical solutions. Luo, Bern, and Chambers (1992) used experimental data from flow loops to determine the parameters of the model. The following controllable variables were found:

- Mud flow rate
- Rate of penetration (ROP)

TABLE 2.15 Angle factors for deviated holes, *AF*

HOLE ANGLE(DEG)	*0*	*25*	*30*	*35*	*40*	*45*	*50*	*55*	*60*	*65*	*70-80*	*80-90*
Angle factors, AF	2.03*	1.51	1.39	1.31	1.24	1.18	1.14	1.10	1.07	1.05	1.02	1.0

* Extrapolated value

TABLE 2.16 Flow rate correction factors for washout holes

8-1/2"		*12-1/4"*		*17-1/2"*	
WASHOUT SIZE, IN.	α	*WASHOUT SIZE, IN.*	α	*WASHOUT SIZE, IN.*	α
9	1.12	13	1.10	18	1.03
10	1.38	14	1.24	19	1.09
11	1.65	15	1.39	20	1.16
12	1.94	16	1.53	21	1.22
13	2.24	17	1.68	22	1.28
14	2.55	18	1.82	23	1.34

- Mud rheology
- Mud flow regime
- Mud weight
- Hole angle
- Hole size

They also defined some uncontrolled variables, such as:

- Drill pipe eccentricity
- Cuttings density
- Cuttings size

These were put into seven dimensionless groups that were fitted to experimental data.

2.4.2.2 Hole Cleaning Model

Figures 2.32, 2.33, and 2.34 provide the results of these studies. The procedure for using these figures is as follows.

1. Enter the Rheology factor chart with the plastic viscosity (PV) and yield point (YP) values and read off the value of the Rheology Factor (RF).
2. Get the angle factor (AF) from Table 2.15.
3. Calculate the Transport Index, TI, based on RF, AF, and MW.

$$TI = RFxAFxMW \tag{2.26}$$

4. Enter the appropriate ROP chart with the TI and the desired (or maximum flow rate), read off the CFR for hole cleaning (or the maximum safe ROP).
5. If the hole is washed out, find the flow rate correction α from Table 2.15. Then, use Equation 2.8 to calculate the CFR for the washout hole section.

$$CFR_{washout} = \alpha x CFR_{gauge} \tag{2.27}$$

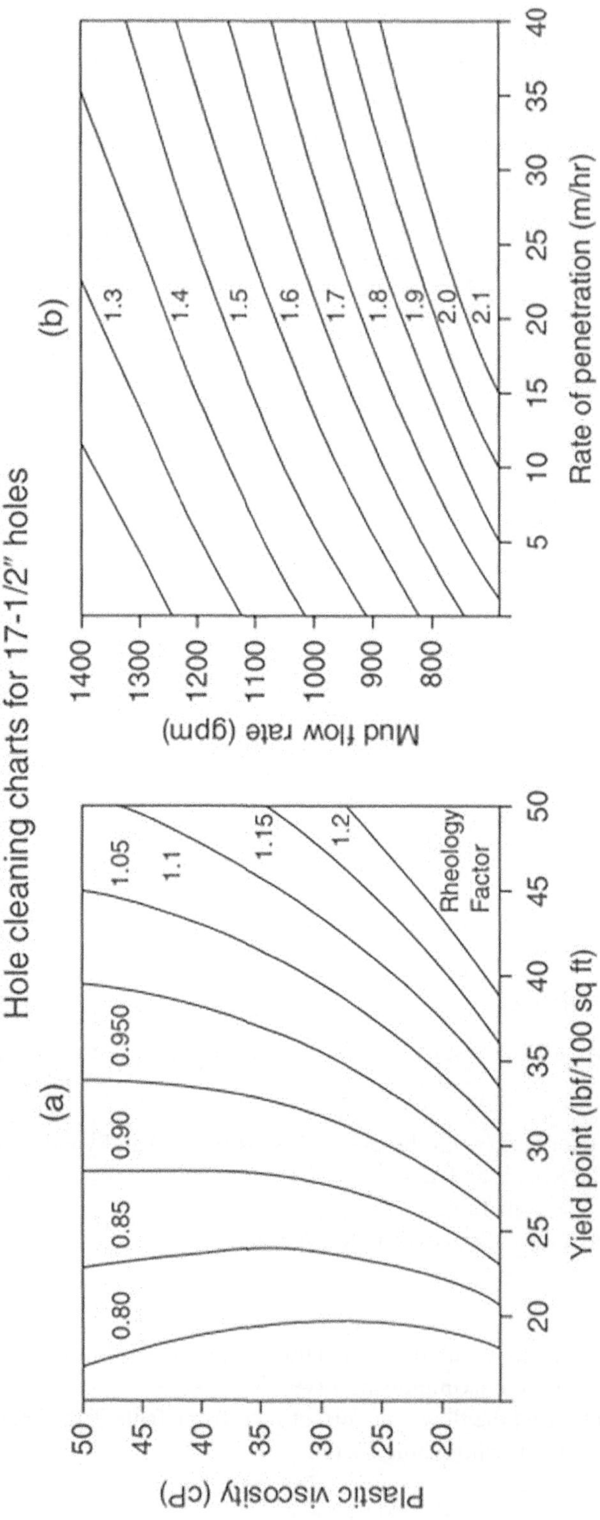

FIGURE 2.32 Rheology and hole cleaning charts for the 17-1/2 in. section.

Source: From Luo, Bern, Chambers, and Kellingray (1994).

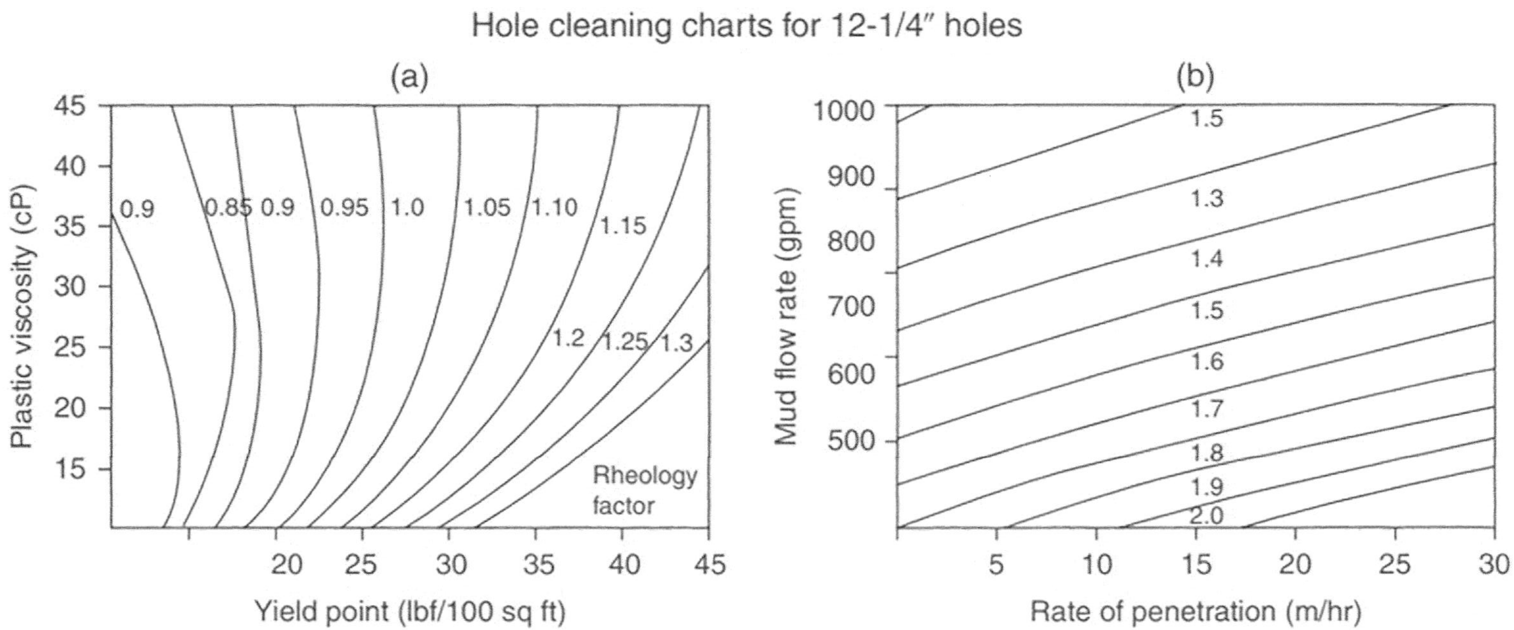

FIGURE 2.33 Rheology and hole cleaning charts for the 12-1/4 in. section.

Source: From Luo, Bern, Chambers, and Kellingray (1994).

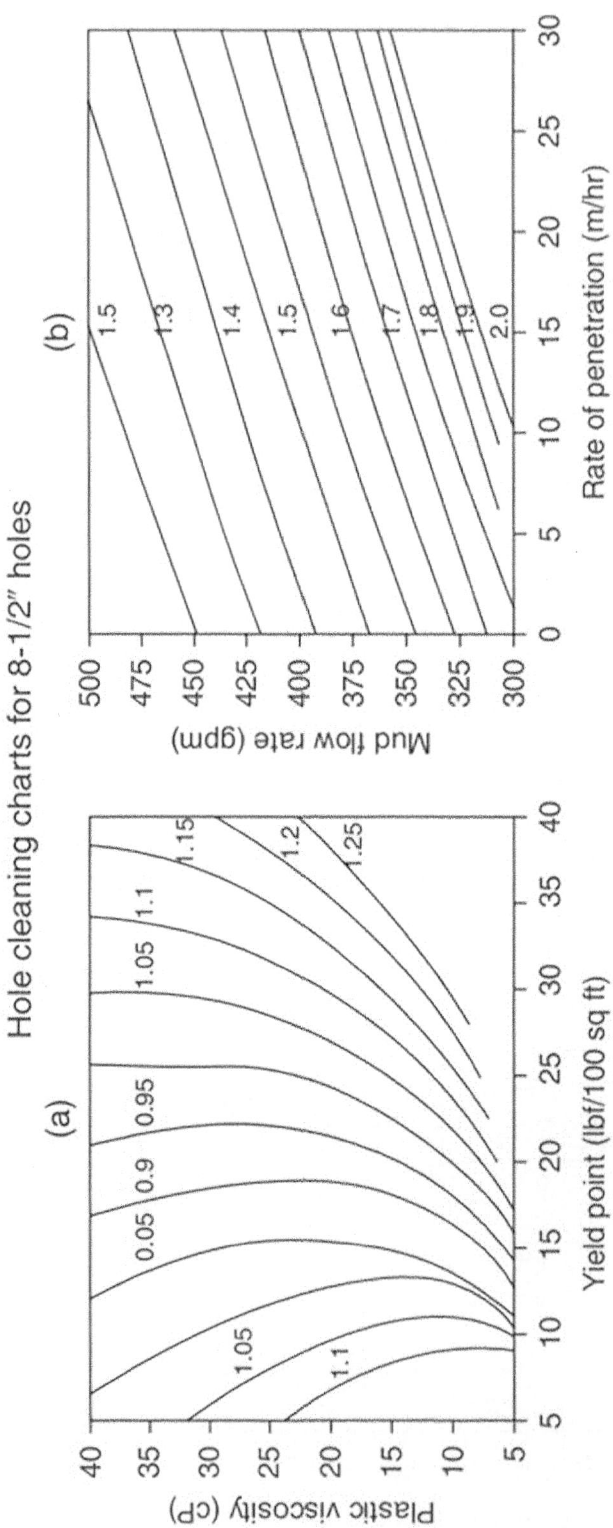

FIGURE 2.34 Rheology and hole cleaning charts for the 8-1/2 in. section.

Source: From Luo, Bern, Chambers, and Kellingray (1994).

The following example is adapted from Luo et al. (1994).

Example. A horizontal 8.5 in. section is to be drilled with a 1.45 s.g. mud where PV=25cP and YP=18 lbf/100ft^2. We want to know:

- What is the maximum safe ROP if the mud pumps can deliver a maximum 450 GPM?
- If it is anticipated that we can drill at an ROP of 20 m/hr, what flow rate will be required to clean the well?
- If we suspect that the hole has been washed out to 10 in., what flow rate should we pump?

Solution

Maximum safe ROP. From the rheology factor chart, Figure 2.34, it may be found that RF=0.91. From Table 2.15, the angle factor AF=1. The transport index becomes from Equation 2.7:

$$TI = 0.91x1.0x1.45 = 1.32$$

Then, from the ROP chart of Figure 2.34 at a TI of 1.32, it can be found that if the maximum achievable flow rate is 450 GPM, the maximum ROP that can be drilled without causing hole cleaning problems is about 23 m/hr.

The flow is at 20 m/hr. If we anticipate we can drill at an ROP of 20 m/hr, then the flow required to clean the hole is 440 GPM.

Washed out hole. However, if the hole is suspected being washed out to 10 in. and we still plan to drill at a rate of 20 m/hr, it may be found from Table 2.16 that the flow rate should be corrected by a factor of α=1.38, i.e.:

$$CFR_{washout} = 1.38x440 = 607GPM$$

Under this circumstance, measures must be taken either to increase the maximum achievable flow rate (e.g., by using larger drill pipes) or to adjust drilling parameters (e.g., mud YP).

The following figures are used to determine the flow rates and rate of penetration.

2.5 HYDRAULIC OPTIMISATION

2.5.1 Introduction

The traditional selection of the hydraulic parameters of a drilling system often involves an optimisation procedure. Typically, the flow rate directly beneath the drill bit is selected to be optimised. Typical optimisation criteria are the maximisation of the hydraulic energy delivered through the bit nozzles or the maximisation of the jet impact force. Although these criteria seem reasonable at first glance, a closer look at the total hydraulic system reveals that they may have limitations.

In this section, we will approach the hydraulic optimisation problem in a non-traditional way. We will use a semi-empirical approach for the pressure drop model. Furthermore, in addition to the classical optimisation methods like maximum hydraulic horsepower and jet impact, new criteria are derived that take the flow rate and cuttings transport into account. The reason is that the classical criteria have not been adequate in the evolution of deeper and longer reach wells.

2.5.2 The Hydraulic System

2.5.2.1 Pressure Losses

We will in this section define some simple equations to perform pressure drop calculations in the hydraulic system. First, we will investigate some properties of the various flow regimes. Bourgoyne et al. (1986) give a good overview of the equations needed to calculate friction losses in tubing and annuli for non-Newtonian fluids.

In general, we deal with two flow regimes. In the laminar flow regime, the fluid moves along defined paths, and the flow equations are determined analytically. In the turbulent flow regime, on the other hand, fluid moves in a chaotic manner. There are no analytical models available for this case, therefore, correlations have to be established using the friction factor concept. In general, we can say that the following relations exist between pressure drop and flow rate for Newtonian fluids:

For laminar flow:

$$P \sim \mu q \tag{2.28}$$

For turbulent flow:

$$P \sim \rho f q^2 \tag{2.29}$$

where: P = pressure drop
q = flow rate
μ = viscosity
ρ = fluid density
f = friction factor

Note that the pressure drop for flow in pipes depends on the flow regime; in laminar flow, the pressure drop is proportional to the viscosity and the flow rate, and in turbulent flow, the pressure drop is proportional to the density and the flow rate squared. Equations (2.28) and (2.29) are valid for Newtonian fluids. For non-Newtonian fluids, more complex relations exist, as described by Bourgoyne et al. (1986). However, the trends are similar, and since we are not going to use these equations in the analysis below, they will not be further addressed here.

Figure 2.35 illustrates the hydraulic system on a floating drilling rig. Inside the drill pipe, the flow velocity is high because of a small cross-sectional area. The velocity increases significantly over the bit nozzles. The inside of the drill string is usually in turbulent flow. In the annulus, the section along the bottom-hole-assembly may be in turbulent flow or in laminar flow, but the rest of the annulus, including the riser, is usually in laminar flow.

Seen in the context of Equations (2.28) and (2.29), we observe that we have a mixture of flow regimes. Therefore, the total pressure loss consists of a mixture of Equations (2.28) and (2.29).

From a functionality point of view, the flow across the bit nozzles shall remove drilled cuttings away from the drill bit. The flow in the annulus has the function of transporting these cuttings up the wellbore to be disposed of on the drilling rig. The pressure drop can be split into two groups:

- The pressure drops across the nozzles, which is aiding the drilling process by providing cleaning and hydraulic power.
- The pressure drops in the rest of the system, or the system pressure drops. This is also called the parasitic pressure drop, as it does not contribute to the drilling process.

If we consider the hydraulic system of Figure 2.35 as a whole, we can split the total pressure drop into a useful and a parasitic group as follows:

$$P_1 = P_2 + P_3 \tag{2.30}$$

where: P_1 = pump pressure
P_2 = pressure drop across bit nozzles
P_3 = parasitic pressure loss or system losses

For a moment, we will consider the parasitic pressure loss. We observe from Equations (2.28) and (2.29) and Figure 2.35 that we have a mixture of flow regimes. Instead of modelling each element of

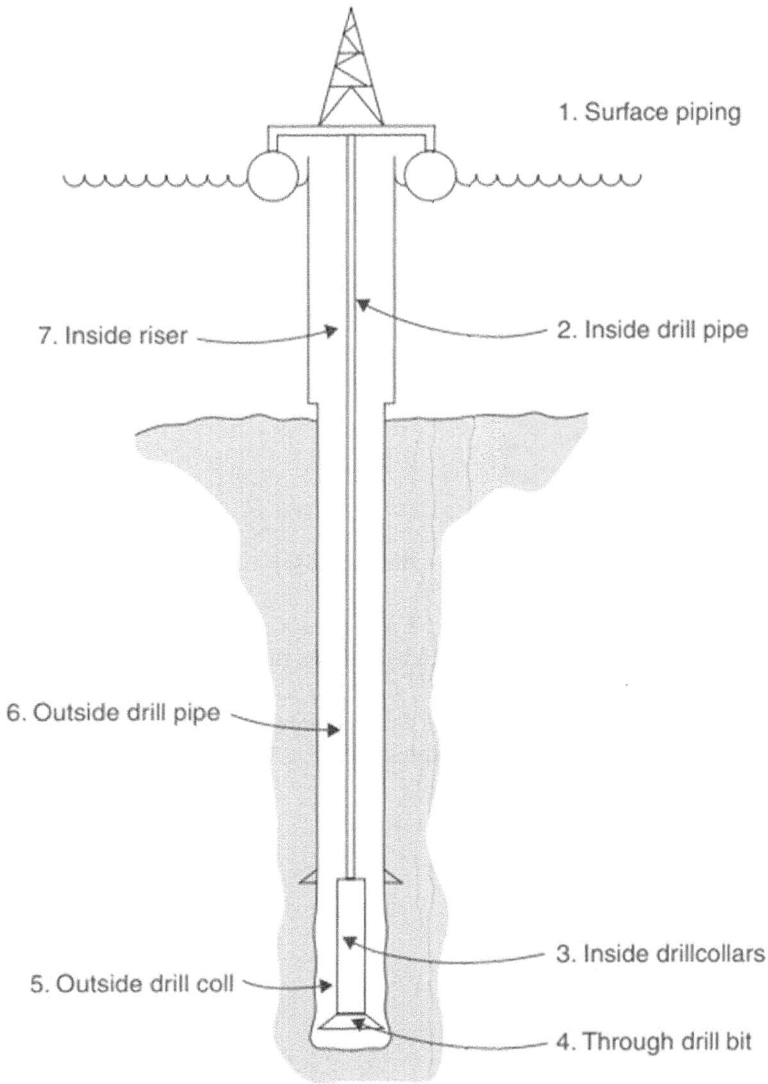

FIGURE 2.35 The hydraulic system.

the system and adding the contributions, we will use one simple equation that describes the whole system.

$$P_3 = Cq^m \tag{2.31}$$

where: C = proportionality constant
m = flow rate exponent

Typically, the pressure losses in the annulus, or the laminar parts of the system, are of the order 10–20 % of the total pressure drop. The losses inside the drill string dominate the parasitic pressure loss. Since this usually is turbulent, Equation (2.29) dominates the process. Therefore, Equation (2.31) is dominated by turbulent flow, which results in an exponent slightly less than two.

The pressure drop across the drill bit nozzles must also be evaluated. The flow rate through the nozzles is given by the continuity equation:

$$q = v_a A_a = v_b A_b = \text{constant}$$

$$\text{or}: v_a = \frac{q}{A_a}, \quad v_b = \frac{q}{A_b} \tag{2.32}$$

where v is velocity and A is area. Using the conservation of energy principle and assuming an incompressible and frictionless system, the pressure drop across the bit nozzles is:

$$\frac{v_a^2}{2} + \frac{P_a}{\rho} = \frac{v_b^2}{2} + \frac{P_b}{\rho}$$

$$P_2 = P_a - P_b = \frac{\rho}{2}\left(v_b^2 - v_a^2\right)$$

where subscript a refers to the drill pipe and b to the nozzles, ρ is fluid density and g is the gravitational constant.

Two simplifications will now be introduced. Firstly, the velocity inside the drill string is negligible compared to the nozzle velocity. Therefore, the drill string flow velocity is neglected. Secondly, from experimental measurements, one has found that the flow is not ideal and is somewhat lower than predicted by the equation above. A discharge coefficient of 0.95 is often used. Introducing these elements and the continuity relation, the above equation can be expressed as:

$$v_o = 0.95\sqrt{\frac{2P_2}{\rho}}$$

or:

$$P_2 = \frac{\rho q^2}{2A^2 0.95^2} \tag{2.33}$$

We have now defined all elements required to use the total pressure drop equation, Equation (2.30). In the following, an example will demonstrate the application.

TABLE 2.17 Parasitic pressure losses

DEPTH (M)	*PRESS. DROP (BAR)*	*FLOW RATE (L/MIN)*
1200	100	2228
	173	3000
2200	103	2000
	218	3000
3200	123	2000
	259	3000

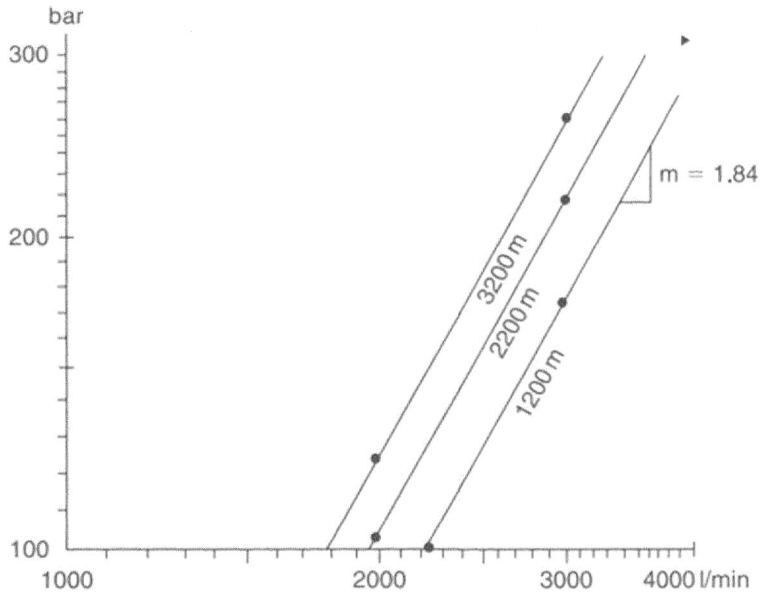

FIGURE 2.36 System pressure losses for various depth intervals.

Example

From an exploration well, the following system losses or parasitic losses have been measured at different well depths (Table 2.17).

It is difficult to show the continuous pressure drop-flow rate behaviour on a plot because we have only a few discrete measurements. However, taking the logarithm on both sides of Equation (2.22), we obtain:

$$lnP_3 = lnC + m\ lnq$$

If the logarithmic relationship given by Equation (2.31) is correct, the data should plot as straight lines on a log-log plot, with a slope equal to *m*. This is performed in Figure 2.36. The numerical value of the slope can be obtained by using the equation above on two data sets and subtracting, as exemplified with the two first entries of the data set above:

$$m = \ln (100/173)/\ln (2228/3000) = 1.84$$

Repeating this process for the two other depth intervals yields the same value. Inserting this value, and a data set from Table 2.17 into Equation 2.31, we obtain an equation for each of the three depth intervals:

$P_3 = 6.92 \times 10^{-5}q^{1.84}$ at 1200 m depth

$P_3 = 8.72 \times 10^{-5}q^{1.84}$ at 2200 m depth

$P_3 = 10.36\times 10^{-5}q^{1.84}$ at 3200 m depth

The three pressure-flow rate functions are shown in Figure 2.36. We see clearly the advantage of using a log-log plot. With only two measurements, the complete pressure range can be established.

We also observe that it is only the scaling factor in front of each of the three expressions above which are different. A plot showed furthermore that this constant is a linear function with depth. Solved as a depth-dependent function, the three expressions above can, therefore, be presented in a single equation as follows:

$$P_3 = (4.86 + 0.00172D)10^{-5}q^{1.84}$$

A few words about the equation above need to be said. If based on a few measurements as in the example, care should be taken to avoid extrapolating beyond the range of the measurements. If measurements are not available, hydraulic modelling can be used.

The equation above is linear with depth. This is expected in general as we typically use one type of bottomhole assembly and only add drill pipes when deepening the well. The pressure drop increase from adding drill pipe is a linear function with a constant flow rate.

Finally, the exponent m, contains both a viscosity and a density element as indicated in Equations (2.28) and (2.29). If significant changes are introduced in the fluid properties or in the equipment, this exponent is no longer expected to stay constant.

We will, in the continuation, use the parasitic pressure loss function above, but point at the fact that the same ideas can be approached with more complex modelling.

2.5.2.2 Classical Optimisation Criteria

Now, we will briefly derive the two classical hydraulic optimisation criteria that have been used for decades: the maximum hydraulic horsepower and the maximum jet impact.

The hydraulic horsepower across the bit nozzles is given by:

$$HP = P_2 q \tag{2.34}$$

We can replace the pressure drop across the bit as the difference between the pump pressure and the parasitic pressure losses, Equations (2.30) and (2.31):

$$HP = (P_1 - Cq^m)\, q$$

To find the maximum hydraulic horsepower across the bit, the equation above is differentiated and set equal to zero:

$$\frac{d(HP)}{dq} = P_1 - C(m+1)q^m = 0$$

$$q^m = \frac{P_1}{C(m+1)}$$

The fraction of parasitic pressure loss over total pump pressure to obtain maximum hydraulic horsepower can be expressed as:

$$\frac{P_3}{P_1} = \frac{1}{m+1} \tag{2.35}$$

This process is repeated for the jet impact criterion. Imagine that a jet impinges the bottom of the hole. If the fluid momentum is destroyed in the impact, then the jet impact force is given by:

$$F_2 = \dot{m}v = \rho q v \tag{2.36}$$

Inserting Equation 2.33, the impact force can be expressed in terms of bit pressure loss and flow rate:

$$F_2 = 1.344\sqrt{\rho q^2 P_2}$$

Again, inserting Equations (2.30) and (2.31), we obtain:

$$F_2 = 1.344\sqrt{\rho q^2 \left(P_1 - Cq^m\right)}$$

Differentiating this equation and setting equal to zero yields:

$$\frac{dF_2}{dq} = 2P_1 - C\left(m+2\right)q^m = 0$$

$$q^m = \frac{2P_1}{C\left(m+2\right)}$$

and the fraction parasitic pressure loss for the maximum jet impact force is:

$$\frac{P_3}{P_1} = \frac{2}{m+2} \tag{2.37}$$

Equations (2.35) and (2.36) define the fraction of parasitic pressure losses that gives the maximum hydraulic horsepower and jet impact beneath the drill bit. An example will demonstrate the application of these concepts.

Example:
Assume that we consider the hydraulic system used in the previous section. The exponent is here $m=1.84$. The fractions of pressure drops are then given by Equations (2.35) and (2.37):

For maximum hydraulic horsepower:

Fraction parasitic pressure drop: $\frac{P_3}{P_1} = \frac{1}{1.84+1} = 0.35$

Fraction pressure drop across drill bit: 1–0.35 = 0.65

For maximum jet impact:

Fraction parasitic pressure drop: $\frac{P_3}{P_1} = \frac{2}{1.84+2} = 0.52$

Fraction pressure drop across drill bit: 1–0.52 = 0.48

When using these criteria, the drill bit nozzles have to be selected to satisfy the fractions defined above.

We observe that there is a considerable spread in the pressure loss fractions depending on the criterion chosen. A pertinent question to ask is what criterion is best. Is there physical significance in these criteria? In all analytical and experimental work seen so far, these two criteria have been used. We will expand with new criteria but will first investigate some cases in an attempt to evaluate shortcomings.

2.5.2.3 Example of Shortcomings with the Classical Approach

The two classical optimisation criteria defined in the previous chapter have been used extensively. A key assumption is that the drill bit works best under these conditions.

We will investigate this further.

A numerical example is shown in Figure 2.37. The equation for parasitic pressure loss from the previous section is plotted. Also, it has been assumed that the pump is working under a maximum constant pressure. This is often the case for the deeper sections of the well.

The total pressure is 300 bar. The difference between the total pump pressure and the parasitic pressure is equal to the pressure drop across the nozzles. We observe that for low parasitic pressure loss, the bit loss is high and vice versa. Also indicated are the maxima for the two previously defined optima. If a high bit nozzle pressure loss is wanted, the flow rate must be low. To summarise, we have so far compared three criteria, the max. hydraulic horsepower, the max. jet impact force, and the max. pressure drop. They all result in different flow rates.

Next, let us consider another element, which is not addressed yet, namely the carrying capacity of the drilling fluid.

Lermo (1993) investigated the relationships between the hydraulic optimalisation criteria and the cuttings transport in the wellbore. Figure 2.38 shows an example from a simulation of a deep well. Shown are two curves

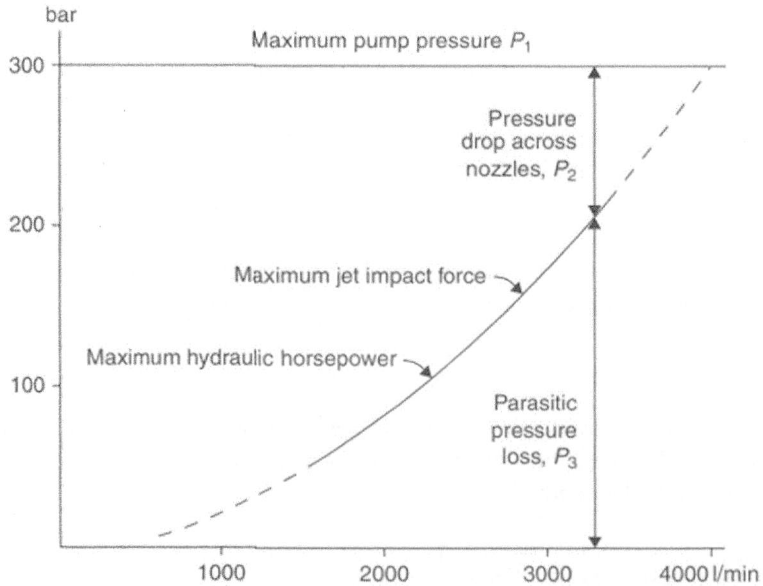

FIGURE 2.37 The total, parasitic and bit pressure loss at 1200 m depth.

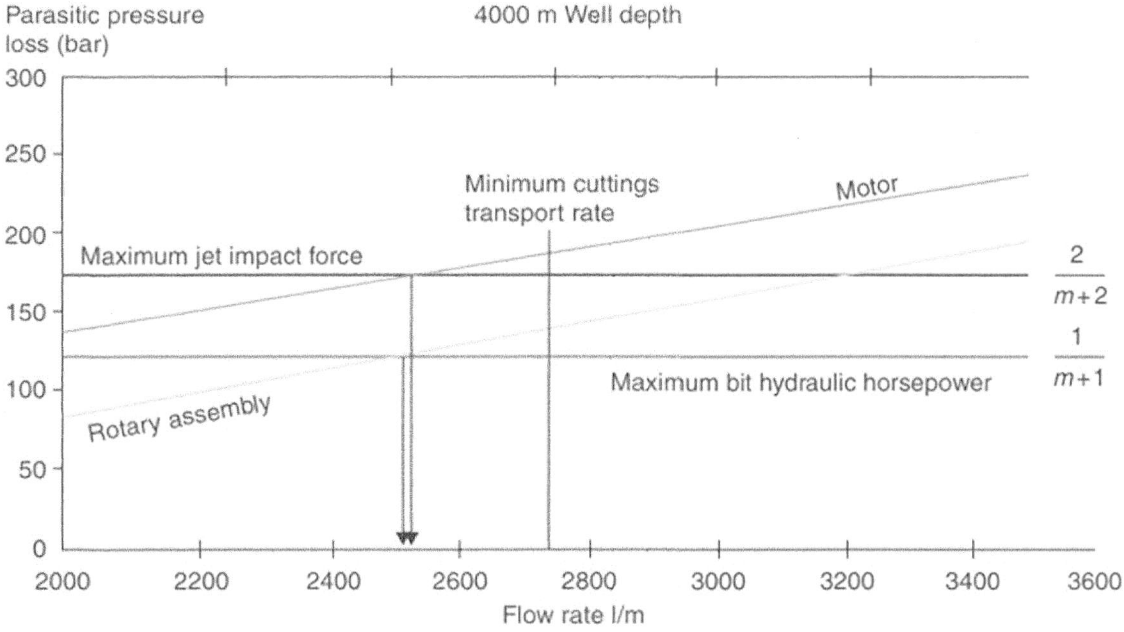

FIGURE 2.38 Example case showing the need for carrying capacity.

TABLE 2.18 Overview over the hydraulic system

POSITION	*FLOW REGIME*	*LIMITATION*	*CRITICAL PARAMETER*
1. Surface piping	Turbulent	Wear	
2. Inside drill string	Turbulent	Wear	
3. Inside drill collars	Turbulent	Wear	
4. Through nozzles	Turbulent	Wear	
5. Outside drill collars	Turb. /Laminar	Washout?	Flow rate
6. Outside drill pipe	Laminar	Cuttings transport	Flow rate
7. Inside riser	Laminar	Cuttings transport	Flow rate

for parasitic pressure losses, one for a standard rotary assembly and one for application with a downhole motor. Also shown are the two pressure levels defining the optimum conditions from the previous discussion. We observe that the max. jet impact force has optima at 2520 and 1800 l/min, while the max. hydraulic power has optima at 2520 and 3190 l/min. However, also shown is a line at 2720 l/min, which is the minimum flow rate to ensure hole cleaning. In this case, the flow rates given by the two classical criteria will be insufficient to clean the well, except for the case with a rotary assembly and max. jet impact criterion.

To summarise this discussion, the following elements may have importance in hydraulic optimisation:

Cleaning or provide impact beneath the drillbit:	maximum hydraulic horsepower, -or maximum jet impact force. -or maximum pressure
Transporting drilled cuttings out of well	-minimum flow rate

Regarding the hydraulic process beneath the drill bit, we will not draw any firm conclusion. Probably some optimum exists regarding the interaction between the rock and the drill bit, but it is not certain which physical mechanism dominates. There are several functions, as pure cleaning beneath the bit, or mechanical work on the rock itself. Experience from metal work suggests also that there may be various optima depending on the rock properties.

Regardless of the criteria chosen for the drill bit, the cuttings must not be allowed to accumulate in the annulus. Therefore, the minimum transport flow rate must be used, even if it violates other hydraulic criteria. For some years, the shortcomings of the classical optimisation criteria have been noted. For deep, inclined, and long reach wells, the flow rates used are often higher than predicted by the classical models. The industry has, however, lacked some systematic way to handle this issue, and in the following a model to solve the problem will be proposed.

The hole cleaning issue will not be pursued further here, but the reader is referred to publications by Zamora & Hanson (1990), Sifferman & Becker (1992), and Hemphill & Larsen (1993).

2.5.3 Hydraulic Optimisation

2.5.3.1 Hole Cleaning

As shown in the previous section, the flow rate might be a limiting factor for cutting cleaning purposes. Before proceeding, let us briefly investigate the hydraulic system on a rig.

With reference to Figure 2.35, we will investigate the various elements of the hydraulic system and the effects of the flow rate on each component.

A brief discussion about the elements above would include the aspect of wear. The majority of the pressure losses occur before the drill bit. A too-high flow rate gives rise to pressures that may promote premature drill string failure or so-called washout. It should, therefore, be mentioned that there could be a penalty for loading the system too much.

The system after the drill bit is mostly limited by the cuttings transport process. This is a central problem and must be properly handled. Therefore, this element must always be satisfied regardless of other optimisation criteria.

The central issue in this section is the new optimisation criteria. However, in the practical application of these criteria, one must use cutting transport models to define the minimum flow rates for a given well. Please refer to Section 2.3 for the assessment of the flow rate.

2.5.3.2 A New Method for Hydraulic Optimisation

Traditionally, performance criteria that relate to the physical process are chosen. Commonly used are the maximum hydraulic horsepower and the maximum jet impact force, as discussed in the foregoing.

The flow rate is believed to be the critical parameter to enhance the transport of drilled cuttings. Recent trends in offshore drilling are also towards higher flow rates than those recommended by the traditional criteria. However, the flow rate itself is not suited as a performance criterion, as it does not take into account other sides of the process.

We need some criteria to select hydraulic parameters in a systematic way in a drilling operation. In the following, another way to approach the problem will be proposed.

Hydraulic horsepower spent across the bit nozzles is equal to the product qP_2, and its maximum is defined by Equation (2.35).

The jet impact force is obtained by multiplying the flow rate with $\sqrt{P_2}$ instead of P_2 as for the hydraulic power, as shown in Equation (2.36).

We will take advantage of the pattern indicated above and define a non-physical variable, which is the product of $q^{n/2}\sqrt{P_2}$, where the parameter n is defined as a performance index.

Differentiating this function and setting it equal to zero gives the optima:

$$\frac{d\left(q^{n/2}\sqrt{P_2}\right)}{dq}=0$$

TABLE 2.19 Overview over optimisation criteria

PERFORMANCE INDEX	*EQUATION*	*CRITERION*	*FRACTION PARASITIC PRESSURE LOSS*	*FLOW RATE*
1	qP_2	Max. HP	$\frac{1}{m+1}$	$\frac{P_1}{C(m+1)}$
2	$q\sqrt{P_2}$	Max. jet impact	$\frac{2}{m+2}$	$\frac{2P_1}{C(m+2)}$
3	$q^{3/2}\sqrt{P_2}$	New A	$\frac{3}{m+3}$	$\frac{3P_1}{C(m+3)}$
4	$q^2\sqrt{P_2}$	New B	$\frac{4}{m+4}$	$\frac{4P_1}{C(m+4)}$
5	$q^{5/2}\sqrt{P_2}$	New C	$\frac{5}{m+5}$	$\frac{5P_1}{C(m+5)}$

Equations (2.30) and (2.31) express the nozzle pressure drop as pump pressure and parasitic pressure:

$$P_2 = (P_1 - Cq^m)$$

Combining the two equations above and solving, we obtain a general equation for optimisation criteria as follows:

$$\begin{aligned} q^m &= \frac{nP_1}{C(m+n)} \\ P_3 &= \frac{P_1 n}{m+n} \end{aligned} \tag{2.38}$$

A number of performance criteria can be derived using the performance indices. Table 2.19 summarises this:

The following steps define the application of the hydraulic optimisation principles outlined here:

- Determine the flow rate q that ensure proper cleaning of the borehole.
- Select performance index.
- Compute system loss P_3 and bit loss P_2.
- Compute nozzle area A.

2.5.4 Field Case

The well that is studied in the field case is vertical and classified as a wildcat well. The problem to be addressed is the design of the hydraulics for the 12-1/4in. section in the interval 1200–3200 m. A PDC bit was chosen to drill this section. This bit had one centre nozzle and five nozzles between the blades.

First, a study was conducted to summarise past experiences with this particular drillbit. Table 2.20 summarises some of the parameters. Of particular interest was the observation that 8 of the 18 drill bits were pulled with plugged centre nozzles. For this reason, the standard 12/32 in. centre nozzle was

TABLE 2.20 Summary of earlier bit runs

BIT NO.	NOZZLES	Q (L/MIN)	ROP (M/HR)	REMARKS
1	5x16, 1x12	2960	1.5	Plugged centre nozzle
2	5x19, 1x12	2660	9.8	
3	5x16, 1x12	2600	13.6	
4	5x19, 1x12	2300	18.2	
5	5x18, 1x12	2400	14.9	Plugged centre nozzle
6	6x12	2600	18.3	
7	5x14, 1x12	2400	15.4	
8	5x15, 1x12	2450	24	
9	5x14, 1x12	2400	4.8	
10	5x14, 1x12	2350	23.8	
11	5x19, 1x12		20	Plugged centre nozzle
12	5x19, 1x12		30	
13	5x18, 1x12		10	
14	5x18, 1x12		22	Plugged centre nozzle
15	5x19, 1x12		7	Plugged centre nozzle
16	5x18, 1x12		27	Plugged centre nozzle
17	5x19, 1x12		16	Plugged centre nozzle
18	5x19, 1x12		19	Plugged centre nozzle

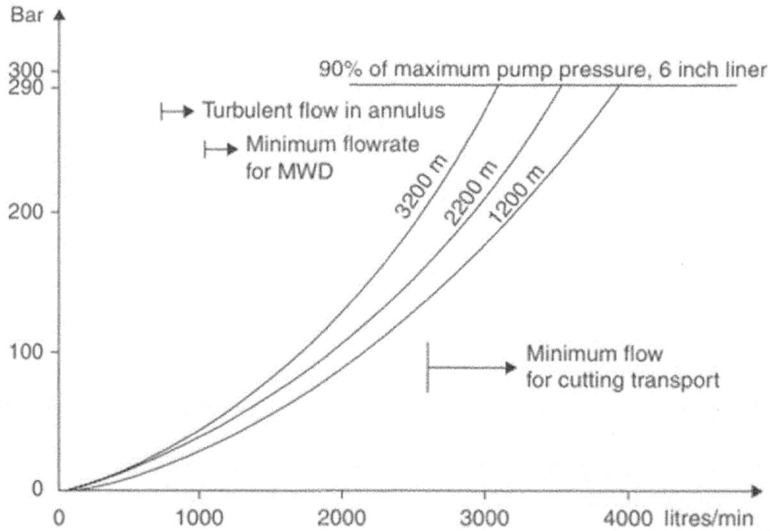

FIGURE 2.39 Parasitic pressure losses and flow rate constraints.

recommended to be increased in size to improve cleaning beneath the bit. There is no correlation between the drilling rate and the number of plugged centre nozzles. The experience is that once the centre nozzle is plugged, the drilling rate decreases significantly, resulting in the pulling of the bit. Table 2.20 can be used to argue that the centre nozzle is particularly sensitive in this drill bit design and that this must be addressed further in the hydraulics design.

Figure 2.39 shows the parasitic pressure losses at 1200 m depth, at 2200 m., and at 3200 m. Also shown is the maximum allowed pump pressure at 290 bars. Three other criteria are shown, namely the minimum flow rate acceptable for the measurement-while-drilling system (MWD), the flow rate at which the flow becomes turbulent around the bottom-hole-assembly, and the minimum flow rate to ensure good

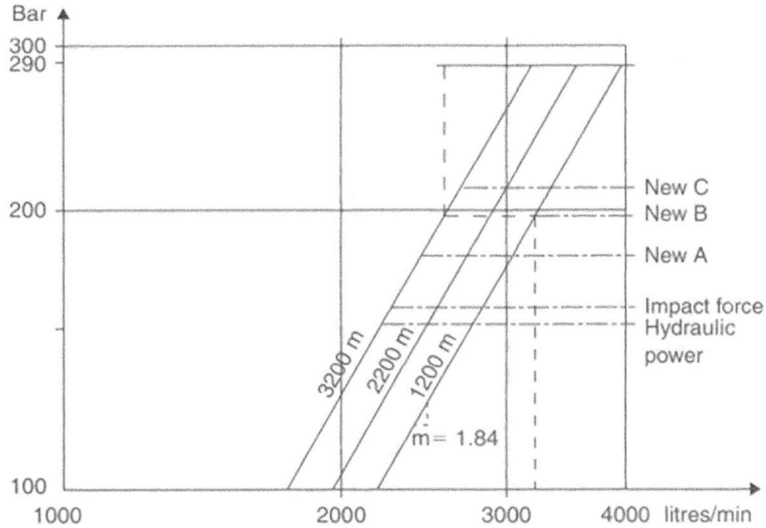

FIGURE 2.40 Determination of flow ranges and optimisation criteria.

TABLE 2.21 Hydraulic parameters for the field case

CRITERION	*PERCENT PARASITIC PRESS. LOSS*	*FLOW RANGE (L/MIN)*
Max. hydraulic power	54	2800–2220
Max. jet impact	52	2850–280
New A	63	3070–2450
New B	69	3250–2580
New C	73	3370–2800

hole cleaning. It is the latter factor that dominates this design, and this is usually the case, therefore, we will accept turbulent flow around the bottom-hole-assembly.

These data are redrawn to a log-log scale in Figure 2.40. For this well, the slope is: m=1.84. Furthermore, the five optimisation criteria of Table 2.5 are shown.

Criteria New B was chosen for the operation, mainly because it was being attempted to keep the flow rate above the critical flow for cuttings transport, 2520 l/min. From Figure 2.40, we also see that the flow rate should ideally vary from 3250 l/min. at 1200 m. to 2580 l/min. at 3200 m., with about 2880 l/min. at 2200 m. The classical criteria recommend a lower flow rate, as shown. These data are summarised in Table 2.21.

The result of the bit run was 1238 m drilled; a company record with this particular bit type. No nozzle plugging was observed. Although this field case is no proof for the new models presented here, it shows that sound pre-planning does give results. In this case, two changes were introduced: a higher flow rate in the whole system and an increased flow rate in the centre nozzle of the bit.

The three lines of Figure 2.40 define the depth levels 1200, 2200, and 3200 m. As the well deepens, the parasitic pressure loss increases due to the added drill pipe. To avoid exceeding the maximum pump pressure of 290 bar, the flow rate has to be gradually decreased. From Table 2.21, and the New B criterion, we see that the proposed flow rate is 3250 l/min at 1200 m depth, decreasing to 2580 l/min at 3200 m. During an actual drilling operation, the flow rate will gradually be reduced by monitoring the pump pressure.

The actual nozzle selection process will be demonstrated with the following example:
At 1200 m depth, the flow rate is 3250 l/min, and the fraction parasitic loss is from Table 2.19:

$$\frac{4}{m+4} = \frac{4}{1.84+4} = 0.69 \quad or, \quad 0.69x290bar = 200bar$$

TABLE 2.22 Optimal nozzle selection for New B criterion

DEPTH (M)	NOZZLES (IN)
1200	five 13/32, one 16/32
2200	five 12/32, one 16/32
3200	five 11/32, one 16/32

and the pressure loss across the nozzles: 290–200 = 90 bar

The nozzle area required can be calculated with Equation (2.33):

$$A = q\sqrt{\frac{\rho}{2P_2}\frac{1}{0.95}}$$

Using the units of density (kg/l), flowrate (l/min), and pressure (bar), the nozzle area in in^2 can be obtained by dividing the equation above with 376. The result is:

$$A = 3250\sqrt{\frac{1.65}{2x90}\frac{1}{0.95x376}} = 0.87in^2$$

Using five 13/32 in. nozzles and one 16/32 in. nozzle, this area is approximately obtained:

$$A = 5\frac{\pi}{4}\left(\frac{13}{32}\right)^2 + \frac{\pi}{4}\left(\frac{16}{32}\right)^2 = 0.85in^2$$

Repeating this process at the two other depths, we can define the nozzle program as follows if we assume a drill bit with six nozzles (Table 2.22).

In a practical application, we have to consider the penetration length of each drill bit, as we can only change nozzles when pulling the drill bit. If we assume one drill bit for each of the intervals of Table 2.18, five 13/32 in. and one 16/32 in. nozzles may be used in the interval 1200–2200 m. However, since we are approaching the critical flow rate for cuttings transport in the bottom interval, we will recommend using the nozzles designed for 3200 m depth.

If more or less bit runs are expected, the nozzle selection must be evaluated accordingly.

2.5.5 Proposed Optimisation Criteria for Various Well Types

Lermo (1993) undertook a larger analysis of the cuttings transport velocity and the optimisation criteria. He evaluated various well types and depths and arrived at the following tabulated recommendation. For the cuttings transport analysis, he used several commercially available simulators, which are considered the state of the art. We will not address this in detail but will show some conclusions of Lermos's work.

Table 2.23 shows some results for the 12-1/4 in. hole section. Although the table defines criteria suitable for the drilling of the wells, the right-hand column proposes a stronger criterion if hole cleaning is

TABLE 2.23 Proposed optimisation criteria for typical 12-1/4 in. hole

HOLE LENGTH	*VERTICAL HOLES*	*DEVIATED WELLS DRILLED WITH MOTOR*	*DEVIATED WELLS WITHOUT MOTOR*	*STRONGER REQUIREMENTS*
Less than 2500 m	Max. HP or Max. Jet Impact	Max. HP or Max. Jet Impact	Max. Jet Impact	New A
2500 – 4000 m	Max. HP or Max. Jet Impact	Max. Jet Impact	New A	New B
Deep (5000 m)	Max. HP or Max. Jet Impact	Max. Jet Impact or New A	New B	New C

Other data: Drill pipe: 675m of 5 in., rest 6-5/8 in. Drill collars: 120 m 8-1/8 in. OD, 2.81 in. ID Mud density: 1.65 s.g. Yield point: 32 lbf/100 sqft. Plastic viscosity: 42 cP.

TABLE 2.24 Minimum flow rate versus drill pipe size

DRILL PIPE SIZE (IN)	*MINIMUM FLOW RATE (L/MIN)*
5	3490
5-1/2	3800
6-5/8	4370

believed to be a problem. If significant hole collapse takes place, for example, a stronger requirement may be applied to ensure good cuttings transport.

For the 17-1/2 in. sections, the cuttings' carrying capacity became more critical. For vertical wells, the criteria of Table 2.19 ensured adequate cuttings transport. However, when wells were inclined with deviation exceeding 45 degrees, a significant increase in flow rate was required. Lermo recommended the use of three mud pumps instead of two, and the optimisation procedure also involved pump liner size selection. In general, the two criteria, New B and New C, were the result of the optimisation process.

The effects of changing drill pipe size were also investigated. By increasing the drill pipe size, the parasitic pressure losses decrease, and the flow velocity in the annulus increases, both improving the total process. Figure 2.41 illustrates this in a parasitic pressure drop-flow rate plot. The three curves represent the system losses for three drill pipe sizes. The criterion new C is applied in each of the cases, and we observe that a higher flow rate results in a larger drill pipe, mainly because of reduced parasitic pressure losses. The following flow rates resulted from this analysis:

By increasing the drill pipe size, we also add flexibility to the hydraulic design.

Hole size:	17-1/2 in.
Drill pipe length:	2323 m
Drill collars:	177 m 8 in. OD, 2.81 in.ID
Mud density:	1.50 s.g.
Yield point:	28 lbf/100 sqft
Plastic viscosity:	29 cP

In this section, we have shown that the two classical hydraulic optimisation criteria may be inadequate for some wells. Since experience shows that poor hole cleaning often is associated with borehole problems, new optimisation criteria are needed. Therefore, in this chapter, new criteria are derived. These are aimed at providing sufficient flow rate to clean the wellbore and should provide a methodology to design the complete hydraulic system.

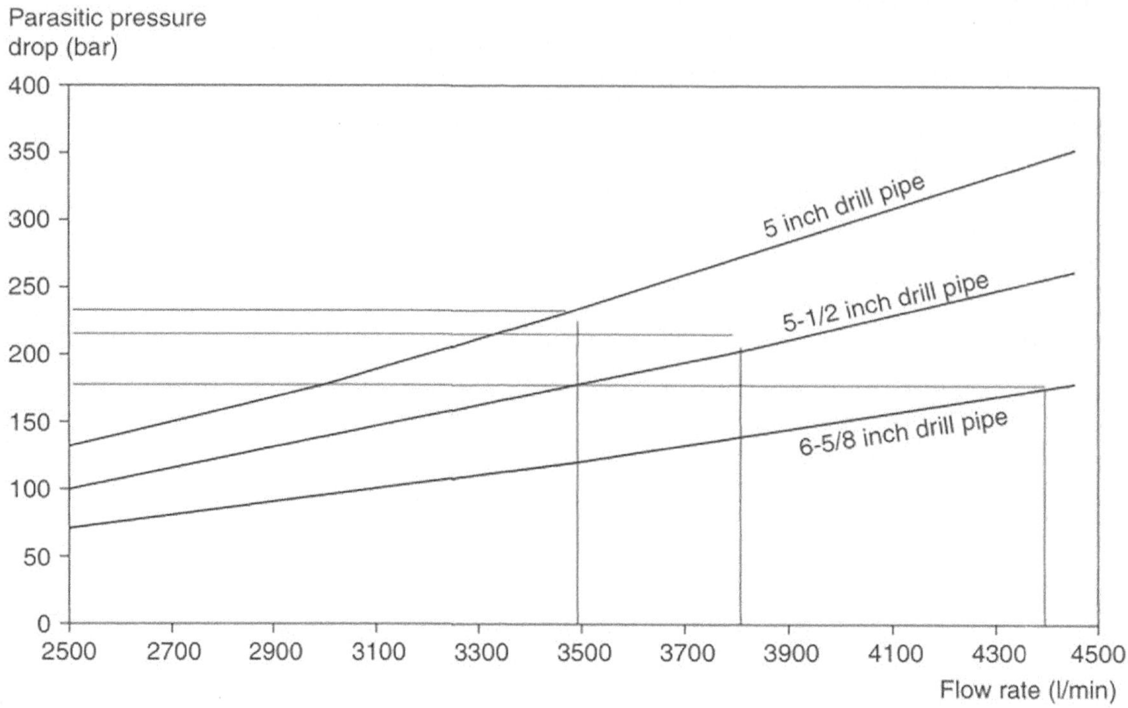

FIGURE 2.41 Pressure losses and flow rates with different drill pipe sizes, using the New C criterion.

Before closing this chapter, a couple of elements related to the hydraulic system should be mentioned. An emerging mechanism is the effect of drill string rotation on the pressure drop. Rotation typically led to increased pressure loss, and the reader is referred to publications by Lockett et al. (1993), Marken and Saasen (1992), Oudeman and Bacarreza (1995), and Cantalos and Dupuis (1993).

Finally, more recently, the understanding has emerged that the element of hole cleaning is an important factor for drill string torque and drag. For long reach wells good hole cleaning must be obtained to be able to reach the target. Aarrestad and Blikra (1994) and Alfsen et al. (1995) give field cases demonstrating this.

Problem

During the drilling of the 12-1/4 in. section of an exploration well, it was decided to use the maximum hydraulic horsepower criterion. The data are:

Flow rate exponent	$m = 1.67$
Mud density	1.25 kg/l
Mud pump pressure	300 bar
Flow rate	2430 l/min

a) Using Table 2.19, determine the parasitic pressure loss.
b) Determine the pressure drop across the bit nozzles.
c) Determine the nozzle sizes, assuming that a 3-cone rock bit is used.

During the operation drilling problems were observed. It was concluded that the hole cleaning was not adequate, and that the flow rate had to be increased to clean the hole. Based on a hydraulic simulator, it was decided to increase the flowrate to 3400 l/min. The parasitic pressure loss was now 196 bar.

d) Calculate the percentage parasitic pressure loss. Using Table 2.19, determine which optimisation criterion which best fits the system now.
e) Determine the pressure drop across the bit nozzles.
f) Determine the nozzle sizes for the 3-cone bit.
g) Compare and discuss the two-optimisation criterion used.

Geomechanical Evaluation 3

3.1 DATA NORMALISATION AND CORRECTION

3.1.1 Introduction

In addition to the general nomenclature defined under Symbols and Units, the following specific nomenclature is used in this chapter:

sf = Sea floor reference level
h_f = Distance drill floor-sea level, m
h_w = Water depth, m
δh = Difference in drill floor elevation between two platforms, m

The petroleum industry uses pressure gradients to a large extent. One key reason is that the hydrostatic mud weight plays a central role in the drilling operation. It is, therefore, simple to refer to a density instead of a pressure, which would vary with depth. By presenting all pressure data as gradients, the element of depth is removed. However, by dividing the pressure with a depth, the actual choice of reference level becomes important. In this chapter, we will, therefore, look into various ways to ensure data consistency.

If we are using data from the same field or platform, we may use them without corrections. However, we are often compiling data from floating rigs, fixed rigs with various drill floor elevations, and data from wells with significant differences in water depths. Also, drilling personnel often use the drill floor as a reference (RKB), while geologists often use the mean sea level (MSL) as a depth reference.

Obviously, using data of various origins may introduce significant errors. In this chapter, we will show how to make the data consistent in a simple manner. However, the key is to select a reference system and to normalise all data to this common reference.

The data we are mainly considering at this point is the mud weight, the frac. gradient, the overburden stress gradient, the horizontal stresses, and the pore pressure gradient. However, all other pressure gradient data should also be corrected to a common reference.

3.1.2 Selecting a Depth Reference

3.1.2.1 Correcting to Mean Sea Level

As already mentioned in the introduction, geologists often adjust the data to the mean sea level. This has the advantage of removing the effects of various drill floor elevations.

Consider a scenario as shown in Figure 3.1. A drilling rig has an elevation h_f from the sea level to the drill floor. Assume that at a given depth D, the pressure is P. This can be either the static mud pressure,

 DOI: 10.1201/9781003618171-3

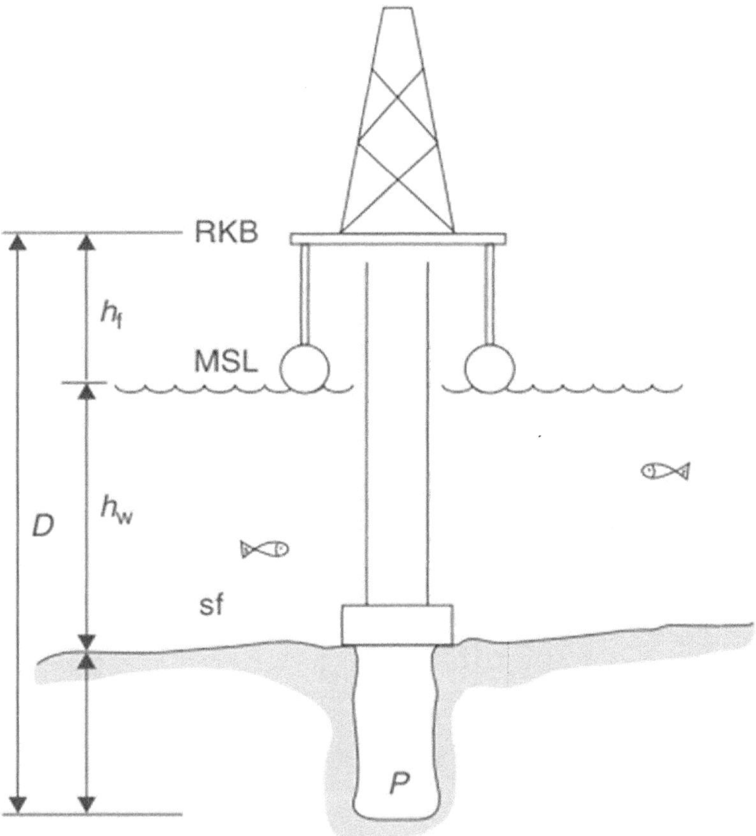

FIGURE 3.1 Definition of references.

a frac. pressure or pore pressure. Expressed as a gradient relative to the drill floor, this pressure can be expressed as:

$$P = 0.098\, d_{RKB} D$$

We want to express this pressure relative to the mean sea level. The pressure must be the same, but the gradient must be different due to a different reference level:

$$P = 0.098 dMSL(D-hf)$$

Equating the two equations results in an expression to correct the RKB data to an MSL reference:

$$d_{MSL} = d_{RKB} \frac{D}{D-h_f} \tag{3.1}$$

If we instead want to convert MSL data to RKB, we can write:

$$d_{RKB} = d_{MSL} \frac{D-h_f}{D} \tag{3.2}$$

In these examples, the reference depth D is chosen from the drill floor. Other reference levels can be used, the important factor is that the bottomhole pressure remains constant regardless of the choice of reference level.

3.1.2.2 Correcting to Another Drill Floor Level

One common problem is that parts of the field data are compiled during exploration drilling and parts during drilling from a production platform. Often, the two have significantly different drill floor elevations. Figure 3.2 illustrates the problem.

Since, for future work, we are most likely going to use the production platform drill floor as a reference (RKB2), we will define this as our reference level. A pressure P at depth D can be expressed as:

$$P = 0.098d_{RKB2}D = 0.098d_{RKB1}(D-\delta h)$$

The correction equation is then:

$$d_{RKB2} = d_{RKB1}\frac{D-\delta h}{D} \tag{3.3}$$

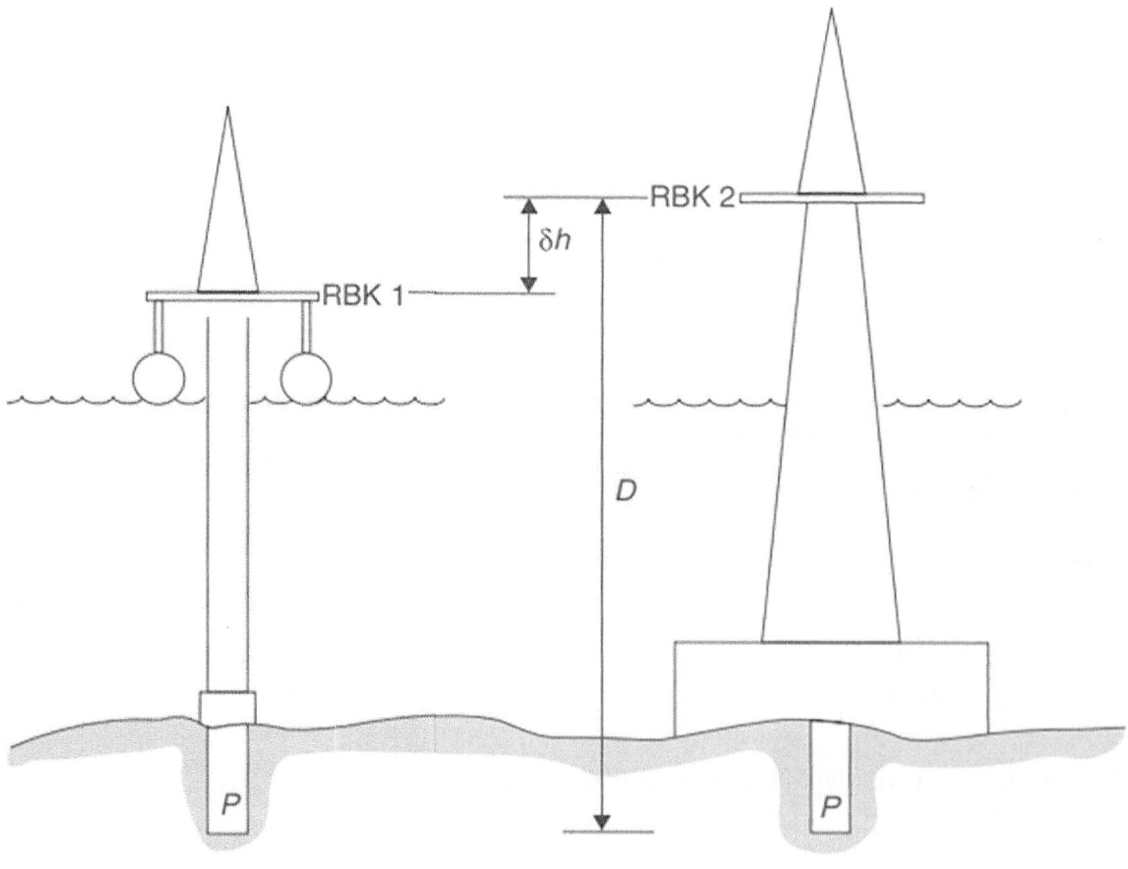

FIGURE. 3.2 Definition of various drill floor elevations.

or, if we want to use RKB1 as a reference:

$$d_{RKB1} = d_{RKB2} \frac{D}{D - \delta h} \tag{3.4}$$

3.1.3 Using the Sea Floor as a Reference

3.1.3.1 Traditional Correction

Again, referring to Figure 3.1, a pressure P at depth D would, if expressed as a pressure gradient, result in:

$$P = 0.098d_{RKB}D = 0.098d_{sf}(D - h_f - h_w)$$

$$d_{sf} = d_{RKB} \frac{D}{D - h_f - h_w} \tag{3.5}$$

This equation is not much used. At shallow depth in deep water, we will obtain very high gradients with little comparison to the mud weights used during the actual drilling operations. Therefore, in the next chapter, we will define another way to handle the conversion.

3.1.3.2 Removing the Water Head

The main purpose of this chapter is to present methods to make leak-off related data consistent. If we compare two shallow leak-off data from very different water depths, the water only has the effect of providing a constant hydrostatic head on top of the rock. Therefore, we may choose to use the sea floor as a reference. By subtracting the water head from each pressure reading, we have removed the effect of water. Again, referring to Figure 3.1, we can express the net pressure at depth D as:

$$P - P_{water} = 0.098d_{RKB}D - 0.098 \times 1.03h_w$$

The gradient from the sea floor is then:

$$P - P_{water} = 0.098d_{sf}(D - h_w - h_f)$$

Equating the two results in:

$$d_{sf} = \frac{d_{RKB}D - 1.03h_w}{D - h_w - h_f} \tag{3.6}$$

This approach is used for small rock penetration only. Note, however, that this approach is not consistent with the other approaches outlined, so the results cannot be mixed. In Sections 3.3.2 and 6.3.1, examples of application are presented.

Problems

In Table 3.1, some data is given for a number of wells. The data are: well no., casing size, depth, water depth, drill floor elevation, leak-off pressure, pore pressure, and overburden stress. Please do the following:

TABLE 3.1 Field data

WELL	*CSG (IN)*	*DEPTH (M)*	*HW (M)*	*HF (M)*	*LOT (SG)*	*PO (SG)*	σ_o*(sg)*
34/7-2	20	848	245	25	1.58	1.06	1.83
	13 3/8	1549			1.69	1.42	2.00
	9 5/8	2031			1.88	1.63	2.00
34/7-8	20	848	286	25	1.62	1.04	1.72
	13 3/8	1859			1.83	1.42	1.93
34/7-14	20	491	148	25	1.49	1.00	1.49
	13 3/8	1559			1.75	1.09	1.70
	9 5/8	1988			1.80	1.53	1.92

a) Correct and plot the data to the mean sea level reference. Remember to plot all data relative to the new depth reference.
b) Assume that we will use the data from a future platform with a 60 m drill floor elevation. Correct all data for this and show them on the same plot.
c) Correct the same data for a sea floor reference and show on the same plot.
d) Correct all data to sea floor by removing the water head and show on a plot.
e) In your own words, compare and discuss the above corrections.

3.2 INTERPRETATION OF FIELD DATA

3.2.1 Introduction

In addition to the symbol list defined in this book, the following specific nomenclature is used in this chapter:

P_o = Pore pressure at point of interest
P_o* = Pore pressure constant
$P_{wf}(\gamma)$= Frac. pressure for a hole inclined an angle γ from vertical
$P_{wf}(0)$= Frac. pressure for a vertical hole
σ_a = Average horizontal stress
σ_H = Maximum horizontal stress
σ_h = Minimum horizontal stress
σ_o = Overburden stress

A common problem in field modelling involving leak-off data is to handle a considerable spread. The task must be to transform these data into a form where interpretations can be made. In this chapter, we will investigate several ways to model and normalise the data, and we will propose ways of interpretation. Also, by introducing several parameters, it is possible to find trends not apparent in the raw data.

Basically, we have the following key parameters:

Leak-off pressure (LOT): This is the key parameter in stress modelling and borehole integrity evaluation. Typically, a leak-off test is performed after each casing is set to ensure integrity before proceeding with the next hole section. Sometimes, especially when using oil-based mud, the hole is not pressurised towards leak-off, and this is called a formation integrity test (FIT). A minifrac.

test is another kind where fluid is injected into the fractured hole. However, although all these tests can be analysed, it is important to use the data consistently; for example, not to evaluate a FIT test as a measure of fracture strength. The leak-off pressure is commonly defined as the critical pressure where fracturing initiates.

Formation pressure P_o: The pore pressure associated with each leak-off test is important to record. However, since most casing shoes are placed in impermeable competent shale sections, there is some inaccuracy associated with the pore pressure prediction, as direct measurements are not available.

Overburden pressure (σ_o): The weight of the overlying sediments is usually obtained from the integration of density logs. In the shallow parts of the hole, density logs are often not run. Here, the density can be obtained from sonic log correlations, lithology, and mineralogical evaluation, or by other methods. The overburden stress is an important parameter since it can be used as a measure of the stress state at any depth.

Lithology: We can group the leak-off data into several groups depending on the lithology of the rock. Usually, the casing shoe is placed in a competent shale section, resulting in leak-off data for an impermeable rock. On the other hand, sometimes we lose circulation in permeable rocks. Therefore, it seems reasonable to group the data into two groups, one high leak-off group for shales and one low leak-off group for other rock types like sand, coal, and chalk.

3.2.2 Definition of Basic Concepts

Process

The process to perform a field analysis is first to collect the raw data. The modelling is, first of all, to normalise these data for known artefacts. Examples are to adjust the data for various drill floor elevations, for various water depths, or for other differences, as described in Section 3.1. Secondly, if the data set consists of data from vertical and inclined boreholes, they could be grouped accordingly. Thirdly, if there are known differences in lithology, the data should preferably be grouped in the two groups defined above.

With a consistent data set at hand, the next step is to model these. Actually, this process is to normalise the data to identify similarities or differences. The examples that follow will demonstrate a number of ways in which this can be performed. We will first look at the raw data, then study the effective stresses, evaluate the horizontal stress field, and remove depth effects. More complex modelling will be discussed later.

Evaluate Leak-Off Data

The leak-off data often shows considerable spread. By evaluating only this single parameter, it is often difficult to obtain good correlations. However, as a starting place for an analysis, one should always start with a plot of the LOT-data versus depth. Remember, the quality of the analysis results mainly from demonstrating an improved degree of correlation when additional factors are considered.

Effective Stresses

The effective stress principle simply says that the stress in the rock matrix is the total stress minus the pore pressure. This is often used in rock mechanics as we are concerned with the rock itself and the strength of the rock matrix. In mathematical terms, the effective stress principle can be expressed as:

$$\sigma' = \sigma - P_o \tag{3.7}$$

Horizontal Stresses

As the LOT test is not a direct measure of the stress state of the rock, we have to develop an expression for the in-situ stresses. We assume that the LOT data are either valid for vertical boreholes or adjusted to represent vertical holes. Furthermore, we assume that the in-situ stress field is oriented in the vertical and horizontal plane, and finally, we assume that the two horizontal stresses are equal. The last condition is often defined as a hydrostatic stress state in the horizontal plane. The coupling between the LOT value, the stress field, and the pore pressure for these conditions is given by Aadnoy and Larsen (1989):

$$LOT = 2\sigma_a - P_o \tag{3.8}$$

The horizontal stress can simply be found by solving for:

$$\sigma_a = \frac{1}{2}(LOT + P_o) \tag{3.9}$$

This equation is very interesting. It simply says that the horizontal rock stresses can be expressed as the average between the leak-off value and the pore pressure. In section §2.2 this concept is defined as the median line principle. We can furthermore evaluate the effective horizontal stress by subtracting the pore pressure as defined in Equation (3.7) resulting in:

$$\sigma_a' = \frac{1}{2}(LOT - P_o) \tag{3.10}$$

Normalising with Overburden Stress

Above, several relationships are shown to evaluate the in-situ stress field from leak-off data.

Now, we will expand on these concepts a little. The leak-off pressures typically increase with depth because the stress state increases with depth. This can be seen from Equation (3.9). The weight of the overburden load also increases with depth. If we assume that for a relaxed depositional basin, the horizontal stress is proportional to the overburden stress, e.g.:

$$\sigma_a = K\sigma_o \tag{3.11}$$

Equations 3.9 and 3.10 can then be expressed as:

$$\frac{\sigma_a}{\sigma_o} = \frac{LOT + P_o}{2\sigma_o} \tag{3.12}$$

$$\frac{\sigma_a'}{\sigma_o'} = \frac{LOT - P_o}{2(\sigma_o - P_o)} \tag{3.13}$$

3.2.3 Field Case

3.2.3.1 Numerical Example

An example will demonstrate the application of the concepts outlined above. Assume that we have the following data set (see Table 3.2).

TABLE 3.2 Field data

WELL	DATASET	DEPTH (M)	LOT (S.G.)	P_o (S.G.)	σ_o (S.G.)
A	1	899	1.46	1.04	1.63
	2	1821	1.74	1.28	1.81
B	3	901	1.55	1.04	1.60
	4	1153	1.56	1.04	1.73
	5	1907	1.81	1.34	1.82
	6	2753	1.95	1.52	1.96

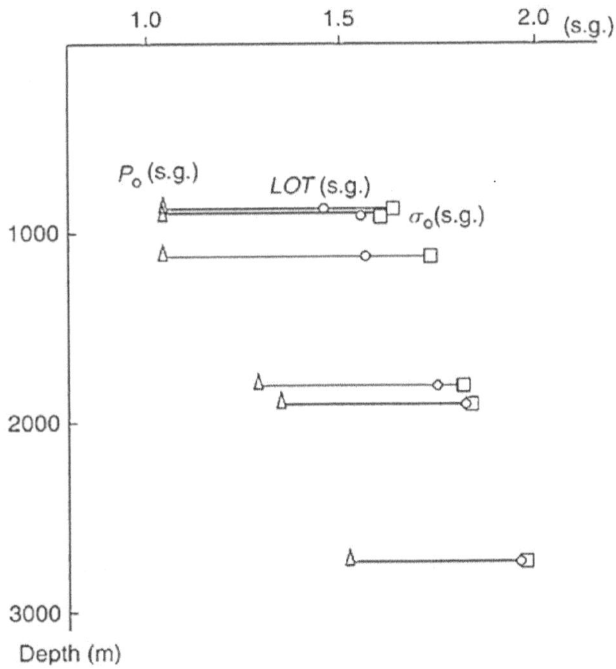

FIGURE 3.3 Leak-off data and the associated pore pressure and overburden stress gradients versus depth.

3.2.3.2 Leak-Off Data Evaluation

The LOT data are shown in Figure 3.3. Also shown are the pore pressure gradient and the overburden stress gradient associated with each leak-off point. We are commonly using gradients instead of pressures because we then have a direct comparison to the mud weight. This is allowed, providing one is consistent. Several observations can be made. First of all, the LOT data show an increase with depth. This is expected since the stress state has to increase with depth, mainly caused by the added weight of the overlying rock mass. The second observation is that the LOT data do not fall on a straight line. In fact, if a best-fit curve is drawn, there is some spread around this curve. Thirdly, the overburden and pore pressure gradients also vary considerably.

Although Figure 3.3 always should be the starting point, further analysis should be performed in an attempt to find geomechanical similarities that could be used to compare various wells or fields and also to provide modelling tools for the planning of new wells.

The leak-off data given in Table 3.2. are converted to pressures and plotted in Figure 3.4. Apparently, a reasonable linear correlation is obtained. In many cases, correlations like this are sufficient, providing that the new well has similar pore pressure and overburden stress states compared to the reference wells.

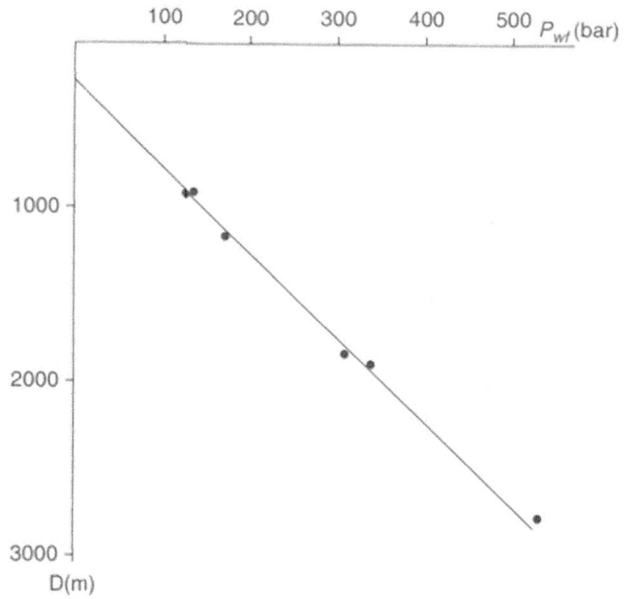

FIGURE 3.4 Leak-off pressure versus depth.

If the new well has a different pore pressure/overburden stress state compared to the reference well, these pressures should be included in the analysis, as demonstrated in the following.

3.2.3.3 Horizontal Stresses

Equation (3.9) gives an expression for the horizontal stresses derived from LOT data. We observe that the estimated horizontal stress for an assumed hydrostatic horizontal stress state simply is the average of the leak-off pressure and the associated pore pressure. In Section 2.2 this concept is simply defined as the "median line principle". By performing the calculation from Equation (3.9), we attempt to evaluate two parameters instead of one as in the previous example. The first entry of Table 3.2 calculated with Equation (3.9) looks as follows:

$$\sigma_a = \frac{1}{2}(1.46 + 1.04) = 1.25 s.g.$$

All datasets from Table 3.2 are computed in this manner. The result is shown in Figure 3.5. We observe that there is still some spread around the trend line, but possibly a little less. However, there is still an increasing trend with depth, as expected. For the data set in Figure 3.3, we have still not arrived at a satisfactory trend correlation and will, therefore, pursue further analysis.

3.2.3.4 Effective Horizontal Stresses

The horizontal stresses generated from Equation (3.9) are the total stresses acting in the horizontal direction. However, the pore pressure gradient provides another variable in this picture. Of main interest is the stresses acting on the rock matrix since fracturing is a phenomenon acting on the matrix itself. The effective stress principle says that the matrix stress is simply the total stress minus the pore pressure. By invoking this principle, Equation (3.10) is generated. Applying Equation (3.10) to the first entry of Table 3.2, we obtain:

$$\sigma_a = \frac{1}{2}(1.46 - 1.04) = 0.21 s.g.$$

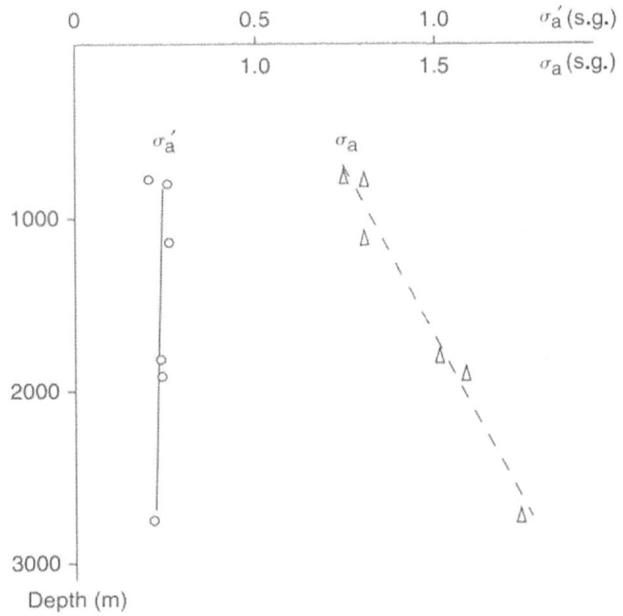

FIGURE 3.5 Average horizontal stress and horizontal effective stress versus depth.

The left-hand side of Figure 3.5 shows the effective horizontal stress generated for Table 3.2. Apparently, we observe an even smaller spread, and the data shows a nearly vertical trend. If we neglect the very top data point, a very good correlation can be obtained. If we decide to use this correlation, an expression for this trend line should be derived, inserted into Equation (3.10) and solved for LOT. However, at this stage, we will investigate a few more solutions.

3.2.3.5 Horizontal Stresses Normalised with Overburden Stress

In the previous analysis, we have investigated possible correlations using one and two parameters. We have, however, three parameters at our disposal as we also have a measurement of the overburden stress. One argument for why there is not a perfect correlation in the previous plots could be that the overburden stresses vary. Thus, by normalising with respect to overburden stress, hopefully, a better correlation could arise. Equation (3.12) gives the horizontal stress normalised with overburden stress. Here, we are using all three parameters. Equation (3.13) gives the effective horizontal stress/overburden stress calculation. Using the data from Table 3.2 in these two equations, letting K=1, the result is shown in Figure 3.6.

The total horizontal stress/overburden stress ratio is shown on the right-hand side. We observe that there is still some spread, and there is still not a vertical trend. Calculating the effective stress ratio, given to the left, the spread is even larger.

3.2.3.6 Summary of the Numerical Example

In the example given above, we have investigated several physical relationships between the frac. pressure, the pore pressure, and the overburden stress gradients. The objective was to look for a trend in the data to provide a more general relationship.

In the raw data plot of Figure 3.3, a straight line through the leak-off data would result in some spread about this line, not a satisfying correlation. Further analysis was conducted to attempt to reduce this spread. In Figure 3.5, the horizontal stress field was investigated. The effective horizontal stress was nearly vertical and showed little spread. By dividing with the overburden stress gradient in Figure 3.6, even more spread about the trend line was observed.

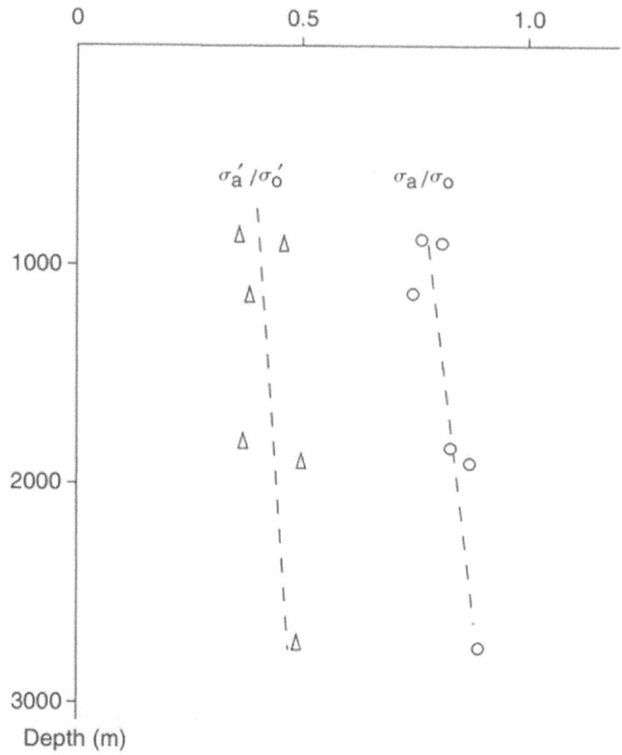

FIGURE 3.6 Horizontal stress normalised with overburden stress.

An overall evaluation of Figures 3.3–3.6 for the particular data from Table 3.2 results in the effective horizontal stress in Figure 3.5 having the least spread around the trend line. Also, this trend is, for this case, nearly vertical, that is, depth independent. For this example, we will derive a simple model. From Figure 3.5 we see that the effective horizontal stress is nearly vertical and equal to 0.23. Setting this value into Equation (3.10):

$$\sigma_a' = \frac{1}{2}(LOT - P_o) = 0.23 \quad or: \quad LOT = P_o + 0.46$$

This is the final model for this example. Figure 3.7 shows the initial data and the modelled data in the same plot. Note, however, that we could have constructed more complex models from, e.g., Figure 3.6, but for this particular case, the spread would increase.

In this example, we started out inspecting the leak-off pressures (Figure 3.4) and concluded the modelling by studying the pressure gradients. Looking for straight-line correlations, one should be aware of some fundamental differences between the two approaches. A straight line in a pressure plot starting at zero would give a constant gradient regardless of depth. If a pressure plot is giving a straight line offset zero, as in Figure 3.4., a gradient plot would actually be given by a non-linear curve. Therefore, if a straight line is not obtained using a pressure or a gradient plot, one should try the other.

Also, we have for simplicity inspected the plots visually for best correlation. With today's availability of computers, one could easily compute all correlations and make comparisons, as in Figure 3.7, to ensure that the best correlation is found.

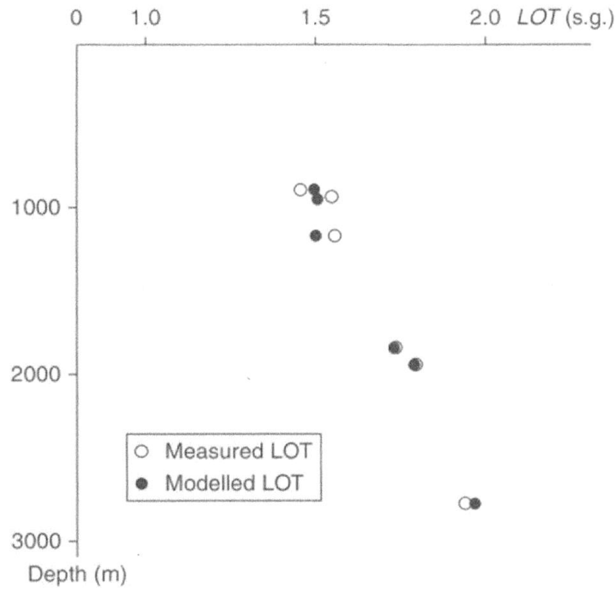

FIGURE 3.7 Comparison of measured vs. modelled leak-off gradients.

3.2.4 More Advanced Modelling

3.2.4.1 Borehole Inclination, Relaxed Depositional Basin

From solid mechanics, it is known that the frac. gradient may depend on the inclination and the azimuth of the borehole. This effect is mainly dependent on the relative magnitudes of the three principal in-situ stresses. For the general anisotropic in-situ stress case, a method of solution will be presented later in this chapter.

In a relaxed depositional basin environment, we typically assume that the two horizontal principal in-situ stresses are equal, or, in other words, we assume a hydrostatic stress field in the horizontal plane. For this case, there will be no dependence on the azimuth angle; in other words, for a given inclination, the frac. pressure will be the same in all geographical directions. Aadnoy and Chenevert (1987) define an equation for this case. In Aadnoy and Larsen (1989) a field method is derived to use this approach. The coupling between the frac. gradient for an inclined and a vertical hole is given by the following expression:

$$P_{wf}(\gamma) = P_{wf}(0) + \frac{1}{3}(P_o - P_o^*)\sin^2\gamma \tag{3.14}$$

This equation can be used in a number of ways. We will illustrate the application by using it to normalise leak-off data for inclined holes to that of equivalent vertical holes.

Example:
Table 3.3 below defines leak-off pressure data from a production field. Assuming a relaxed depositional environment, we will normalise the data by calculating the frac. gradients for vertical holes.

Rearranging the equation as follows, and assuming a pore pressure constant of 2.09 s.g., the third entry is:

$$P_{wf}(0) = 1.63 - \frac{1}{3}(1.44 - 2.09)\sin^2 24° = 1.67 s.g.$$

TABLE 3.3 Field data

DEPTH (M)	*FRAC.GRAD. (S.G.)*	*P_O (S.G.)*	*σ_O (S.G.)*	*INCLINATION(°)*
1610	1.66	1.08	1.79	0
1671	1.64	1.32	1.80	10
1792	1.63	1.44	1.81	24
1914	1.58	1.51	1.83	46

Both the raw data and the inclination normalised data are shown in Fig. 3.8. We observe that the normalised data shows a better trend than the raw data. Also, when corrected toward vertical, the frac. pressure gradients increase with depth as expected.

In the previous discussion, Equation (3.14) was actually used to normalise both the pore pressure and the inclination of the borehole. As will be seen later in the chapter on compaction modelling, the scaling factor of 1/3 is consistent with the scaling factor for the pore pressure effect. In the following we will develop an equation which is useful for normalisation of the borehole inclination.

In the following analysis, we assume equal horizontal principal stresses and a relaxed depositional basin environment. From Aadnoy (1990), Appendix B, we can express the fracture gradient in terms of stresses on the borehole wall:

$$P_{wf} = 3\sigma_y - \sigma_x - P_o \tag{3.15a}$$

The stress components can further be expressed with the above assumptions as (Aadnoy and Chenevert, 1987):

$$\sigma_x = \sigma_a cos^2\sigma + \sigma_o sin^2\sigma \tag{3.15b}$$

$$\sigma_y = \sigma_a \tag{3.15c}$$

Combining the expressions above, we may obtain an expression for the fracture gradient for any inclination as:

$$P_{wf} = 2\sigma_a - P_o - (\sigma_o - \sigma_a)sin^2\gamma \tag{3.16}$$

Now let us assume that we have fracturing data for inclined boreholes, which we want to make comparable by calculating the equivalent fracturing gradient for a vertical hole. By setting up two equations (Equation 3.16) one for an inclined hole and one for a vertical hole and combining, the result is:

$$P_{wf}(0) = P_{wf}(\gamma) + (\sigma_o - \sigma_a)sin^2\gamma \tag{3.17}$$

Evaluating field data, we see that we need an estimate for the average horizontal stress to use the equation above. However, setting up an equation for fracturing of a vertical hole, the horizontal stress can be eliminated in Equation (3.17). The result is:

$$P_{wf}(0) = \frac{P_{wf}(\gamma) + \left(\sigma_o - \frac{1}{2}P_o\right)\sin^2\gamma}{1 + \frac{1}{2}\sin^2\gamma} \tag{3.18}$$

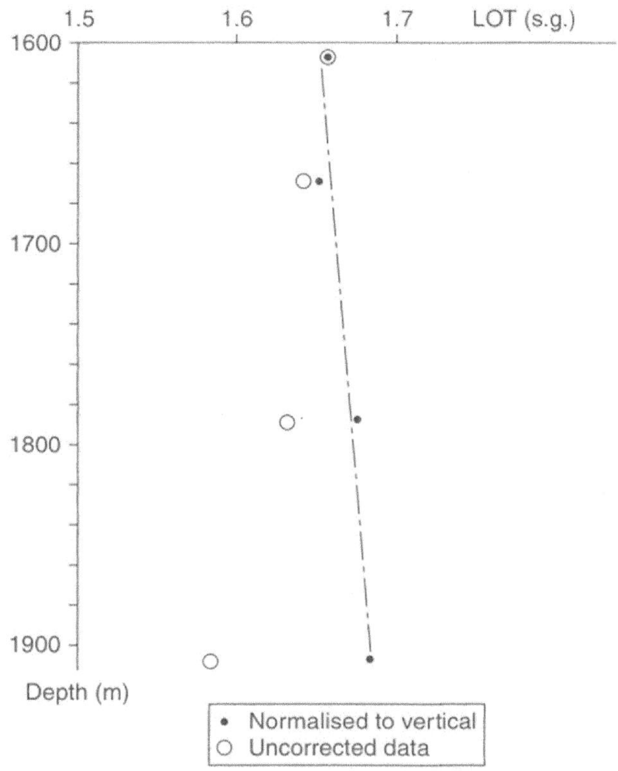

FIGURE 3.8 Leak-off data corrected for borehole inclination.

Equation (3.18) is the final result. We will in the following demonstrate its application.

Example

You are asked to normalise the data of the previous Table 3.3 towards that of a vertical hole.

The result from Equation (3.18) on the fourth entry is:

$$P_{wf}(0) = \frac{1.58 + \left(1.83 - \frac{1}{2}1.51\right)\sin^2 46^\circ}{1 + \frac{1}{2}\sin^2 46^\circ} = 1.70 s.g.$$

The above example demonstrates an alternative normalisation if we assume no changes in pore pressure. Also, the overburden stress is here coupled into the fracture equation by the stress transformation equations. The significance of establishing the horizontal and the vertical stress ratio is also clear from the above example. The above normalised data can be plotted into Figure 3.8 and evaluated as an exercise.

3.2.4.2 Compaction Model

The modelling shown in the previous chapters are mainly based on the effective stress principle. This principle simply decomposes the total stress into a rock matrix stress and a pore pressure component. This principle basically determines the stresses with no reference to history and have therefore limitations.

By introducing a simple compaction model, we will have a tool to perform stress history calculations. If the pore pressure has changed over time, we can estimate what effect this will have on the fracturing pressure. One example is to normalise all leak-off data to a given reference pore pressure. If all data then form a trend, we may interpret this as that all data have the same origin. In this chapter we will first define a simple compaction model, then use an example to illustrate the application.

Crockett et al. (1986) gives a more general derivation of the so-called "backstress" concept, which is the influence of charged pore pressure on the fracturing pressure. Morita et al. (1988) also address the same problem. Aadnoy (1991) derives this concept in a simpler way. Figure 3.9 a and b show a rock before and after the pore pressure has been changed.

Assuming that the overburden stress remains constant, and that no strain is allowed on the sides of the rock, we can calculate the changes in the horizontal rock stress. Since the overburden stress is constant and the pore pressure is, e.g., lowered, the rock matrix must take the load held by the initial pore pressure. This increased vertical matrix stress will via the Poisson's ratio also increase the horizontal stress. This horizontal stress increase is (Aadnoy, 1991):

$$\Delta\sigma_a = \Delta P_o \frac{1-2\nu}{1-\nu} \tag{3.19}$$

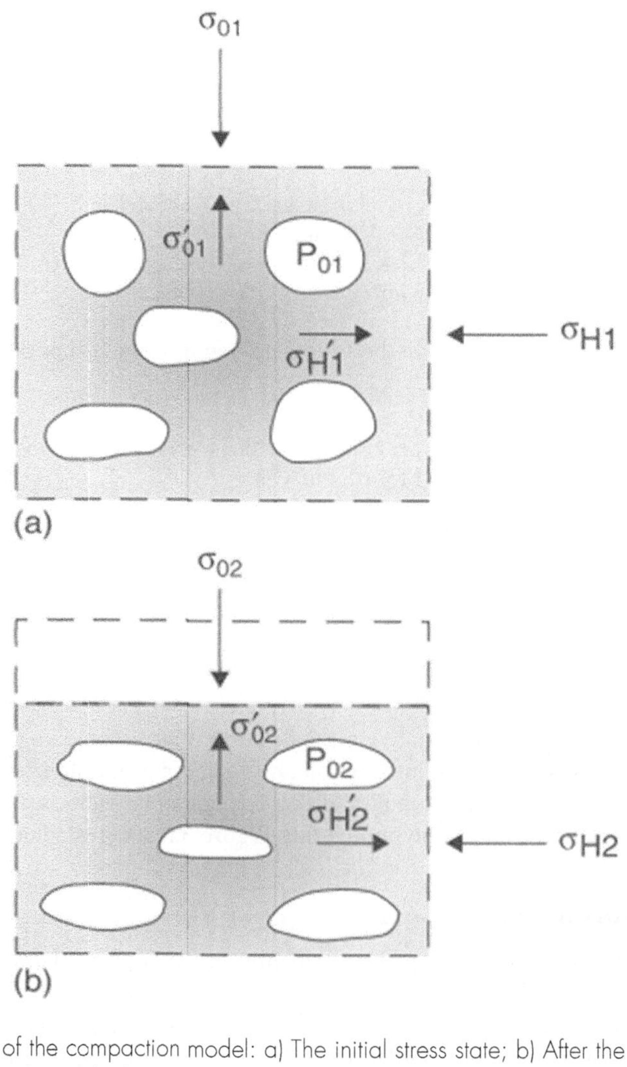

FIGURE 3.9 Illustration of the compaction model: a) The initial stress state; b) After the pore pressure is reduced increases the horizontal stress, while the overburden remains constant.

Inserting this matrix stress change into the general frac. equations, the corresponding change in frac. pressure can be calculated. The following equation results:

$$\Delta P_{wf} = \Delta P_o \frac{1-3\nu}{1-\nu} \tag{3.20}$$

As an example, Table 3.4 shows some leak-off data and their associated pore pressure gradients. If we decide to normalise the data to the same pore pressure gradient, arbitrarily chosen to be 1.80 s.g., and with a Poisson's ratio of 0.25, the third entry looks like:

$$P_{wf} - 1.98 = (1.80 - 1.44)\frac{1-3x0.25}{1-0.25}$$
$$P_{wf} = 2.10 s.g.$$

The raw data and the pore pressure-corrected data are shown in Figure 3.10. It is observed that the raw data shows hardly any correlation. The compaction-corrected data, on the other hand, falls nearly on a line, with a reasonable correlation. One interpretation is that the data are very similar or have the same origin. The initial stress state for the four wells was possibly similar, but the variations observed in the frac. pressures are mainly dependent on local variations in the pore pressures. In other words, the pressure depletion history is also reflected in the frac. pressure.

Assuming that the vertical trend line of Figure 3.10 (at 2.11 s.g.) reasonably well describes the frac. pressure, the compaction equation can be expressed as:

$$2.11 - P_{wf} = (1.80 - P_o)\frac{1-3x0.25}{1-0.25}$$

or rearranged the frac. equation looks like:

$$P_{wf} = 1.51 + \frac{1}{3}P_o$$

If Table 3.4 is used to reconstruct the data with this equation, the result is very close to the initial raw data.

Another example is that in a field the initial frac. gradient during exploration drilling was 1.70 s.g. After the field has been put into production, the overall reservoir pressure gradient declines from 1.5 to 1.3 s.g. What is the expected change in the horizontal stresses and the frac. pressure in the reservoir?

The change in the horizontal stress gradient would be:

$$\Delta\sigma_a = (1.5 - 1.3)\frac{1-2x0.25}{1-0.25} = 0.13 s.g.$$

TABLE 3.4 Field data

DEPTH (M)	*LOT (S.G.)*	*PO (S.G.)*
3885	2.10	1.79
3821	2.13	1.84
3818	1.98	1.44
3914	2.06	1.58

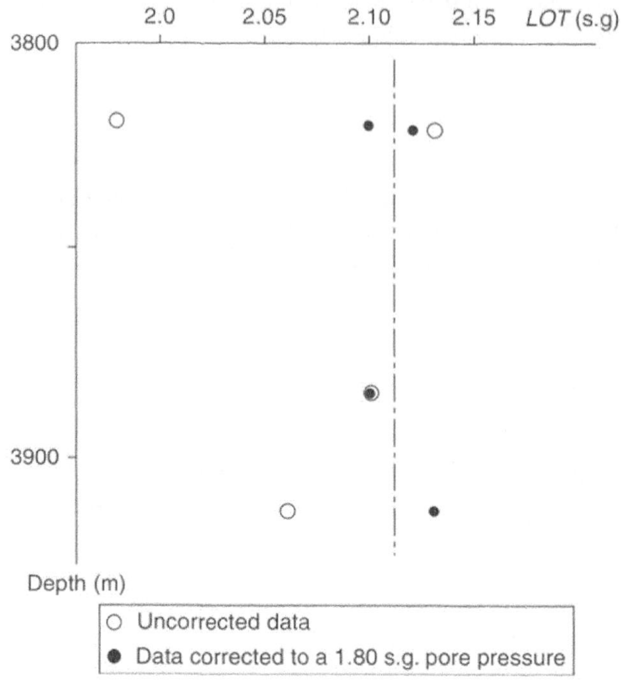

FIGURE 3.10 Frac. data shown uncorrected and corrected with a compaction model.

If an infill well was drilled at a later stage, the estimated frac. pressure gradient would be:

$$P_{wf} = 1.7 - (1.5 - 1.3)\frac{1 - 3x0.25}{1 - 0.25} = 1.63 s.g.$$

3.2.4.3 Anisotropic Horizontal Stresses

In a relaxed depositional environment, we often neglect tectonic effects, and assume that the horizontal in-situ stress field is due to compaction only. It is often called a hydrostatic or isotropic stress field in the horizontal plane. That implies the same horizontal stresses in all directions. If deviated boreholes are drilled, there are no directional abnormalities for the same wellbore inclination, and the same leak-off value is expected in all geographical directions. Since the horizontal stresses in a relaxed depositional environment are lower than the overburden stress, the fracture gradient will decrease with hole angle as illustrated in Figure 3.11. This situation has been addressed earlier in this book, and is relatively simple to analyse. However, even though a relaxed environment exists, this ideal stress situation is not always the case. Often a more complex situation exists. However, one should always start a field analysis with the simple approaches shown earlier in this chapter. If the correlations are not satisfactory, more complex approaches should be used.

The horizontal stress field often varies with direction, and we have two different horizontal stresses. This stress state is called *anisotropic*. This may be caused by global geologic processes as plate tectonics or due to local effects like salt domes, topography, or faults. The resulting stress state varies over the area. Figure 3.12 shows an example from a field in the North Sea. Two immediate observations are: there is a considerable spread in the leak-off data, and there is no apparent trend with respect to wellbore inclination. It is obvious that the isotropic model is not adequate for this case. The answer is that there are several stress states for the wells of Figure 3.11.

A method is derived to model the anisotropic stress field (Aadnoy, 1990). The so-called leak-off inversion technique uses leak-off pressure, pore pressure, overburden stress, borehole inclination, and borehole azimuth to estimate the principal stress field. The method can use any number of data sets. This model was used in the following example in the following way.

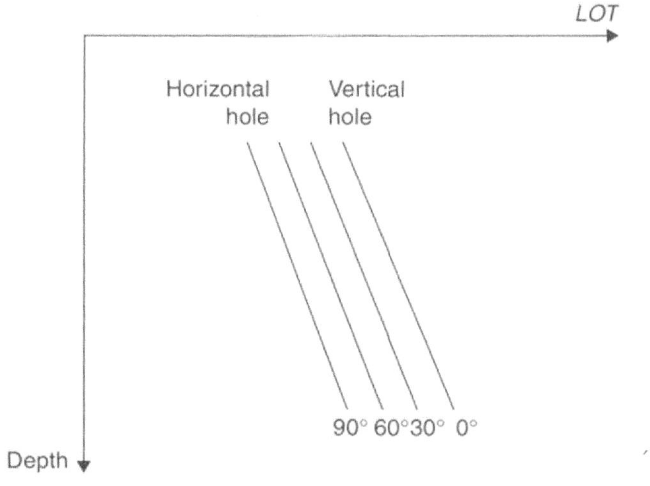

FIGURE 3.11 Expected leak-off behaviour for a relaxed depositional basin.

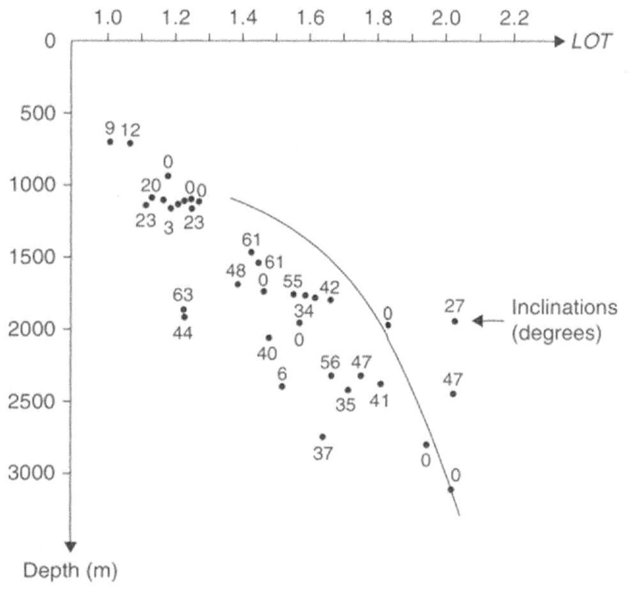

FIGURE 3.12 Snorre leak-off data and wellbore inclinations, showing the effect of stress anisotropy.

First, all data were used (42 data sets) to estimate the average stress level. Then, the data were grouped both depth-wise and sector-wise to determine local stress fields. The result is a stress model that varies with depth and over the field. In this manner, we were able to model the leak-off data from Figure 3.11 with a reasonable degree of accuracy.

Example

The following data set gives leak-off data for three wells and some data for a new well. The objective is to use the inversion technique to determine the in-situ stresses and to use these to predict the leak-off values for the new well.

TABLE 3.5 Data for anisotropic case

DATA SET	WELL	CASING	DEPTH (M)	LOT (S.G.)	P_o (S.G.)	σ_o (S.G.)	γ	φ
1	A	20	1101	1.53	1.03	1.71	0	0
2		13 3/8	1888	1.84	1.39	1.82	27	92
3		9 5/8	2423	1.82	1.53	1.89	35	92
4	B	20	1148	1.47	1.03	1.71	23	183
5		13 3/8	1812	1.78	1.25	1.82	42	183
6		9 5/8	2362	1.87	1.57	1.88	41	183
7	C	20	1141	1.49	1.03	1.71	23	284
8		13 3/8	1607	1.64	1.05	1.78	48	284
9		9 5/8	2320	1.84	1.53	1.88	27	284
10	New	20	1100	?	1.03	1.71	15	135
11		13 3/8	1700	?	1.19	1.80	30	135
12		9 5/8	2400	?	1.55	1.89	45	135

TABLE 3.6 Results from simulations

DEPTH INTERVAL (M)	σ_H/σ_o	σ_h/σ_o	DIRECTION	LEAK-OFF, NEW WELL
1100-1148	0.754	0.750	44	1.53
1607-1812	0.854	0.814	96	1.71
2320-2423	0.927	0.906	90	1.86

The data from Table 3.5 were simulated in various combinations, resulting in a reasonable field model. The estimated stresses were as follows:

The results can be interpreted as follows. Table 3.6 shows the ratio of the two horizontal stresses divided by the overburden. In addition is the direction shown. The first entry is the maximum horizontal in-situ stress pointing at 44 degrees north-east, or 44 degrees clockwise as per convention in surveying. The stress field is increasing with depth as expected. Furthermore, it is clear that the stress field is anisotropic. The shallow stress model shows a nearly isotropic stress model, as expected from younger sediments. However, the two lower stress levels show anisotropy with the maximum horizontal principal stress pointing approximately East–West. The program also estimated the leak-off gradients for the new well. These data will be used in the design of the new well.

Figure 3.13 shows the location of the wells in the field. Figure 3.14 shows the estimated stress fields at the various depth levels. This example shows only a moderate degree of stress anisotropy. In many fields, the anisotropy can be much stronger. In certain fields, the maximum horizontal stress can even exceed the overburden stress. For further details on application, Aadnoy et al. (1994) give a detailed stress analysis of the field using the leak-off inversion technique.

Problems

Problem 1. Table 3.7 gives the field data from some exploration wells.

In order to evaluate and model the data from Table 3.7, you are asked to do the following:

a) Plot the leak-off data versus depth.
b) Plot both the horizontal stresses and the effective horizontal stresses versus depth.
c) Normalise the horizontal stresses from b) with overburden stress and plot versus depth.
d) Evaluate the correlations on each plot and propose the best correlation.
e) Using the best correlation, derive an expression for the frac. pressure. Make a plot comparing the raw data against the modelled data.

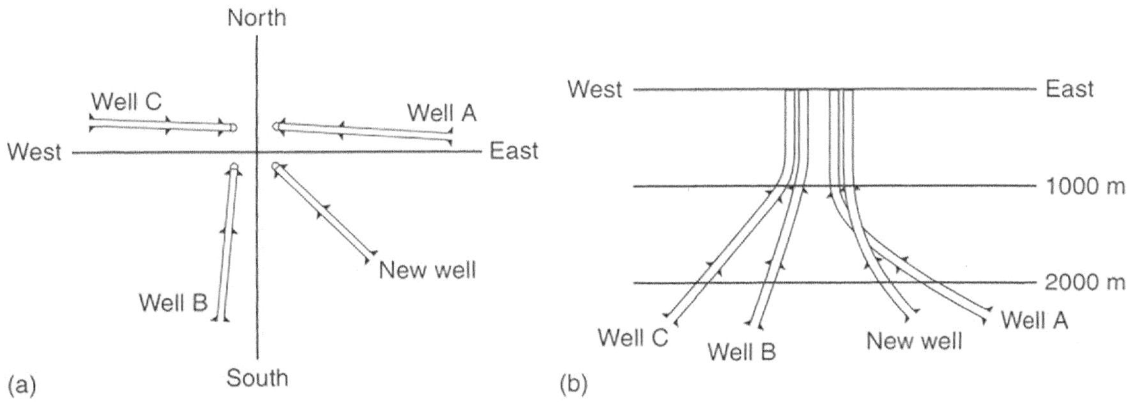

FIGURE 3.13 Field layout, for example. a) Placement of wells in the field; b) vertical west-east projection of the wells.

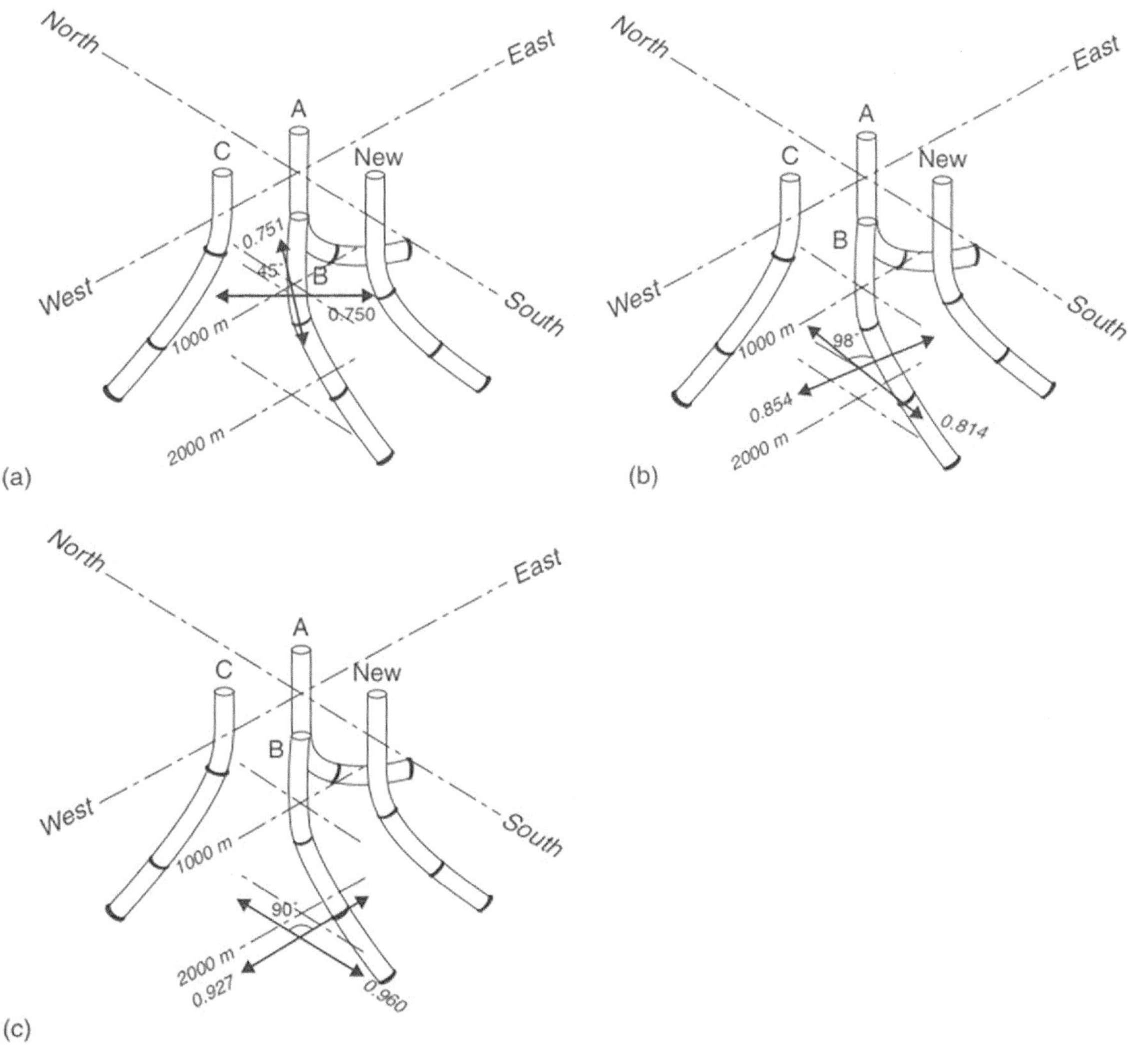

FIGURE 3.14 The three stress levels and their direction. a) predicted stress field at 11–1148 m; b) predicted stress field at 1607–1812 m; c) predicted stress field at 2320–2428 m.

TABLE 3.7 Leak-off data from some exploration wells

WELL	CSG	DEPTH (M)	LOT (S.G.)	PO (S.G.)	σ_O (S.G.)
34/7-1	20	1071	1.57	1.03	1.62
	9 5/8	1844	1.93	1.29	1.81
34/7-2	20	848	1.58	1.03	1.41
	13 3/8	1549	1.69	1.13	1.63
	9 5/8	2031	1.88	1.48	1.75
34/7-3	20	1153	1.56	1.03	1.73
	13 3/8	1922	1.77	1.36	1.83
	9 5/8	2753	1.95	1.51	1.96
34/7-4	20	953	1.53	1.04	1.63
	13 3/8	1915	1.78	1.35	1.92
	9 5/8	2745	1.90	1.52	1.95
34/7-5	20	913	1.52	1.03	1.60
	9 5/8	1844	1.80	1.30	1.81
34/7-6	20	935	1.55	1.04	1.55
	13 3/8	1876	1.91	1.33	1.82
	9 5/8	3000	2.01	1.40	1.84
34/7-7	20	969	1.50	1.04	1.65
	13 3/8	1862	1.90	1.32	1.82
	9 5/8	2750	1.85	1.51	1.95

Problem 2. Using the data from the table in the previous problem, you are now asked to normalise the data with respect to compaction. Specifically, normalise each leak-off data to a pore pressure gradient of 1.03 s.g. for the shallow data, 1.30 s.g. for the intermediate data, and 1.5 s.g. for the deep data.

Plot the compaction-corrected data on the same scale as the raw data. Compare the two. If an improved correlation exists at any depth level, derive a correlation equation.

Problem 3. Assume that the total overburden weight is given by the equation:

$$\sigma_o = 0.098\left\{1.03h_w + \rho_{rock}\left(D - h_w - h_f\right)\right\}$$

The density gradient from the drill floor is $\sigma_o = 0.098 d_o D$. The average rock density is 2.0 s.g.

a) Plot the overburden gradient for a land well.
b) Plot the overburden for a land well where the drill floor is 10 m above the surface.
c) Plot the overburden gradient for an offshore well where the water depth is 200 m and the drill floor is 25 m above sea level.
d) Plot the overburden gradient for an offshore well in a water depth of 500 m and a drill floor at 25 m above sea level.

3.2.5 Interpretation of Leak-Off Tests

3.2.5.1 Introduction

The first part of this chapter presents fracturing from *a mechanistic* perspective. This will include repeated fracturing as used during an extended leak-off test (XLOT). However, many companies use a more correlation-based approach, and each company has its own "best practices". Wellbore stability analysis is often a mixture of correlation-based and mechanistic approaches. The first part will address the fundamentals, and the second part current applications.

This section presents a model for leak-off interpretation that includes evaluation after the borehole is fractured. The Kirsch equation is valid up to the conventional leak-off point, but beyond leak-off, the hole is fractured, and an elastoplastic bridge model applies to ultimate failure. Also presented is an evaluation of in-situ stress interpretations from leak-off tests.

Findings of the new model are that the leak-off point using Kirsch is correctly defined by the in-situ stress state and the rock tensile strength. The model also explains the circulation loss problem with oil-based drilling fluids where the wellbore strength reduces to the minimum horizontal stress level.

The models presented (in this subsection and Aadnøy et al., 2009) apply to a typical drilling operation, using particle-laden drilling muds with filtrate control, performing leak-off tests with a large annular volume and relatively small fractures opening during the tests. The model is not applicable for massive hydraulic fracturing with clean penetrating fluid, as performed during well-stimulation operations. This is an important definition as many publications either confuse or use a model for penetrating clean fluid instead of a non-penetrating model for drilling mud.

Performing numerous fracturing experiments on hollow concrete cores, Aadnoy and Belayneh (2004) found that for most drilling fluids, the Kirsch equation underpredicted the ultimate fracture pressure significantly. Figure 3.15 shows three fracture tests. Mud B and C are common commercial water-based drilling fluids. For these, the ultimate fracture pressure is two to three times higher than predicted by the Kirsch equation. Mud A is a special mud with fewer particles but a particular chemistry. It is in accordance with Kirsch.

The results of Figure 3.15 also show that the ultimate fracture pressure varies for different drilling fluids. This indicates that in the future, we should be able to design the mud to avoid circulation losses. This paper is one more step towards this goal.

3.2.5.2 Fracturing of the Borehole: Experiments with Continuous Pumping

The results, as seen in Figure 3.15, led to a search for an explanation. As reported in Aadnoy and Belayneh (2004), the explanation is that during pressure testing, the rock actually fractures. When this fracture opens, solid particles form a bridge that prevents more fluid from entering the fracture. From this perspective, the fracturing process consists of two steps, a pre-failure, and a post-failure phase.

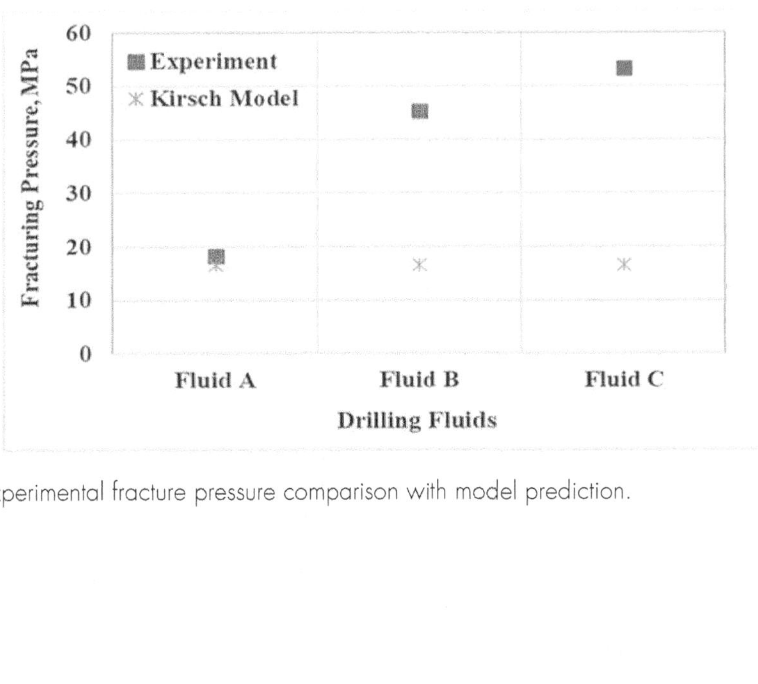

FIGURE 3.15 Experimental fracture pressure comparison with model prediction.

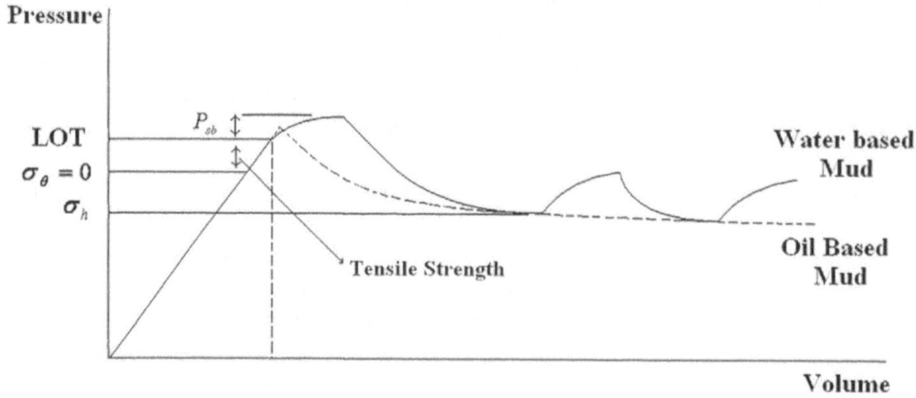

FIGURE 3.16 Fracture pressure with continuous pumping.

The pre-failure phase is valid until the borehole fractures. This phase can be described by the Kirsch equation, which is a continuum mechanics approach. Assuming a simple scenario with a vertical well subjected to anisotropic horizontal stresses, the leak-off pressure is:

$$LOT = 3\sigma_h - \sigma_H - P_o + \sigma_{tensile} \tag{3.21}$$

As shown in Figure 3.16, leak-off tests are usually defined as the point where the pressure plot deviates from a straight line. This is the point where the borehole actually fractures. Also shown in Figure 3.16 is the tensile strength of the rock. The tensile strength must be exceeded for the first fracturing cycle of the borehole. A well-defined LOT is therefore defined in terms of the in-situ stress state and the rock tensile strength.

A good estimate of the in-situ stress state can be obtained from Equation (3.21). Assuming equal horizontal stresses, the estimate is:

$$\sigma_h = \frac{1}{2}\left(LOT + P_o - \sigma_{tensile}\right) \tag{3.22}$$

The *post-failure phase* is after the borehole is fractured. Solid particles in the mud form a bridge across the fracture and allow the pressure to increase further. At the maximum pressure P_{sb}, this bridge fails, and the pressure drops as mud invades the fractures.

Figure 3.16 is valid for continuous pumping. Following the failure of the bridge, the pressure drops towards the minimum normal stress, σ_h. Continuing pumping, particles try to form bridges as shown in Figure 3.17 but break down at a given pressure. This is the "self-healing" effect of water-based drilling fluids.

Oil-based drilling fluids behave differently. Often, a more abrupt breakdown is seen, and the propagation pressure is constant with continuous pumping. This effect is well known from drilling operations, where it is known that with oil-based muds it is often difficult to cure circulation losses. One mechanism attributed to this effect is the wettability contrast between the rock and the drilling fluid, leading to low filtrate losses. We have, however, an additional plausible mechanism. To form a stable bridge of mud particles, a certain amount of friction is required to make the bridge stable. In other words, too much lubricity between the particles lets them slide instead of locking up as a bridge. It may be pointless to decrease lubricity to oil-based muds. Instead, we may use more angular particles, which may lock up more easily.

Before addressing the actual LOT tests, here is a brief summary of the fracturing research:

- The LOT is a correct stress indicator, provided that the tensile rock strength is taken into account.
- The LOT is defined by the Kirsch equation, as defined by Equation (3.21). This defines the minimum fracture resistance before fracturing. The maximum fracture resistance P_{sb} varies and depends on the quality of the mud. The LOT and the P_{sb} define the hole strength interval of the unfractured borehole.
- After failure, the fracture resistance is given by the stress bridge and σ_h, a reduced hole strength. For water-based muds, the strength can be partially recovered, but for oil-based muds, the fracture strength usually remains low.

3.2.5.2.1 What Happens at Failure?

The description above is based on results from numerous fracturing tests that have been conducted at the University of Stavanger since 1996. During the research, a mechanistic model was developed to better explain the post-failure behaviour. Figure 3.17 shows what we believe happens. At the LOT point the borehole fractures. By further pumping some mud enters the fracture. However, near the fracture entrance a stress bridge builds up that prevents further fluid flow into the fracture, and that allows borehole pressure to increase. At the ultimate pressure, this stress bridge collapses, and the wellbore pressure drops.

Mechanical equilibrium demands that this stress bridge form a curved shape as indicated. Because it consists of individual particles, inherent material stiffness does not exist, and geometric stability is the natural way such a mechanical barrier can be established. The appendix at the end of this section is a short version of the model fully presented by Aadnoy and Belayneh (2004). It also shows that the maximum stress bridge pressure is proportional to the compressive strength of the particles. In other words, strong particles may give higher ultimate fracturing pressures than weak particles.

Figure 3.18 shows the results of experiments using particles of different hardness, that is, different mechanical strengths, tested at the Fracturing lab at the University of Stavanger. Calcium carbonate is a weak material and creates a weak stress bridge. Quartz is stronger, and Feldspar is the strongest material, giving the highest fracture pressure. This may also explain why sometimes the fracture pressure increases when the well has been drilled for some time. Particles from drilled cuttings may help in establishing a stronger stress bridge.

The post-failure fracture pressure can be formulated as:

$$P_{wf} = P_{sb} + f(\sigma, E, P_o, ...) \tag{3.23}$$

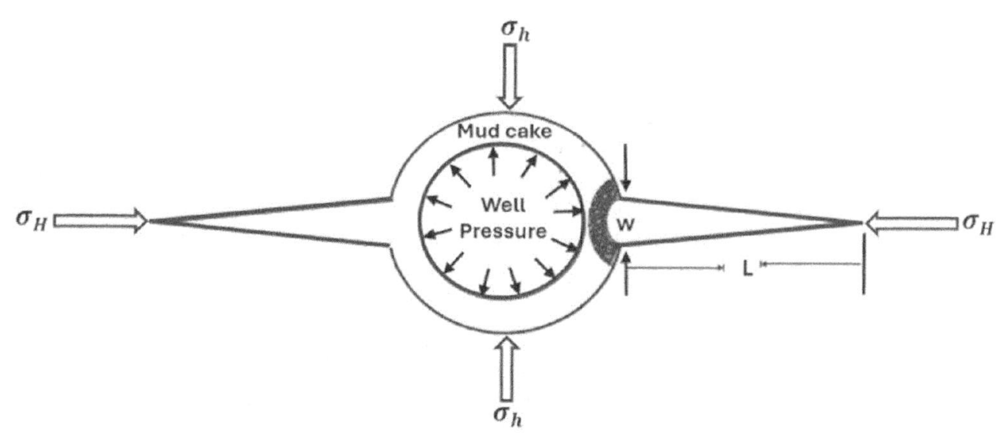

FIGURE 3.17 Post-failure establishment of stress bridge.

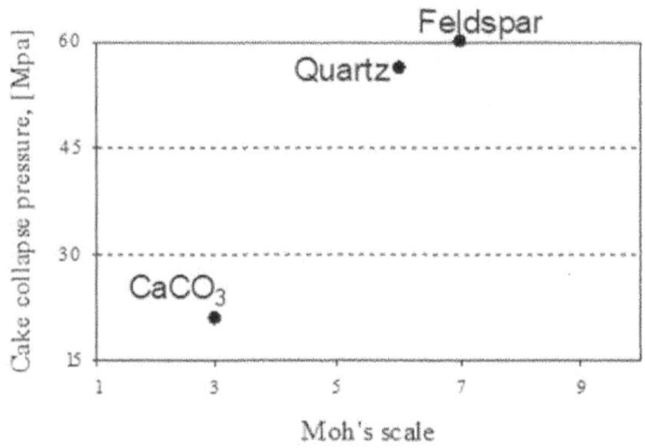

FIGURE 3.18 Fracture pressure versus particle hardness.

A model of a fracture growing through a rock is complex. It is related to the stress state, the elastic rock properties, crack tip propagation conditions, and other parameters. The minimum value of the fracture propagation pressure is the in-situ stress normal to the fracture surface. We will use this minimum value in our analysis.

$$P_{wf} = P_{sb} + \sigma_h \tag{3.24}$$

We will use the above definition for the minimum fracture propagation pressure. Further research may modify this by including crack tip propagation conditions and other effects.

3.2.5.3 Interpretation of the LOT

In the following, the previous understanding will be applied to a traditional extended LOT or ELOT. This is a LOT that is pumped beyond breakdown and is often repeated one or more times.

Figure 3.19 shows a typical ELOT. The lower curve shows the pump cycles. Often, the pump is stopped at or past the ultimate pressure. The pressure then drops towards a stable value, and the test is repeated one or more times.

In the following, the common ELOT interpretation will be discussed in light of the previous model.

- The LOT value is today used as a stress indicator assuming zero tensile rock strength. For this assumption to hold true, the direction of a preexisting crack must coincide with the fracturing stress conditions. This appears unlikely, and we will, therefore, argue that the tensile rock strength has to be overcome for the first leak-off test cycle.
- The pressure decline is monitored after the pump is stopped. For massive hydraulic fracturing with penetrating fluid (ref. e.g. De Bree and Walters 1989), the slope change of the pressure decline may define the in-situ stress. For these operations, the injected fluid is often penetrating. Applying these pressure decline methods to LOT tests during drilling raises some questions. One is the presence of particles and fluid loss control. During flowback, stress bridges may arise inside the fracture, giving an incorrect fracture closure pressure. Particles can also keep the fracture open, allowing seepage below the fracture closure pressure. Also, the well volume is often several orders of magnitude larger than the mud volume entering the fractures. The compressibility effects of the mud may dominate the flowback volume, making it difficult to discriminate between mud compressibility effects and fracture closure effects. More research is needed with drilling muds to confirm if clean fluid models can be applied to drilling muds.
- The minimum pressure, P_e, is sometimes used as an indicator for the minimum horizontal stress. We do not see the physical connection to the in-situ stress because the pump is stopped. Possibly, the minimum pressure is more related to the static weight of the mud in the well.

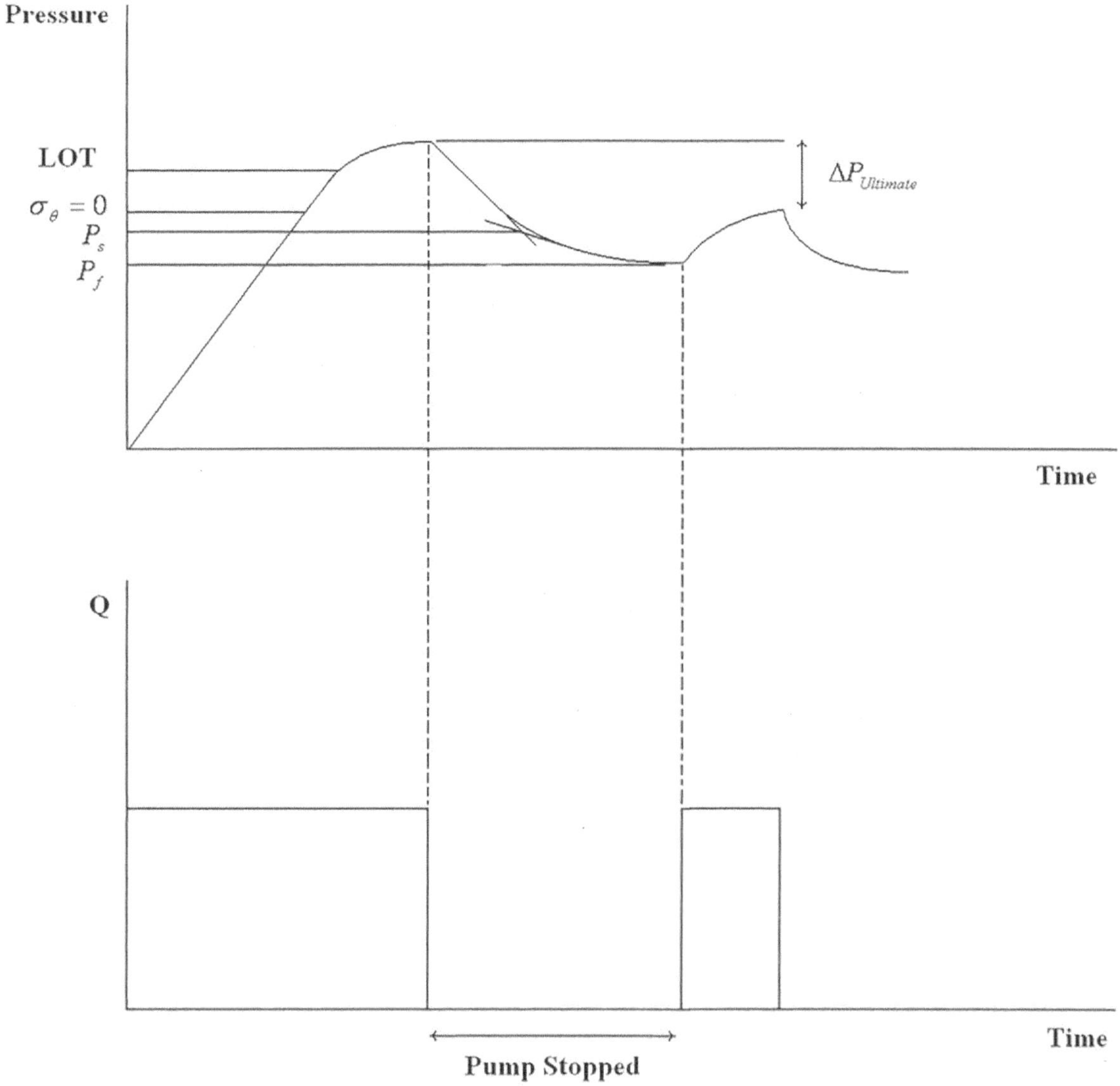

FIGURE 3.19 Extended LOT test.

- The difference between the first two repeated ELOT cycles is often interpreted as a measure of the rock's tensile strength. This is only correct if the two stress bridges are identical. Our experiments indicate that the first stress bridge pressure is often larger than the second one, which has to bridge an existing fracture. From this observation, the ΔP_{ult} may overestimate the tensile rock strength.

Before proceeding, a summary of current ELOT interpretation practice is:

- For stress evaluation of the first cycle of the ELOT, we should include the tensile rock strength. For repeated ELOT cycles, tensile strength is zero.
- Pressure decline after the pump is stopped may have no unique coupling to the in-situ stress state for large-volume wells with drilling fluids. Pressure decline, flowback, and minimum pressure assessment may, therefore, not give unique information about the in-situ stress magnitudes.

For the most part, a LOT is used to qualify the well for further drilling. Figure 3.20 shows a typical LOT test. Usually, the test is terminated just beyond the LOT point, and the pump is stopped. In light of

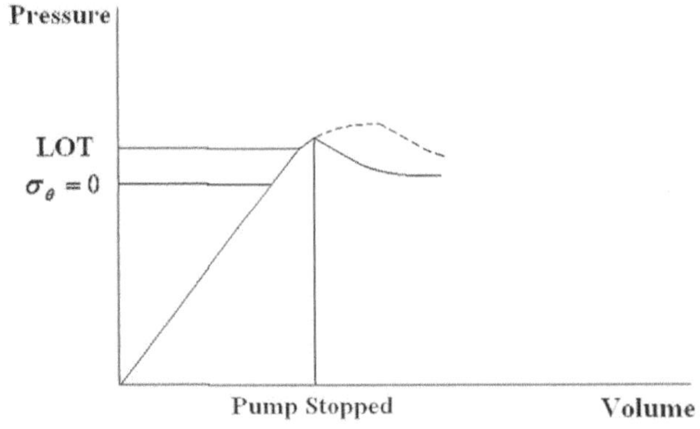

FIGURE 3.20 Typical LOT test.

TABLE 3.8 LOT data for North Sea well

LOT (s.g.):	1.90
P_o (s.g.):	1.08
$\sigma_{tensile}$ (s.g.):	0.1

the previous discussion, the in-situ stress information is in the LOT point. After the pump is stopped, we cannot see any in-situ stress information.

From the discussion above, the following fracturing limits are seen in petroleum wells during drilling: Upper and lower limits of fracture pressure for an *unfractured* well:

$$P_{wf}(upper) = 3\sigma_h - \sigma_H - P_o + \sigma_{tensile} + P_{sb}$$
$$P_{wf}(lower) = 3\sigma_h - \sigma_H - P_o + \sigma_{tensile} \tag{3.25}$$

Upper and lower limits of fracture pressure for a *fractured* well:

Example A: LOT from Vertical Well

Table 3.8 defines data from a vertical well in the North Sea area.

The caliper logs of the well show washouts over large depth intervals but basically circular shapes. The interpretation of this is that the two horizontal stresses are equal. From Equation (3.22) an estimate for the horizontal stress state is:

$$\sigma_h = \frac{1}{2}(LOT + P_o - \sigma_{tensile}) = \frac{1}{2}(1.9 + 1.08 - 0.1) = 1.44\text{s.g.}$$

Example B: Losing Circulation

Using water-based mud, circulation is often reestablished by reducing wellbore pressure or by pumping lost circulation material. The post-failure strength is 1.44 s.g. With water-based muds some of the strength can be recovered by the stress bridge. Using oil-based muds the expected hole strength is now only 1.44 SG.

3.2.5.4 Further Comments on Wellbore Fracturing

From this discussion, it is apparent that fracturing is a non-reversible process. Once a borehole is fractured, the tensile strength cannot be recovered. This applies basically at high wellbore pressures where the tangential stress goes into tension ($P_w > \sigma_h$). At high wellbore pressures where the tangential stress goes into tension, the Kirsch equation no longer applies.

At low wellbore pressures ($P_w < \sigma_h$), the tangential stress is always in compression. The fracture may not have an impact on the stresses but may have affected the compressive rock strength locally. For this case, the continuum mechanics approach given by the Kirsch equation still applies.

Some mud may be left in the fractures, leaving a small but permanently expanded fracture. One hypothesis used in later years is that this leads to an increased hoop stress. From a mechanistic point of view, this hypothesis is questionable. During pressurisation towards the LOT, a radial and tangential strain is introduced on the borehole wall. This strain is coupled to the change in radial and tangential stresses as defined by the Kirsch equation. However, at failure, the Kirsch equation is no longer valid, and the pressure resistance is given by the least in-situ stress σ_h. Because now we have a long fracture crossing the borehole on both sides, we no longer have a radial stress/strain situation but a linear stress-strain in an infinite medium. As the pressure now is approximately equal to the least in-situ stress, the strain is very small. On reducing the pressure in the well, in-situ stress conditions still exist.

One key point in this discussion is that in order for an increase in stress to occur, a corresponding increase in strain is required. Another key point is that after failure, the rock wall becomes discontinuous, leading to a change in stresses.

A related question is if it is possible to create a small fracture that does not penetrate the stress concentration region around the borehole (less than 5 radii distance). Hypothetically, this may create a stress concentration as a tensile tangential stress still exists. Experiences from fracturing of concrete cores indicate an abrupt failure often associated with a noise. A partial fracture was never observed. If the fracturing is an unstable process that always penetrates the stress concentration, it is not likely to increase stress due to the filling of fractures. Future research may show if it is possible to create short partial fractures that actually increase the tangential stress.

3.2.5.5 Future Work

The subject treated in this paper is very important for the oil industry. We have looked at the dominating mechanisms and have suggested alternative research paths. Further work is needed on:

- More experimental work is needed on stress bridges. This would include particle concentrations, particle sorting, stability testing, and obtainable pressure magnitudes.
- Both theoretical and experimental work on tensile rock fracturing. This will include research on compressibility effects, bridging conditions, and fracture propagation conditions.
- Tables for rock tensile strength should be established for analysis of LOT tests. Possibly correlations between rock tensile strength and uniaxial compressive strength could be established for use in both fracturing and collapse analysis.
- Full-scale tests can be carried out to determine if in-situ stress information can be found after the pump stops. In addition to downhole pressure readings, downhole flow measurements should also be collected.
- The model presented may suggest that increased intra-particle friction may help reduce circulation losses with oil-based drilling fluids. One way to do this is to increase the angularity of the mud particles.

Summary

The paper presents a fracture model for the leak-off testing using drilling muds with filtrate control. The model consists of a continuum mechanics model up to the leak-off point and a fracture evaluation for the post-failure pressure behaviour.

The most important findings are:

- The leak-off pressure accurately represents the in-situ stress state and the tensile rock strength.

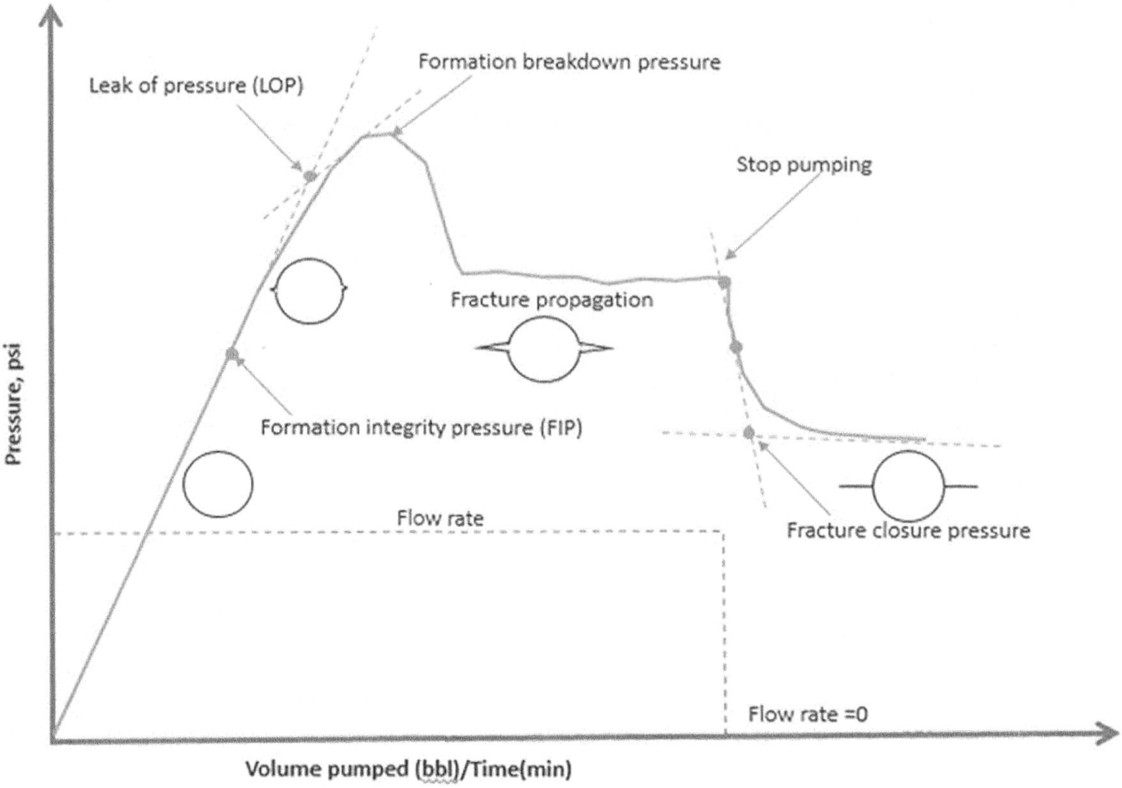

FIGURE 3.21 Illustration of leak-off and formation integrity tests.

- During the first fracture cycle, the tensile rock strength must be overcome. To determine the in-situ stresses from the test, the tensile rock strength must be known.
- After the borehole is fractured, the hole strength consists of the stress bridge and the least in-situ stress.
- The minimum horizontal stress can be estimated from the bottom level of the pressure curve during continuous pumping.
- Decline curve analysis, flowback, and minimum pressure evaluation performed after the pump is stopped may not give unique information of the in-situ stresses using drilling muds.
- The tensile strength obtained from extended LOTs most likely overpredicts the real rock tensile strength.

3.2.5.6 *Applied Conventional Well Stability Modelling Workflow*

Well integrity is essential throughout the life of a petroleum well. Maintaining the well pressure between the collapse and fracture gradients during drilling is critical to prevent well failure.

The NORSOK D-010 standard outlines minimum formation integrity requirements for both new and existing wells, utilising formation tests such as the Fracture Integrity Test (FIT), Leak-Off Test (LOT), and Extended Leak-Off Test (XLOT). Figure 3.21 illustrates these tests, which help estimate fracture closure pressure (FCP). According to the standard, the minimum formation stress/fracture closure pressure must exceed the maximum wellbore pressure at the formation depth. However, the standard does not address collapse-related issues or provide generalised well-stability models within the wellbore.

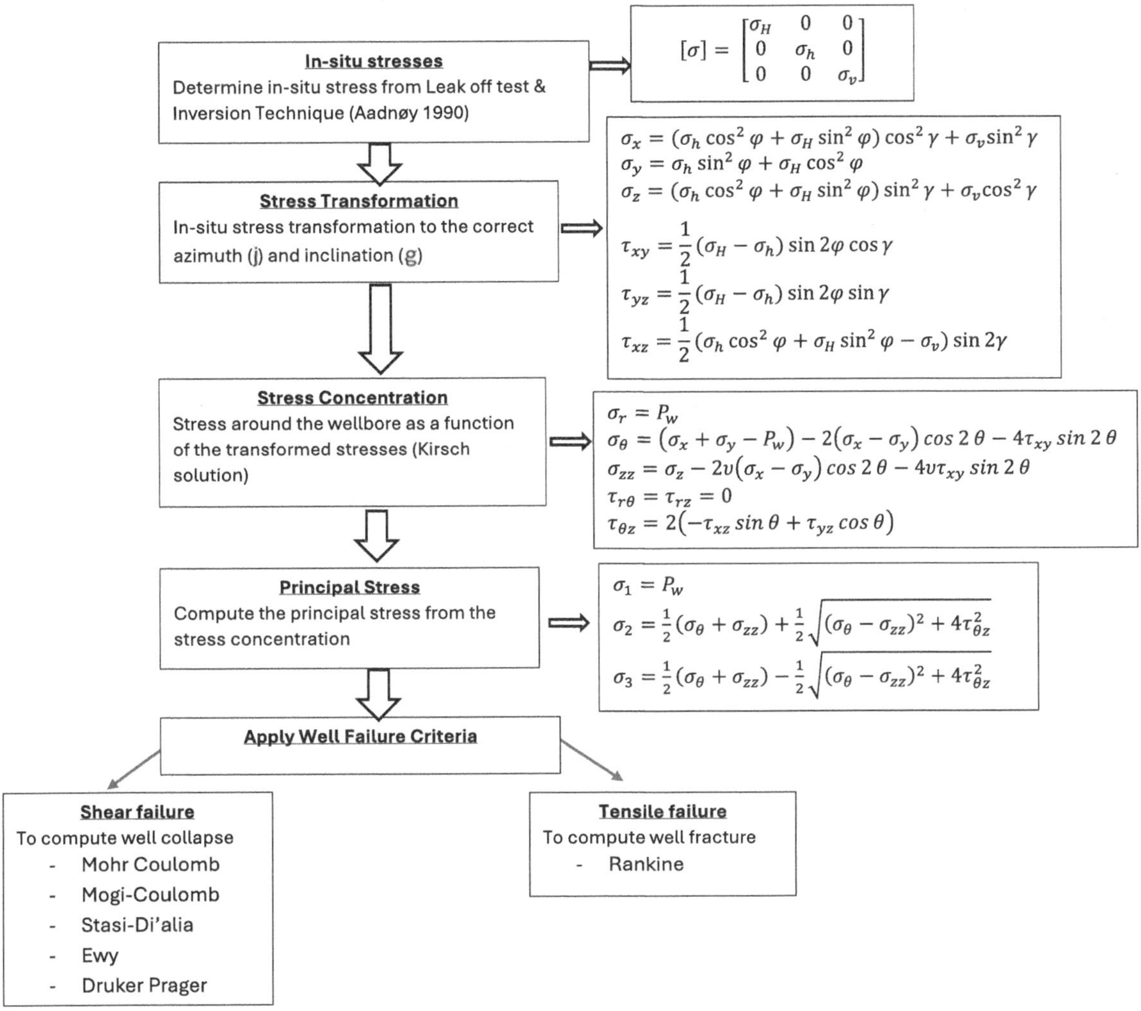

$$[\sigma] = \begin{bmatrix} \sigma_H & 0 & 0 \\ 0 & \sigma_h & 0 \\ 0 & 0 & \sigma_v \end{bmatrix}$$

$$\sigma_x = (\sigma_h \cos^2\varphi + \sigma_H \sin^2\varphi)\cos^2\gamma + \sigma_v \sin^2\gamma$$
$$\sigma_y = \sigma_h \sin^2\varphi + \sigma_H \cos^2\varphi$$
$$\sigma_z = (\sigma_h \cos^2\varphi + \sigma_H \sin^2\varphi)\sin^2\gamma + \sigma_v \cos^2\gamma$$
$$\tau_{xy} = \frac{1}{2}(\sigma_H - \sigma_h)\sin 2\varphi \cos\gamma$$
$$\tau_{yz} = \frac{1}{2}(\sigma_H - \sigma_h)\sin 2\varphi \sin\gamma$$
$$\tau_{xz} = \frac{1}{2}(\sigma_h \cos^2\varphi + \sigma_H \sin^2\varphi - \sigma_v)\sin 2\gamma$$

$$\sigma_r = P_w$$
$$\sigma_\theta = (\sigma_x + \sigma_y - P_w) - 2(\sigma_x - \sigma_y)\cos 2\theta - 4\tau_{xy}\sin 2\theta$$
$$\sigma_{zz} = \sigma_z - 2\upsilon(\sigma_x - \sigma_y)\cos 2\theta - 4\upsilon\tau_{xy}\sin 2\theta$$
$$\tau_{r\theta} = \tau_{rz} = 0$$
$$\tau_{\theta z} = 2(-\tau_{xz}\sin\theta + \tau_{yz}\cos\theta)$$

$$\sigma_1 = P_w$$
$$\sigma_2 = \frac{1}{2}(\sigma_\theta + \sigma_{zz}) + \frac{1}{2}\sqrt{(\sigma_\theta - \sigma_{zz})^2 + 4\tau_{\theta z}^2}$$
$$\sigma_3 = \frac{1}{2}(\sigma_\theta + \sigma_{zz}) - \frac{1}{2}\sqrt{(\sigma_\theta - \sigma_{zz})^2 + 4\tau_{\theta z}^2}$$

FIGURE 3.22 Chart showing the general wellbore stability modelling workflow.

For wellbore integrity analysis, it is crucial to consider both well collapse and fracture. The *general wellbore stability* workflow is outlined as follows:

- As shown in Figure 3.22, the workflow begins by collecting pore pressure, well inclination and azimuth, overburden, and leak-off data from three or more wells. Using these data and a fracture model, the inversion technique introduced by Aadnøy (1990) is applied to invert the minimum and maximum horizontal in-situ stresses.
- The in-situ stress tensor (σ_h, σ_H, σ_v) is then transformed to account for the correct well inclination and azimuth.
- During drilling, stresses are concentrated around the wellbore, as solved by Kirsch (1898). The stress concentration tensors are functions of the transformed in-situ stresses.
- Finally, the *principal stress* is calculated from the stress concentration stress tensor.

- The well *collapse and fracture gradients* are then derived using *failure criteria* and the principal stresses.
  - Several shear failure criteria are utilised to determine the collapse pressure gradient of wells, resulting in varying predictions for minimum mud weight. The typical shear failure criteria employed in the petroleum industry include Drucker-Prager, Ewy, Modified Ewy, Stassi Di'Alia, Mogi-Coulomb, and Mohr-Coulomb.
  - For tensile failure, the *Rankine failure* criterion is typically used.

Example 1: Application of Well Fracture Criteria for Well Fracture Pressure
The borehole will fracture when the *minimum effective principal stress* reaches the tensile rock strength σ_t. Assuming that the *shear stress* is small, which will be neglected, the *hoop stress* at θ =0 becomes the minimum, and applying the *Rankine failure* criterion is expressed as:

$$\sigma'_\theta \leq \sigma_t \tag{3.26}$$

$$\sigma_\theta - P_o \leq \sigma_t \tag{3.27}$$

Inserting the hoop stress in the absence of shear stress, the fracturing equation based on the transformed stress becomes (Aadnoy and Looyeh, 2019):

$$P_w = 3\sigma_x - \sigma_y - P_o - \sigma_t \text{ for } \sigma_x < \sigma_y, \text{ and } \theta = 90° \tag{3.28}$$

$$P_w = 3\sigma_y - \sigma_x - P_o - \sigma_t \text{ for } \sigma_y < \sigma_x, \text{ and } \theta = 0° \tag{3.29}$$

These equations say that a fracture will initiate normal to the least stress and propagate in the direction of the largest normal stress.

- The in-situ stress tensor (σ_h, σ_H, σ_v) is then transformed to account for the correct well inclination and azimuth.

Please also observe that, assuming *maximum and minimum normal* stress to the borehole wall and vanishing shear stresses, the general fracturing equation becomes:

$$P_w = 3\sigma_{min} - \sigma_{max} - P_o - \sigma_t \tag{3.30}$$

Numerical Example C
Assume that the following data exist for a well:

Overburden stress: σ_v = 100 bar

Horizontal stress: $\sigma_H = \sigma_h$ = 90 bar

Pore pressure: P_o = 50 bar

Borehole inclination: γ = 40°

Borehole azimuth: ϕ = 165°

Task: Determine the fracture pressure for a ***vertical well*** and the ***deviated well*** given above.

- For the **vertical well**, the in-situ stresses relate directly to the borehole direction and become:

  $\sigma_x = \sigma_y = 90$ bar. The fracture pressure is determined directly by Equation (3.30):

  $P_{wf} = 2\sigma_x - P_0 = 2x90 - 50 = 130 bar$

- For **the inclined well**, the stresses must first be transformed to the orientation of the wellbore. The result is:

  $\sigma_x = 94.13$ bar

  $\sigma_y = 90$ bar

  $\sigma_z = 95.87$ bar

  $\tau_{xz} = - 4.92$ bar, $\tau_{yz} = \tau_{xy} = 0$

$$\sigma_r = P_w \quad (3.31)$$

$$\sigma_\theta = 2\sigma_h - P_w \quad (3.32)$$

$$\sigma_z = \sigma_v \quad (3.33)$$

These data are again inserted into the equations for the borehole stresses, which become:

Radial stress: $\sigma_r = P_w$
Tangential stress: $\sigma_\theta = 184.13 - P_w - 8.26\cos(2\theta)$
Axial stress, plane strain: $\sigma_z = 95.87 - 2.07\cos(2\theta)$
Shear stress: $\tau_{\theta z} = -9.84\sin\theta$

The angle θ from the x-axis at which the fracture starts must be determined. Equation (3.30) results for these particular data in: θ = 0°, and the fracturing pressure becomes now:

$$P_{wf} = 3x90 - 94.13 - 50 = 125.9 bar$$

We observe that the *fracturing pressure decreases with increased borehole inclination*. This is believed to be a general trend. However, for anisotropic stress states, this behaviour may differ.

Example 2: Application of Well Collapse Failure Criteria for Well Collapse Pressure
For an illustration, let us consider the Mohr-Coulomb failure criterion given as:

$$\sigma_1 - \alpha P_o = C_o + (\sigma_3 - \alpha P_o)\tan^2\beta \quad (3.34)$$

where α is the Biot-coefficient (0.8–1), P_o is the pore pressure and C_o is the uniaxial compression strength. The uniaxial compressive strength can be calculated as:

$$C_o = \frac{2\tau_o \cos\varphi}{1 - \sin\varphi} \quad (3.35)$$

Uniaxial compressive strength can also be estimated from the sonic log as proposed by the empirical models, τ_o is the inherent shear strength (cohesion) and φ is the friction angle calculated by Equation (3.36).

$$\tan^2\beta = \frac{1+\sin\varphi}{1-\sin\varphi} \tag{3.36}$$

Example D: Derive the Well Collapse Pressure

For the sake of simplicity, let us assume a vertical well and isotropic stress state where the horizontal stresses are equal *and no shear stress.* The stress concentrations at the inner wall of the wellbore will be reduced to:

$$\sigma_r = P_w$$

$$\sigma_\theta = 2\sigma_h - P_w$$

$$\sigma_z = \sigma_v$$

where:

- σ_r = radial stress
- σ_θ = Hoop stress
- σ_z = axial stress
- P_w = well+ pressure
- σ_v = Overburden stress
- σ_h = Horizontal stress

Solution:
Failure criteria

- Use the failure criteria given in Equation (3.34).

Assume that the stress state is in the order of:

- $\sigma_\theta \geq \sigma_z \geq \sigma_r$

The maximum and the least principal stresses are σ_θ and $\sigma_{r,}$ respectively.
Inserting these in the Mohr-Coulomb failure criteria, we get:

$$\sigma_1 - \alpha P_o = C_o + (\sigma_3 - \alpha P_o)\tan^2\beta$$

$$\sigma_\theta - \alpha P_o = C_o + (\sigma_r - \alpha P_o)\tan^2\beta$$

$$2\sigma_h - P_w - \alpha P_o = C_o + (P_w - \alpha P_o)\tan^2\beta$$

Solving the well pressure, P_w, we get:

$$P_w \leq \alpha P_0 + \frac{2(\sigma_h - \alpha P_0) - C_0}{1 + \tan^2 \beta}$$

Problem:
Assuming the stress fields ($\sigma_z \geq \sigma_r \geq \sigma_\theta$) have pore pressure, P_o, your task is to derive the well collapse pressure by using the Mohr-Coulomb failure criteria (Equation 3.34).

Nomenclature

a,b	inner and outer radius of core sample
c	elastic/plastic zone of core sample
t	plastic zone thickness
q	plastic/elastic interface pressure
d,f	inner and outer radius of cylindrical mud cake
e	elastic/plastic zone of core sample
U	yield stress for stable stress bridge
P_w	well pressure
P_{wf}	fracture pressure
P_e	final pressure after pump is stopped
P_{ult}	ultimate well pressure before stress bridge fails
σ_h	minimum horizontal in-situ stress
σ_H	maximum horizontal in-situ stress
$\sigma_{tensile}$	tensile strength
σ_v	vertical stress
P_o	pore pressure
LOT	leak-off-test
ELOT	extended leak-off-test
Q	flow rate

3.3 FRACTURING PRESSURES FOR SHALLOW PENETRATION

3.3.1 General

The nomenclature for this chapter is as defined in Section 3.1. and is shown in Figure 3.23.

There exists little compiled information for fracture strength in shallow layers. The reason is mainly that the shallow casing strings are usually not critical and that the shallow holes are often drilled without blow-out preventers, thereby not allowing for pressure integrity testing. Yet, there is a need from a well design point of view to establish fracture gradient curves in order to optimise casing setting depths, especially with respect to shallow gas zones.

Aadnoy et al. (1991) showed that the spread in fracture gradients usually decreases with depth. For depths down to 1000 meters, the spread is considerable, and it has so far not been possible to model this total span. We usually assume a pore pressure equal to that of hydrostatic sea water, so a pressure normalisation provides no improvement. Using this assumption, we have no benefit of the effective stress principle. Figure 3.23 defines the physical setting of the problem. Figure 3.24 shows some of the shallow fracture gradients for the Sleipner field (Aadnoy et al., 1991). The before-mentioned spread is very pronounced, and several models can be defined, e.g., giving average, low, or high leak-off values.

At very shallow depths, pressure control is of little concern, and the high frac. gradient has little interest aside of providing an excess integrity for the mud weight. The average value could be of more practical use. However, using the average value, one will usually not know whether the actual leak-off values for future wells become lower or higher than the average. If a significantly lower value occurs, the casing design may have to be altered. The approach taken here will be to define the low frac. gradient with reasonable accuracy. There is little likelihood of obtaining an even lower value in future wells.

From a practical drilling point of view, a reasonable accurate prediction of the low side of the frac. gradient plot would have significance, as a more marginal casing program could be developed with potential for optimisation and with a small likelihood for mud loss problems. Therefore, in this analysis, we will attempt to model lost circulation behaviour; that is, the high fracture points are neglected, and the low points are attempted to be modelled.

A similar collection and evaluation of leak-off data covering deeper depths was done by Breckels and van Eeckelen (1982). Also, Daines (1982) developed a correlation method for deeper intervals.

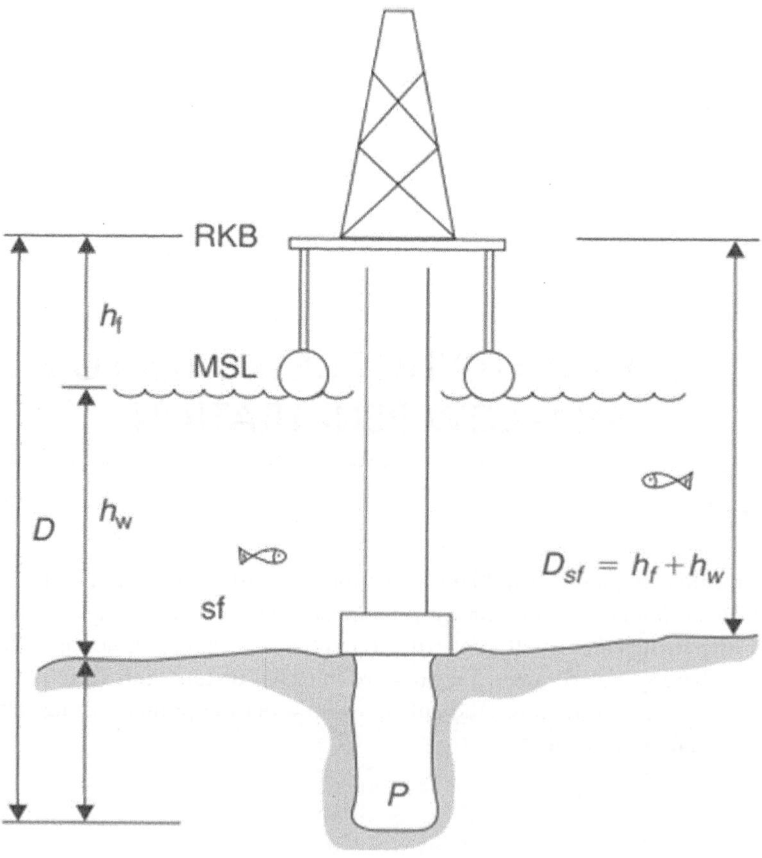

FIGURE 3.23 Definition of references.

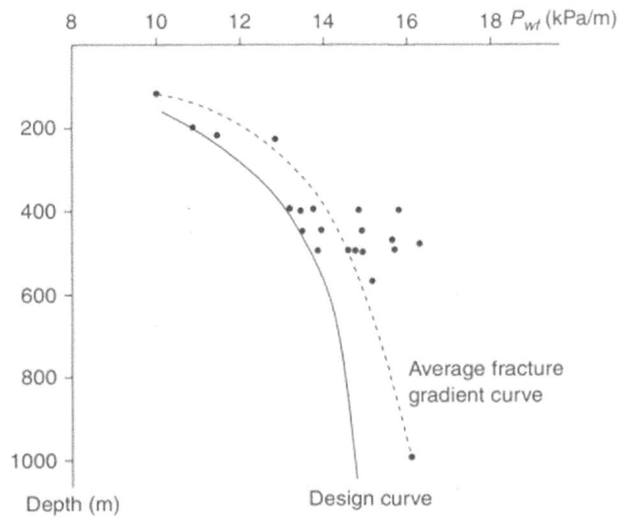

FIGURE 3.24 Example of shallow fracture gradients from the Sleipner field. Observe the large variation and the design curve that defines the low points.

3.3.2 Depth-Normalised Shallow Frac. Data

Figure 3.24 shows data from the Sleipner field only. The water depth is here 118 m. The shallow fracture data from other fields are recorded for various water depths. To make the data consistent, they are therefore normalised to the seabed by subtracting the water pressure, following the principles outlined in Section 3.1.3. Furthermore, we have collected four data sets and used only the low side of these to approach the low frac. gradient, or critical lost circulation gradients, as stated above. The data are:

- Saga Petroleum seabed diverter obtained data.
- General soil strength data from various platform studies.
- Shallow leak-off data from unidentified exploration wells.
- Low frac. data from a study of the Sleipner field (Aadnoy et al., 1991).

The result is shown in Figure 3.25. We observe that there seems to be consistency in the data. However, one pertinent question is whether these can be used in a general sense since they come from various sources. Another, but opposite, argument is that the very shallow deposits are very young and unconsolidated and have similar properties regardless of location, which is the approach we will take. We will not accept this without concern, but we will use the following model as a design tool.

The generalised frac. curve of Figure 3.25 can be expressed as equations for an arbitrary water depth and height of drill floor, referred to as RKB. Here, we have added a general water pressure and used the drill floor as the reference level instead of the seabed.

$$d = 1.03\frac{D_{sf} - h_f}{D} + 1.276\frac{D - D_{sf}}{D} \quad \text{for}: 120\text{m} > \text{D} - \text{D}_{\text{sf}} > 0 \tag{3.37a}$$

$$d = 1.03\frac{D_{sf} - h_f}{D} + 1.541\frac{D - D_{sf}}{D} - \frac{33.16}{D} \quad \text{for}: 600\text{m} > \text{D} - \text{D}_{\text{sb}} > 120\text{m} \tag{3.37b}$$

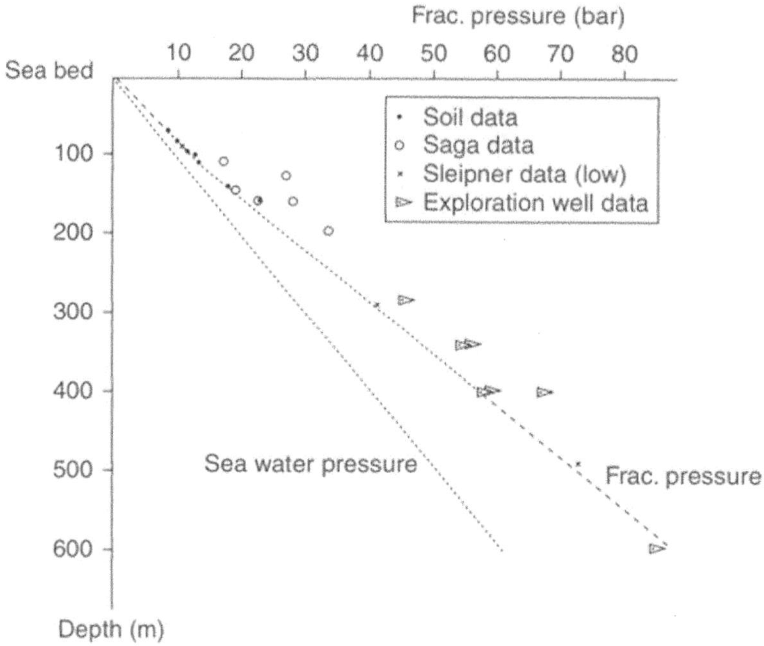

FIGURE 3.25 Low shallow frac. data normalised to seabed level, and by subtracting the pressure of sea water.

For shallow formations, we assume a hydrostatic pore pressure. This implies that the pressure gradient is equal to 1.03 s.g. both in the sea water and inside the formation. If the reference level is the sea level, the pore pressure would simply be obtained by using the saltwater gradient. However, the drilling rigs always have the drill floor above sea level, providing a different reference point. To use gradient plots, we have to correct for drill floor elevation. Assume a pressure P at depth D from the drill floor. If the drill floor is used as a reference, the pressure can be expressed as:

$$P = 0.098 d_{RKB} D$$

The same pressure can be expressed from the mean sea level as:

$$P = 0.098 d_{MSL}(D - h_f) = 0.098x1.03(D - h_f)$$

Equating the two expressions results in an expression for the normal pore pressure gradient from any elevation:

$$d_{RKB} = d_{MSL} \frac{D - h_f}{D} = 1.03 \frac{D - h_f}{D} \tag{3.38}$$

3.3.3 Estimation of Shallow Frac. Grad. for a Semi-Sub and a Jack-Up Rig

In the example shown in Figure 3.26, we assume a water depth of 68 m. In addition, we are considering two drilling rigs: a semi-submersible rig with the drill floor 26 m above sea level and a jack-up rig with

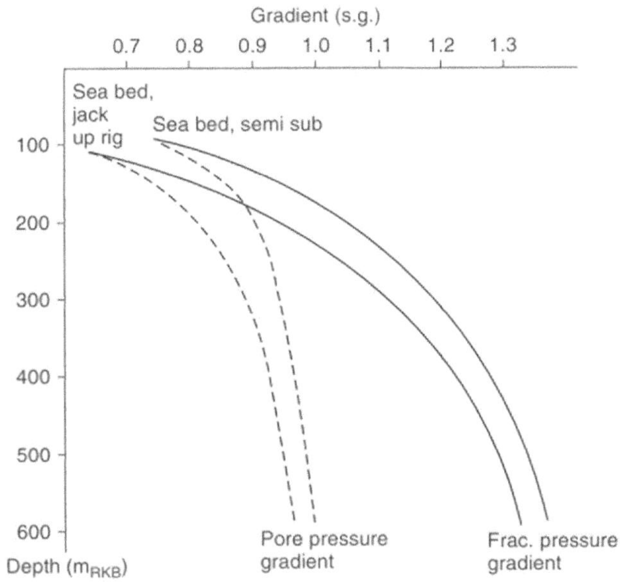

FIGURE 3.26 Fracture and pore pressure gradients for two drill floor elevations. Example: water depth 68 m, semi-sub elevation 26 m, and jack-up elevation 42 m from sea level to drill floor.

the drill floor 42 m above sea level. Generating gradient curves, this air gap reflects the results. Using Equations (3.37) and (3.38), Figure 3.26 was made. The frac. gradient curves for the two rig types are shown. In addition, the gradient curves for the pore pressure are shown. We have assumed the static weight of sea water according to Equation (3.38).

Problem

You are planning two wells which are going to be drilled with the semi-submersible drilling rig Wildcat. The drill floor elevation is 22 m. One well will be drilled in 56 m water depth while the other will be drilled in 172 m water depth.

a) Use Equations (3.37) and (3.38) and make a plot for the shallow frac. gradient and pore pressure gradients down to 600 m below seabed for both wells.
b) Make a plot of the frac. gradients and pore pressure gradients from a mean sea level reference.
c) Plot the difference between the pore pressure gradients and the frac. gradients for the cases above.

3.4 FIELD EVALUATION OF BOREHOLE COLLAPSE

3.4.1 Introduction

The following specific nomenclature is used in this chapter:

$\sigma_r = P_w$ = Borehole pressure gradient
σ_θ= Tangential stress gradient

σ_a = Estimated average horizontal stress gradient
σ_v = Overburden stress gradient
P_o = Pore pressure gradient
P_{wf} = Fracturing pressure gradient
P_{wc} = Critical collapse pressure gradient
D_h' = Bit diameter

Borehole stability is a very central issue, as the wells in later years have become longer and with higher inclination. For years, people have estimated the time loss associated with unexpected borehole stability problems to account for 10–15% of the time required to drill a well. Since the rig time is the major cost factor in a drilling operation, we understand that borehole stability problems are very costly for the industry.

Problems encountered may be unexpected circulation losses that take time to cure or have to be handled by using contingency casing strings. However, another class of problems is associated with borehole collapse. Very often, a tight hole occurs, which requires frequent wiper trips or reaming. This can, in certain wells, lead to stuck drill strings or difficulties in landing the casing string. There are many reasons for tight hole, for example, dog-leg severity can contribute, or simply inward creep of the borehole wall, also aided by shale swelling.

Most boreholes will enlarge over time. This is often a time-dependent collapse phenomenon. Problems caused by hole enlargement are difficulties in removing rock fragments and drilled cuttings from the borehole or a reduced quality of the cement placement behind casing strings. It is important to understand that tight hole and borehole collapse are similar events; in one case, the hole may yield, while in the latter case, an abrupt failure may occur. Borehole collapse is possibly the most costly single problem encountered during the drilling of a well, and there does not exist a simple cure for the problem.

3.4.2 Collapse Mechanisms

The concept of borehole collapse will be used to demonstrate the borehole problems and their dependence on borehole pressure. There are two aspects, namely, mechanical failure and chemical effects. We will discuss the former.

Field personnel have for many years known that borehole collapse is most severe at low mud weights. Yet, low mud weight has often been used for reasons of pore pressure estimation and maintaining the drilling rate. These concepts have been discussed in Section 2.1 on mud weight selection. Although a high mud weight may not eliminate hole enlargement, we believe that enlargement can be minimised by maintaining the mud weight above a critical level.

We will briefly look at the stresses acting on the borehole and show the effects of the mud weight on these. The most important stress component is hoop stress or tangential stress. On the borehole wall, we have three normal stress components, which in the simplest form can be defined as follows for a vertical well:

Radial stress: $\sigma_r = P_w$ = borehole pressure

Tangential stress: $\sigma_\theta = 2\sigma_a - P_w$ (3.39)

Vertical stress: σ_v = overburden = constant

Pore pressure: P_o

In geosciences, we must also take the formation pressure into account. The effective stress principle simply says that the total stress is the sum of the pore fluid pressure and the stress taken up by the rock matrix itself. We are here considering the failure of the rock matrix and must therefore use the effective stresses, which can be written as:

Radial effective stress: $\sigma_{r'} = P_w - P_o$

$$\text{Tangential effective stress: } \sigma_\theta' = 2\sigma_a - P_w - P_o \tag{3.40}$$

$$\text{Vertical effective stress: } \sigma_v' = \sigma_v - P_o,$$

Figure 3.27a shows these stresses on the borehole wall, and Figure 3.27b shows a numerical example. Field data are inserted into Equation (3.40), and the three stress components are shown as a function of borehole pressure. Three borehole pressure levels are shown. Table 3.9 gives the stress magnitudes at these borehole pressures.

We observe from the example above that the vertical (axial) stress component is constant regardless of the borehole pressure, while the radial and tangential stress varies with the borehole pressure. Of course, the radial stress is given by the borehole pressure itself, so the key parameter is the tangential stress, or the hoop stress. In the following, we will illustrate how these stresses relate to the failure of the borehole.

Figure 3.27b shows these borehole pressures. Common for these is the difference between the radial and the tangential effective stresses. This stress contrast gives rise to shear stresses. We see that for the

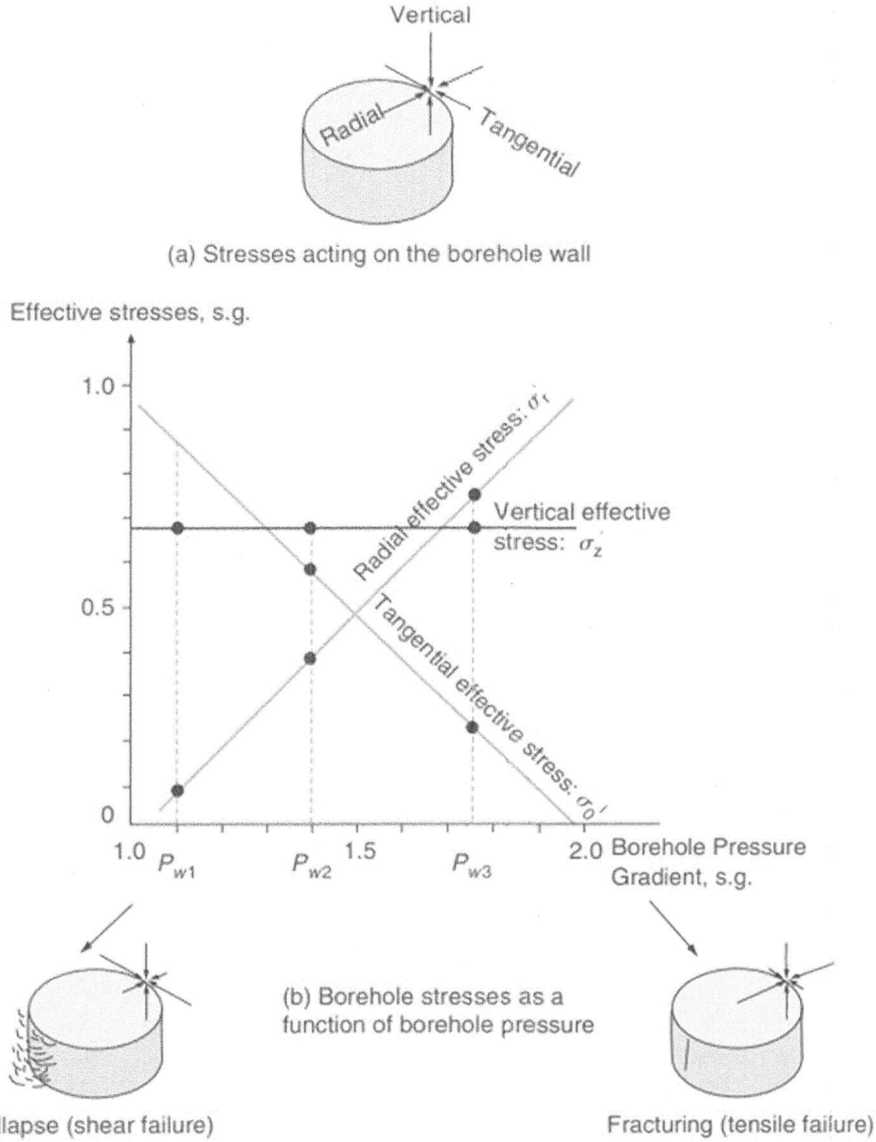

FIGURE 3.27 Example of wellbore stresses. Data: σ_a = 1.5 s.g., σ_v = 1.7 s.g., P_o = 1.03 s.g.

lowest borehole pressure gradient (P_{w1}) this difference is the largest. At P_{w2} the difference is smaller, but we observe that at high pressures P_{w3} the stress difference again increases.

The three cases from Figure 3.27b are shown in a Mohr-Coulomb plot in Figure 3.28. The straight line is a failure envelope obtained from the tri-axial testing of cores. We see that P_{w2} and P_{w3} are well below the failure envelope, and the hole should, therefore, be stable for these borehole pressures. However, P_{w1} approaches the failure envelope, and at this point, the stress loading is equal to the strength of the rock. The borehole will fail at this point.

From the two previous figures, we see that by lowering the borehole pressure, the tangential stress will increase, ultimately resulting in borehole failure. Also, from Figure 3.27, we observe that the stress contrast can be minimised by selecting a proper mud weight. This is addressed in Section 2.2 about mud weight selection. Furthermore, we see that high shear stresses arise also for high mud weights. This is not commonly considered today but should be kept in mind. The *median line principle* derived in Section 2.2 takes this into account.

In this section, we will address borehole collapse. Since we have presented field data, we may briefly also evaluate the fracturing pressure for the example. Fracturing is a tensile failure defined as taking place when the effective tangential stress of Equation (3.40) equals zero. Using the numbers given in Figure 3.19, we find a fracturing pressure gradient of 1.97 s.g.

We will not pursue this modelling approach further but move into the main topic of this section, the evaluation of field data. For more details on modelling, please refer to Aadnoy and Chenevert (1987).

3.4.3 Interpretation of Caliper Logs

We have just shown that borehole pressure plays a key role in the collapse failures of boreholes. Often, the remedy is to increase the borehole pressure, that is to lower the shear stress, as shown in Figure 3.28. We have also briefly introduced the elements of stresses and rock strength. Unfortunately, we often do not have sufficient stress and rock strength information to perform a reliable collapse analysis.

However, we have many caliper logs from boreholes showing hole enlargement. In this section, we will take advantage of these to analyse the field data in an empirical way. First, we will analyse a vertical exploration well in the southern North Sea.

3.4.3.1 Exploration Well Example

During the drilling of a well, some borehole collapse took place. Although the collapse was not very pronounced, one had difficulty landing the production casing and the liner. Remedial work of reaming, hole cleaning, and under-reaming took considerable time before the well was successfully finished.

Figure 3.29 shows the stresses and pressures in the well. We observe that the 12-1/4" hole section is abnormally long, 2000 m. Also, the pore pressure is hydrostatic sea water down to 2800 m. The well has a low pore pressure gradient profile. Also shown are the mud weights used, the overburden stress gradient obtained from the integration of density logs, and the measured leak-off values below each casing shoe.

We have, in the foregoing, seen that the well will reach a critical pressure and collapse as the mud weight is lowered, and we have seen that the tangential (or hoop) stress is the key parameter. We will now calculate the hoop stress for this well and correlate it with the caliper log. We will use the results of an analysis of the well, which concluded that the horizontal stress level is about 80% of the overburden stress for this well. The effective stresses from Equation (3.40) can then be written:

$$\sigma_{r'} = P_w - P_o$$

$$\sigma_\theta' = 2\sigma_a - P_w - P_o = 2x0.8x\sigma_v - P_w - P_o \tag{3.41}$$

$$\sigma_v' = \sigma_v - P_o$$

TABLE 3.9 Numerical example of wellbore stress gradients

BOREHOLE PRESSURE GRADIENT (S.G.)	*PW1 = 1.1*	*PW2 = 1.4*	*PW3 = 1.75*
Radial effective stress	0.07	0.37	0.72
Tangential effective stress	0.87	0.57	0.22
Vertical effective stress	0.77	0.77	0.77

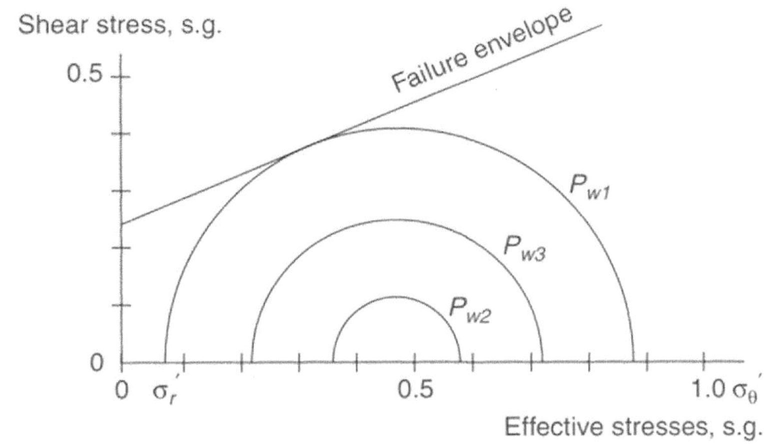

FIGURE 3.28 Mohr-Coulomb representation of the three cases in the numerical example.

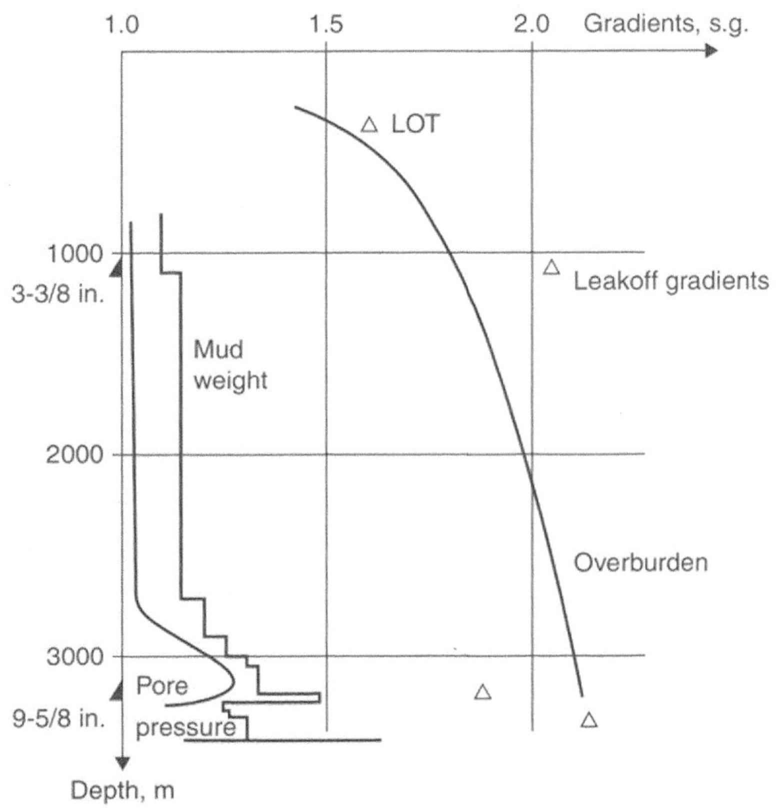

FIGURE 3.29 Pressure gradients for the exploration well.

We want to depth-normalise the data. If we, for example, calculate the effective hoop stress and divide by the effective overburden stress, we expect a vertical trend (see Section 3.2.2). This ratio can be obtained from Equation (3.41) as follows:

$$\frac{\sigma_\theta'}{\sigma_v'} = \frac{1.6\sigma_v - P_w - P_o}{\sigma_v - P_o} \tag{3.42}$$

Figure 3.30 shows Equation (3.42) plotted using the field data. Also, the caliper log is shown. We observe that there is a very good correlation as the hole is collapsed for $\sigma_\theta'/\sigma_v'>1$, but basically in-gauge if this ratio is lower. At the bottom of the 12-1/4" section of Figure 3.29 we observe that the mud weight has been increased. The result is an in-gauge hole. However, in the 8-1/2" section below, the mud weight was temporarily reduced, and some collapse took place. Maintaining a higher mud weight for the remainder of the drilling operation resulted in an in-gauge hole.

From the evaluation of Figure 3.30, we can state that the critical collapse pressure was reached when $\sigma_\theta'/\sigma_v' = 1$. Solving Equation (3.42) for this condition results in:

$$P_{wc} = 0.6\sigma_v \tag{3.43}$$

Actually, the solution to this evaluation of this particular well is that the mud weight should be higher than 60% of the overburden stress. This curve is shown in Figure 3.31. It is seen that only a slight increase in mud weight would probably have resulted in a more in-gauge hole. Also, from Figure 3.31, we see that the overburden stress gradient is systematically increasing with depth, which is also reflected in the horizontal stress state. The mud weight, however, is kept nearly constant, which implies that the stresses acting on the hole wall are increasing with depth. The optimal choice of mud weight has been addressed in Section 2.2.

The lithology of the collapsed intervals of Figure 3.31 covers the Upper Jurassic to the Upper Cretaceous. A comparison between the caliper log and the lithology did not result in a correlation.

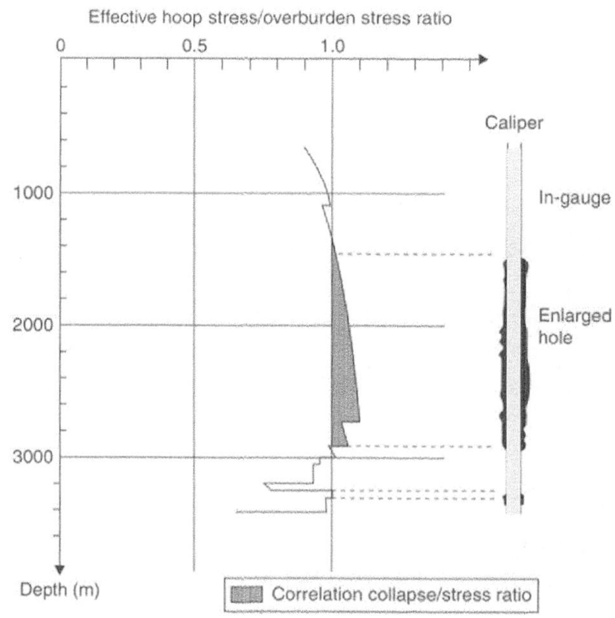

FIGURE 3.30 Correlation between depth normalised tangential stress and caliper log. Horizontal stress equal to 80% of overburden stress.

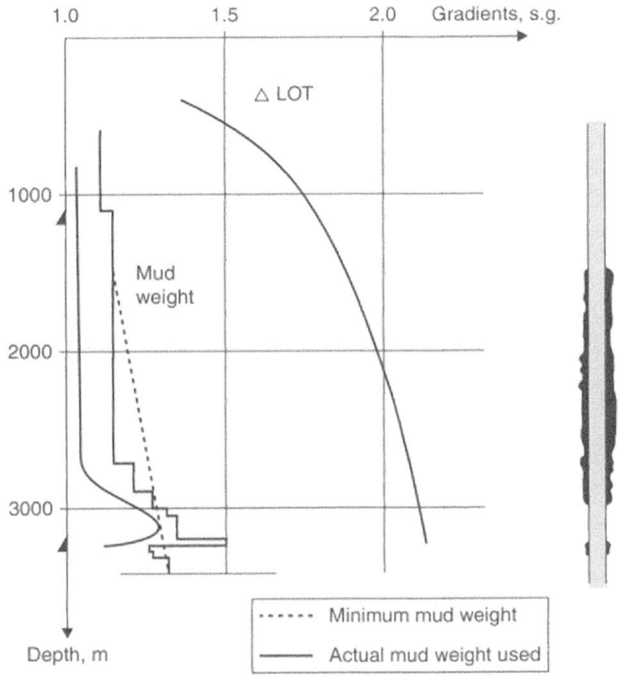

FIGURE 3.31 Results of the analysis showing critical mud weight to avoid collapse.

3.4.3.2 Deep Well Example

We will now look at another case, which is a little more complicated but can be analysed using the same approach.

The lower portion of the 12-1/4 in. section of a deep well experienced a considerable collapse. The depth interval is 4200–4630 m, well below where we expect reactive clays. We believe, therefore, that it is a mechanical hole collapse and that the remedy would have been to use a slightly higher mud weight. Figure 3.32 shows the pressure gradients, and Figure 3.33 shows the resulting collapse from two different caliper log runs. Figure 3.34 shows the open hole exposure time.

In order to perform modelling, the in-situ stress state must first be established. A study showed that the fracture gradient could be approximated as equal to the overburden stress gradient at this depth ($P_{wf} = \sigma_v$). Using this equation and setting the effective tangential stress equal to zero (Equation (3.40)), the horizontal stress is estimated from:

$$\sigma_\theta' = 2\sigma_a - P_{wf} - P_o = 0$$
$$= 2\sigma_a - \sigma_v - P_o = 0$$
$$\text{or: } 2\sigma_a = \sigma_v + P_o \quad (3.44)$$

Following the process derived earlier, we will compute the effective hoop/overburden stress ratio by using the horizontal stress from Equation (3.44).

$$\frac{\sigma_\theta'}{\sigma_v'} = \frac{2\sigma_a - P_w - P_o}{\sigma_v - P_o}$$
$$= \frac{\{(\sigma_v + P_o) - P_w - P_o\}}{\sigma_v - P_o}$$

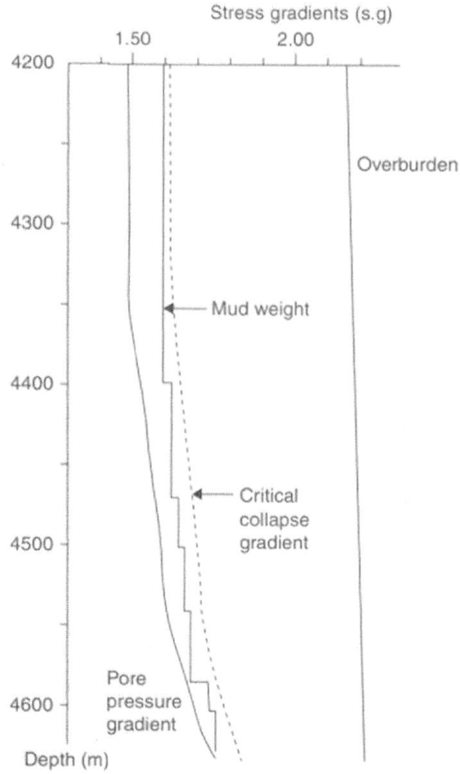

FIGURE 3.32 Stress and pressure gradients for the deep well.

$$\frac{\sigma_\theta'}{\sigma_v'} = \frac{\sigma_v - P_w}{\sigma_v - P_o} \tag{3.45}$$

Figure 3.33 shows this ratio computed for both caliper log runs. We have a problem, however. No sections of the caliper are in-gauge, so we have no reference depth with transition for in-gauge/collapsed hole. However, by comparing Figures 3.33 and 3.34 we may get a clue. In Figure 3.34, we observe that the open hole exposure time is about 200 hours more for the second logging run above 4500 m. Yet, the two caliper logs are nearly identical. One interpretation of this may be as follows:

When drilling below 4500 m, the mud weight is gradually increased. This means that the section from 4200 to 4500 m is now exposed to a higher pressure than when the section was drilled. Since no additional collapse has taken place in the upper part, possibly the mud weight was above a critical level. In other words, for the hole section above 4500 m the initial mud weight was too low (caliper log 1), while the mud weight used when drilling the last part was above the critical collapse level (caliper log 2). However, during drilling of the section from 4500–4632 m, the mud weight was too low for this bottom part. This is a zone of increasing pore pressure. At about 4420 m, both caliper curves show a minimum collapse. This may suggest that here we are close to the critical collapse pressure.

If we use the definitions above, then the correlation of the effective stress ratio of Figure 3.32 will be between the two curves. Since the effective stress ratio is depth normalised, we choose a vertical trend of ratio 0.82 as our design line, which crosses the second stress ratio curve at 4220 m. Inserting this value into Equation (3.45) yields the following equation for the critical collapse pressure:

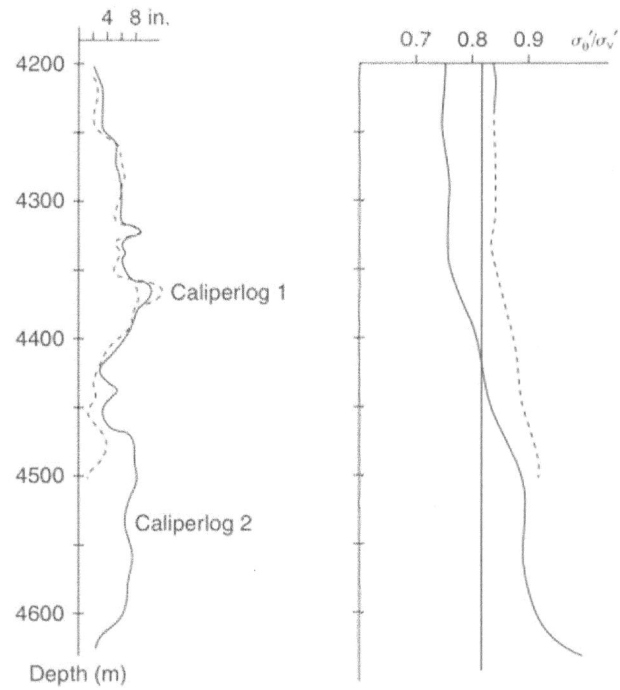

FIGURE 3.33 Caliper logs and the computed stress ratio.

$$\frac{\sigma'_\theta}{\sigma'_v} = \frac{\sigma_v - P_{wc}}{\sigma_v - P_o} = 0.82$$

$$P_{wc} = 0.18\sigma_v + 0.82P_o \tag{3.46}$$

Equation (3.46) defines the critical mud weight. This is indicated in Figure 3.32. If another well will be drilled in the same area, this critical mud weight should be used as a minimum value. Note, however, that this equation is valid only for the conditions met in this well. If for example, another pore pressure regime is met in another well, the in-situ stress state should probably be normalised for those conditions. We have not shown the lithology in Figure 3.33, but the rock above about 4450 m is chalk, and below it is shale.

In this chapter, we have demonstrated that even if in-gauge portions of the caliper log are missing, we may still be able to estimate the critical collapse pressure. In this case, we used a second caliper log to identify that the collapse had stopped. This problem also leads us towards time-dependent borehole collapse, which will be discussed in the following chapter.

3.4.4 Time Dependency

In the previous example, we have only considered the stress state versus failure. From field experience, we know, however, that borehole stability problems are clearly time dependent. If a well can be drilled and cased off in a short time, we usually have no problems. If, on the other hand, the borehole is allowed to stay open for some time, problems like difficulties in landing the casing string may arise.

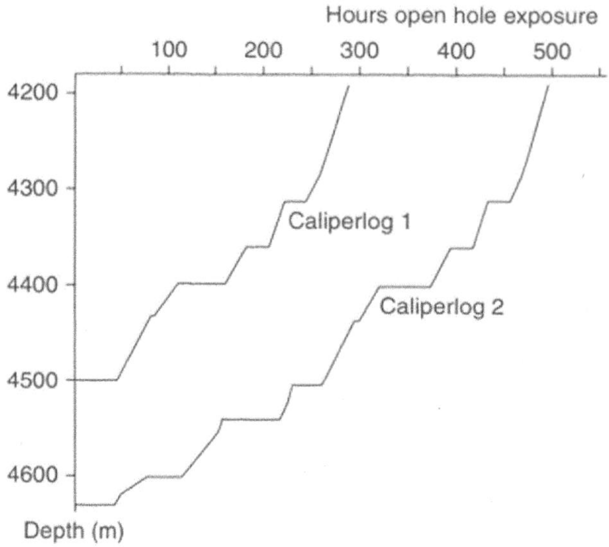

FIGURE 3.34 Open hole exposure times.

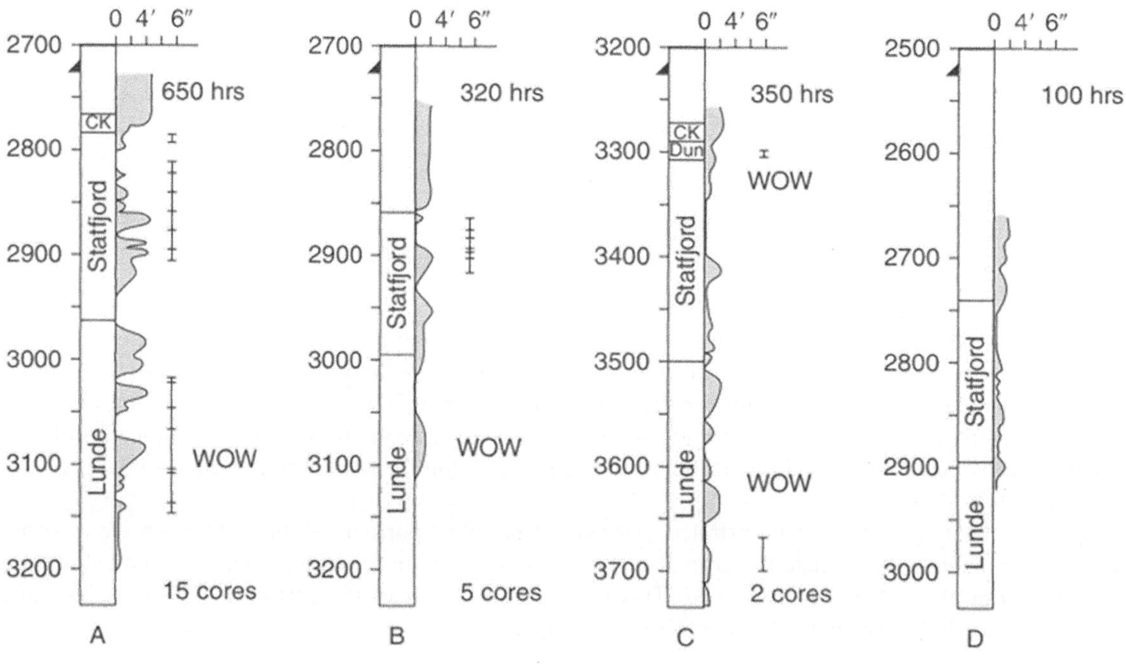

FIGURE 3.35 Caliper logs for four production wells.

Figure 3.33 demonstrates the time dependency on hole enlargement below 4420 m. We will, in the following, compare four production wells.

Figure 3.35 shows four wells. The open hole exposure time is different for these wells mainly because the coring programs were different. Figure 3.36 shows the time curves for the various parts of the boreholes, which is the time from drilling to logging.

From Figure 3.36, we see that the bottom section of each well has stayed open approximately the same time, from 30 to 60 hours. This is consistent with the caliper logs of Figure 3.35, which shows similar, but little, collapse for all four wells. Further up, the open hole exposure times are quite different since various

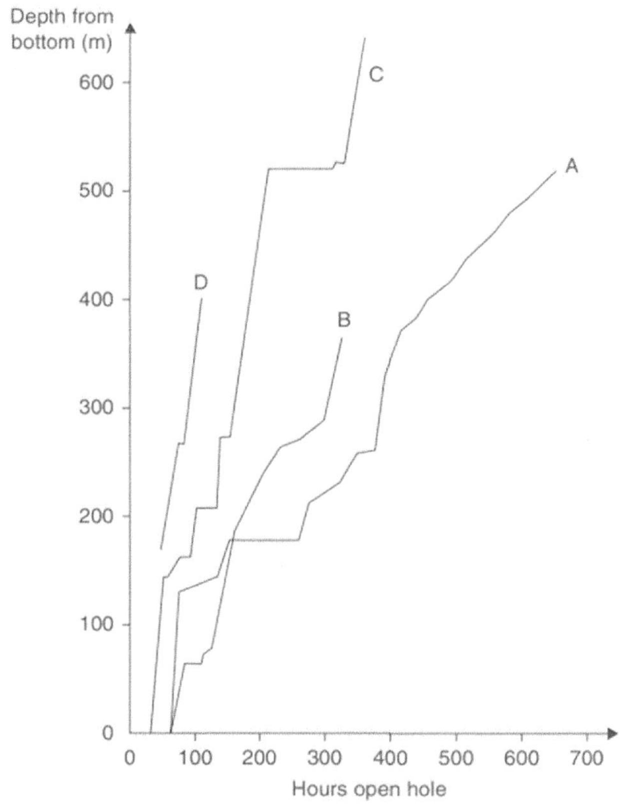

FIGURE 3.36 Open hole exposure time for the four wells.

coring programs have been implemented, and the wells have also been temporarily abandoned because of bad weather conditions, as indicated with waiting on weather (WOW) in Figure 3.35. Well A has been open for 650 hrs with a maximum collapse of 4.5 in. Wells B and C have been open for 320–350 hrs, giving a maximum collapse of more than 2 in., while well D has been open 100 hrs, with a collapse of less than 2 in.

The above example illustrates the time dependency of borehole collapse.

Berland (1993) performed an analysis of time-dependent borehole collapse of a North Sea oil field. The following main elements were investigated:

- the anisotropic stress field
- the overburden stress
- the time of open hole exposure
- the depth
- the lithology
- the effects of KCl inhibition
- effective stresses of the elements above

Since 41 different analyses were performed, we will not discuss all aspects of the modelling, only point at some of the observations, which can be given as:

- Although the use of anisotropic horizontal stress for the field gave the best results, a reasonable correlation was obtained using the overburden stress as external loading.

- The collapse is clearly lithology dependent, with the following order from most collapse to less: paleosol, mudstone, siltstone, and sandstone.
- A new parameter was found that dominated the extent of the collapse. The collapse extent seemed to be proportional to the thickness of each lithological group.

The last element requires some explanation. Figure 3.37 shows a portion of a caliper log. Also shown are the lithological groups. We observe that the sandstone sections are basically in-gauge, while the clay sections are collapsed. Berland (1993) observed that the collapse was proportional to the thickness of the lithological unit. Or, to put it another way, if sand stringers occur within a clay sequence, the effect is that the collapse is reduced. The mechanisms are not fully understood.

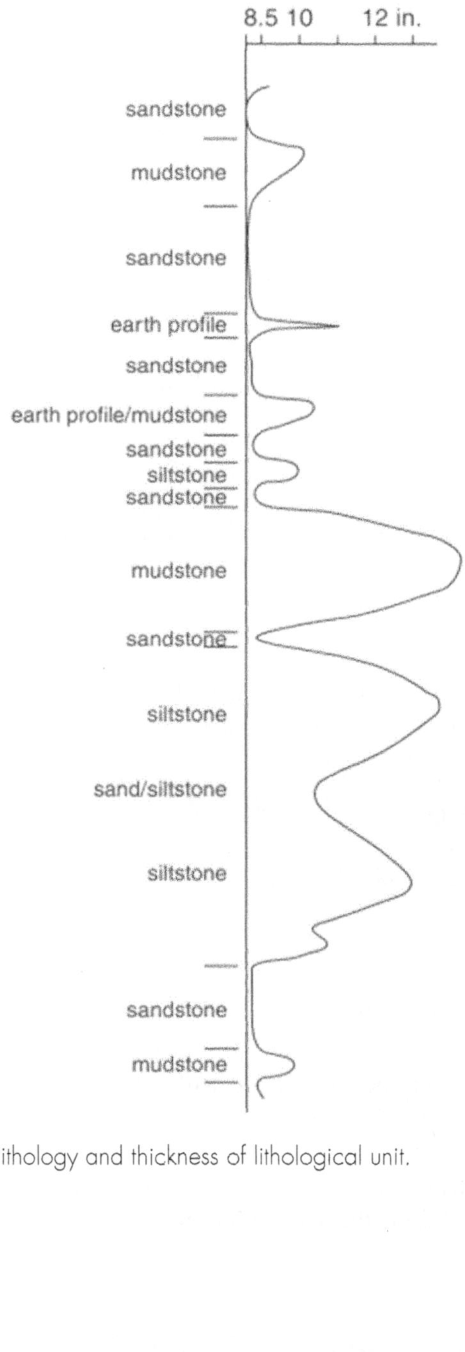

FIGURE 3.37 Collapse versus lithology and thickness of lithological unit.

Source: Berland (1993).

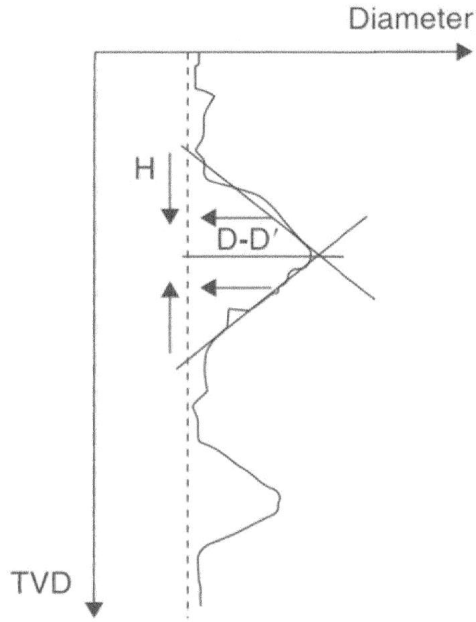

FIGURE 3.38 Model of the collapse within a lithological unit.

To resolve the problem with unit thickness, Berland introduced a depth-dependent collapse model. The principle is shown in Figure 3.38. For a clay sequence, the upper and lower sandstone boundaries are marked out. The collapse in between is represented by an ellipse or some other function. In our example, we have used two straight lines. The collapse within this triangle can be calculated from the following equation:

$$X = \frac{D_h - D_h^{'}}{H} \tag{3.47}$$

Using this model, Berland derived a time-dependent collapse model of the following form:

$$\frac{D_h}{D_h^{'}} = X\sigma_v\left(a + bD_{TVD}\right)\ln(t) \tag{3.48}$$

where the following notations are used:

D_h = hole diameter, where $D_h = D_h(t)$
D_h' = bit diameter
D_{TVD}= true vertical depth of well
X = geometrical factor
t = time
a,b = constants

Using this approach, Berland managed to model the collapse quite well.

It must be pointed out that Berland (1993) performed both modelling and hypothesis testing. The data had possibly too little variations to develop a general model. Future work will refine this.

We have now shown that borehole collapse or borehole enlargement is a complex issue. Although simple stress and material models describe how the borehole behaves with respect to varying borehole pressures, they are often inadequate to simulate real borehole behaviour. We have shown that the open hole exposure time may be a governing parameter and also that the unit thickness of the rock may be a governing parameter.

However, we have demonstrated that simple correlations can be established between the caliper log, and the mud weight used to establish critical mud weights. These can be important tools for field applications. Furthermore, this analysis can be expanded when other data becomes available.

3.5 DRILLABILITY EVALUATION

3.5.1 Introduction

The nomenclature used in this section is:

ROP = Rate of penetration of the drill bit
WOB = Weight applied on the drill bit
N = Rotary speed
d_e = Drillability called the d-exponent
d_l = Linear drillability
d_{ln} = Normalised linear drillability
D_h' = Diameter of the drill bit
σ_v = Overburden stress gradient
D = Well depth
t = Time

In petroleum engineering, we have relatively few data available compared to the vast coverage of our wells. Logging of the well is important to determine the petrophysical parameters for further evaluation. One drawback is that the logs are always delayed in time; even the measurement-while-drilling tool (MWD) is lagging 5–10 meters behind the drill bit. However, the drill bit is at the very bottom of the hole. If we can utilise the drill bit information, we will instantly know changes as they occur. Also, the drilling information is an under-utilised source of information with a large potential for correlations. In the future, a high-quality drillability log will be generated which is used with the electrical logs. At this time, improvements in drilling data are being made, which will give us a higher quality drillability log in the future. Kyllingstad et al. (1993) addresses some of the recent improvements of the drilling data.

The d-exponent is actually a drillability, which is used for pore pressure estimation. The interpretation techniques have not advanced significantly in the past 20 years. The common interpretation is to look for deviations from a straight line, which is often interpreted as an indicator for increasing pore pressure. The d-exponent is a logarithmic function, as illustrated below:

$$ROP = N\left(\frac{WOB}{D_h'}\right)^d \quad \text{or :}$$

$$d_e = \frac{\log\left(\frac{ROP}{N}\right)}{\log\left(\frac{WOB}{D_h^{'}}\right)} \tag{3.49}$$

We know that the drilling rate depends on the loading and the rotary speed. The simplest drilling model can then be expressed as:

$$ROP = d_1 \frac{WOBxN}{D_h^{'}}$$

Here, we have used d_l to define the drillability, which is actually a scaling factor for the coupling between the rock and the drill bit. From the drilling data, all parameters are available except the drillability. This is expressed as:

$$d_1 = ROP \frac{D_h^{'}}{WOBxN} \tag{3.50}$$

We see that both Equations (3.49) and (3.50) actually define drillabilities. The drawback of the d-exponent is that the logarithmic scale is non-linear. There is also little evidence that this logarithmic dependence is related to the real physics of the drilling process. In the following examples, we will use only Equation (3.50).

3.5.2 Clay Diapir Example

During the drilling of production wells in a North Sea oil field, loss of circulation was experienced in the 1100–1200 m depth interval. An investigation concluded that from the evaluation of cuttings and electrical logs, the loss interval is identical to the rocks above and below, that is silty clay. There were no indicators to distinguish the loss interval, as also faulting and tectonics were ruled out.

A drillability analysis was performed for four of the wells. Figure 3.39 shows the result. The drillability logs are generated using Equation (3.50). It was found that the drillability was higher in the loss interval in all four wells considered. In Figure 3.39a, we observe that the drilling rate increases by a factor of 5–10 in the interval 1138–1212 m. In Figure 3.39b, this effect is less pronounced, but an increased drillability is seen in the interval 1167–1232 m. Figure 3.39c shows an increased drillability in the interval 1095–1185 m, and Figure 3.39d shows an increase in the interval 1162–1210 m. In the last well, a sand stringer is penetrated just below the interval investigated. Often, sand stringers have higher drillabilities than shale intervals in this field. Please observe that these wells are inclined and that the depth shown is along the well path.

The problem observed was that weaker rocks existed in the middle of a clay sequence. From cutting measurements and electrical logs, we were unable to distinguish this interval from the surroundings. However, the drillability analysis showed that this interval drilled easier than the surroundings. It was also found that a clay diapir existed at this depth interval. It is, therefore, reasonable to tie this weaker zone to the properties of the diapir. It is believed that all formations at this shallow depth are normally pressurised. Since the fracturing pressure was also reduced in this interval, possibly a reduced horizontal stress state is the primary mechanism causing the problem. Another explanation is that the clay diapir has higher water content than the surroundings,

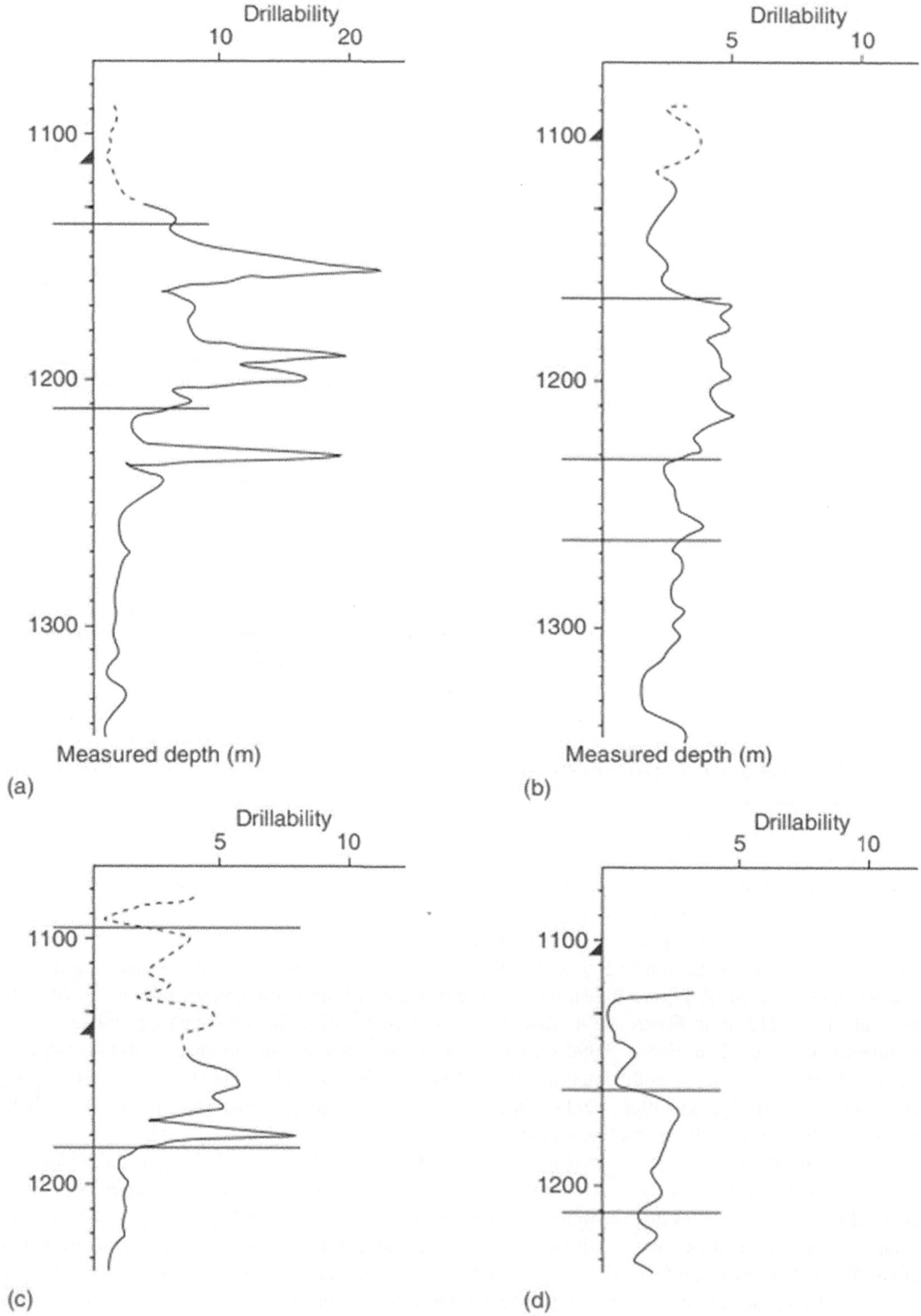

FIGURE 3.39 Drillability logs for four production wells. a) Well A; b) well B; c) well C; and d) well D.

causing a higher porosity. However, at this stage, we do not have a full understanding of the problem but have demonstrated the potential of using the drillability logs for empirical correlations.

3.5.3 Relief Well Example

The underground blowout in problem well 2/4-14 took nearly a year to control. A relief well, 2/4-15s, was drilled in 1989 to nearly 5000 m depth, and communication was established with a successful killing operation as a result. Rock mechanics considerations are discussed by Aadnoy and Bakoy (1992). In particular, the mechanism of the breakthrough was identified as a collapse rather than a fracture, which was the conventional belief at that time.

At the time of the breakthrough, well 2/4-14 had been subjected to an underground flow, which was estimated to be 18,000 barrels of oil/condensate per day, for nearly a year. The reservoir below had, therefore, a reduced pressure, and the rock around the blowing wellbore had been subjected to high hoop stresses due to the bottom hole drawdown pressure. In other words, stress and pressure changes had taken place in the neighbourhood of the well.

As part of the post evaluation, drillability logs were made for both wells. Near bottom, the two wells were just a few meters apart. Therefore, it was assumed that the virgin rock was identical around the two wellbores and that any difference should be attributed to effects of the underground blowout.

Figure 3.40 shows the resulting drillability curves. The dotted line shows the drillability for the initial well. The drillability was approximately constant down to 4714 m, where the weaker Mandal sand was penetrated. The drillability increased here by a factor of six. During further drilling, the drillability again increased when the reservoir was penetrated at 4732 m, causing an underground blowout to the Mandal formation and also to a shallower sand stringer at about 900 m depth.

The solid curve shows the drillability for the adjacent relief well. Down to 4673 m, the drillabilities of the two wells are nearly identical. However, at this depth, the drillability of the relief well increases significantly over the curve for the initial well. 4673 m defines the depth at which there are changing properties due to the underground blowout. At this depth, the formations and the pressure regime have undergone alterations. At 4673 m, the distance between the two wells was about 6.4 m. The conclusion from Aadnoy and Bakoy (1992) is that the underground flow that took place over a year caused changes which penetrated 30 borehole radii away from the wellbore, which was by far exceeded the 5 borehole radii which was predicted by rock mechanics theory. This is important in planning the target size in relief well drilling.

These two examples demonstrate that the drillability log can provide significant information in rock mechanics evaluations. Other aspects, like pore pressure estimation, will not be addressed at this time. A refined drillability log may also, in the future, be able to provide better information about formation tops and cap rock tagging.

3.5.4 Drilling Time Curve

The top hole is drilled with a larger drill bit than the deeper parts of the well. Yet the top layers are easier and quicker to drill. The rock is more compacted and has undergone diagenetic changes with depth, which makes it harder. The in-situ stress also increases with depth. There is, of course, a coupling between the stress state and the rock properties.

Figure 3.41a illustrates a drilling time curve with a significant increase in time with depth. The flat sections represent operation with no drilling like casing landing and cementing. If we remove the casing operations time, we obtain an effective drilling time curve, as indicated in Figure 3.41b. We observe that the shape of this curve can be approximated as some power of depth.

That the drilling time increases with depth means that the drilling rate decreases with depth. The following coupling between drilled depth, drilling rate, and time can be established:

$$\delta D = ROP\ \delta t \tag{3.51}$$

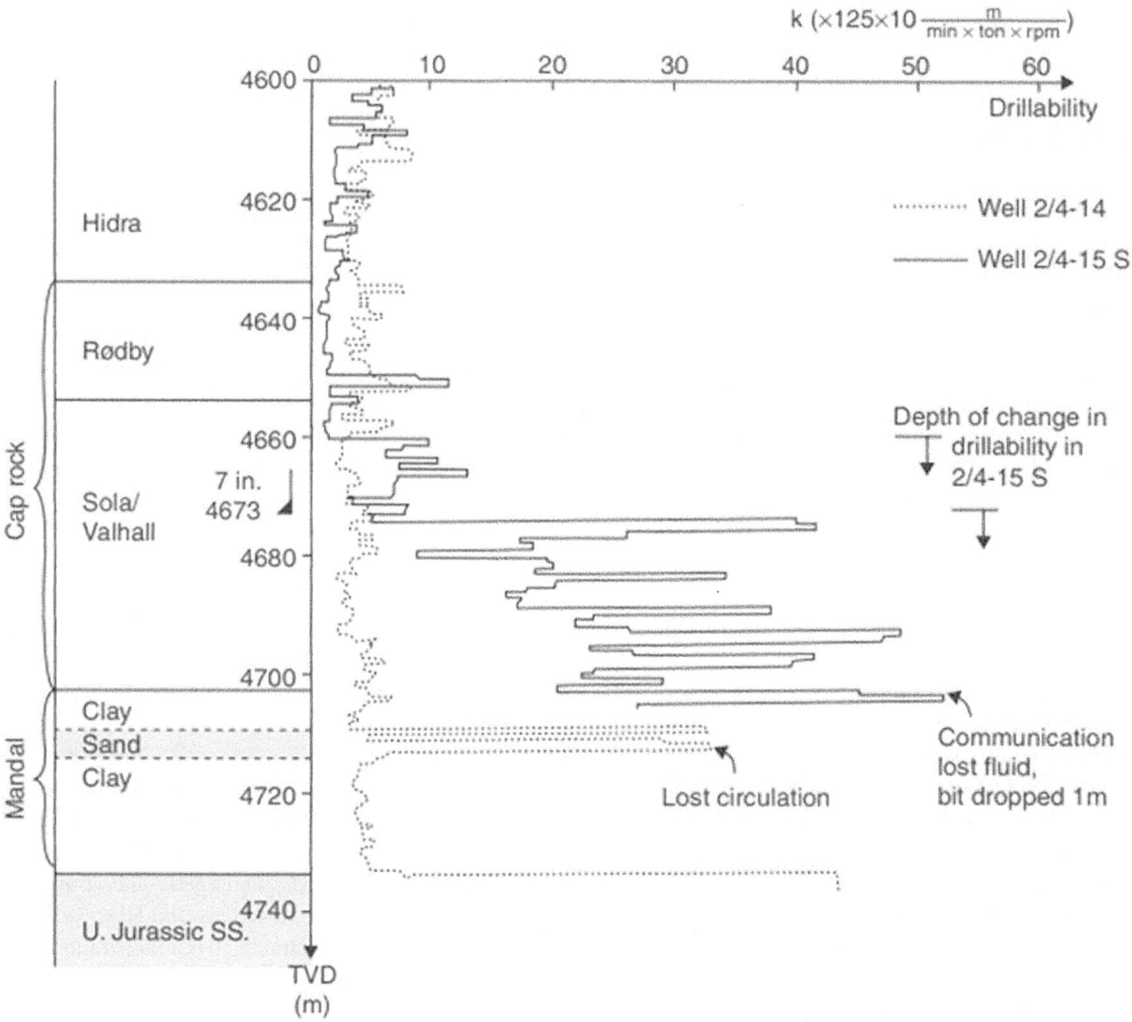

FIGURE 3.40 Drillabilities for the blowing well (2/4-14) and the relief well.

Source: Aadnoy and Bakoy (1992).

We will now assume that the overburden stress gives a reasonable magnitude of the general stress state at any depth, in the absence of more detailed stress information. We usually have an overburden stress curve for all wells. Figure 3.41c shows a typical overburden stress curve. Although some deviations occur in surface layers, it can be approximated as a straight line, that is, a nearly constant gradient.

We observe that the overburden stress increases with depth while the drilling rate decreases. Therefore, we will assume that an inverse relation exists. (The overburden stress is the gradient multiplied by the depth).

$$ROP = 1/\sigma_v D \tag{3.52}$$

Eliminating the drilling rate by combining the two equations above results in:

$$\delta t = \sigma_v D \delta D$$

The time required to drill a given distance can be obtained by integrating this equation.

$$t = \int_{D_0}^{D_1} \sigma_v D dD \tag{3.53}$$

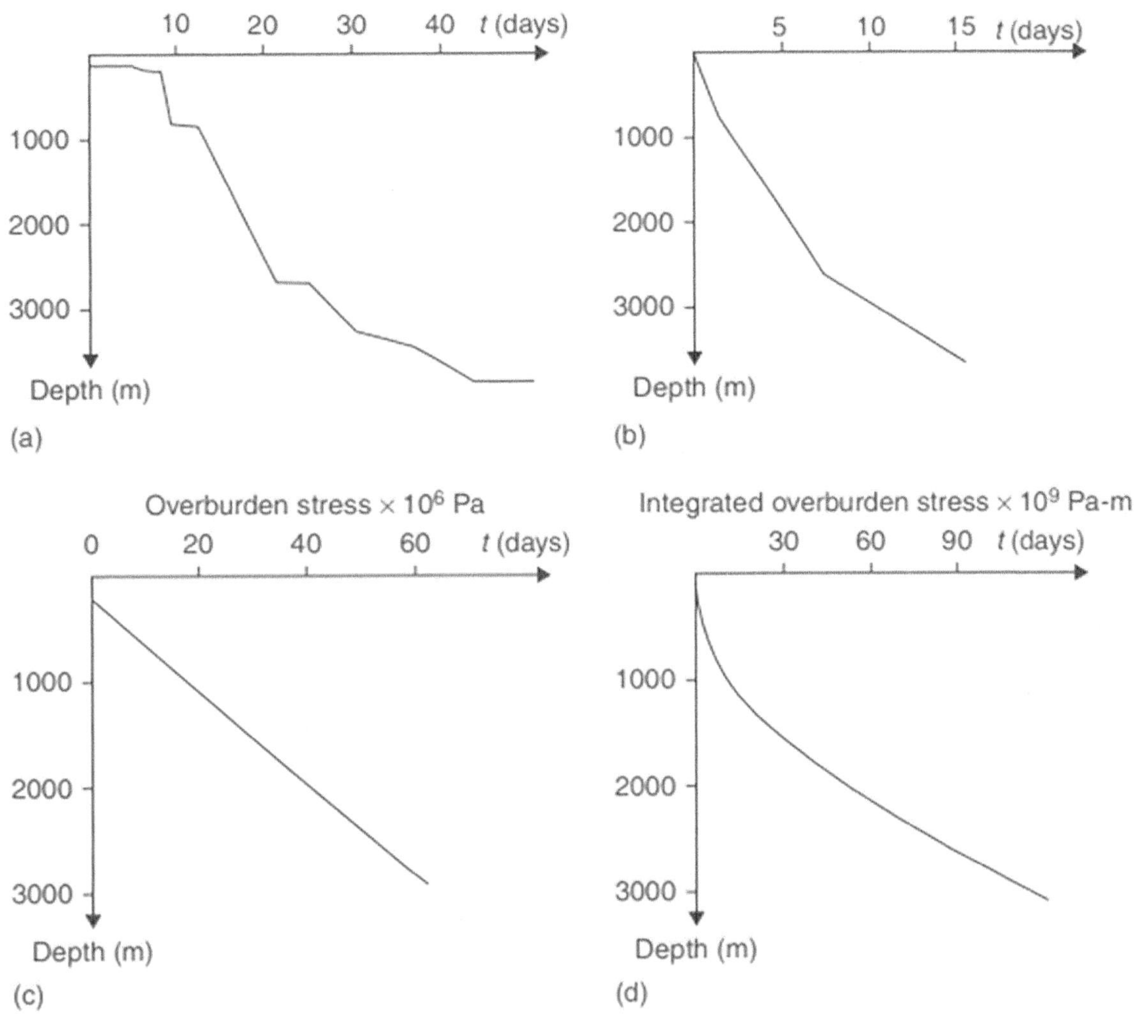

FIGURE 3.41 Visual comparison between drilling time and integrated overburden time.

If we assume a linear overburden stress function, then the drilling time becomes proportional to the squared depth, i.e.:

$$t = \sigma_v D^2 \tag{3.54}$$

Figure 3.41d shows the integrated overburden curve. Comparing the effective drilling time curve in Figure 3.41b with the integrated overburden stress curve of Figure 3.41d, we see a good resemblance. Equation (3.53) can, therefore, be used to model drilling time.

Equation (3.53) can also be described by an analogy description.

The integration of the drilling time from t_0 to t_1 is equivalent to the drill bit working through the stress field from D_0 to D_1. The mechanical work spent is equivalent to an integration in time.

3.5.5 Depth Normalisation of Drillability

We have just demonstrated that there is an apparent correlation between the overburden stress and the drilling time. One application of this concept is to establish correlations for drilling time modelling and optimisation. Another aspect will be briefly outlined in the following.

Equation (3.52) defines an inverse relationship between the drilling rate and the overburden stress. This will be used to depth normalise the drilling rate. We will use this relation to define a pressure and depth normalised drillability:

$$d_{ln} = d_l \sigma_v D$$

If we insert this expression into Equation (3.50), the normalised drillability becomes:

$$d_h = ROP \frac{\sigma_v DxD_h'}{WOBxN} \tag{3.55}$$

In traditional drillability evaluation, e.g. when calculating the d-exponent, there is a declining trend with depth. By stress and depth normalisation, we expect a straight vertical trend, this means that the normalised drillability from the top hole can be compared against the normalised drillability at depth. This will, therefore, provide a framework with increased potential for interpretation.

Readers are also referred to Hareland and Hoberock (1993) for drillability interpretation.

3.6 A GENERAL FRACTURING MODEL

3.6.1 Introduction

In Section 3.1, general normalisation equations are derived for arbitrary drill floor height. In Section 3.3, an empirical frac. model is derived based on shallow frac. data. In the application of the latter, a normalisation procedure is applied for varying seawater depth.

The above normalisation concepts have been extended to deep water wells. Aadnoy (1998) showed that the fracture pressure is basically related to the effective overburden stress and presented a general model which gave good results when applied to wells in various parts of the world, such as the North Sea, Gulf of Mexico, Brazil, Angola, and so on. It was, therefore, termed a "worldwide model". In particular, it was found that the model works for any water depth, deep or shallow. The major aspect of the model is to properly normalise for the water depth. Kaarstad and Aadnoy (2008) summarise this model. In the following, a general presentation is given as well as several examples of application.

3.6.2 Development of the Model

3.6.2.1 The Overburden Stress

The fracture pressure of a borehole depends on the in-situ stress state, which is defined in terms of a three-parameter tensor: the overburden stress σ_o and the two horizontal stresses σ_H and σ_h. The overburden stress is defined as the cumulative weight of sediments above at a given depth. In integral form:

$$\sigma_o = g \int_0^{D_w} \rho_{sw} dD + g \int_{D_w}^{D} \rho_b dD \tag{3.56}$$

A constant seawater density is a good approximation, whereas for the bulk rock density, it may not always be the case. For example, if constant densities are assumed, Equation (3.56) can be formulated as:

$$d_{ob} = 1.03\frac{D_w}{D} + \rho_b\left(1 - \frac{D_w}{D}\right) \tag{3.57}$$

Figure 3.42 shows overburden gradient curves for various water depths using Equation (3.57). It is observed that the overburden gradient reduces with increased water depths because of the low density of water.

Kaarstad and Aadnoy (2008) showed that when applied on deepwater wells, there is a strong correlation between the fracture pressure and the overburden stress. This correlation will be used in the following to derive general normalisation equations. The purpose is to be able to use data from one water depth to predict the fracture pressure at another water depth.

3.6.2.2 The Fracturing Equations

Reference is given to Section 3.2 where the elementary equation relating to borehole fracturing are presented. The following derivations are valid for:

- Relaxed depositional basin environments or fields with equal horizontal in-situ stresses.
- Normal pore pressure.
- Abnormal pore pressure, but the same pore pressure in the two cases considered.
- Vertical boreholes. Inclined boreholes can be handled by first deriving the solution for vertical holes, then transforming the solution to the wellbore direction of interest.

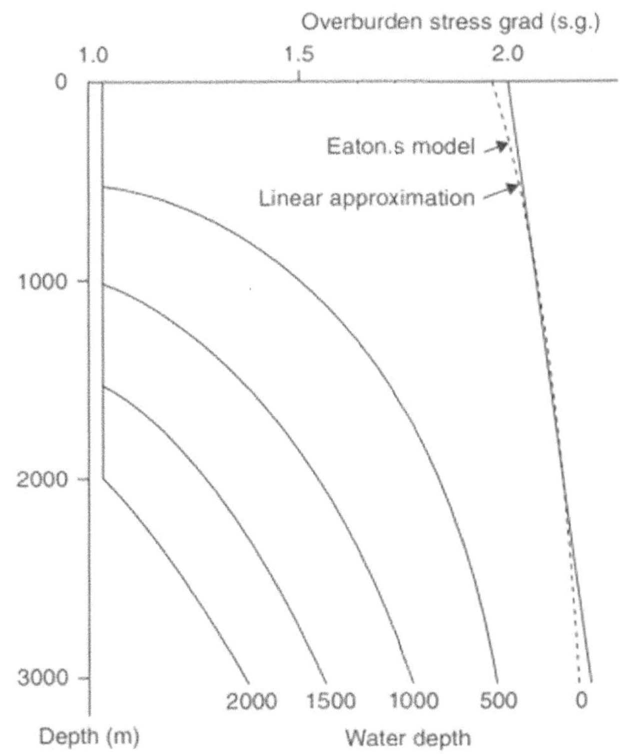

FIGURE 3.42 Overburden stress gradient curves for various water depth.

3.6.2.3 Normalisation of Frac. Pressures

The fracturing pressure is normalised to seabed and correlates to the respective overburden pressure for relaxed depositional basin environments. Figure 3.43 shows normalised pressure from seabed of two offshore locations. Seawater pressure is subtracted, and the depth is also calculated from seabed.

Subscript 1 refers to reference, while subscript 2 refers to the prognosis.

The zero reference is the drill floor. The following definitions apply:

h_f = Air gap from drill floor to sea level (m)
h_w = Water depth (m)
D = Depth (m)
D_{sb} = Depth of rock below seabed (m)
P = Pressure (MPa)
d = Gradient with reference to water (s.g.)
d_b = Bulk density gradient (s.g.)
d_{sw} = Relative density of sea water (s.g.)

In the general case, a new well (index 2) will have different rock penetration below the seabed, varying bulk densities, different water depth, and different air gap to drill floor compared to a reference well (index 1). Applying the direct correlation between fracture pressure and overburden yields:

$$P_{wf} = cons\tan t.\sigma_o \tag{3.58}$$

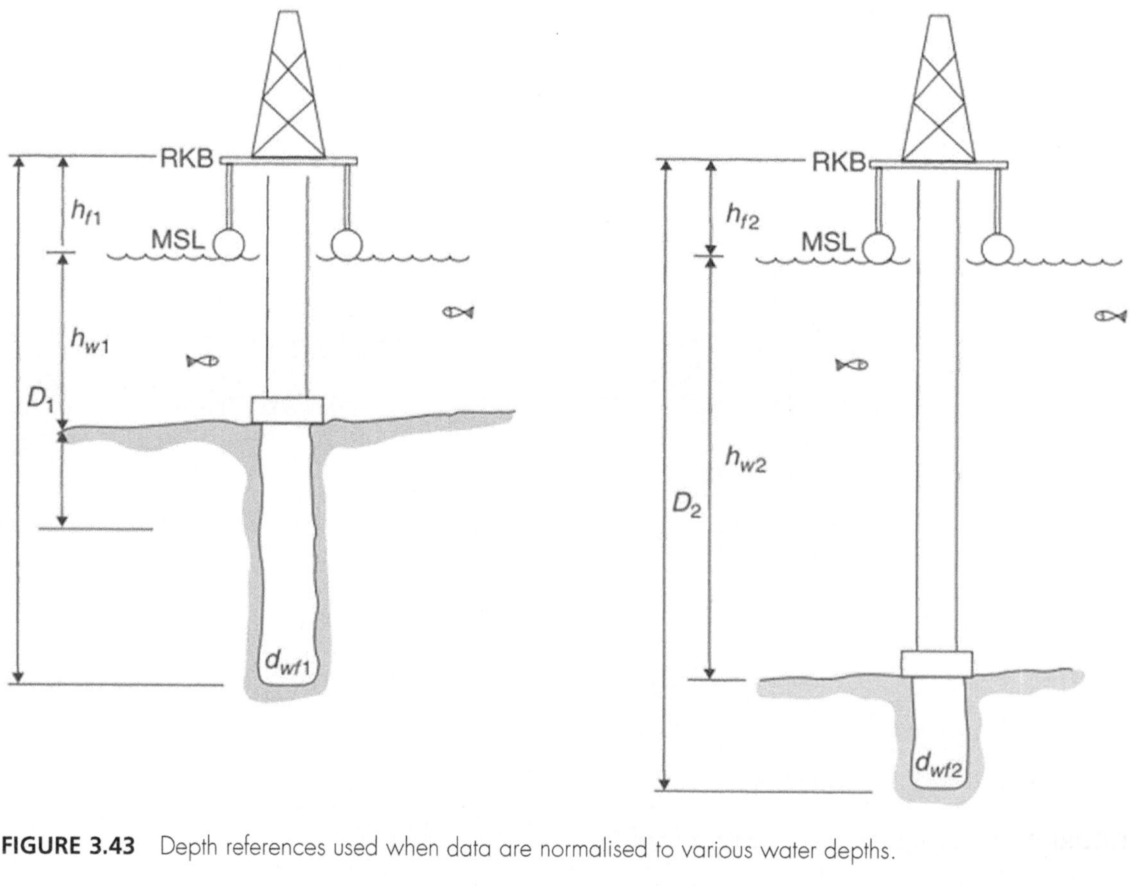

FIGURE 3.43 Depth references used when data are normalised to various water depths.

When solving for the fracture pressure gradient, the normalisation equations for the prognosis become, assuming a variable bulk density:

$$
\begin{aligned}
D_{wf2} &= D_{wf1} + \Delta h_w + \Delta h_f + \Delta D_{sb} \\
d_{wf2} &= d_{sw}\frac{h_{w2}}{D_{wf2}} + \left(d_{wf1}\frac{D_{wf1}}{D_{wf2}} - d_{sw}\frac{h_{w1}}{D_{wf2}} \right) \frac{\int^{D_{sb2}} d_{b2}\,dD}{\int^{D_{sb1}} d_{b1}\,dD}
\end{aligned}
\tag{3.59}
$$

This equation requires detailed information about the bulk density profile and is applicable when significant reference data exist. However, simplifying assumptions often apply, and can be categorised as follows:

3.6.2.4 Different, but Constant Bulk Densities

Applying constant bulk densities to Equation (3.59), the integrals are reduced, and the normalisation equations can be expressed as follows:

$$
\begin{aligned}
D_{wf2} &= D_{wf1} + \Delta h_w + \Delta h_f + \Delta D_{sb} \\
d_{wf2} &= d_{sw}\frac{h_{w2}}{D_{wf2}} + \left(d_{wf1}\frac{D_{wf1}}{D_{wf2}} - d_{sw}\frac{h_{w1}}{D_{wf2}} \right) \frac{d_{b2} D_{sb2}}{d_{b1} D_{sb1}}
\end{aligned}
\tag{3.60}
$$

3.6.2.5 Constant Bulk Density

For wells in the same area, it may often be assumed that the bulk density is equal for the different wells. Equation (3.59) is then further simplified, and the normalisation equations become:

$$
\begin{aligned}
D_{wf2} &= D_{wf1} + \Delta h_w + \Delta h_f + \Delta D_{sb} \\
d_{wf2} &= d_{sw}\frac{h_{w2}}{D_{wf2}} + \left(d_{wf1}\frac{D_{wf1}}{D_{wf2}} - d_{sw}\frac{h_{w1}}{D_{wf2}} \right) \frac{D_{sb2}}{D_{sb1}}
\end{aligned}
\tag{3.61}
$$

These equations are used to normalise between varying water depths, platform elevations, and rock penetrations.

3.6.2.6 Same Rock Penetration Below Seabed for Data and Prognosis and Constant Bulk Density

When setting $\Delta D_{sb}=0$, the following normalisation equations result:

$$
\begin{aligned}
D_{wf2} &= D_{wf1} + \Delta h_w + \Delta h_f \\
d_{wf2} &= d_{wf1}\frac{D_{wf1}}{D_{wf2}} + \frac{d_{sw}\Delta h_w}{D_{wf2}}
\end{aligned}
\tag{3.62}
$$

These equations are used to normalise between varying water depths or platform elevations. The overburden pressure at depth D_l is given by:

$$P_{wf1} = 0.098 d_{sw1} h_{w1} + 0.098 d_{b1} \left(D_{wf1} - h_{w1} - h_{f1} \right) \tag{3.63}$$

or

$$P_{wf1} = 0.098 d_{wf1} D_{wf1} \tag{3.64}$$

The depth of the rock below the seabed is:

$$D_{sb1} = D_{wf1} - h_{w1} - h_{f1} \tag{3.65}$$

3.6.3 Field Cases

Five deep-water wells and one shallow-water well offshore Norway are analysed in detail in Kaarstad and Aadnoy (2008). The lithology and the bulk densities were analysed to provide an accurate overburden stress curve in each well. Because the overburden stress serves as a fracturing gradient reference, it is important to obtain as accurate bulk density data as possible.

Analysing leak-off data and overburden stress gradient for the five deep-water wells gave a fracture prognosis of 98% of the overburden stress gradient with a standard deviation of 0.049, and errors ranging from 0 to 5%. The overburden, leak-off data, and resulting fracture pressure gradient curve for one of the wells are shown in Figure 3.44.

3.6.3.1 Example of Use of Normalisation Methods

Data normalisation is an indispensable method to compare data sets with different references. Equation (3.59) defines the general normalisation equations used to compare pressures (e.g., overburden, leak-off pressure, in-situ stresses, etc.) with differences in bulk density, rig floor height, water depth, and depth of penetration. To demonstrate the application, we present two examples:

Example 1

The reference well is drilled in 400 m of water. The following data will be used to derive a prognosis for a well in 1100 m water depth. Assume that the rig floor height, the bulk density, and the penetration depth remain unchanged.

- Drill floor height: $h_f = 25$ m
- Total depth of well 1: $D_l = 900$ m
- Water depth for well 1: $h_{wl} = 400$ m
- Leak-off pressure for well 1: $d_l = 1.5$ s.g. @ 900 m
- Water depth for well 2: $h_{w2} = 1100$ m
- Seawater density: $d_{sw} = 1.03$ s.g.

Given the assumptions of this example, we can apply Equation (3.60) to calculate the prognosis for the leak-off pressure gradient for the new well. First, we calculate the new depth reference:

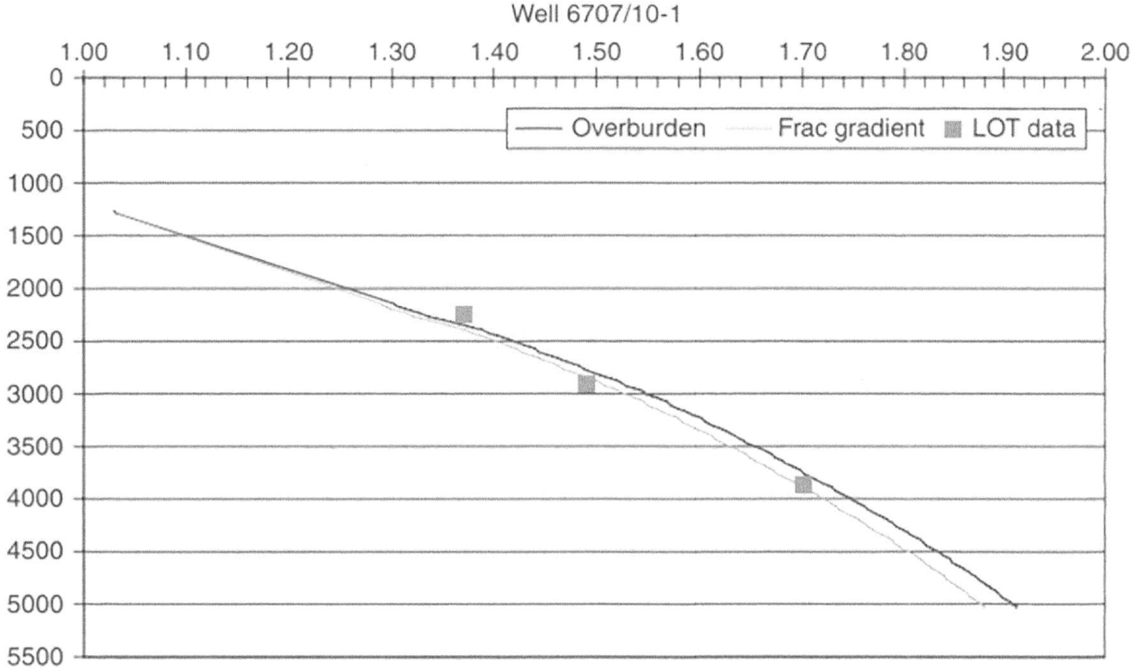

FIGURE 3.44 Example of application of generalised frac. model.

$$
\begin{aligned}
D_2 &= D_1 + \Delta h_w + \Delta h_f \\
&= 900m + (1100m - 400m) + (25m - 25m) \\
&= 1600m
\end{aligned}
$$

Next, we can calculate the prognosis for the leak-off pressure gradient:

$$
\begin{aligned}
d_2 &= d_1 \frac{D_1}{D_2} + \frac{d_{sw} \Delta h_w}{D_2} \\
&= 1.5 \text{ s.g.} \frac{900m}{1600m} + 1.03 \text{ s.g.} \frac{700m}{1600m} \\
&= 1.29 \text{ s.g.}
\end{aligned}
$$

In this example, the increase of the water depth from 400 m to 1100 m resulted in a decrease in leak-off pressure gradient from 1.5 s.g. to 1.29 s.g.

Example 2

Unless the wells are very close to each other, it is reasonable to assume that there are differences in the bulk densities. Changes in lithology may have a significant effect on the overburden stress gradient. Therefore, the normalisation should take into account differences in bulk density.

In this example, we want to show the effect of differences in bulk density between the two wells. We consider the same wells as in Example 1, with the following additional information for the new well:

- Bulk density gradient for reference well: $d_{b1} = 2.05$ s.g.
- Bulk density gradient for new well: $d_{b2} = 1.85$ s.g.

Equation (3.59) must be applied to normalise the data. The new depth reference becomes:

$$\begin{aligned} D_2 &= D_1 + \Delta h_w + \Delta h_f + \Delta D_{sb} \\ &= 900 + (1100 - 400) + 0 + 0 \\ &= 1600m \end{aligned} \tag{3.66}$$

The new leak-off pressure gradient is:

$$\begin{aligned} d_2 &= d_{sw} \frac{h_{w2}}{D_{wf2}} + \left(d_{wf1} \frac{D_{wf1}}{D_{wf2}} - d_{sw} \frac{h_{w1}}{D_{wf2}} \right) \frac{d_{b2}}{d_{b1}} \\ &= 1.03 \frac{1100}{1600} + \left(1.50 \frac{900}{1600} - 1.03 \frac{400}{1600} \right) \frac{1.85}{2.05} \\ &= 1.24 \text{ s.g.} \end{aligned} \tag{3.67}$$

We observe that the lower bulk density in well 2 leads to a decrease in overburden stress, resulting in a lower leak-off prognosis.

We also observe that water contributes significantly to the total overburden stress. The result is that with the same penetration depth, an increase in water depth gives a decrease in overburden stress and fracture pressure. If we look at the curves in Figures 3.1–3.6 and observe the overburden stress gradient at, e.g., 150 m below the seabed, we see that the largest water depths give the lowest overburden gradient.

APPENDIX: MODEL FOR THE STRESS BRIDGE

During fracturing the fracture opens up with increasing pressure. A stress bridge forms across the fracture preventing mud to enter. From a mechanistic perspective this bridge must have a curved shape to give a force equilibrium between the mud and the borehole wall. Figure A.1 shows a half-cylindrical model. We assume that the inner part of the bridge is elastic, with a plastic layer in the mud cake on the outside. When the entire stress cylinder is plastified, the stress bridge fails, and mud enters the fracture. In the following a failure model for the stress bridge will be presented. More details can be found in Aadnoy and Belayneh (2004).

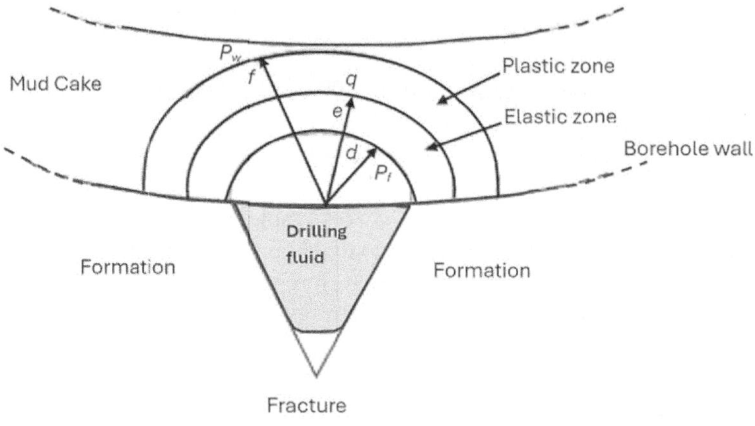

FIGURE A.1 Model of stress bridge.

Figure A.1 shows a stress bridge across a fracture which is analogous to a cylinder subjected to an external load, or a collapse load. The following is a short presentation of the bridge stress model.

The stress bridge arises in the mudcake and consists of a plastic outer zone and an elastic inner zone. Assuming a very thin elastic zone, $e \approx d$, the mud cake strength for small arguments can estimated as:

$$P_{mc} = \frac{2U}{\sqrt{3}}\left(\frac{f}{e} - 1\right) + P_o \qquad \text{(A1)}$$

Because U is the compressive strength of the cylinder, Equation (A1) shows that the maximum pressure of the stress bridge is directly proportional to the yield strength of the particles in the stress bridge. In other words, if a high borehole pressure is required, a strong particle must be used. We refer to Aadnoy and Belayneh (2004) paper with more detail. It is also important to point out that the elastoplastic model is not routinely used in the oil industry today.

Well Design Premises

4

4.1 WELL INTEGRITY

4.1.1 Definitions

During drilling of a well the design must ensure that the well can withstand abnormal events. Two incidents that can create severe problems are significant loss of mud return, and taking a high pressure kick.

Loss of circulation may halt the operation since it usually has to be cured either by plugging with lost circulation material (LCM) or by cementing, or by sealing off the loss zone with a casing string or a liner. Usually loss of circulation delays further drilling since a hydrostatic pressure balance is difficult to obtain in the annulus. In extreme cases, loss of circulation may initiate a well pressure control problem for the same reason. From a casing design point of view the major impact of circulation losses is an increased collapse loading on the casing.

A kick can develop either as a consequence of circulation losses, by reduced hydrostatic head in the annulus, or by drilling into a high-pressure reservoir with insufficient mud weight. As long as the well is open there will be low loading on the casing. However, if the well is shut in, when partially or fully filled with gas, a significant pressure may develop in the shallower parts of the well. In the following we will describe three different scenarios that may arise after a gas-filled well is shut in.

Figure 4.1 illustrates a well which is filled with gas and shut in. At the bottom the formation pressure is acting. Since the formation fluid is often light, the wellhead is exposed to a considerable pressure, which is the formation pressure minus the weight of the gas column.

Figure 4.1a illustrates the full well integrity case. The burst loading of the casing is the inside pressure minus the outside pressure. Here the burst strength is exceeding the load. Also, the critical fracturing pressure below the casing shoe is higher than the pressure inside the well. In other words we have full well integrity over the casing and also over the open hole interval.

Figure 4.1b illustrates the same loading scenario, only in this case the open hole below the casing is too weak, and the well may fracture. In this case we have a reduced well integrity, since a flow may take place between the reservoir and the casing shoe further up. This scenario is called an underground blowout, and even if it is not wanted, it may be acceptable temporarily as long as the flow is confined to the casing shoe and if the drilling fluid level above the shoe is maintained. However, in some instances, a leak may occur behind the casing, or a fracture may move towards the surface, leading to more severe problems.

a) Full well integrity. Both open hole below casing and the casing can handle gas filled well.
b) Reduced well integrity. Casing can handle gas filled well, but formation below is too weak.
c) Reduced well integrity. Now the open hole below the casing has sufficient strength, but the casing is too weak resulting in burst below the wellhead.

DOI: 10.1201/9781003618171-4

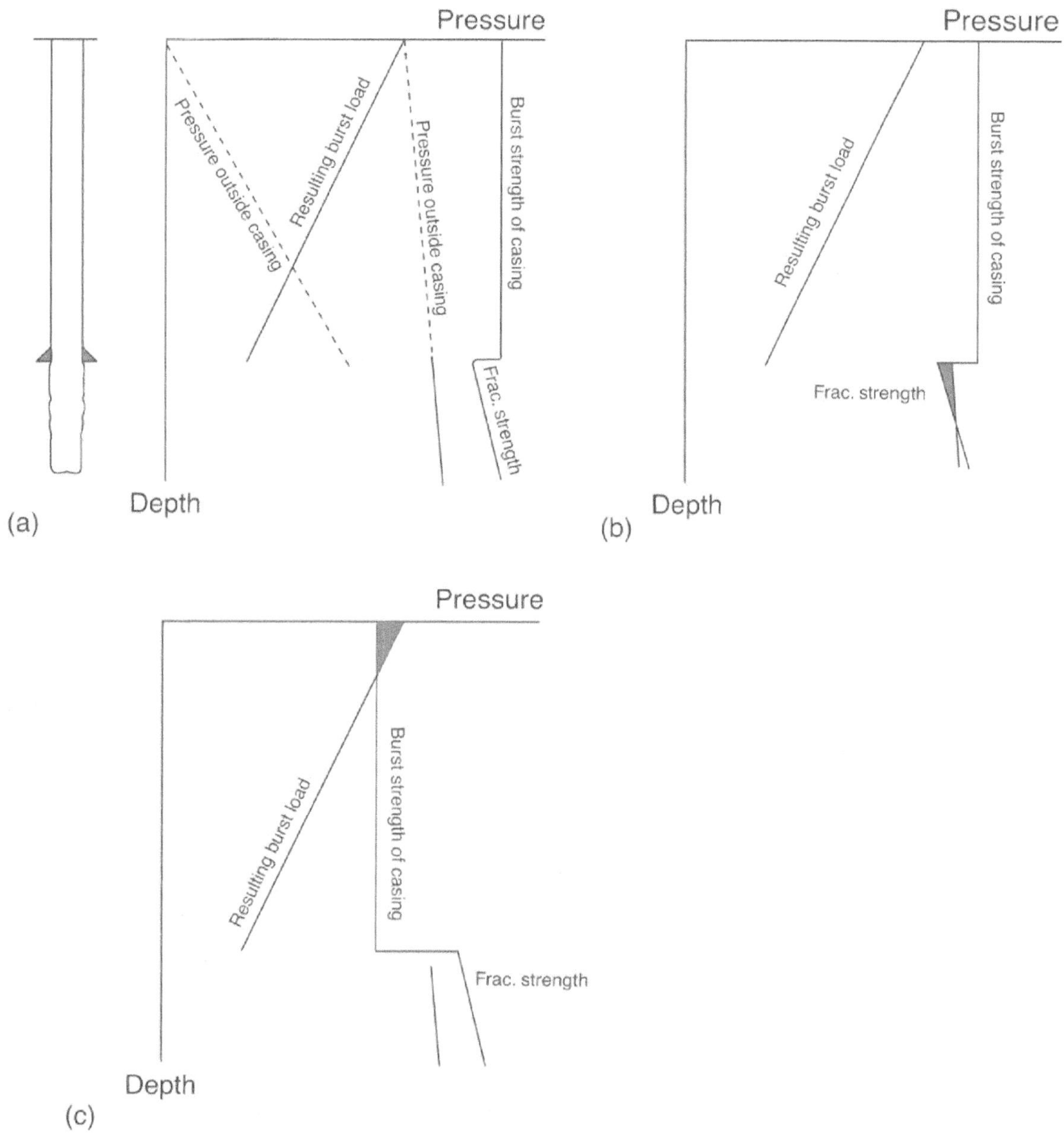

FIGURE 4.1 Definitions of well integrity.

Figure 4.1c also shows the same scenario as before. In this case the casing is too weak, and it will burst just below the wellhead when the well is shut in. The critical fracture pressure below the shoe may or may not exceed the annulus pressure, but is higher than the strength of the casing. This situation is completely unacceptable, since a failure below the wellhead will almost certainly result in a full blowout, a possible disaster both for personnel and equipment. In fact, as shown later we will always design the well such that the upper part of the casing can withstand the full reservoir pressure. If a weak point is required, this must always be below the casing shoe.

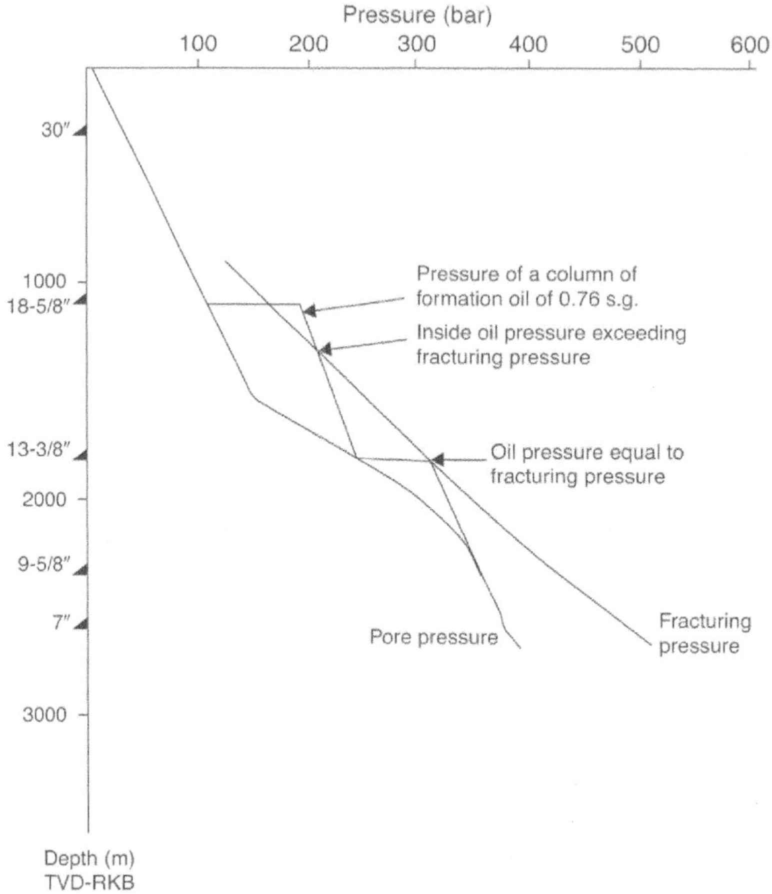

FIGURE 4.2 Examples of full and reduced well integrity for a shut-in oil-filled well.

Figure 4.2 further illustrates the integrity issue. This shows the frac. pressure, the pore pressure, and the pressure inside the borehole if the well is filled with formation fluid and shut in. We observe that the 7″ and 9 5/8″ intervals have full integrity, as the hole will not fracture if the well is shut in. The 13 3/8″ hole interval, on the other hand, has reduced well integrity as the inside oil pressure exceeds the fracturing pressure at a depth of 1300 m.

4.1.2 Well Integrity

4.1.2.1 Full Well Integrity

The production casing, which is the last casing string installed before installing the production tubing always need full well integrity. (Figure 4.1a). During production, a leak in the top of the production tubing may give the production casing a loading corresponding to Figure 4.1c with full well integrity; both the casing and the open hole can withstand gas-filled casing exposed to full reservoir pressure. For this case, the following design premise must be met:

- The minimum frac. gradient required to ensure full well integrity and to reach the end of the next open hole section.

4.1.2.2 Reduced Well Integrity

All casing strings, except the production casing, may be designed for reduced well integrity. For this case, the well cannot handle a shut-in when the well is completely filled with gas. An important condition is that the casing (which usually is weakest just below the wellhead), must not burst. In other words, one must ensure that the open hole below the casing represents the weakest point. A failure in the top of the casing can be disastrous for personnel and equipment, while a failure in the rock below the casing may result in an underground blowout with less impact on the surface. For the reduced integrity case, the following design conditions must be established:

- The minimum frac. gradient required to reach next casing setting depth.
- The maximum allowable frac. gradient to ensure that the weak point stays below the casing.
- The maximum kick size that can be handled without fracturing below the casing.

In other words, provided that the maximum kick size is not exceeded, full well integrity is obtained.

4.1.3 Example

Figure 4.3 shows the pressure prognosis for a well under planning. Also shown are the leak-off values for the reference wells. The problem at hand is to determine the deepest setting point for the 18 5/8″ casing shoe, using the concept of reduced well integrity. Also the kick margin will be calculated.

Solution:
Inspection of Figure 4.3 shows that the open hole below the 18 5/8″ shoe is planned to 3400 m, which is the setting depth for the 14″ casing. At this depth the pore pressure is:

$$0.098 \times 1.5(\text{s.g.}) \times 3400(\text{m}) = 500\text{bar}$$

Figure 4.4 shows the pressure profile in the well if it is shut-in when filled with 0.188 s.g. methane gas. The surface pressure will be:

$$500 \text{ bar} - 0.098 \times 0.188(\text{s.g.}) \times 3400(\text{m}) = 437 \text{ bar}$$

Also shown in Figure 4.4 is the burst strength of the 18 5/8″, X-70, 84.5 lbs/ft casing chosen. Using a 10% safety factor, maximum allowable surface pressure in this casing is:

$$197 \text{ bar} / 1.10 = 179 \text{ bar}$$

As this allowable pressure is significantly lower than the gas-filled casing pressure, we have a reduced integrity case. At a pressure of 179 bar at the wellhead, the following pressure will be seen at the shoe:

$$179 \text{ bar} + 0.098 \times 0.188(\text{s.g.}) \times 1000(\text{m}) = 198 \text{ bar}$$

This pressure is equivalent to a gradient of:

$$198\text{bar} / (0.098 \times 1000(\text{m})) = 2.02\text{s.g.}$$

If the actual frac. gradient is lower than 2.02 s.g., the open hole below the casing will fracture, which is the acceptable solution. A frac. gradient higher than 2.02 s.g. will result in a burst casing below the

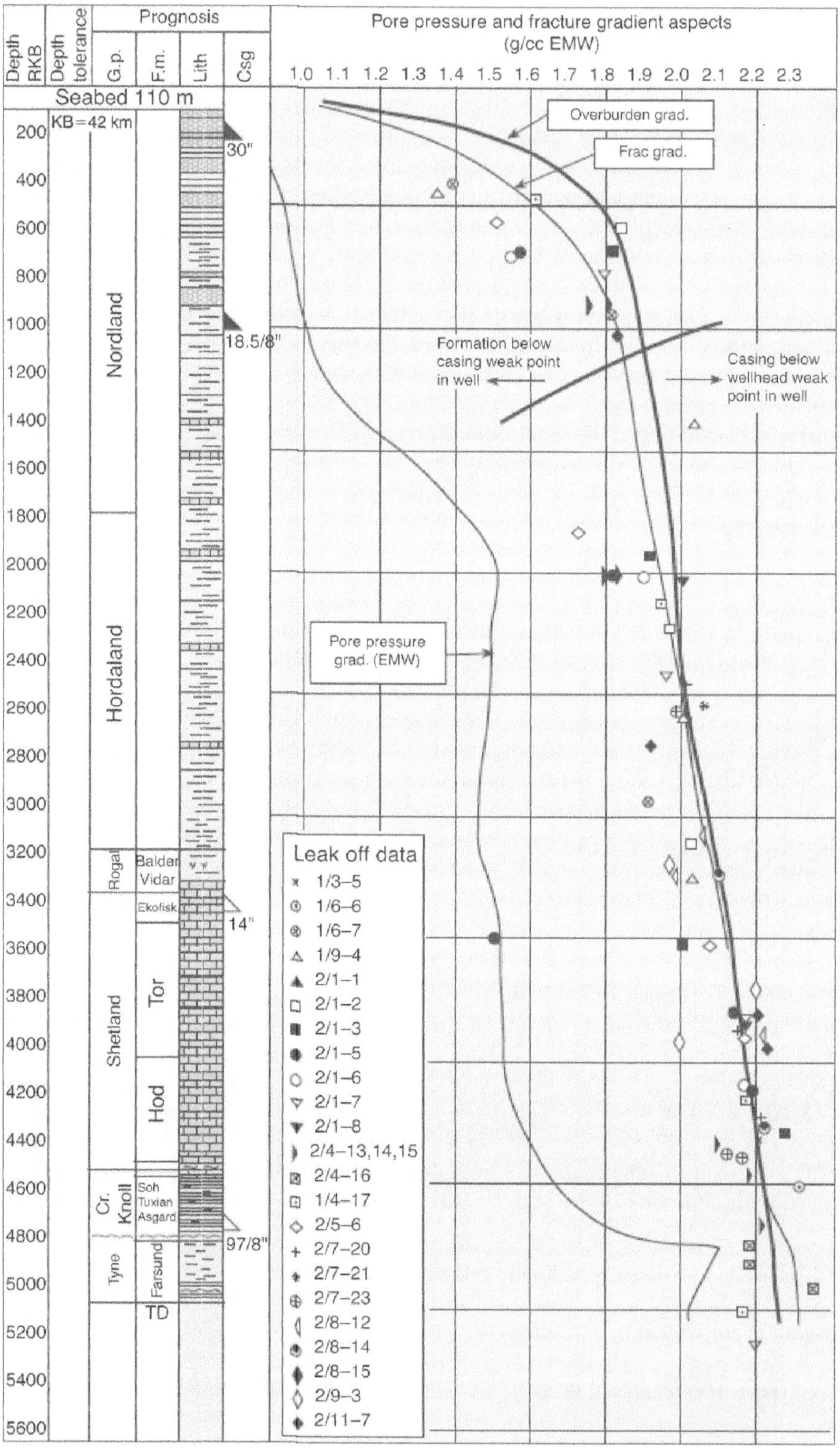

FIGURE 4.3 Example case, selecting casing setting depth from reduced well integrity considerations.

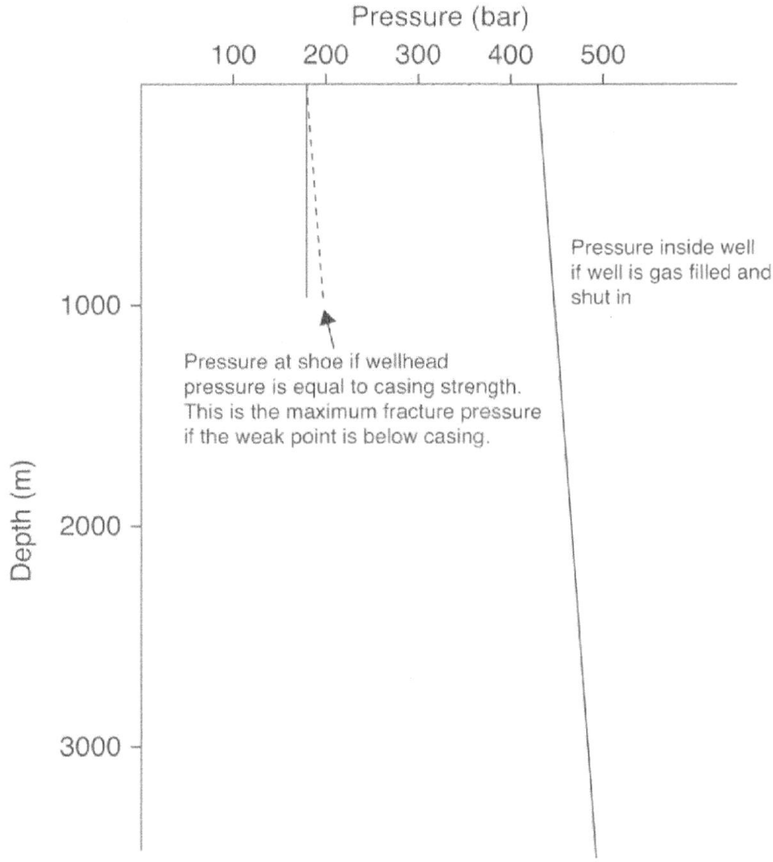

FIGURE 4.4 Design pressures for a reduced integrity case.

wellhead, which is not acceptable. Since we want to investigate the integrity over a depth interval, this calculation is repeated for 1200- and 1400-meters depth resulting in:

$$179 + 0.098 \times 0.188 \times 1200 / (0.098 \times 1200) \quad = 1.71 \text{ s.g.}$$
$$179 + 0.098 \times 0.188 \times 1400 / (0.098 \times 1400) \quad = 1.49 \text{ s.g.}$$

These three points are plotted and connected in Figure 4.3. This curve may be interpreted as follows. If the actual frac. gradient is lower than this line, the formation below the casing string will represent the weak point and the design is acceptable. Conversely, if the actual frac. gradient is higher, the casing just below the wellhead is the weak point, and this is not acceptable.

At 1100 m, this design line crosses the frac. gradient curve at a value of 1.86 s.g. However, at this depth the frac. gradient may exceed the design curve, resulting in an unacceptable design. We will therefore place the casing at 1000 m depth, where the weak point will stay below the casing shoe provided that the frac. gradient is less than 2.02 s.g.

The actual kick margin depends on the actual frac. gradient obtained. We will assume the design gradient of 1.83 s.g. at 1000 m. The fracturing pressure at 1000 m then becomes:

$$0.098 \times 1.83 \text{ (s.g.)} \times 1000(\text{m}) = 179 \text{ bar}$$

Now assume that the well has been shut in with a gas bubble at the bottom. At a critical gas height h, the frac. pressure will be reached at the shoe, as determined from the pressure balance:

$$179 \text{ bar} = 500 \text{ bar} - 0.098 \times 0.188 \times \text{h} - 0.098 \times 1.70(3400 - 1000 - \text{h})$$
$$\text{or} : \text{h} = 533 \text{ m}$$

Assuming a 16″ hole and a 5″ drill pipe the annular capacity is 0.117 m^3/m, which yields a kick margin of 62.4 m^3.

To summarise, the results of the reduced well integrity design are:

- Setting depth of casing: 1000 m.
- Maximum allowable frac. gradient to ensure that the formation below the casing shoe represent the weak point: 2.02 s.g.
- Minimum frac. gradient to be able to drill next open hole section to 3400 m: 1.50 s.g.
- Kick margin, maximum allowable gas influx to avoid fracturing below casing shoe: 62.4m^3.

Problems

Problem 1. You are asked to perform a reduced integrity design for the next casing string, the 14″ casing (Figure 4.3) that will be set in the interval 3000–3400 m. The open hole below the 14″ string is planned to extend to 4730 m. The 14″ casing chosen is a grade P-110, 86 lbs/ft, with a burst pressure of 569 bar. Using a 10% safety factor, perform the reduced well integrity design for the two cases:

a) Assuming methane gas with density of 0.267 s.g.
b) The geologists have convinced you that only a heavier formation fluid of 0.547 s.g. will be encountered.

Problem 2. The 17-1/2 in. section of a well is just drilled. Due to operational problems, combined with a gas-bearing permeable formation, a kick occurs at the bottom of the well. Well data are:

Depth of well	1700 m
Pore pressure	1.56 s.g. assuming weightless gas
Previous casing	20 in. set at 1100 m. Burst strength 190 bar
Leak-off below	20 in. 1.62 s.g.
Mud density	1.60 s.g.

The 20 in. casing is cemented to the wellhead; we will therefore assume that effective casing back-pressure is equivalent to seawater gradient.

a) Prepare a figure over the well.
b) Compute bottom-hole pressure, the leak-off pressure and effective burst pressure for the casing at the surface and at the casing shoe.
c) Determine if the well has reduced integrity. Where is the weak point, if any?
d) If it has reduced integrity, determine the kick margin.

4.2 CASING SIZES AND SETTING DEPTH

4.2.1 Introduction

In this section we will first define some requirements for the casing, then look at the process of setting depth selection. A field example will be used to demonstrate the process. In particular, we will show that the casing seat selection is mainly dependent on the pore pressure, the fracturing pressure, and the mud weight. Handling of well kicks imposes little constraint in the process, as we will use the concept of kick margins.

4.2.2 Casing Strings

4.2.2.1 Types

The following standard types of casing strings are commonly used:

• Conductor casing:	30 in. diameter
• Surface casing:	20 in. diameter
• Intermediate casing:	13-3/8 in. diam.
• Production casing:	9-5/8 in. diam.
• Production liner:	7 in. diameter

Other sizes may also be used. Examples are:

• Surface casing:	18-5/8 in. diam.
• Intermediate casing:	16 in. diam.
• Production casing:	10-3/4 in. diam.
• Production liner:	5-1/2 in. diam.

With the increased attention to slim hole drilling and other means of cost reduction, we will probably see a trend towards smaller strings or elimination of, for example, the intermediate string.

4.2.2.2 Functional Requirements

The most important requirements for each string can be identified as follows.

Conductor casing:

- Isolate unconsolidated layers below seabed.
- Support template and marine riser on floating rigs, and mudline suspension/riser system on jack-up rigs. Extend the well to deck on fixed platforms.
- Support surface casing and wellhead.
- Sufficient internal diameter to accommodate surface casing and provide efficient cement displacement.
- Outer diameter or connections shall allow for installation through the rotary table.
- Conductor shall be deep enough and sufficiently strong to safely handle a gas kick if a diverter system is planned to be used.

Surface casing:

- Isolate weak formations down to a depth where the formation integrity is sufficient to ensure proper control of abnormally pressured formations in the hole below.
- Support the wellhead and the blow-out-preventer.

- Isolate the formations down to any potential shallow gas zone or isolate such zones in order to establish integrity for further drilling.

Intermediate casing:

- Isolate all formations up to the surface casing shoe so that the next hole section can be drilled safely and efficiently through the pay zone.
- Due to abnormally pressurised zones, weak zones, and zones with lack of stability, more than one intermediate casing may be required. The sizes of the casings must therefore be selected such that the pay zone later can be isolated by a casing with sufficient internal diameter, even if a second casing is used.
- Give sufficient well integrity for drilling the pay zone or any abnormally pressurised zones as requested by the drilling program.
- Fulfil production casing design requirements if only a production liner is planned below.

Intermediate liner:

- Has the capability of being set between two ordinary strings of casing if hole conditions call for an intermediate isolation of the wellbore.
- Dimensions that give sufficient clearances for the previous and the next casing strings.

Production casing:

- Shall isolate the productive zones.
- Ensure proper cementing of the annulus across the productive zones, so that fluid cannot migrate along the wellbore.
- Be able to withstand mechanical and chemical wear from formation- and completion-fluids over the planned production lifetime of the well.
- Be designed to maintain well integrity during all planned production and workover periods.
- Shall be designed and set to allow for further deepening of the hole if specified in the drilling program.

Production liner:

- Shall isolate the productive zones if a production casing is not used, or if the production casing is set only to the top of the reservoir.
- All casing strings and liners exposed to production activities shall fulfil the production casing requirements with respect to well integrity during all phases of the productive life of the well.

Tieback casing:

- The tieback casing has the same functional requirements as the production liner except that the axial load from testing is not present.
- The tieback casing is used to increase the well pressure integrity, often in connection with options such as flow testing of the well. Also, it may be installed to increase the corrosion resistance if H_2S and CO_2 gases are present.

4.2.2.3 Dimensional Requirements

The casing clearances are always governed by the connector/coupling configuration. Strength requirements may require coupling with larger outer diameter and smaller inner diameter than the casing itself. We will in the following define some important dimensional requirements.

The conductor casing dimensions:

- Outer diameter relates to the rotary table opening, opening of the temporary guide base if used, and flow area and top hole diameter.
- Inner diameter relates to flow area between conductor and surface casing.
- Wall thickness relates to stiffness for support of wellhead, blow-out-preventer, and marine riser, to resist buckling if drive pipe is used, and to handle fatigue stresses for subsea production wells due to bending and thermal expansion. Thick wall conductors are often required for the upper joints.

The surface casing dimensions:

- Outer diameter relates to the flow area between the casing, connectors, and the hole.
- Inner diameter relates to clearance for next drill bit, and outside diameter of a possible intermediate liner hanger system.
- Wall thickness of upper joints relates to fatigue stresses for subsea production wells due to bending and thermal expansion.

The intermediate casing dimensions:

- Outer diameter relates to the flow area between casing, connectors, and hole, and the smallest inner diameter in the intermediate liner system if installed.
- Inner diameter relates to the next drill bit size, and the outside diameter of a possible liner hanger system.

The intermediate liner dimensions:

- Same requirements as for the intermediate casing.

The production casing and liner dimensions:

- Outer diameter relates to the flow area between casing, connectors, and hole. Annular space for proper cementing of the productive interval, and the smallest inner diameter in the intermediate liner system if such is installed.
- The inner diameter relates to the next drill bit size if further drilling will be performed. Clearance required for logging, perforating, and well testing equipment to be run, and clearance required for completion- and possible gravel packing equipment installation.

The dimensions of any other type of string:

- If casing strings or liners other than those mentioned above are to be used, considerations shall always be given to diameters and clearances.

4.2.3 Setting Depth

4.2.3.1 Factors to be Evaluated

The following main elements should be evaluated and used as a basis for the casing design. The list may be changed according to particularities in each individual well design.

Hole stability:

- unconsolidated formations
- swelling clays
- fractured formations

- collapse/washout
- fluid loss zones
- plastic formations
- subsidence
- zone isolation

Formation pressure and integrity:

- high integrity formations
- low integrity formations
- high pressure formations
- charged formations
- highly permeable formations
- well control integrity and margins

Drilling fluids,, hole cleaning, and cementing precautions:

- pressure losses, circulation densities, and pump performance
- hole cleaning capabilities
- cementing of permeable intervals
- H_2S and CO_2 bearing intervals
- formation temperature
- mud system chemical/physical tolerances
- differential sticking
- reservoir invasion and damage

Hole curvature:

- kick-off points
- drop-off points
- hole angles
- dog leg severity
- build-up/drop-off rates
- potential side tracks

Mechanical equipment:

- drilling rig hoisting/rotating capability
- drill string and bottom-hole-assembly capability
- casing tensile strength, burst, and collapse capability
- mechanical wear on previous casing
- equipment availability

Economy:

- equipment cost
- penetration rate
- pilot holes
- time versus depth profile
- probability and consequences of hole problems
- primary and secondary objectives
- formation evaluation and geological markers

The evaluation of risk is an important element when performing the well design, in which the casing setting depth evaluation is essential.

4.2.3.2 Setting Depth Evaluation

The process of evaluating casing setting depths is briefly outlined below:

- Determine primary and secondary well objectives.
- Identify factors to be given special emphasis.
- Determine cementing requirements.
- Determine objectives for each proposed hole interval based on the factors listed above.
- If the hole cannot be drilled with standard sizes of casing strings, an extra intermediate or production liner can be installed.
- If the above-mentioned liner does not solve the problem, then consider a non-standard casing program.
- Perform the design.
- If the casing design results in unfavourable casing qualities, then re-evaluate the setting depths prior to confirming the required qualities.

An evaluation of the well control capabilities and kick margins should always be performed for each hole interval below the surface casing after the setting depths have been determined.

A check on the availability of the selected casing should always be performed immediately after the casing has been designed. Lack of availability may lead to a re-evaluation of the setting depths or the use of higher quality casing than strictly required by the design.

4.2.4 Example of Setting Depth Evaluation

In this example calculation we will investigate various ways to evaluate the setting depth of each casing string. The two key parameters are the fracturing pressure and the pore pressure. The relationship between these two parameters determines the maximum length of each open hole section.

Given a pressure prognosis, the casing seats are always determined starting at the bottom of the well and working upwards. In this manner the minimum number of casing strings will be determined. First, we will look at the simplest case.

4.2.4.1 Setting Depth Limited by Mud Weight

Figure 4.5 shows the pressure gradients for a well. At about 2400 m the reservoir is penetrated, causing the regression on the pore pressure gradient. In the fracture gradient curve two potential loss zones are shown, these are the Utsira sand and the Balder tuff. We will assume that the 9-5/8 in. production casing is set just above the reservoir, and that the 7 in. liner will be set through the reservoir.

At 2400 m the minimum mud density is 1.6 s.g. A vertical line upwards crosses the fracture gradient prognosis at 1300 m, which is the shallowest setting depth of the 13-3/8 in. casing. The two shallower casing strings can be set over a wide range, as there is a wide margin between the two gradient curves. From Figure 4.5 we have determined the setting depths as follows:

CASING SIZE (INCH)	*DEPTH (M)*	*MUD WEIGHT (S.G.)*
7	2700	1.60
9 5/8	2400	1.60
13 3/8	1300	1.30
18 5/8	700	1.20
30	400	Sea water

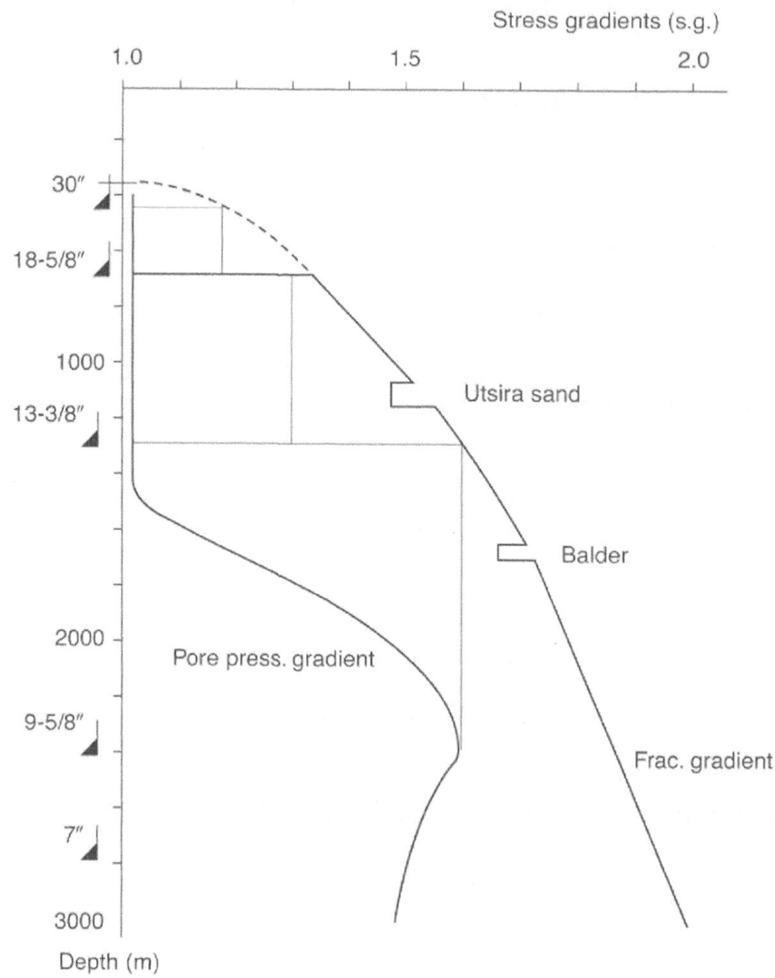

FIGURE 4.5 Casing depth limited by mud weight.

Please observe that even if the Balder zone locally is weaker, it is not critical in this design. The top of each open hole section is critical, that is at 1300, 700, and 400 m.

4.2.4.2 Setting Depth Limited by Mud Weight and Riser Margin

If we decide to drill from a floating drilling rig, we are required to take the riser margin into account. The purpose is to keep the well stable if, for example, the riser has to be disconnected in an emergency situation, such as in bad weather conditions. Figure 4.6 illustrates the problem.

During the drilling operation, drilling mud is filling the hole all the way to the drill floor as shown in Figure 4.6a. However, during disconnect, the hydrostatic head caused by the elevation from the drill floor to the sea level is lost, and the mud inside the marine riser is now replaced by sea water, as shown in Figure 4.6b.

During disconnect, the pore pressure must be balanced by the total hydrostatic pressure, which is:

$$P_o = 0.098\left\{\rho\left(D - h_f - h_w\right) + 1.03h_w\right\} \tag{4.1}$$

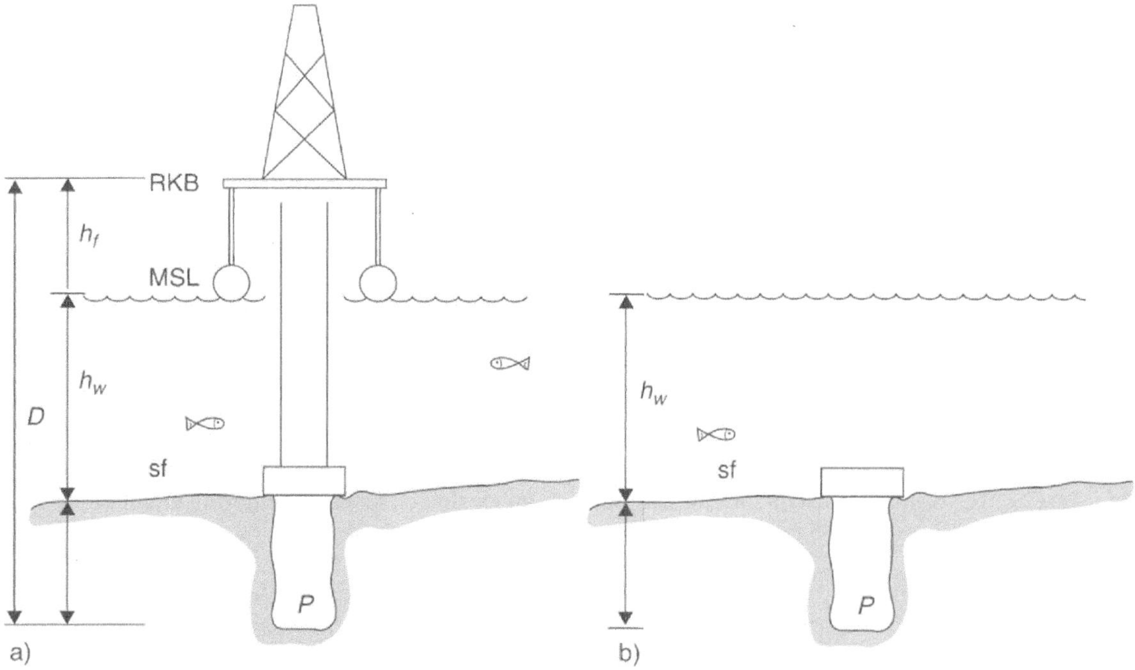

FIGURE 4.6 Drilling and abandonment scenarios to define the riser margin. a) During drilling and b) after abandonment.

Because of this loss of hydrostatic head, we will during regular drilling use an over pressure, called the riser margin. The bottomhole pressure is then:

$$P = 0.098\rho \times D \tag{4.2}$$

and the pore pressure at depth D can be expressed as:

$$P_o = 0.098\, d_p D \tag{4.3}$$

Now we want to determine the minimum mud weight used to ensure that there is a riser margin. If we equate Equations (4.1) and (4.3), we can calculate this minimum mud weight, which is expressed as:

$$\rho = \frac{d_p D - 1.03 h_w}{D - h_f - h_w} \tag{4.4}$$

Figure 4.7 shows the same data as Figure 4.5, except that the minimum mud weight curve calculated from Equation (4.4) is shown. The mud weight used must at least have this value to include the riser margin. The casing evaluation process is now repeated. Since the minimum mud weight allowed at 2400 m now is 1.69 s.g., the previous open hole section must now at least pass 1700 m. From Figure 4.7 the following setting depths are suggested:

CASING SIZE (INCH)	*DEPTH (M)*	*MUD WEIGHT (S.G.)*
7	2700	1.69
9 5/8	2400	1.69

(continued)

CASING SIZE (INCH)	*DEPTH (M)*	*MUD WEIGHT (S.G.)*
13 3/8	1700	1.40
18 5/8	900	1.20
30	440	Sea water

4.2.4.3 Setting Depth Limited by Kick Criteria

Kick control is a critical element in all casing design, and also when it comes to setting depth. When doing this evaluation it is important to use pressures, and not pressure gradients. Remember that the well is exposed to a pressure. The gradient is introduced as some kind of depth normalisation. If a gas is allowed to migrate up the hole with constant pressure, the gradient is going to increase.

Figure 4.8 shows a pressure plot for the same well. From the reservoir geology section, we have learned that the formation fluid expected is a condensate with density 0.76 s.g., and calculations have shown that the pressure will exceed the bubble point at all phases of the circulation of a kick. The density can therefore be assumed constant, and we assume that no expansion takes place during circulation.

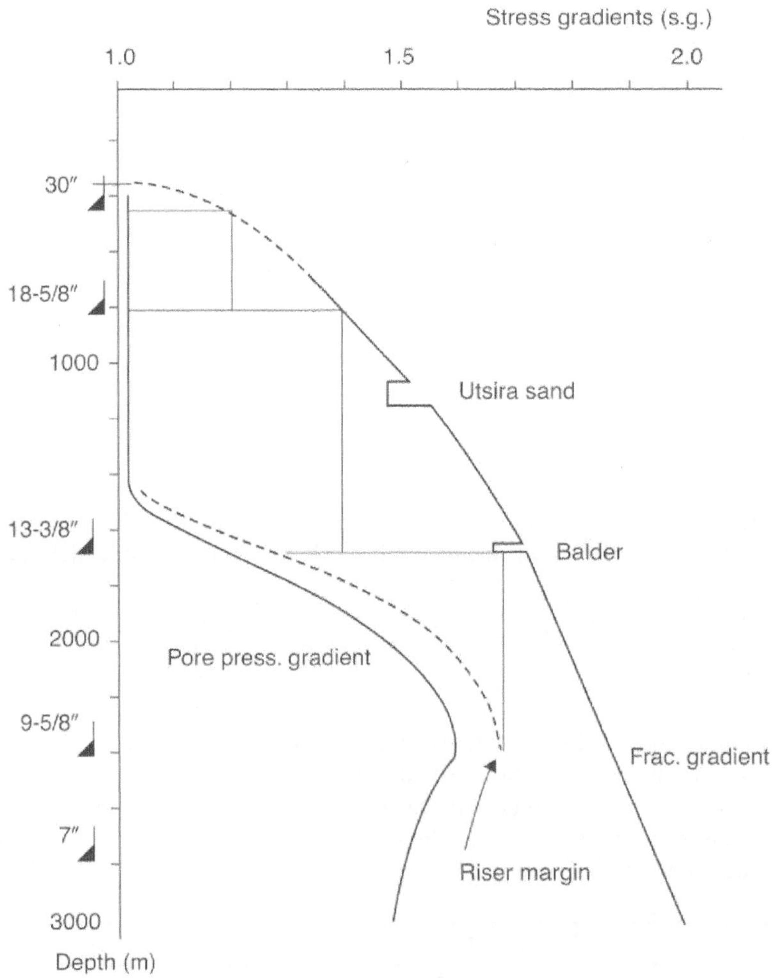

FIGURE 4.7 Casing depth limited by mud weight and riser margin.

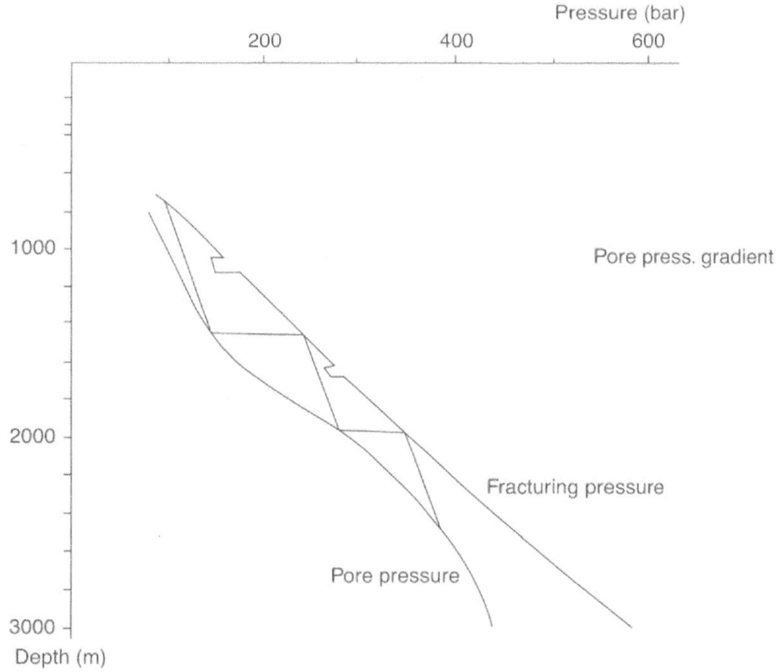

FIGURE 4.8 Setting depth from kick criteria.

We can assume that the 12-1/4 in. section has been drilled to 2400 m, and the well takes a kick. The well is filled with 0.76 s.g. condensate, and pressure up the hole is reduced by the weight of this fluid. A gradient corresponding to the condensate density is drawn from 2400 m upwards, and it crosses the fracture pressure line at 1960 m. Repeating this process upwards yields the minimum setting depths of:

CASING SIZE (INCH)	DEPTH (M)
7	2700
9 5/8	2400
13 3/8	1960
18 5/8	1450
30	750

We observe that if we want to ensure full well integrity in all hole sections, we end up with an unacceptably long casing string. This will be further discussed in the following.

4.2.4.4 Summary of Example Case

The first and second cases above are similar, except that the first case applies to jack-up rigs and fixed installations, while the second case applies to mobile drilling rigs. The depths determined from mud weight considerations must always be considered. However, if we considered a kick scenario, we found that we obtained unrealistically long casing strings. To overcome this problem, we will therefore use the definitions of well integrity defined in Section 4.1. The casing seat selection would look as follows:

The casing seats are selected from mud weight considerations as shown in the examples above.

- The production casing must always have full well integrity, that is the capability of handling a kick.

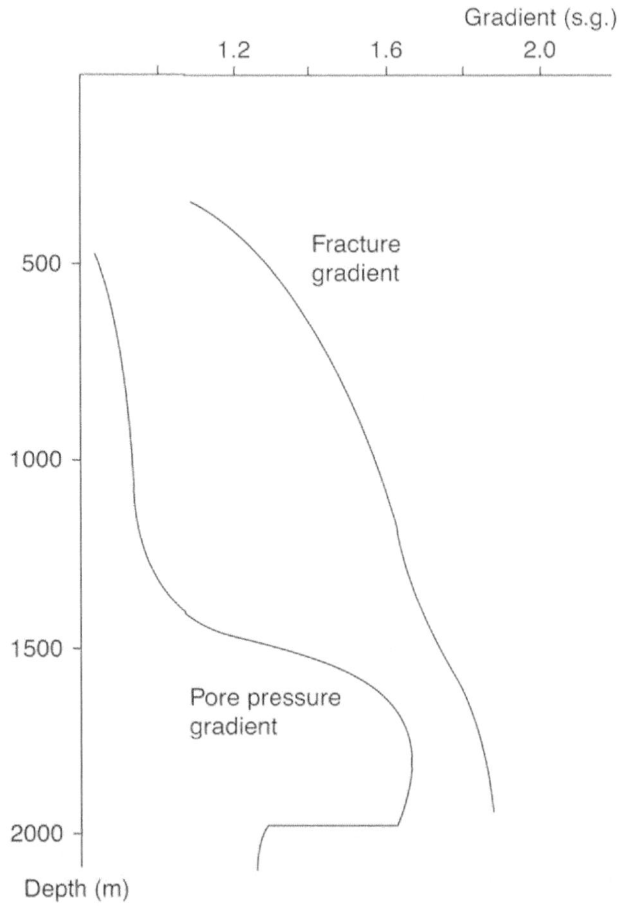

FIGURE 4.9 Problem 1 of Section 4.2.4.

- The shallower casing strings will not be designed for full well integrity because this requires unacceptably long casing strings. Instead, we will here use the concept of reduced integrity, defined in Section 4.1. Reduced well integrity means that a kick margin is introduced. If this kick volume is not exceeded, the well will remain stable.

Problems

Problem 1. In Figure 4.9 is a prognosis for an exploration well. You are asked to determine the setting depth of each casing string. The 9-5/8 in. casing is set at 1900 m, and a 7 in. liner is set through the reservoir.

The water depth is 150 m, and the drill floor elevation is 25 m above sea level. Perform another casing seat evaluation by including the riser margin.

Using the "median line principle" from Section 2.1, propose a mud weight schedule for this well.

Problem 2. In Figure 4.10 is a prognosis for a high pressure well in the Central Graben area. The 9-5/8 in. casing will be set at 4730 m, and a 7 in. liner will be set through the reservoir. Perform the following setting depth evaluations:

a) Using the fracture gradient curve.
b) Using the frac. gradient curve for weak zones.
c) Using the frac. grad. curve and the median line for the mud weight.

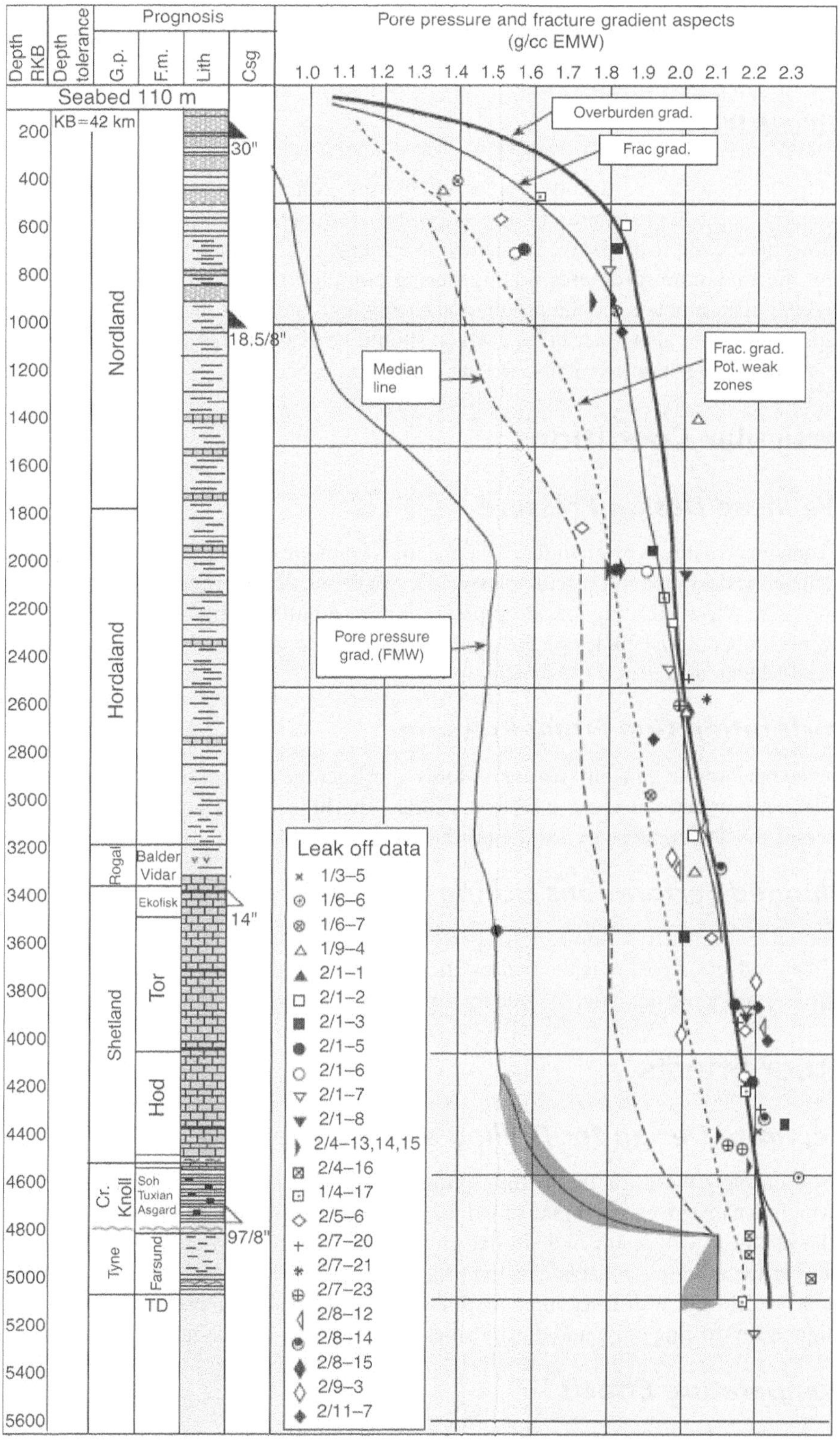

FIGURE 4.10 Problem 2 of Section 4.2.4.

4.3 COMPLETION AND PRODUCTION REQUIREMENTS

4.3.1 Introduction

A number of pressure tests are usually performed during completion and workover operations. The objective is usually to check production tubing and packer for leaks. Also, plugged perforations and fracturing operations may result in excessive pressures. These may exceed the pressures the casing is designed for. Therefore, the maximum pressures arising during completion and workover operations should be established as design parameters for the production casing, and included in the design.

This chapter addresses also other effects which should be evaluated in the well design process, in an attempt to bring more long-term elements into the evaluation process.

4.3.2 Particular Conditions

4.3.2.1 Wellhead Design Pressure

The wellhead design pressure is often higher than the maximum pressures expected in the field. One reason is that the wellhead is designed in pressure classes, i.e. 5000 psi. Another reason to use a higher wellhead design pressure is to allow for a higher pump pressure during bullheading and fracturing operations. For these reasons it is often practice to design the production casing to withstand the wellhead design pressure, even if it exceeds the pressures in the formation.

4.3.2.2 Perforation Gun Firing Pressure

Tubing conveyed perforation guns often use pressures to trigger the charge. In some instances this pressure exceeds the design pressure for the production casing. The highest physical pressure occurring should always be used as the design pressure for the production casing.

4.3.2.3 Plugged Perforations During Bullheading

When bullheading the well after a flow test, the perforations may become plugged. In that case, the pressure acting below the production packer is the hydrostatic head inside the string plus the surface pump pressure. If this scenario is considered realistic, the maximum permissible pump pressure should be evaluated.

4.3.3 Other Effects

4.3.3.1 Separate Design for Drilling and Well Testing

The production casing design for the drilling phase involves a gas-filled casing scenario for the burst evaluation, which usually gives the highest load just below the wellhead. In case of well testing, a tubing leak below the wellhead will result in a similar pressure in the wellhead area, but also to a considerable pressure on top of the production packer, as the production casing has a significant head of mud.

In exploration drilling, well tests are sometimes not performed. In these cases the production casing could be designed for drilling only, and during an eventual well test a tieback string could be installed.

4.3.3.2 Temperature Effects

During well testing and production a considerable amount of heat is transported from the reservoir up to the wellhead, resulting in a higher temperature profile. This increased temperature results in volume expansion of the fluids behind the casing strings. If these annuli are closed, a considerable pressure can develop, resulting in collapse or burst of one of the casings. Usually this problem is avoided by ensuring that there always is some open hole exposure in each annulus. In other words, keeping the cement level below the previous casing shoe.

For wells where the bottomhole temperature exceeds 100°C, a derating of the casing strength is required. The derating should follow the casing manufacturer's recommendations. When derated field values are determined, derating of the collapse rating can be done as described in API bulletin 5C3(1990).

4.3.3.3 Long Time Effects

The main emphasis in this book is exploration drilling, where the time frame is short. In production wells, which may have a life of 20–30 years, long time effects should be considered, especially for the production casing. Some factors are:

- Particle settling behind the casing.
- Subsidence and compaction.
- Corrosion below the production packer.
- Pore pressure reduction over time.

4.3.4 Examples

4.3.4.1 Test Pressures During Completion and Workover

During design of a well, the drilling department used a reservoir pressure of 194 bars as a design parameter. The wellhead had a 5000 psi (345 bar) rating. It was questioned whether one could install weaker casing that satisfied drilling requirements, or whether the wellhead design rating should be used. The completion group evaluated various test scenarios, and concluded that the full 345 bar design pressure was required. The elements considered were:

1. Weakness evaluation of casing and tubing.
2. Pressure testing of tubing before setting production packer.
3. Pressure testing after setting production packer.
4. Performing scale inhibitor squeeze, or fracturing.
5. Temperature expansion.

We will in the following perform a brief discussion about each element, and perform simple calculations.

Scenario 1: Weakness evaluation of casing and tubing. Figure 4.11 defines the data used in the example case. We are mainly concerned about the integrity of the production casing. If excessive pressures arise inside the production casing, then the casing may burst, or the production tubing may collapse. In our case the tubing collapse pressure is 433 bar and the casing burst pressure is 473 bar, resulting in the tubing being the weakest element. The production tubing may in other words act as a safety element for the casing. The exception is when the pressure inside the tubing is higher than the pressure inside the casing.

Scenario 2: Pressure testing of tubing before setting packer. This scenario is just a test of the production tubing integrity. We assume that if leakage occurs, the operation will be stopped before higher pressures build up inside the production casing.

Scenario 3: Pressure testing after setting production packer. During pressure testing of the integrity of the packer, on the other hand, the production casing will be pressured. Figure 4.11 illustrates the problem, and we will in the following perform simple calculations:

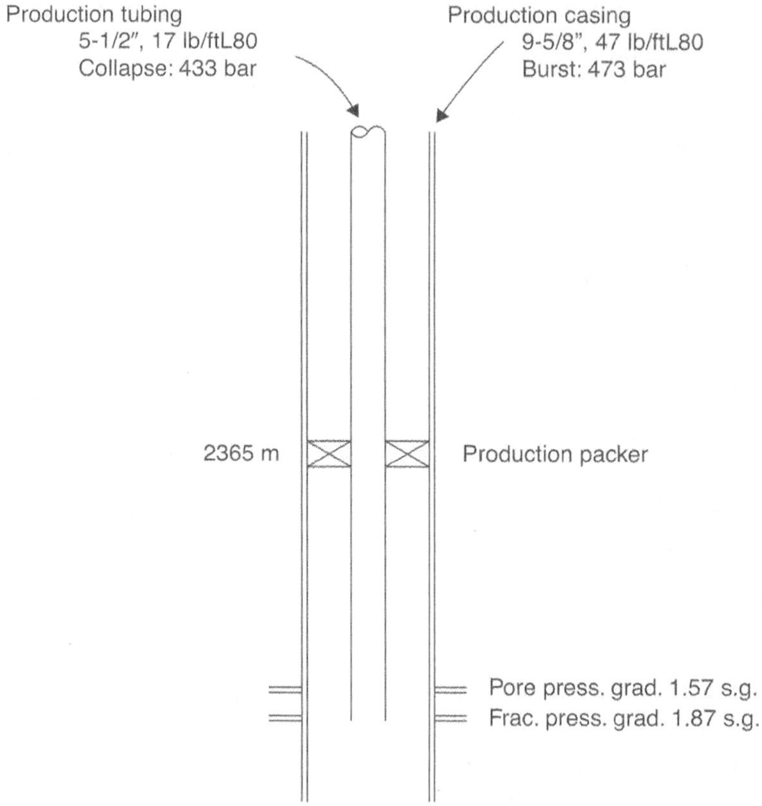

FIGURE 4.11 Illustration of completed well.

The actual pressures at the depth of 2365 m are:

$$\begin{aligned} \text{Pore pressure:} &\quad 0.098 \times 1.57 \times 2365 &= 364 \text{ bar} \\ \text{Frac. pressure:} &\quad 0.098 \times 1.87 \times 2365 &= 433 \text{ bar} \\ \text{Pressure behind casing:} &\quad 0.098 \times 1.03 \times 2365 &= 239 \text{ bar} \end{aligned}$$

We have assumed that the effective pressure behind the casing is hydrostatic sea water, since this is the only mobile phase when particles have settled and the cement hardened. However, we are fully aware that this is a conservative assumption.

If the production casing above the packer is filled with sea water, maximum differential pressure across the packer is:

$$0.098(1.57 - 1.03)2365 = 125 \text{ bar}$$

If we instead assume reservoir fluid with density 0.76 s.g., maximum differential pressure across the packer is:

$$0.098(1.57 - 0.76)2365 = 188 \text{ bar}$$

and if we assume gas-filled annulus above the packer, maximum differential pressure is:

$$0.098(1.57-0.20)2365 = 317 \text{ bar}$$

The test pressure is defined as 50 bar more than the maximum pressure occurring. The first case above is most realistic, but if we assume the third case is unrealistic then the second case represents the worst case. The maximum test pressure required is:

$$188+50 = 238 \text{ bar}$$

The conclusion is that the packer does not have to be tested to 345 bar from a formation pressure point of view. However, the packer is usually tested to full design pressure, that is 345 bar, because well operations may give pressures higher than the formation pressure. Both when testing from the upper and lower side, the pressure is 345 bar, which exceeds the requirement of the highest physical pressure + 50 bar, which is 238 bar. The design pressure is significantly lower than the burst strength of the casing, which is 473 bar.

Scenario 4: Performing scale inhibitor squeeze, or fracturing. During this process we are pumping fluids down the production string and into the reservoir. There are two main sources of excessive pressures, the frictional losses call for a higher surface pressure, and if the perforations become plugged, higher pressures arise under the packer.

Assume that the production tubing is filled with 0.76 s.g. fluid, and we pump to fracture behind the perforations. The perforations become plugged, and we pump until the casing burst below the packer with 473 bar. The static surface pressure is then:

$$473 \text{ bar} = 0.098 \times 0.76 \times 2365 + P$$

$$P = 297 \text{ bar}$$

Assuming the same fluid density on both the inside and outside of the production tubing, we have a considerable margin against burst above the packer, but below the packer the casing may burst. We have not evaluated frictional pressure losses, but this is often used as an argument to design the casing for full well-head design pressure.

Scenario 5: Temperature expansion. This element will be addressed in more detail in the next chapter. During production, heat will be transported upwards inside the well. The annulus between the tubing and the production casing is actually a closed chamber. When heating occurs, expansion takes place, causing pressure to build up. On jack-up rigs and on production platforms, this annulus can be vented to relieve the pressure, but on present sub-sea wellheads used in exploration drilling, there are no venting capabilities.

In scenario 1 we found that if a high annulus pressure arises, the tubing will collapse before the casing bursts. This is the preferred solution if that occurs.

To summarise, we observe that most scenarios give lower pressures than the well-head design pressure. However, during injection and other operations, failures like plugged perforations, combined with high frictional losses may result in considerable pressures. Therefore, it is conventional practice to design the production casing of the well for full wellhead design pressure.

4.3.4.2 Temperature-Induced B-Annulus Pressure

During well testing and production a significant amount of heat is transported up the wellbore. The temperature will increase throughout the well. If closed annuli are present, the temperature expansion of fluids

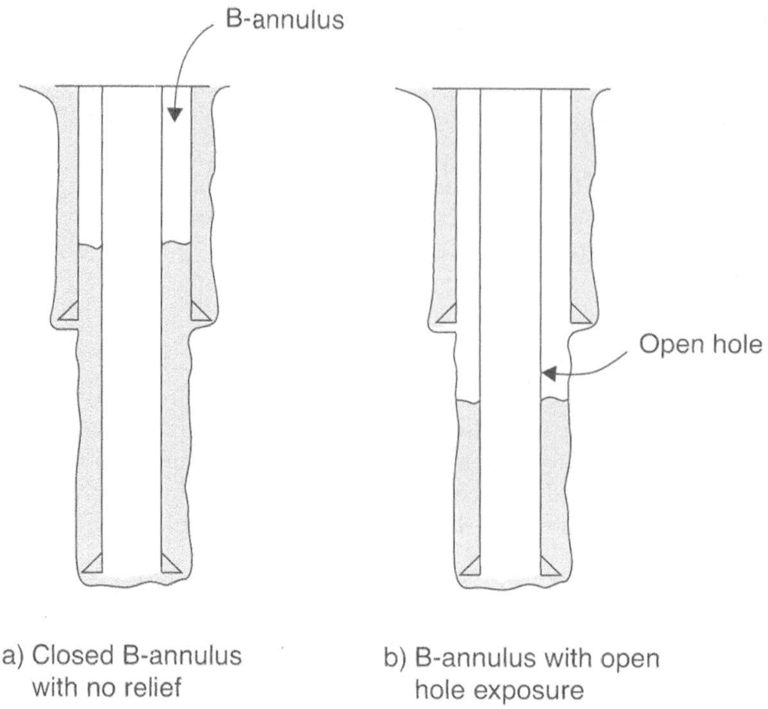

FIGURE 4.12 Situations where temperature induced pressure may/may not cause casing failure.

inside these can cause significant pressure rise. In extreme cases, temperature expansion can cause burst or collapse of casing strings.

Figure 4.12 illustrates the problem. In Figure 4.12a the inner casing string is cemented above the previous casing shoe. Provided a high-quality cement job is performed, this annulus (B-annulus) is considered closed. In sub-sea exploration wells there are usually no venting possibilities. Therefore, one must always check the expected pressures and the casing strength in these cases.

The common way to reduce temperature-induced pressures is shown in Figure 4.12b. The cement for the next casing string is not covering the previous shoe. It is believed that the exposed open hole section may allow for a small fluid loss, which will lead to a reduced pressure. The pressures should also be checked against the fracturing gradient of the section, which should act as a safety valve. In the following we will demonstrate the temperature-induced pressures with an example.

Figure 4.13 shows the temperature profiles during casing installation, and also the profile during a well test. These are considered the two extreme cases. Assuming a linear temperature profile, the changes in temperatures can be expressed as:

$$\begin{aligned} &\text{at well head:} \quad \Delta T_1 = T_3 - T_1 \\ &\text{at well bottom:} \quad \Delta T_2 = T_4 - T_2 \end{aligned} \tag{4.5}$$

Having a total volume V in the annulus, free expansion of this volume subjected to the temperature change of Equation (4.5) yields:

$$\frac{\Delta V}{V} = \frac{1}{2}\alpha\left(\Delta_1 + \Delta T_2\right)$$

or if we insert the actual temperatures from Equation (4.5), we obtain:

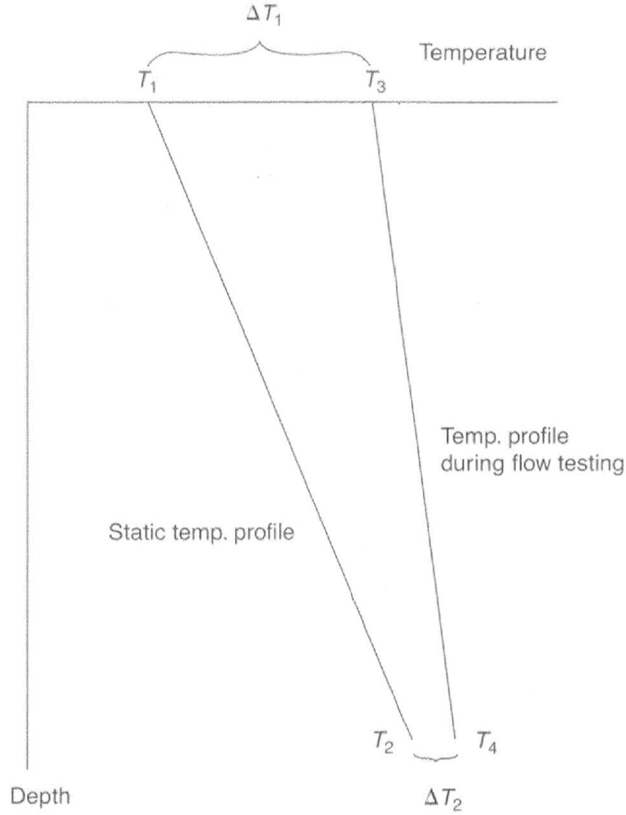

FIGURE 4.13 Simplified temperature profiles in a well during various phases.

$$\frac{\Delta V}{V} = \alpha \left\{ \frac{T_3 + T_4}{2} - \frac{T_1 + T_2}{2} \right\} \tag{4.6}$$

We observe that the volume change is proportional to the average temperature change.

To consider the pressure element, first imagine that the fluid is allowed to expand freely according to Equation (4.6). Then, the fluid is compressed back to its initial volume. The pressure required is:

$$\Delta P = \left(\frac{-1}{c} \right) \frac{\Delta V}{V}$$

or by inserting Equation (4.6),

$$\Delta P = \left(\frac{-\alpha}{2c} \right) \left\{ (T_3 + T_4) - (T_1 + T_2) \right\} (3.37\text{b})(3.37\text{b})(3.37\text{b}) \tag{4.7}$$

where: c = compressibility of the fluid
α = heat expansion coefficient

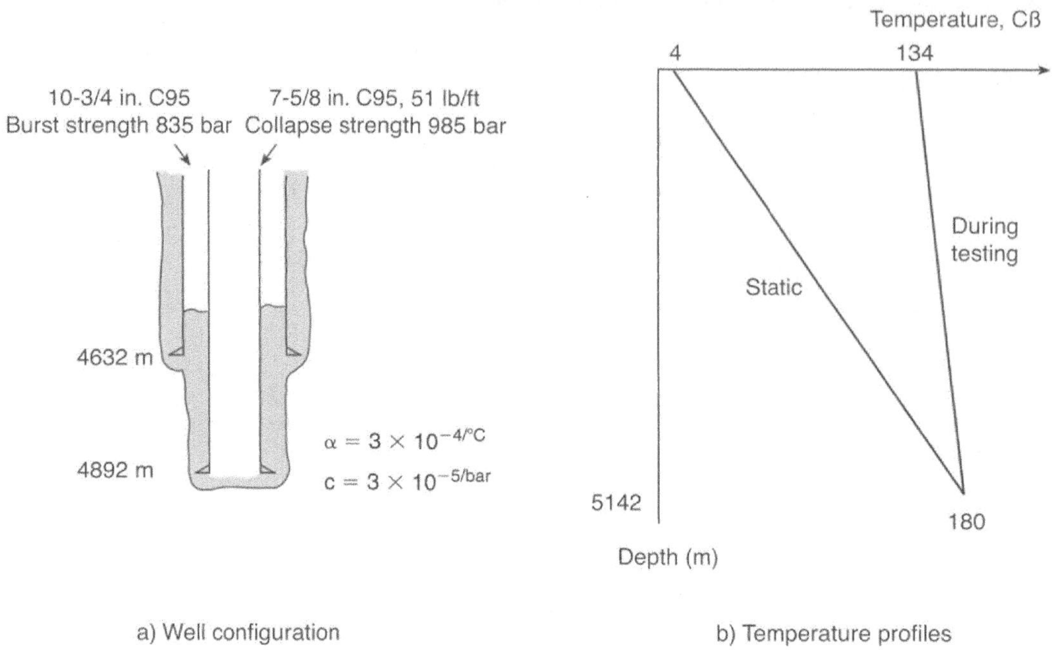

FIGURE 4.14 Data for an exploration well.

Equation (4.7) gives an estimate for the temperature-induced pressure in a closed annulus, assuming that the mass of the fluid remains constant. In the example above, only the fluid behaviour is studied. Although the casing and the rock have less effect, their compressibilities could be included as well.

Figure 4.14a shows an example from an exploration well. Figure 4.14b shows the estimated temperature profiles before and after testing. The expected pressure increase from Equation (4.7) is:

$$\Delta P = \frac{3\times10^{-4}}{2\times3\times10^{-5}}\{(134+180)-(4+180)\} = 650 \text{ bar}$$

Comparing this excessive pressure to the strength of the casings, results in:

Collapse strength of 7-5/8 in casing:	985 bar > 650 bar
Burst strength of 10-3/4 in. casing:	835 bar > 650 bar

We conclude that the design is acceptable because both casing strengths exceed the expected pressure increase. Also, the 7-5/8 in. production casing is considered most critical, and we observe that failure would in any case occur in the outer intermediate string before the production casing.

Halal and Michell (1994) propose a casing design procedure for trapped annular pressure buildup.

Casing Design

5

5.1 DESIGN CRITERIA

The selection of the design criteria is the most critical activity in the design of a well. All relevant criteria should be evaluated and used in the design. Often several criteria can be used for each design, and from these the most realistic scenario must be established.

The casing design basically involves strength assessment of burst, collapse, and tensional loads on the string. In the following each of these elements will be discussed.

In addition to the general nomenclature, the following specific symbols are used in this chapter:

σ_t	Tangential stress on casing due to inner pressure
σ_a	Axial stress on casing due to inner pressure
L	Length of casing
P_{burst}	Critical inner pressure when casing will fail in tension
$\sigma_{tensile}$	Tensile strength of casing material
D_i, D_o	Inner and outer diameter of casing
$P_{collapse}$	Critical outer pressure when casing will collapse
d_{sw}	Specific gravity of sea water
d_{ce}	Specific gravity of cement
d_{mw}	Specific gravity of drilling mud

5.1.1 Burst Mechanism and Design Criteria

The pipe body will have a tendency to burst when the difference between the internal and the external pressures exceeds the mechanical strength of the pipe. Burst is a tensile failure, resulting in rupture along the axis of the pipe. Figure 5.1 illustrates a thin-walled pipe, and the following example shows the failure mechanism.

A pipe, or a casing string can be considered as a thin-walled vessel. Figure 5.1a shows such a vessel. It consists of a tube, with both ends closed. Imagine that this vessel is being pressurised in the interior. We will now briefly investigate the stresses that arise during pressurisation. We will investigate the stresses in the two directions, axially, and circumferentially (the tangential or hoop stress).

Figure 5.1b shows the tangential stress case, and we have indicated a plane which we will investigate. The total force acting on this plane is the internal pressure P multiplied with the projected area, or: $F_t = PD_i L$. The area that takes this force is given by the wall thickness on both sides, or: $A_t = 2tL$. The stress acting on the tube wall is:

$$\sigma_t = \frac{F_t}{A_t} = \frac{1}{2} P \left(\frac{D_i}{t} \right) \tag{5.1}$$

DOI: 10.1201/9781003618171-5

Figure 5.1c shows the axial stress case, and now we will investigate a plane cutting across the vessel. The force acting on this plane is equal to the end loads of the vessel, or: $F_a = P\pi D_i^2/4$, and the area that holds this load is the total wall thickness: $A_a = \pi Dt$. The axial stress in the vessel is:

$$\sigma_a = \frac{F_a}{A_a} = \frac{1}{4}P\left(\frac{D_i}{t}\right) \tag{5.2}$$

Taking the ratio of these two stresses, we obtain:

$$\sigma_t = 2\sigma_a \tag{5.3}$$

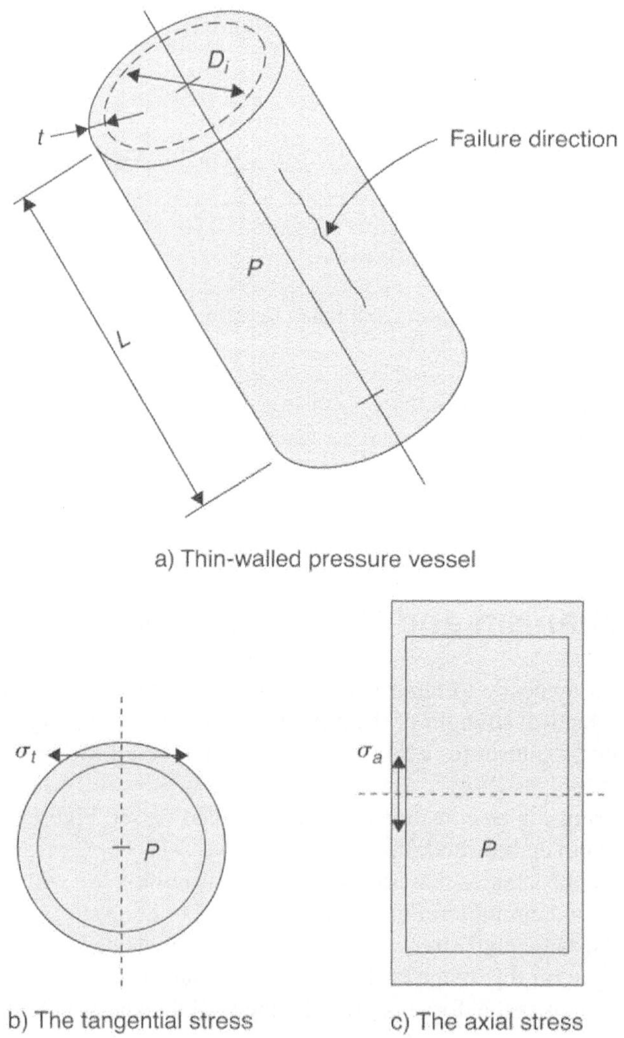

FIGURE 5.1 Stresses and failure of thin-walled vessel pressured from the inside.

This is a very interesting equation. If a thin-walled vessel is pressurised from the interior, the stress acting around the circumference is twice the axial stress. This is well known in pressure vessel technology, and if such a vessel fails, it usually opens up along the axis as indicated in Figure 5.1a.

What we have described above is the tensile failure mechanism which in petroleum terminology is called bursted casing. A number of complications can be introduced, such as added axial load due to string weight, bi-axial or tri-axial stress conditions, thick-walled stress analysis, elastoplastic analysis, and so on. Payne and Swanson (1990) gives a probabilistic approach, while Kastor (1986) and Johnson et al. (1987) perform triaxial casing design. However, we will in this chapter use the simple definition of Equation (5.1), which results in the following burst equation if the tangential stress is set equal to the tensile material strength:

$$P_{burst} = 2\sigma_{tensile}\left(\frac{t}{D_i}\right)$$

or, if we use the outer diameter:

$$P_{burst} = 2\sigma_{tensile}\left(\frac{t}{D_o}\right) \tag{5.4}$$

Example
Given the following data for a casing (IFP, 1991):

Size:	9-5/8 in. outer diameter (244.5 mm)
Inner diameter:	226.6 mm
Wall thickness:	8.94 mm
Grade:	C95
Weight:	36 lbs/ft
Cross-sectional area:	66.23 cm^2
Burst strength:	419 bar
Collapse resistance:	169 bar
Yield strength:	433,000 daN

Calculate the burst strength with Equation (5.4) and compare with manufacturer's data.

Solution
The yield strength of the pipe is:

$$\sigma_{tensile} = \frac{F}{A} = \frac{433000 \text{ daN}}{66.23 \text{ cm}^2} = 6538 \text{ bar}$$

Equation (5.4) gives:

$$P_{burst} = 2 \times 6538 \text{ bar}\left(\frac{8.94 \text{ mm}}{244.5 \text{ mm}}\right) = 478 \text{ bar}$$

The manufacturer defines the burst strength of this casing to be 419 bar, and we observe that Equation (5.4) gives a higher burst value. We have also used the outer diameter of the casing in the calculation above, to reduce the difference. However, for practical applications we are always using the manufacturer's data. Equation (5.4) will be used to adjust these for wear and corrosion, and is considered sufficiently accurate for that purpose.

As an example, assume that the same casing has been installed in a well. However, during drilling of the next section, a considerable wear arises. A casing caliper log shows that the thinnest section of the casing has been reduced from the nominal value of 8.94 mm to 5 mm. Estimate the reduced burst strength of the weakest point in the casing.

We observed in the previous example that a perfect match was not obtained. Since the manufacturer's recommendation is the base value, we will simply use Equation (5.4) as a proportionality relation, that is:

$$\frac{P_{burst}}{t} = \frac{2\sigma_{tensile}}{D_o} = \text{constant (3.37b)(3.37b)(3.37b)}$$

$$P_{burst} = 235 \text{ bar}$$

There are numerous situations where pressure conditions arise that can result in a bursted pipe. The following list identifies some:

a. The hydrostatic mud pressure inside the casing exceeds the formation pressure or the hydrostatic pressure outside the casing.
b. During well shut-in, the differential borehole pressure allows formation fluid to enter the wellbore.
c. A gas bubble, caused by a kick, is allowed to migrate up the casing with or without limited expansion.
d. During kick circulation.
e. The casing is filled with gas migrating up the wellbore during a temporary abandonment or disconnect in an emergency situation.
f. During testing or production a leak occurs in the tubing just below the wellhead.
g. Temperature expansion of fluid in closed annuli between casing strings.
h. When squeeze cementing.

Although the situations above are all different, the pressure picture is similar for several of the operations. From a design point of view, the following main categories can be established:

1. *Gas-filled casing.* This is a conservative design criterion. It assumes that the well is completely filled with gas or formation fluids, and then shut in. Figure 5.2a illustrates this. The inside pressure below the wellhead will be the formation pressure minus the weight of the gas column, and the outside pressure on the casing will usually be the hydrostatic weight of the fluids behind the casing string. Also shown is the difference between these two pressures, which gives the load on the casing string. The gas-filled casing criterion must be applied to the production casing since the well is tested and produced, and therefore potentially is exposed to a gas-filled case. For shallow or intermediate casing strings, with no flow test options, the gas-filled casing criterion is too conservative. Here we introduce the concept of reduced well integrity, which sets an upper limit for the size of an eventual kick allowing us to use weaker casing strings in the design.
2 *Leaking tubing criterion.* During well testing or production a leak may occur at the top of the production tubing just below the wellhead. Figure 5.2b illustrates this. Now we have added another component to the system, namely the production string, which at bottom usually is installed on a packer. This packer isolates the annuli above or below. First, we will study the pressure across the tubing wall. The inside pressure is gas-filled similar to case 5.2a, but outside the tubing the

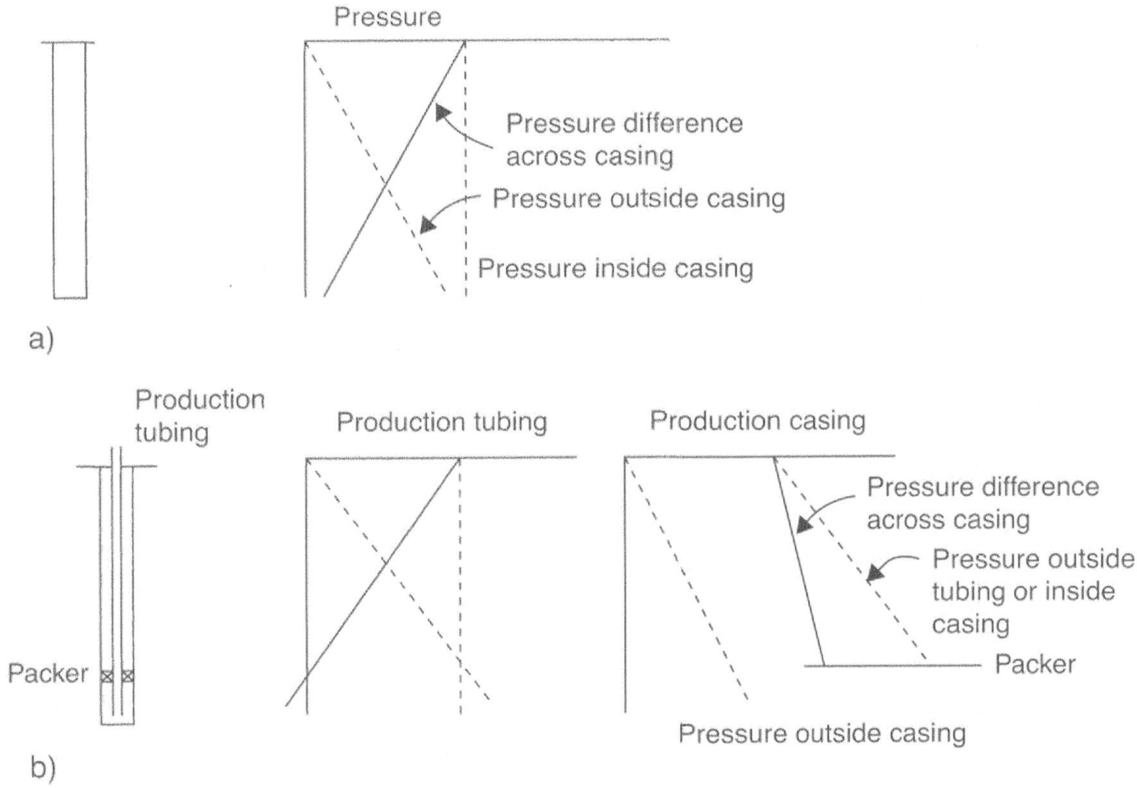

FIGURE 5.2 Pressure regimes causing gas filled casing and leaking tubing failures.

pressure is caused by a completion fluid. If a leak occurs in the top of the tubing, this inside tubing pressure is superimposed on top of the casing/tubing annulus. Therefore, due to the hydrostatic head, a significant pressure arises in the casing annulus. The top of the packer usually becomes the critical element under these circumstances. Also note that the leaking tubing criterion also implicitly includes the gas-filled casing criterion.

3. *Maximum gas kick.* This criterion is based on the largest gas influx volume which can enter the wellbore at next casing setting depth and be circulated out without fracturing the formation at the previous casing shoe. In Section 4.1 the concept of reduced well integrity was defined. The design premise results in the following constraints:
   - The minimum leak-off value to reach next casing shoe.
   - The maximum acceptable leak-off value to ensure that weakest point in the well stays below the casing shoe.
   - The maximum influx size that can safely be handled without fracturing the hole.

This criterion is commonly used for casing strings which are not intended to be production casings. In other words, this criterion is valid for all strings except for the production casing. A maximum kick size must be chosen. The following guidelines are based on the kick volume detection accuracy on present-day drilling rigs:

- Floating units, kick size: 1–8 m^3.
- Fixed installations, kick size: 0.2–4 m^3.

Furthermore, to avoid having the weakest point below the wellhead, a maximum leak-off value is determined. This maximum leak-off value depends directly on the casing strength chosen.

5.1.2 Collapse Mechanism and Design Criteria

When a tubing collapses due to external loading, it changes shape from circular to elliptical or some other non-circular form. The main problem is that equipment may no longer pass through the interior of the pipe.

The casing will have a tendency to collapse when the external pressure acting on the casing body exceeds the internal pressure. The external pressure is caused by pore pressure, drilling fluid pressure, or temperature expansion, and the internal pressure is defined equal to the hydrostatic pressure exerted by a mud or saltwater column. The collapse is a deformation of the casing, and is a geometric failure rather than a materials failure.

The collapse is actually a stability problem. At a critical pressure, only a slight imperfection in circularity or loading leads to deformation and a shape change. The derivation of the collapse equations is a formidable task. Also there are complications such as elastoplasticity. We will use the following linear elastic equation, which is considered sufficiently accurate for our purposes (Holmquist and Nadia, 1939; API Bulletin 5C3, 1990):

$$P_{collapse} = \frac{2CE}{1-v^2}\left\{\frac{1}{\left(\frac{D_o}{t}-1\right)^2 \frac{D_o}{t}}\right\} \tag{5.5}$$

Actually the diameter/wall thickness ratio determines the particular collapse type, which are defined as yield collapse, plastic collapse, transitional collapse, and elastic collapse. Equation (5.5) is called elastic collapse and is valid for large *D/t* ratio. For more details on collapse performance, the reader is referred to a publication by Fowler et al. (1983).

As steel usually has the same elastic properties regardless of quality, we observe that the collapse resistance largely depends on the diameter/thickness ratio. An example illustrates this:

Example

Assume the same casing that was studied under the previous burst discussion. This casing had a collapse pressure of 169 bar given by the manufacturer. Determine the constant C from Equation (5.5). Assume further that: $E = 2.07 \times 10^6$ bar, $v = 0.3$.

Inverting Equation (5.5):

$$C = P_c \frac{1-v^2}{2E}\left\{\frac{D_o}{t}-1\right\}^2 \frac{D_o}{t}$$

$$= 169\,\text{bar}\frac{1-0.3^2}{2\times 2.07\times 10^6\,\text{bar}}\left\{\frac{244.5\text{ mm}}{8.94\text{ mm}}-1\right\}^2 \frac{244.5\text{mm}}{8.94\text{mm}}$$

$$C = 0.705$$

Assume furthermore that Equation (5.5) is valid also for estimating effects of casing wear. What will the collapse resistance be if the casing locally is worn from 8.94 mm to 7 mm?

$$P_{collapse} = \frac{2\times 0.705\times 2.07\times 10^6\,\text{bar}}{1-0.3^2}\left\{\frac{1}{\frac{244.5\text{mm}}{7\text{mm}}-1}\right\}^2 \frac{1}{\frac{244.5\text{mm}}{7\text{mm}}}$$

$$P_{collapse} = 80\,\text{bar}$$

We observe that casing wear can have considerable impact on the collapse resistance. The application of Equation (5.5) implicitly assumes that the wall thickness is evenly reduced. It is not known what the reduction in collapse resistance would be if a smaller portion is worn, but it is reasonable to expect that the reduction is less severe.

Examples of situations that may lead to casing collapse are:

a. The mud level inside the casing drops due to lost circulation, which can be caused by high mud weights, natural fractures, or extremely high permeability.
b. During cement squeeze jobs through perforations, high pressures may arise behind the casing.
c. During primary cementing of the casing, the high cement density on the outside exceeds the internal pressure of the displacing fluid, resulting in collapse.
d. During drilling through salt sections, the plastic properties of salt may cause a significant load on the casing, which may collapse.
e. If the casing is emptied, the external pressures may exceed the collapse resistance. One example is if the well is swabbed in for production, another example is if serious circulation losses take place.
f. The casing string is not properly filled with mud. This can create high collapse loading in deep water.
g. Temperature expansion on closed liquid-filled annuli between casing strings.

We will in the following define two criteria, which incorporate most examples above:

1. *Mud losses to a thief zone.* During drilling, sometimes mud losses occur unexpectedly. In severe cases, the fluid level in the annulus may drop. If a permeable formation is exposed, a well control problem can evolve as well. When mud is lost, the pressure outside the casing remains constant, but the inside pressure decreases, developing a collapse pressure. Many criteria have been proposed to model this situation. We will use a simple definition for all cases, but this should of course be modified if other information becomes available. Please note that there are several realistic mud loss scenarios, some of which are:

   - Losing mud at the bottom of the present hole section. This could be related to the drilling or the cementing operation.
   - Losing mud at the bottom of the next openhole section. This is a post-installation scenario where the annulus level may drop and cause collapse loading. For this case we may assume that the cemented part of the casing is mechanically fixed, and assume that a collapse eventually arises in the fluid-filled annulus above.
   - Penetrating an overpressurised reservoir, we may also assume that the mud losses stabilise when the annulus pressure equals the pore pressure.
   - Other realistic scenarios can be developed and used as the basis for the collapse design of the casing. In the following field examples only a few of these are applied.

From practical experience in the North Sea, pore pressures lower than the hydrostatic saltwater pressure are seldom reported. This will be used as a lower limit for the mud loss pressure. In other words, if mud losses occur, the annulus level will stabilise when the bottomhole pressure is equal to the weight of a saltwater column to that depth. This is illustrated in Figure 5.3a.

The pressure of this saltwater column is:

$$P = 0.098 d_{sw} \left(D - D_a \right)$$

We have here assumed a floating rig with a drill floor elevation above the sea level. If a mud loss situation occurs, the pressure at the bottom is:

$$P = 0.098 d_{mw} (D - h)$$

Equating the two pressures, the depth to the mud level in the annulus is:

$$h = D - (D - D_a)\frac{d_{sw}}{d_{mw}} \tag{5.6}$$

Assuming drilling mud both outside and inside the casing, Fig. 5.3a illustrates the collapse pressure that arises. The equations governing these pressures are as follows:

Outside pressure:

$$P_o = 0.098 d_{mw} D$$

Inside pressure:

$$\begin{aligned} P_i &= 0.098 d_{mw} (D - h) \\ &= 0.098 d_{sw} (D - D_a) \quad \text{for } D > h \\ &= 0 \qquad\qquad\qquad\quad \text{for } D < h \end{aligned}$$

The collapse loading is:

$$\begin{aligned} P_{collapse} &= P_o - P_i = 0.098\{d_{mw} D - d_{sw}(D - D_a)\} \quad \text{for } D > h \\ &= 0.098 d_{mw} D \qquad\qquad\qquad\qquad\qquad\quad \text{for } D < h \end{aligned} \tag{5.7}$$

From this evaluation, the collapse pressure is constant in the interval (h, D).

2. *Collapse during cementing*. Immediately following cementing, there is a hydrostatic wet cement pressure along the whole surface casing. The external pressure is then the sum of the weight of the sea water to seabed and the cement below in the annulus. This situation is shown in Figure 5.3b.

Let us now for simplicity assume that the outside of the annulus is completely filled with cement, and the inside is filled with displacing mud. The pressures are then:

Outside pressure:

$$P_o = 0.098\{d_{sw}(D_{sb} - D_a) + d_{ce}(D - D_{sb})\}$$

Inside pressure:

$$P_i = 0.098 d_{mw} D$$

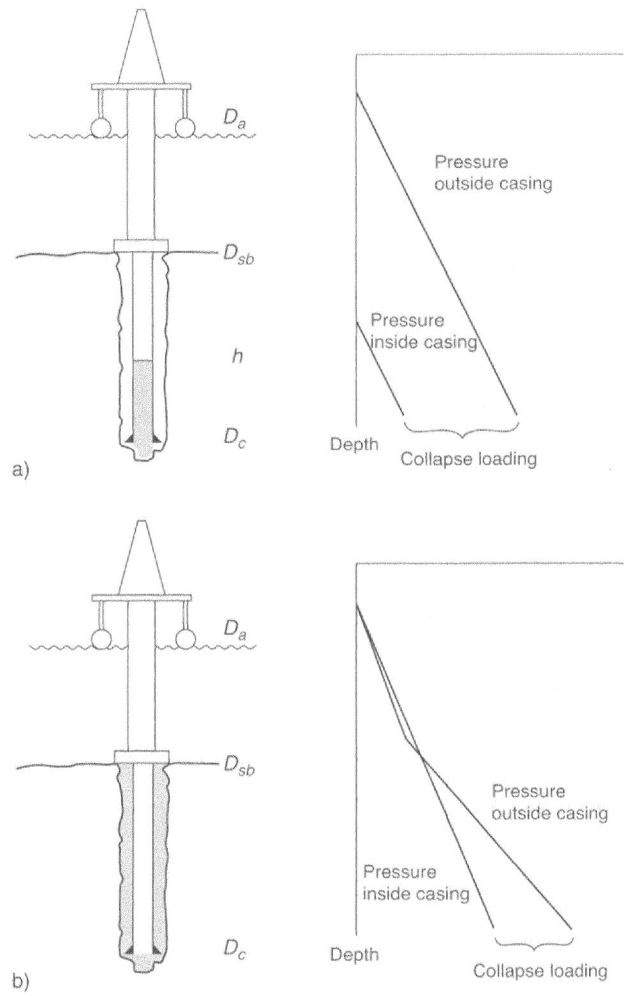

FIGURE 5.3 Two central casing collapse scenarios. a) Collapse loading due to mud losses to a thief zone and b) collapse loading due to cementing.

Collapse loading:

$$P_{collapse} = P_o - P_i = 0.098\{(d_{sw} - d_{ce})D_{sb} + (d_{ce} - d_{mw})D - d_{sw}D_a\} \tag{5.8}$$

The maximum collapse load will typically arise at the casing shoe for this example. Usually this criterion applies for casing strings that are cemented all the way to the seabed, like the conductor casing and the surface casing.

5.1.3 Tension Mechanism and Design Criteria

Tension failure occurs when the axial loads exceed the material strength, resulting in a parted pipe or connection. Tension loads will be imposed on the casing at any time by some or all of the following:

a. Dynamic forces or shock loads. While lowering the casing through tight spots or dog-legs, stick-slip effects on the casing string can induce tensional shock loads. Abrupt changes in the running speed will also induce similar loads. Such loads may be excessive on floating rigs when the casing string is supported on the drill floor when the rig heaves.
b. Movements to free a differentially stuck pipe may induce considerable tensional loads.
c. When pressure testing, high internal pressure will induce tensional stress loads in the casing string, as will be the case when pressure testing the casing after a cementing job, before the cement sets.
d. The static weight of the casing string.
e. Buoyancy, that reduces the effective weight of the string, reduces tension.
f. Bending loads due to dog-leg severities.
g. Drag forces.

The maximum load during installation of a casing string should be evaluated. This must also take into account drag forces if the casing will temporarily be pulled. The following defines typical forces considered:

- Weight in air minus buoyancy force, plus bending forces, plus drag forces, plus load from pressure testing; or
- Weight in air minus buoyancy force, plus bending force, plus drag force, plus shock load.

5.1.4 Temperature Effects

The strength of the casing decreases as the temperature increases. Figure 5.4 shows a curve for a C 110 material as supplied by the manufacturer. For shallower wells with a maximum temperature in the order of 80–100°C, no correction is usually applied. In deeper, high-pressure wells, the margins are often small, and a temperature strength reduction is called for. For these wells a strength-depth curve is made, showing the local strength at each depth interval. To obtain this, temperature curves are used, one showing the static well temperature, and one higher curve showing the temperature profile during testing and production.

Another problem caused by temperature change is pressure build-up in closed annuli. This has been discussed in Section 4.3: Completion and production requirements.

5.1.5 Biaxial Loading

There are three main stresses acting on the casing string, the axial load, the radial load, and the tangential load (the hoop load). We have studied each of these in the foregoing. However, we also know that all of these are present simultaneously in a casing string. In this chapter we will briefly investigate the coupled behaviour. We know that real materials often yield before failure. For example, a casing subjected to external loading may start to yield on the inner surface before it collapses.

The classical Hencky-von Mises maximum distortion energy theory says that a critical yield limit exists in the casing, regardless of direction. Mathematically it can be expressed as:

$$(\sigma_1 - \sigma_2)^2 + (\sigma_1 - \sigma_3)^2 + (\sigma_2 - \sigma_3)^2 = 2\sigma_{yield}^2 \tag{5.9}$$

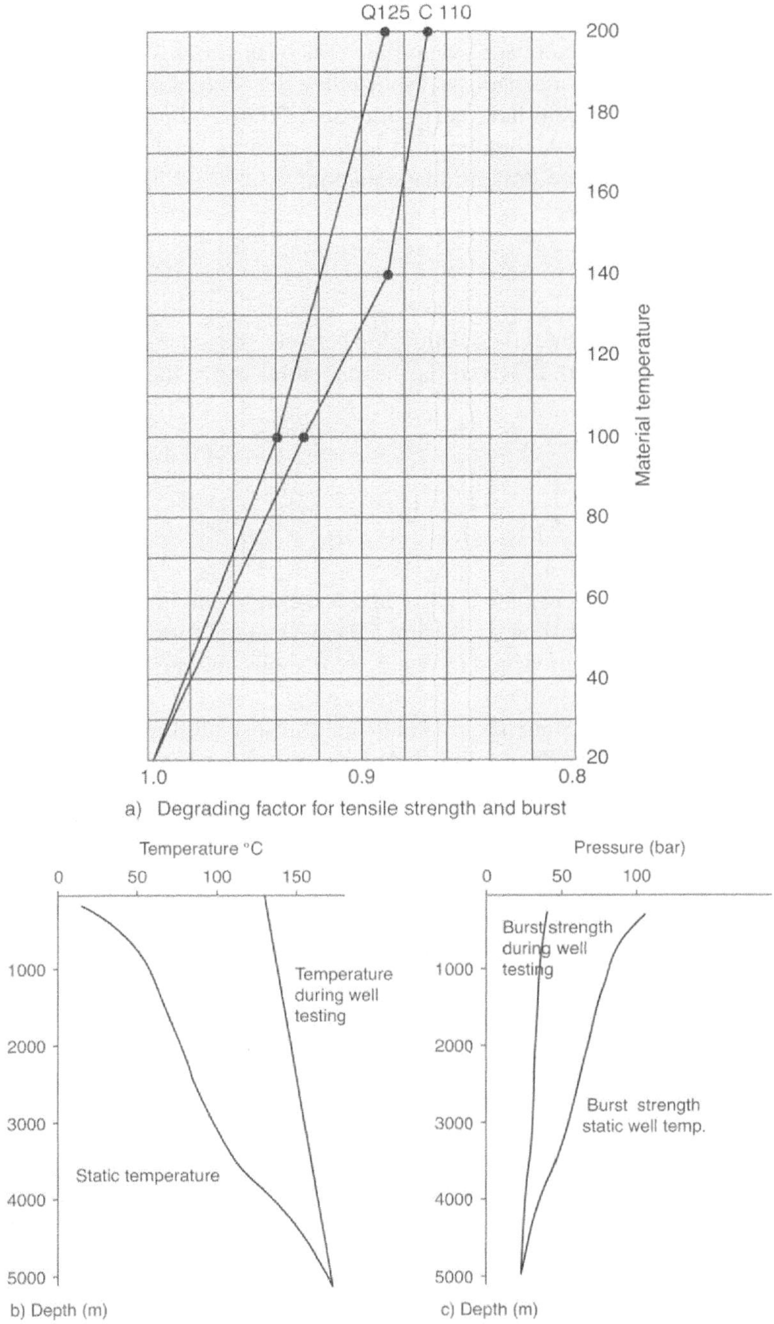

FIGURE 5.4 a) Elements of strength reduction due to well temperature. b) Static temperature profile and design temperature for well testing. c) Resulting burst strength plot as a function of depth.

Here the three subscripts refer to the axial, tangential, and radial stress components. The equation above simply says that the material will start to yield when the shear, or the deviatoric loading of the material has reached a certain value. The casing material has been uniaxially tested, and has a yield strength $\sigma = \sigma_{yield}$.

Looking back on the earlier definitions, we remember that the tensional strength depends on the axial tension, while both the burst and the collapse were functions of the hoop stress. In other words, the hoop and axial stresses govern all failure mechanisms evaluated so far. The radial stress is of course a loading factor, but no radial failure mechanisms have been addressed. From a failure point of view we can neglect radial failure. Usually the radial stress component is neglected in the analysis, as it has little effect. This results in the following simplified two-dimensional equation:

$$\sigma_t^2 + 2\sigma_t\sigma_a + \sigma_a^2 = \sigma_{yield}^2 \tag{5.10}$$

This is an elliptic equation which defines the relationship between the tangential and axial stress conditions and the yield strength of the material. If solved for the tangential stress, the equation results in:

$$\frac{\sigma_t}{\sigma_{yield}} = \frac{1}{2}\frac{\sigma_a}{\sigma_{yield}} \pm \sqrt{1 - \frac{3}{4}\left(\frac{\sigma_a}{\sigma_{yield}}\right)^2} \tag{5.11}$$

The above equation is now in a dimensionless form, and one ellipse can in a generalised form be used for all loading conditions. This ellipse is shown in Figure 5.5a in a general form. The quadrant of most interest in our well design applications is reproduced in Fig. 5.5b. It shows the reduction in collapse strength as a function of axial tension.

It must be emphasised that in casing design practice, the ellipse of plasticity cannot be applied unless the assumption of a yield-strength mode of failure is known to be valid. If a very high-grade casing is used, the string may fail in a brittle manner, before any significant yield takes place.

An example of using the biaxial ellipse follows:

A 18-5/8 in., X-70, 84.5 lbs/ft casing has a collapse resistance of 43 bar and a yield strength of 800 × 10^3 daN. What will the collapse strength be if the casing is loaded axially to 400 × 10^3 daN?

The ratio between the axial load and the yield strength is 400 daN/800 daN = 0.5.

Entering Figure 5.5b, we read a scaling factor for collapse of 0.65. The reduced collapse resistance under this axial load is therefore 0.65 × 43 bar = 28 bar.

5.1.6 Sour Service

The topic of corrosion is a complex issue. Here we will only briefly discuss two aspects of sour service, that is, presence of corrosive gases in the borehole. These are:

- corrosion, that lead to long-term material failure, and
- hydrogen embrittlement, that may lead to failure on a short-term basis.

If a well is planned for production over a 20-year period, the production casing should last the whole period. Typically a production packer is set just above the reservoir, isolating the annulus above. The fluid in the annulus above the packer is usually non-corrosive. The casing below the packer, on the other hand, is exposed to reservoir fluids and therefore subjected to corrosion. If corrosive elements are expected, typically the bottom part of the production casing is made of a stainless steel quality, and often with an increased wall thickness as a corrosion allowance. There are no simple methods to design the production casing for corrosive environments. Judgement and experience often form the basis for materials selection.

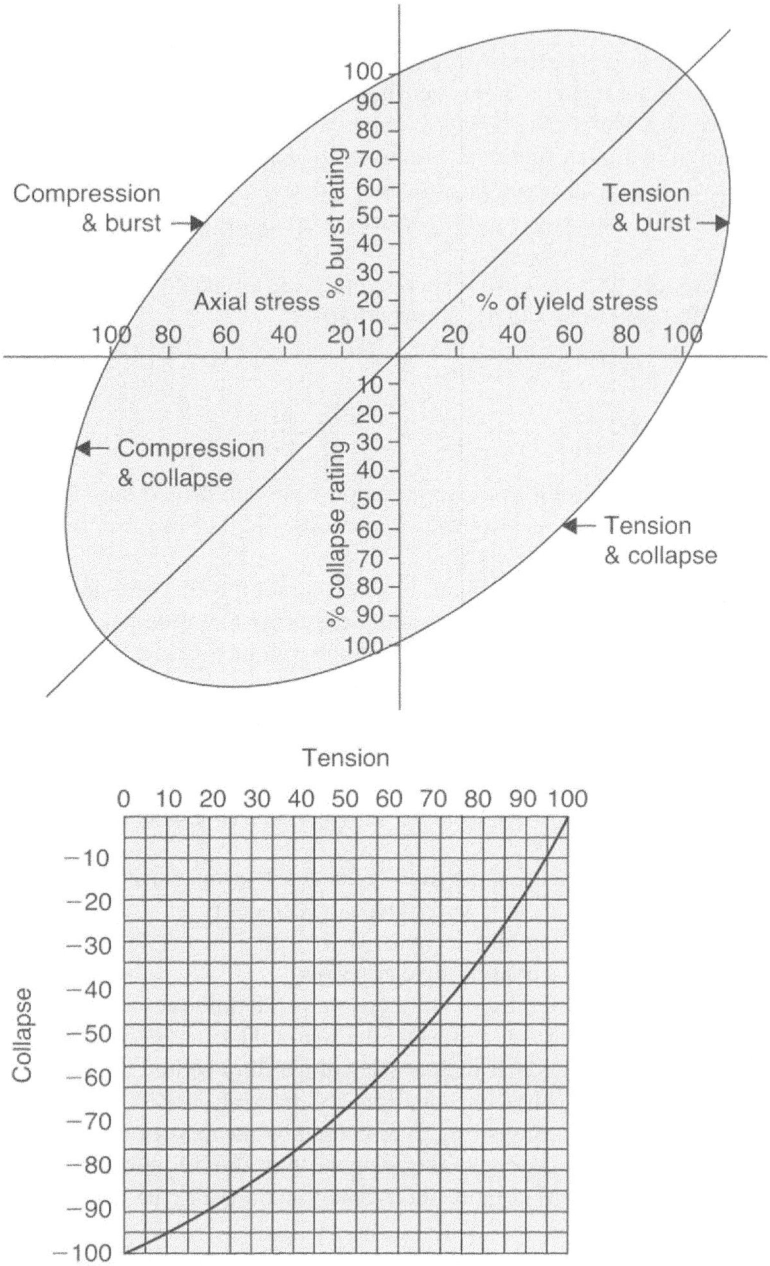

FIGURE 5.5 Ellipse of plasticity.

However, the correct choice of casing is important, because a premature failure of the production casing is very costly to repair.

The other element of sour service discussed is hydrogen embrittlement. This is an important element during drilling of the well. Sour gases are natural gases with a defined hydrogen sulphide content (H_2S). Dry sour gases are normally not corrosive, but as soon as hydrogen sulphide or carbon dioxide gas is introduced into water, its pH value drops, which means that the water becomes acidic or sour. When exposed to aqueous solutions, carbon and low-alloy steels undergo weight loss corrosion. Additional damage in the form of crack formation in the material microstructure can only occur in hydrogen sulphide water.

In general, soft steel casing is not as susceptible to hydrogen embrittlement because of its ductility. High grade casing, on the other hand, is highly sensitive to embrittlement, and exposure to hydrogen sulphide water must be avoided. Wells have been lost due to total failure of the casing.

Very little presence of hydrogen sulphide is required to generate potential for embrittlement. In practical well applications, as little as 50 ppm is sufficient. However, this problem is usually present at low temperatures. Above 80–100°C, hydrogen embrittlement is usually discarded because the casing becomes more ductile.

Some selected references relating to sour service are given by Ikedo (1992), Marshall et al. (1994), Monrose and Boyer (1992), and Stair and McInturff (1986).

5.1.7 Time Scenario

We have in general two different well types. Exploration wells are drilled and abandoned within a few months, while production wells may be designed for a lifetime of 20 years or more. The time frame of the wells should be addressed in the design criteria.

One aspect of time is the corrosion problem. In an exploration well, corrosion is normally not a concern beyond the sour service requirements discussed in Section 5.1.6. A production well, on the other hand, may be exposed to corrosive fluids over decades. In order to minimise costly workover operations, proper corrosion allowance should be taken into account. However, remember the functionality requirements of each casing string. The conductor casing is usually giving support to the wellhead equipment. The production casing provides the pressure integrity throughout the life of the well. The surface and intermediate casing strings have a more temporary function, namely to provide integrity while drilling the well. Corrosion allowance is usually included in the production casing design, but often not for the other casing strings.

Another element of time is the state of the fluid behind each casing string. For burst design, the fluid density may be a critical element for certain wells. Once the cement is set, the only mobile phase is water. Mechanical support is usually not taken into account, as any deficiency inside the cemented interval may be a weak spot. We recognise this as a conservative criterion.

Exploration wells have a short time frame. In these wells the fluid pressure behind the casing strings consists of drilling mud above the cement, and water density in the cemented interval.

In production wells, it is expected that the particles in the fluid behind the casing strings will settle, leaving a water phase on top. Typically one neglects mechanical support from settled particles, and considers the pressure behind each casing string to be caused by the density of the water phase. Since the production casing has the long-term integrity requirement, this should be designed taking settling into account. The other casing strings, with a more short-time function, may be designed using the full mud density.

From the discussion above we understand that the design premises are an important input. In optimisation one should always re-evaluate these conditions. If changes can be justified, these may become important for the casing design optimisation process.

5.1.8 Casing Wear

Casing wear may become important in some cases. One example is an HPHT well where wear may reduce the burst strength of the production casing. It is difficult to predict wear, but it is often related to the number of hours the drill string is being rotated inside the casing. For further discussion about casing wear, the reader is referred to publications by White & Dawson (1987) and Schoemaker (1987).

TABLE 5.1 Common design criteria for exploration wells

CASING	DESIGN CRITERIA: BURST OUTSIDE	 INSIDE	DESIGN CRITERIA: COLLAPSE OUTSIDE	 INSIDE	DESIGN CRITERIA: TENSION
Conductor	None*	None*	None*	None*	None*
Surface	Sea water	Fracture pressure at casing shoe	Hydrostatic sea water plus mud, wet cement	Thief zone drains mud to balance sea water gradient, mud	Buoyed weight, running resistance, testing
Intermediate	Sea water column	Reservoir pressure, next hole section	Mud column	Thief zone drains mud to balance sea water gradient	Buoyed weight, running resistance, testing
Production casing without liner	Sea water column	Leak in test string or gas filled casing	Mud column or cushion fluid	Thief zone or well flow	Buoyed weight, running resistance, testing
Production casing with liner	Sea water column, one string*	Leak in test string or gas filled casing, one string**	Mud column or cushion fluid, one string**	Thief zone or well flow, one string**	Buoyed weight, running resistance, testing

*Unless particular conditions arise. **Production casing and liner should be regarded as one string.

5.1.9 Summary of Design Criteria

The different casing strings have different functions as discussed previously. The conductor casing has the function of supporting the wellhead. The other casing strings have a more temporary function, except for the production casing, which must ensure the integrity of the well over its lifetime. In other words, the intermediate strings have no functionality once the next string is set.

In the following will we list the most common design criteria for well design. Of course, these must be changed if specific conditions arise.

Problem

During drilling of a production well a severe dog-leg occurred about 200 m above the shoe when a casing was set. A combination of high load due to the well profile, and extensive drilling in the following openhole section resulted in considerable casing wear. A wear log was run, and it indicated a reduction in wall thickness of 15% of the casing. Because this is a production casing you are requested to evaluate the remaining strength of the casing. Casing data are:

9-5/8 in. P110	61.1 lbs/ft
Burst strength	862 bar
Collapse resistance	747 bar

Assume that the collapse strength reduction is approximately to the third power of the remaining wall thickness.

a) Compute the burst and the collapse strength of the casing at the position of excessive wear.
b) Discuss the consequences of the reduced collapse strength.
c) Discuss the consequences of the reduced burst strength.
d) Propose adjustments to the completion design to compensate for the reduced casing strength.

5.2 CASING TEST PRESSURE

5.2.1 General Test Requirements

The casing must be tested after it is cemented. There are two purposes for this, namely the casing itself must be tested to withstand the design pressure, while the quality of the cement on the outside and the formation strength is tested with a so-called leak-off test.

The casing integrity can be tested either after the cement is set, or during cementing when the cement plug lands in the float collar. The latter is called bumping of plug. According to regulations (NPD, 1991) the test pressure should:

- Be equal to the maximum burst pressure defined in the test scenario.
- Not exceed 85% of the internal yield pressure of the casing string.

In the following, we will briefly outline the test pressure design process. Since we have different fluids in the well during testing compared to when a well control situation arises, we also need to perform a test design. We will investigate two common scenarios to illustrate the process.

5.2.1.1 Test Pressure of Surface Casing

The surface casing is typically installed before the marine riser is installed. In the following we will determine the surface test pressures for two cases, during cementing and after the cement is set.

If the cement plug is landed (bumped) during cement placement, the casing can be tested immediately. The advantages are saving time, and obtaining a better test since the cement is not set, creating a well-defined hydrostatic pressure profile. However, the large diameter surface casing gets a high tensional load during pressure testing. Therefore, testing is often delayed until the cement is set.

Figure 5.6a illustrates the situation when testing a surface casing from a floating rig. During plug bumping, the external hydrostatic pressure is:

$$P_o = 0.098\left\{d_{sw}\left(D_{sb} - D_a\right) + d_{ce}\left(D - D_{sb}\right)\right\}$$

The inside pressure is:

$$P_i = 0.098 d_{mw} D$$

If a test pressure is applied inside the casing, the burst loading is:

$$P_{burst} = P_t + P_i - P_o = P_t + 0.098\left\{\left(d_{mw} - d_{ce}\right)D + \left(d_{ce} - d_{sw}\right)D_{sb} + d_{sw}D_a\right\}$$

Often, sea water is used as a displacing fluid, reducing the above equation to:

$$P_{burst} = P_t + 0.098\left\{\left(D - D_{sb}\right)\left(d_{sw} - d_{ce}\right) + d_{sw}D_a\right\}$$

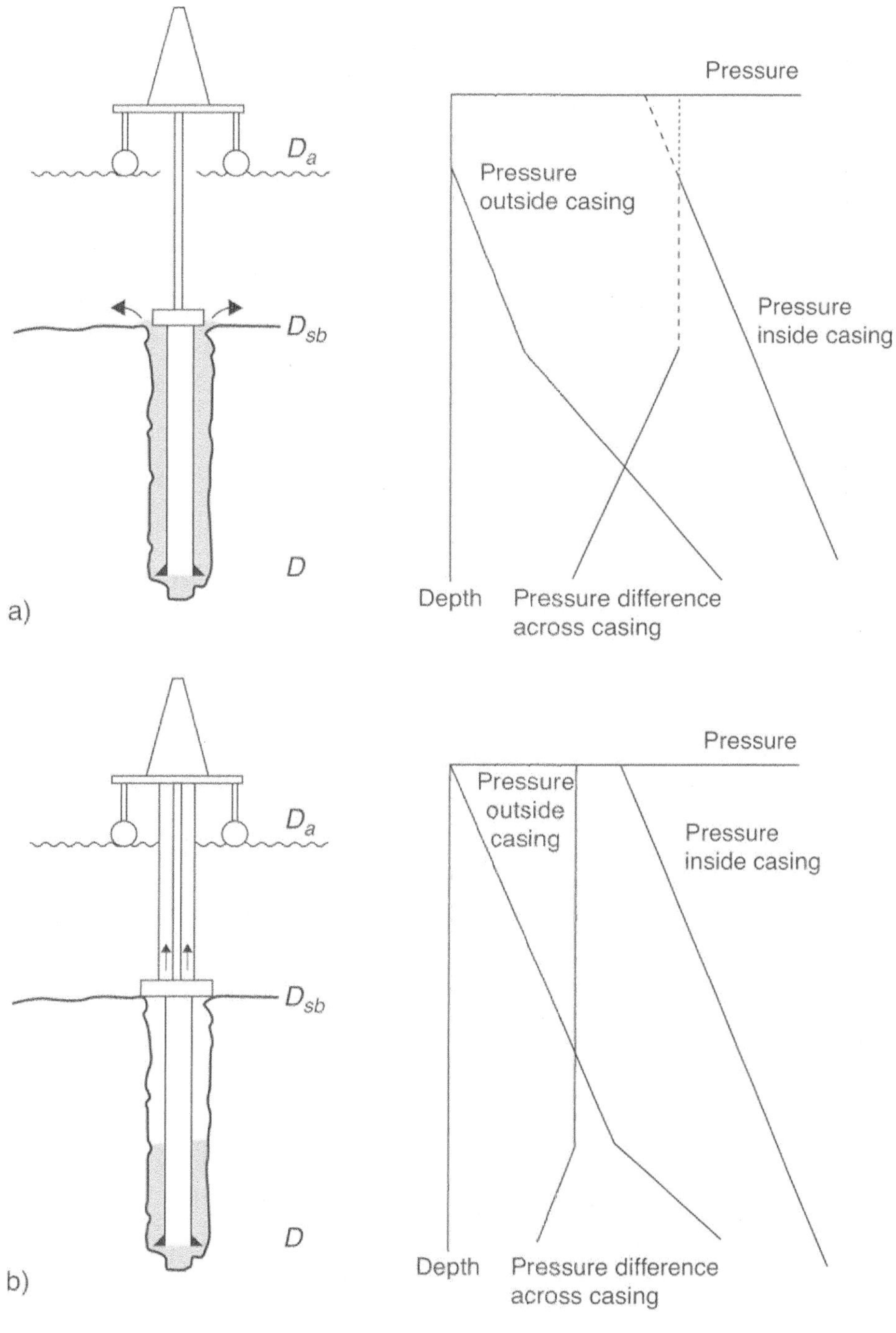

FIGURE 5.6 Test scenarios for surface casing and deeper casing strings.

Inspection of this equation reveals that the test pressure is first reached under the wellhead, resulting in a test pressure of:

$$P_t = P_{burst} - 0.098 d_{sw} D_a \tag{5.12}$$

Now we will investigate the test pressure after the cement is set. The surface casing is typically cemented along its whole length. After the cement is set, its weight is no longer providing hydrostatic pressure. However, some water exists in the cement, and possibly also in voids and cracks around the wellbore. We may therefore imagine that a saltwater pressure is the only mobile component that may transmit pressure after the cement is set. The pressures then become:
Outside pressure:

$$P_o = 0.098 d_{sw} (D - D_a)$$

Inside pressure:

$$P_i = 0.098 d_{mw} D$$

If a test pressure is applied inside the casing, the burst loading is:

$$P_{burst} = P_t + P_i - P_o = P_t + 0.098\left\{(d_{mw} - d_{sw}) D + d_{sw} D_a\right\} \tag{5.13}$$

Inspection of this equation reveals that for this case, the burst pressure is first reached at the casing shoe. Again, if sea water is used as the displacing fluid, the test pressure becomes identical to that of the bumping plug case. The only difference is that the test pressure is first reached in the top of the casing when bumping the plug, but at the bottom of the casing when testing after the cement is set.

5.2.1.2 Test Pressure of Deeper Casings

The deeper casing strings are set through a marine riser, as shown in Fig. 5.6b. They are not cemented to the wellhead, but have a drilling fluid or a completion fluid above the top of the cement. In this demonstration we assume that the mud outside and inside the casing string have the same densities.

When bumping the plug, the following pressures arise:
Outside pressure:

$$P_o = 0.098\left\{d_{mw} D_{ce} + d_{ce} (D - D_{ce})\right\} \quad \text{in the cemented interval}$$
$$P_o = 0.098 d_{mw} D \quad \text{above the cemented interval}$$

Inside pressure:

$$P_i = 0.098 d_{mw} D$$

If a test pressure is applied inside the casing, the burst loading is:

$$P_{burst} = P_t + P_i - P_o = P_t + 0.098\{(d_{ce} - d_{mw})(D_{ce} - D) \quad \text{in the cemented interval} \tag{5.14}$$

$$P_{burst} = P_t \quad \text{above the cemented interval} \tag{5.15}$$

If the cement is heavier than the mud, the critical burst pressure will first be reached at the wellhead, which gives a surface test pressure of:

$$P_{burst} = P_t \tag{5.16}$$

Finally, we will investigate the pressure testing of deeper casing strings after the cement is set. In this case we assume drilling mud over the cement, and a hydrostatic sea water pressure as the only mobile phase inside the cement itself. The pressures are:
Outside pressure:

$$P_o = 0.098\left\{d_{mw}D_{ce} + d_{sw}\left(D - D_{ce}\right)\right\} \quad \text{in the cemented interval}$$
$$P_o = 0.098 d_{mw} D \quad \text{above the cemented interval}$$

Inside pressure:

$$P_i = 0.098 d_{mw} D$$

If a test pressure is applied inside the casing, the burst loading is:

$$P_{burst} = P_t + P_i - P_o = P_t + 0.098\{(d_{mw} - d_{sw})(D - D_{ce}) \quad \text{in the cemented interval} \tag{5.17}$$

$$P_{burst} = P_t \quad \text{at the wellhead} \tag{5.18}$$

We observe that for deeper casing strings the design test pressure was identical for the bumping plug and cement set cases, and both tests resulted in the test pressure being reached at the wellhead.

This chapter shows the methodology when designing test pressures. We observe that if different conditions exist, the test pressure must be redesigned. Examples are fixed platforms and cement programs involving several cement densities.

Another aspect is shown with the examples above. The kick scenario in the casing design results typically in the top or the bottom of the string becoming the most loaded position. In this chapter we have calculated the test pressure without comparison to the design scenario. We will find in some well designs, that the casing is not fully tested at its most severe position according to the design. This seems to be accepted practice in the industry. In the next chapter we will address this further.

5.2.2 Special Pressure Tests for Critical Wells

It is clear that the casing will burst when the pressure exceeds the strength at any depth. In the previous discussion we calculated a test pressure, but did not compare this to the casing design scenario. In this chapter we will enlarge on this by designing a test pressure that simultaneously tests the whole casing string to its design pressure.

The problem is the different fluids in the well during testing and during an eventual kick situation. For critical wells we may require a fully tested casing. The following example demonstrates this process.

Example

In an exploration well the 9-7/8 × 9-5/8 in production casing was considered very critical since it had to handle the drilling of a high-pressure reservoir below. Also, this casing is considered the most critical issue in high-pressure (HPHT) wells. It was decided to test it above the design pressure along the whole length of the casing. To obtain this, the following scenarios were investigated:

1. To bump the cement plug during cementing. Since this is a 5000 m deep well, the differential pressure across the cement plug would be in the order of 700 bars. No supplier could guarantee that the plug would withstand such a pressure, so this concept was abandoned.
2. The design assumed a saltwater pressure across the cemented interval. Although many people claim that this is a very conservative criterion, there were no rational arguments to deviate from this assumption, which therefore was maintained.
3. A packer could be set in the middle of the well, and the over- and underside tested in two sequences. The operational staff were afraid of damaging the casing in this process, so it was abandoned.
4. A back-pressure could be established behind the casing. This is a real possibility, but one would have to compare the pressures to the strength of the formations exposed behind the casing. It was not performed because a simpler solution came up.
5. During testing, the upper half of the casing was evacuated to sea water, while the bottom part contained mud. This is the solution that was chosen.

The data are as follows:

Casing:	9-7/8 in SMC 110, 66.4 lbs/ft in the interval 0–2400 m
	Burst strength: 889 bar
	9-5/8 in. Q 125, 53.5 lbs/ft in the interval 2400–4730 m
	Burst strength: 854 bar

Reservoir pressure gradient: 2.11 s.g. at 5042 m

Formation fluid density: 0.547 s.g.

Maximum reservoir pressure at 5042 m: 0.098 × 2.11 × 5042 = 1043 bar

Pressure of column of reservoir fluid in well: 0.098 × 0.547 × 5042 =270 bar

Wellhead pressure if gas filled casing: 1043–270 = 773 bar

These are the design pressures that the casing will be tested for, and we will in the following calculate the test pressures.

We assume that there is no pressure behind the casing just below the wellhead. At 3692 m the transition between mud and cement is giving a pressure of:

$$0.098 \times 1.9 \times 3692 = 687 \text{ bar}$$

From 3692 to 4700 there is cement. Assuming that the effective pressure in this interval is equal to sea water weight after the cement is set, this pressure amounts to:

$$0.098 \times 1.03(4700 - 3692) = 102\text{bar}$$

The pressure on the outside of the casing at the shoe is then: 687 + 102 = 789 bar. Figure 5.7a shows these pressure profiles. Also, the difference between the inside and the outside pressure is shown over the well depth.

This difference curve is the burst design pressure for the casing string. The most severe loaded position is just below the wellhead. The loading decreases with depth, because of the increased outside pressure.

Figure 5.7b shows the burst strength for the casing strings in question. The right-hand curve shows the burst strength given by the manufacturer. Because of increased temperatures with depth, this curve is temperature-corrected for the static well temperature. Finally, the solid line shows the 85% level of this temperature-corrected burst strength curve, which must not be exceeded according to regulations (NPD, 1991). This is the design strength curve we will use.

Next, assume that we perform a pressure test with a 1.9 s.g. drilling mud, and apply a surface pressure equal to the surface design pressure of 773 bar. The pressure inside the casing at the shoe is:

$$773 + 0.098 \times 1.9 \times 4700 = 773 + 875 = 1648 \text{ bar.}$$

We subtract the external pressure and obtain the test load curve as shown in Figure 5.8a. Two observations are clear: 1) the design strength exceeds the design load curve throughout the well, so the casing is sufficiently strong; and 2) the test load exceeds the design strength throughout the interval, so the test cannot be performed.

The task at hand is to design a test which exceeds the design curve in the whole well, but which does not exceed the design strength of the casing. This will be performed in the following.

From Figure 5.8a we observe that the largest difference between test pressure and design strength is at the casing shoe. Actually here the difference is 200 bar. Now we want to pump in some saltwater in the top of the well to reduce the bottomhole pressure with 200 bar. The minimum depth of this saltwater pill is:

$$200\,\text{bar} = 0.098(1.9 - 1.03)\text{h} \quad \text{or :} \quad \text{h} = 2346\,\text{m}$$

For design purposes we decide to set the saltwater pill to 2500 m, and the 1.9 s.g. drilling mud from 2500 m to 4700 m. The inside pressure at 2500 m is:

$$773 + 0.098 \times 1.03 \times 2500 = 1025 \text{ bar}$$

and the inside pressure at 4700 m is:

$$1025 + 0.098 \times 1.9(4700 - 2500) = 1435\,\text{bar}$$

Again subtracting the outside pressure from the inside pressure, the test design curve is shown in Figure 5.8b. We observe that the test curve falls between the design curve and the strength curve over the whole depth range. This will therefore test the casing fully.

Before leaving this subject one must be aware that by evacuating the upper part of the production casing to sea water, the collapse load on the casing is increased. At 2500 m depth, a pressure difference exists:

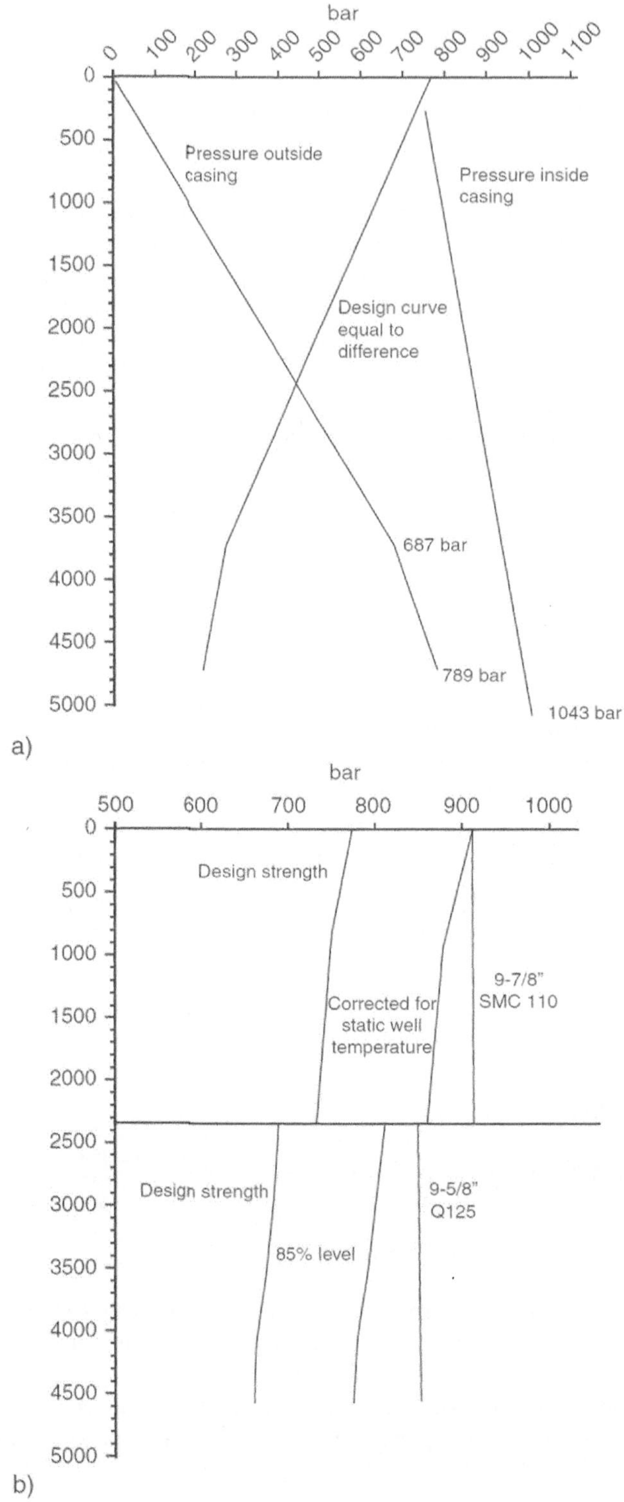

FIGURE 5.7 Defining design load and design casing strength. a) Design pressure for gas filled casing. b) Allowed burst loading, temperature corrected, and reduced to 85% level.

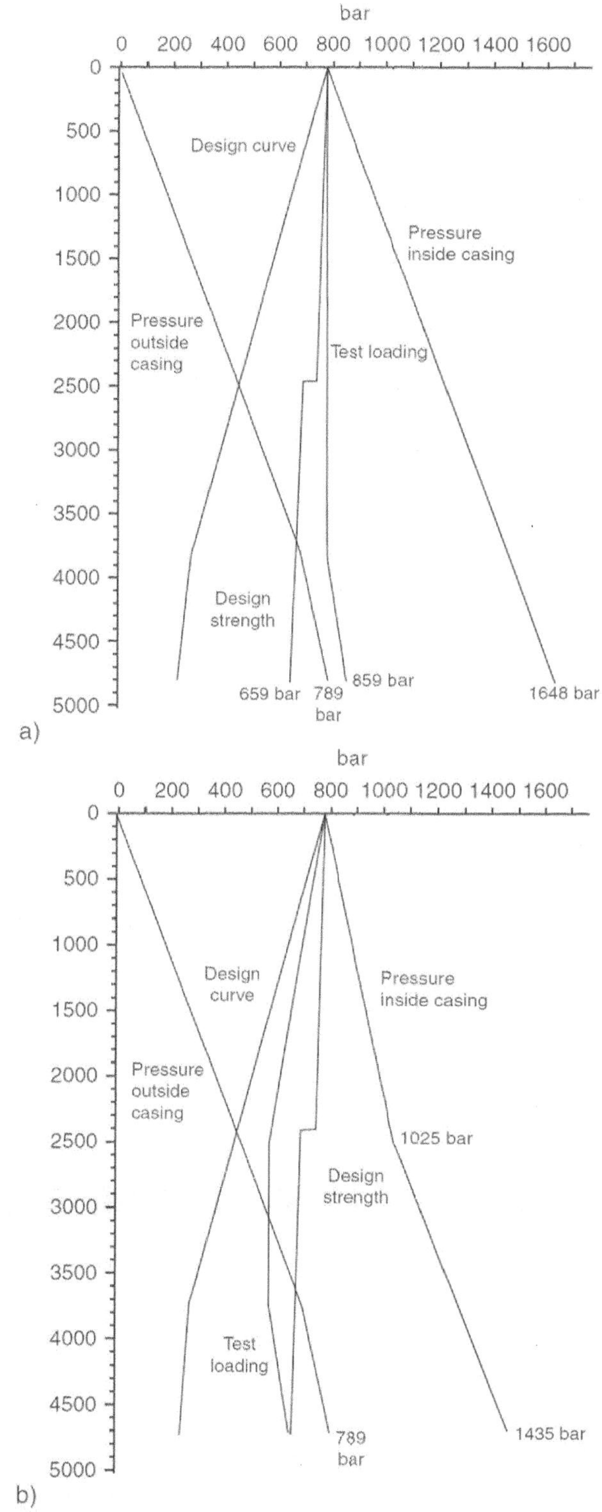

FIGURE 5.8 Test pressures with mud and sea water/mud inside casing. a) Test pressure with 1.9 s.g. mud. b) Test pressure with sea water to 2500 m and 1.9 sg mud from 2500 m to 4700 m.

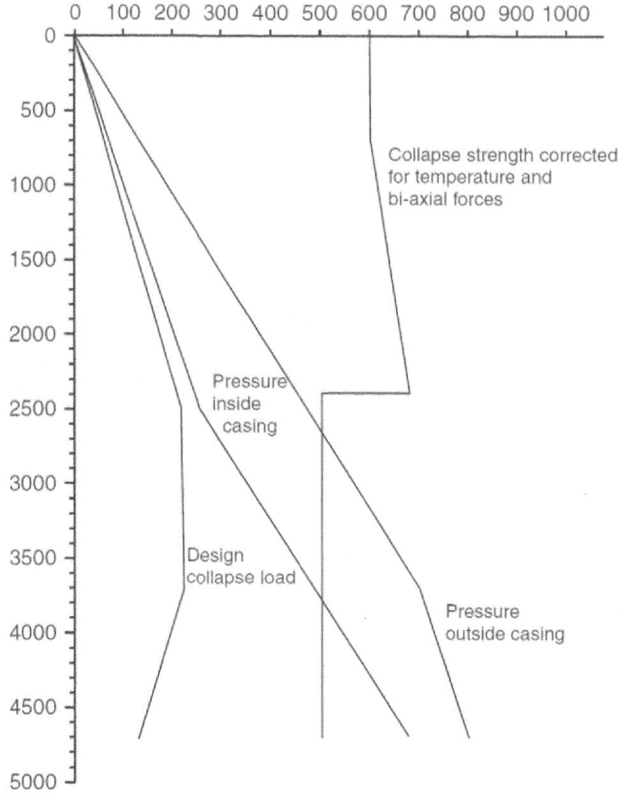

FIGURE 5.9 Evaluation of collapse loading during test.

$$0.098(1.9-1.03)2500 = 213 \text{ bar}$$

The collapse loading is shown in Figure 5.9. Also shown is the temperature-adjusted collapse strength curve. At 2500 m the collapse strength-collapse load ratio is 499 bar/213 bar = 2.34. There is sufficient margin to perform the pressure test.

5.3 CASING DESIGN EXAMPLE

5.3.1 Summary

The example presented is for a subsea well: see Figure 7.1.

The enclosed casing design for a small field results in the following design:

- The 30 in. conductors will be batch set to 300 m.
- The 18-5/8 in., X-70, 84.5 lbs/ft. casing will be set in the Hordaland group at approximately 1100 m depth.
- The 13-3/8 in., P-110, 72 lbs/ft. casing will be set into the Lista/Sele formation at approximately 1820 m depth.

- The 9-5/8 in., L-80, 47 lbs/ft. casing will be set through the reservoir in the production wells, and above top reservoir in the injection wells. This is at approximately 2365 m depth for both cases.
- A 7 in. L-80, 32 lbs/ft. liner will be set in the injection wells at 2625 m depth. The liner is also a contingency option for the production wells if problems arise during drilling.

5.3.2 Selection of Casing Setting Depths

The methodology of casing setting depth evaluation is described elsewhere in this book. Here we will present the results, and some of the arguments.

Figure 5.10 shows the setting depths selected for the field. There are two designs, one for production wells and one for injectors. Furthermore, Figure 5.11 shows the pressure prognosis for the field. Since all wells are drilled from one template location, the top layers are nearly identical. However, at reservoir level, the formation tops are located at different depths. This is illustrated with a range in the formation types. The base design uses the deepest depths shown.

Specific points in selection of casing setting depths are as follows:

30 in. casing. The main functional requirements for the conductor casing are covered by a 75-m-long string.

18-5/8 in. casing. The purpose of the surface casing is to seal off the Utsira formation (Figure 5.11), which is a potential lost circulation formation, to provide sufficient integrity for the next hole section. The leak-off value below the surface casing should yield a value exceeding 1.57 s.g., while

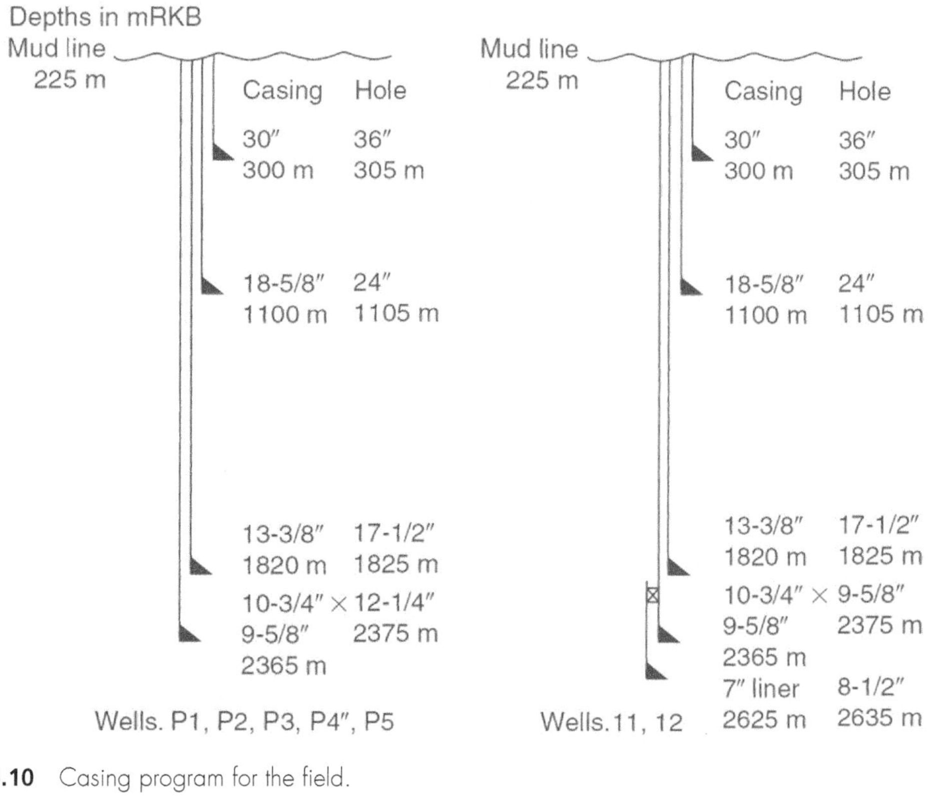

FIGURE 5.10 Casing program for the field.

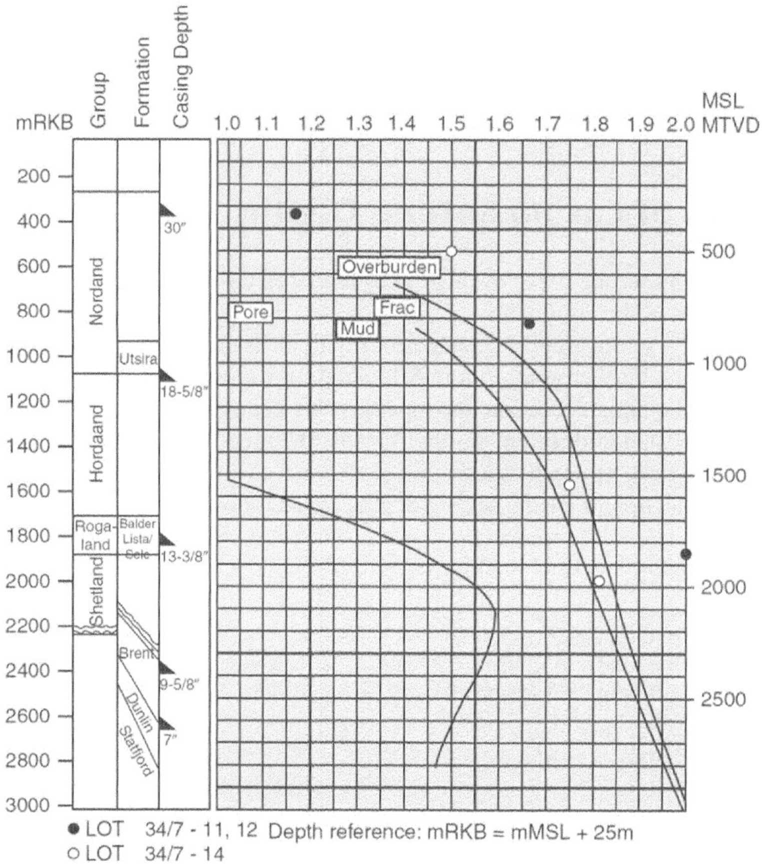

FIGURE 5.11 Geology and pressure gradient plot.

the maximum mud weight required in the next hole section will be 1.50 s.g. The design setting depth is 1100 m.

13-3/8 in. casing. The purpose of the intermediate casing is:

- to seal off the Hordaland group which is normally pressurised but contains reactive clays.
- to seal off weak sand stringers and the potentially weak Balder formation.
- to provide sufficient shoe integrity for the drilling of the 12-1/4 in. section.

The selection of the intermediate casing setting depth is not critical with respect to casing design loads. At the recommended setting depth at 1820 m, the prognosed leak-off value is 1.77 s.g.

9-5/8 in. casing. Current planning is to drill all reservoir sections in the production wells with 12-1/4 in. holes, and to install a 9-5/8 in. casing. In the injection wells, this casing string will be set above the reservoir, and a 7 in. liner set through the reservoir. The 9-5/8 in. casing is intended to isolate the Shetland and Cromer Knoll formations prior to penetrating the reservoir for wells with a liner. For all wells, the 9-5/8 in. casing will isolate the Shetland formation. The design setting depth is 2365 m.

Special casings are required below the production packer. In the two injection wells, L80 grade will be sufficient, while in the production wells 13 Cr. casings will be installed. There is a corrosion allowance added for all wells.

7 in. liner. The liner is intended to seal off the reservoir sections in the deeper injection wells. The design setting depth is 2625 m. Also, the liner will serve as a contingency string if hole problems arise during drilling of the production well.

5.3.3 Design Basis

30 in. casing. Collapse, burst and compression, and bending forces are usually not a problem for the conductor casing. Therefore, no design calculations will be performed.

18-5/8 in. casing. The collapse design criteria are as follows:

Collapse during cementation is evaluated by assuming cement along the outside and sea water inside. The wet cement consists of 100 m of 1.90 s.g. tail cement slurry, with the remaining annulus to seabed filled with 1.45 s.g. lead cement slurry. Sea water is assumed to be the displacing fluid.

Fluid level drop assuming a thief zone, after the casing string is landed, is also checked. The inside fluid level drops until the bottomhole pressure is equivalent to a sea water column. The fluid outside is mud up to seabed. Since both collapse scenarios take place prior to drilling out, the collapse calculations have been adjusted for biaxial stress, but not for wear, because this will occur in the following drilling.

The burst design criteria are as follows:

It is assumed that the maximum possible internal pressure is equivalent to the maximum pore pressure in the next open hole section. Although there are no reservoir zones in the following 16 in. section, an oil gradient has been assumed. Since the burst has a post-installation perspective, it is assumed that the cement is set and that the particles in the mud behind the casing are settled. The only mobile phase is water (see production casing scenario, section 5.1.7), and we therefore assume hydrostatic weight of saltwater behind the whole casing string. The burst calculations include an allowance for casing wear.

The tension design criteria are:

The weight of the casing in mud is the key criterion. In addition are bending effects included plus the forces arising when pressure testing the casing. Maximum tensile loading occurs during installation, so the calculations have not been adjusted for later wear.

The 13-3/8 in. casing. Mud loss to a thief zone is considered the worst case scenario for casing collapse. The inside fluid level drops until the bottomhole pressure is equivalent to a sea water column. The fluid outside the casing is mud. Again the calculations allow for biaxial stresses, but not for casing wear since this is an installation scenario.

The burst design is based on an *oil-filled casing* scenario. The calculations are adjusted for later casing wear. The external pressure is assumed to be caused by a sea water gradient, according to the production casing scenario discussed in Section 5.1.7.

The tension design criteria are the weight of the casing in mud, plus the bending, plus the forces created during pressure testing of the casing. Maximum tensile loading is expected to occur during installation, so the calculations have not been adjusted for later wear.

9-5/8 in. casing. The collapse design involves a thief zone scenario as for the previous casing strings. The fluid outside the casing is assumed to be mud. Again the calculations allow for biaxial stresses but not for casing wear since this is an installation scenario.

In addition, *collapse potential due to plugging of perforations* during well flow is checked. During production, the perforations are assumed to be plugged by particles coming from the formation. There will be a pressure difference across the casing. The external pressure is the formation pressure in the reservoir interval. The internal pressure is assumed to be caused by the hydrostatic weight of formation fluids inside the production tubing plus the separator pressure at surface. Since this is a post-drilling scenario, the collapse calculation has been adjusted for corrosion.

The burst design criteria are:

The oil-filled casing scenario assumes that the reservoir pressure reduced by a hydrostatic column of oil creates a maximum burst loading below the wellhead. The external pressure is assumed to be a hydrostatic column of sea water. The burst calculations include corrosion and wear allowance.

The tubing leak criterion assumes a leak in the production tubing just below the wellhead. This creates a maximum burst loading at the production packer. There is assumed to be oil in the tubing and a 1.10 s.g. packer fluid in the annulus above the packer. A sea water gradient is assumed on the outside of the casing, according to the production casing scenario of Section 5.1.7. The calculations include corrosion and wear adjustments.

The tension design criteria are weight in mud, plus bending, plus forces created during pressure testing. Maximum load occurs during installation, so the calculations have not been adjusted for wear and corrosion.

The 7 in. liner. The *collapse design scenario* involves reservoir pressure acting on the liner due to plugged perforations during well flow. The inside of the liner is assumed to be filled with produced oil, which gives a hydrostatic pressure added to the separator pressure. The outside is assumed to be reservoir formation pressure. The collapse calculations have been adjusted for estimated corrosion.

The burst design criterion for the liner below the production packer is based on a *bullheading operation*. The internal pressure is defined by the maximum expected fracturing pressure in this interval. The external pressure is assumed to be a sea water gradient all the way. The calculations are adjusted for estimated corrosion, but not for wear since no drilling will take place after the liner is installed. Since the production packer is planned set in the 9-5/8 in. casing above the liner, leaking tubing criterion is not used for the liner.

The tension design criteria are the weight in mud plus bending. Maximum tensile loading occurs during installation, so corrosion adjustment is not included.

Corrections for wear and corrosion. After the casing strings are installed, continued drilling may result in wear. Long time exposure with various fluids may also result in corrosion damage. A scenario is designed to take wear and corrosion into consideration in the design, resulting in derating of the casing strengths as follows:

- 18-5/8 in. casing, derated in burst strength for 10% wear.
- 13-3/8 in. casing, derated in burst and tension for 10% wear.
- 9-5/8 in. casing above production packer, derated for 10% wear in tension, and in burst.
- 9-5/8 in. casing below production packer and 7in. liner, derated for 10% corrosion in burst and collapse.
- 7 in. liner in injectors, derated for 10% corrosion in burst and collapse.

5.3.4 Casing Design

In this section we will perform design calculations for each of the casing strings of the field case. The calculations for each string will be performed separately.

The 18-5/8 in. Surface Casing

The design parameters are as follows:

Depth of casing:	1100 m
Depth to seabed:	225 m
Depth to sea level:	25 m
Depth to top of tail cement:	1000 m
Depth to top lead cement:	225 m

Depth next open hole section:	1820 m
Design fracture gradient at casing shoe:	1.57 s.g.
Pore pressure gradient, casing shoe:	1.03 s.g.
Pore pressure gradient, next section:	1.40 s.g.
Formation fluid density:	0.76 s.g.
Mud density:	1.20 s.g.
Mud density, next open hole section:	1.50 s.g.
Lead cement density:	1.45 s.g.
Tail cement density:	1.90 s.g.

The casing data are:

18-5/8 in. grade X-70, 84.5 lb/ft. casing	
Weight:	186 kg/m
Cross-sectional inner area:	1527 cm^2
Burst strength:	197 bar
Collapse resistance:	43 bar
Pipe body yield strength:	800×10^3 daN

Note that both the 30 in. and the 18-5/8 in. hole sections are drilled without riser. The BOP and the marine riser are installed after the 18-5/8 in. casing is installed.

First, we will calculate the *collapse loading during cementation*. The external load is cement all along the outside of the casing and sea water as a displacing fluid inside the casing. The pressures are:

External casing pressure	at drillfloor:		= 0
	at seabed:	0.098 × 1.03(225–25)	= 20 bar
	at top of tail cement:	20 + 0.098 × 1.45(1000–225)	= 130 bar
	at casing shoe:	130 + 0.098 × 1.90(1100–1000)	= 149 bar
Internal pressure	at drillfloor:		= 0
	at shoe:	0.098 × 1.03 × 1100	= 111 bar

The pressure profiles are shown in Figure 5.12a. We observe that the maximum collapse load occurs at the shoe, and is 149–111 bar = 38 bar.

The collapse strength is derated due to bi-axial forces. As shown later, the axial design load becomes 398×10^3 daN and the axial load/strength ratio is 395/800 = 0.49. From Figure 5.5b, this gives a collapse strength reduction factor of 0.66, and the derated collapse pressure is 0.66 × 43 = 28 bar. At the casing shoe, there is no axial load, and no correction applies. Since the axial load is a linear function with depth, a straight line connects the two extreme collapse strength points. The most severe loading is at the casing shoe (Figure 5.12a), where the collapse strength-load ratio is: 43 bar/ 38 bar = 1.13.

Also the *fluid level drop* case must be checked. We here assume a thief zone at the bottom of the well. If losses occur to a thief zone, the inside mud level may stabilise at a depth equal to a hydrostatic sea water pressure, or:

$$0.098 \times 1.03(1100 - 25) = 0.098 \times 1.20(1100 - h)$$

The height of mud from the drill floor is 177 m. The maximum collapse load at this position is equal to the external pressure, since the inside is air. The collapse load is:

0.098 × 1.03 × 177 = 18 bar, which is less than in the previous case.

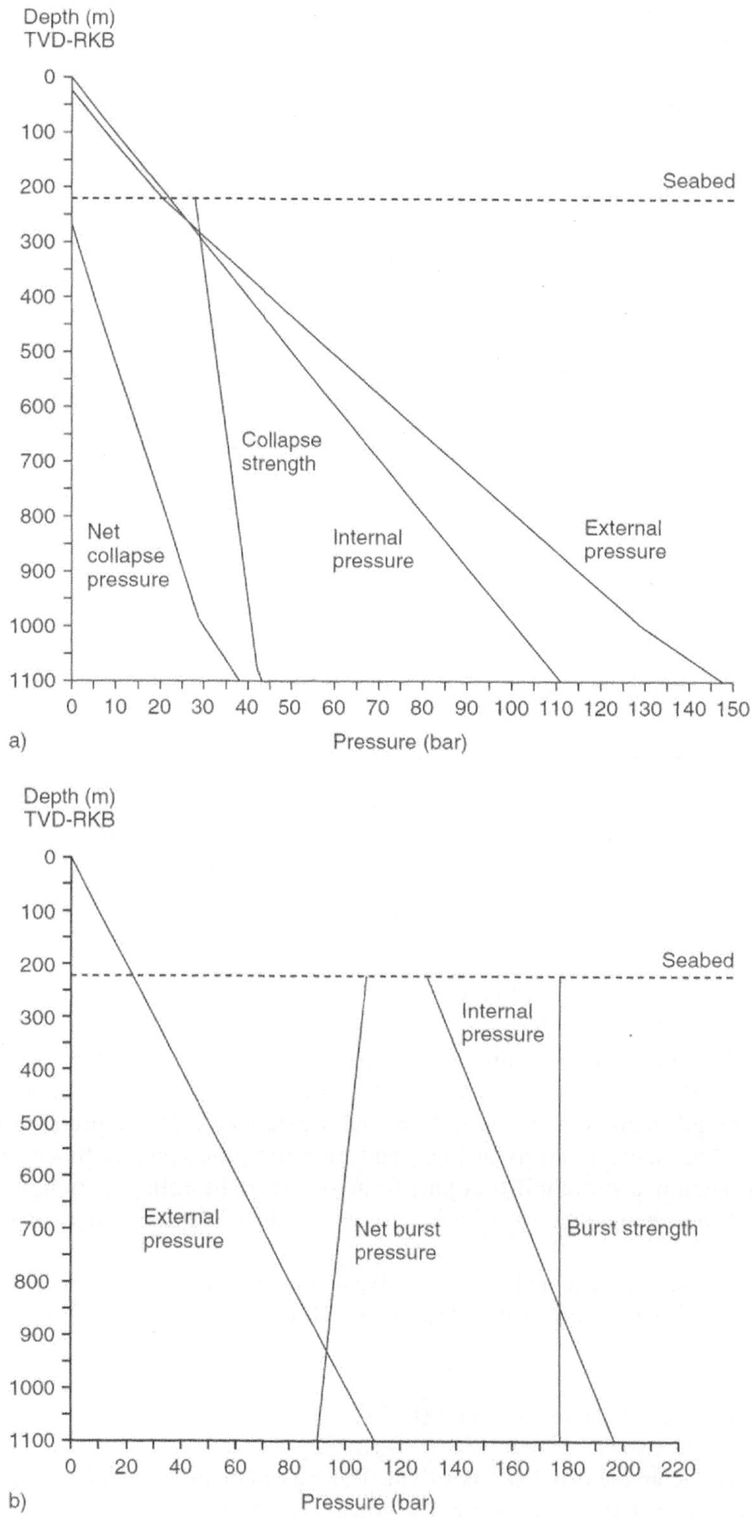

FIGURE 5.12 (a) 18-5/8 in. casing collapse design. (b) 18-5/8 in. casing burst design.

For burst design, we assume a production casing scenario with sea water behind the casing and obtain the following pressures:

External pressure at wellhead:	0.098 × 1.03 × 225 = 23 bar
at the casing shoe:	0.098 × 1.03 × 1100 = 111 bar
Pressure at bottom of next open hole section:	0.098 × 1.40 × 1820 = 250 bar

Assuming oil-filled casing, the internal pressures are:

at casing shoe:	250–0.098 × 0.76(1820–1100) = 196 bar
at wellhead:	196–0.098 × 0.76(1100–225) = 131 bar

The pressure profiles are shown in Figure 5.12b. We have also shown the burst strength, which is corrected for 10% wear by using Equation (5.4). Since all elements of this equation are constant except the burst pressure and the wall thickness, a proportionality relation gives us: $P_{burst} = 0.9 \times 197 = 177$ bar, which also is shown in Figure 5.12b. We observe that the highest burst load occurs below the wellhead, with a burst load of 131–23 = 108 bar. This furthermore results in a strength/load ratio of 177/108 = 1.84.

For the tensional load design, the weight of the string is:

$$186\,\text{kg/m}\left(1100-225\right)\text{m}\times 0.981\,\text{daN/kg} = 160\times 10^3\,\text{daN}$$

The casing is landed in a 1.20 s.g. mud, providing buoyancy. The net weight in mud is:

$$\left(1-\frac{1.2\ \text{s.g.}}{7.8\ \text{s.g.}}\right)160\times 10^3\ \text{daN} = 135\times 10^3\,\text{daN}$$

Additional tension due to bending is estimated from other wells to 260 × 10^3 daN.

Based on the definitions in Section 5.2 the casing test pressure for both the bumping plug case and the cement set case is:

$$108-0.098\times 1.03\times 25 = 106\ \text{bar}$$

This test pressure causes tension in the casing, which is a closed chamber during testing. The axial force is:

$$106\ \text{bar}\times 1527\,\text{cm}^2 = 162\times 10^3\ \text{daN}$$

The design tensional load is the sum of the weight in mud, the bending, and the pressure testing, or:

$$\left(135+260+162\right)10^3 = 557\times 10^3\ \text{daN}$$

The strength/load ratio in tension is then: 800 × 10^3 daN/557 × 10^3 daN = 1.44

For the bi-axial collapse correction, only the buoyed weight and bending tension applies, giving a design load of:

$$(135+260)10^3 = 395\times10^3\,\text{daN}$$

Finally, an evaluation of well integrity is required. Following Section 4.1 we will consider the integrity of this section of the well.

If the well is filled with oil and shut in, the pressure at the 18-5/8 in. casing shoe will be equal to 196 bar, as calculated above. This pressure is equal to a gradient of 196 bar/0.098 × 1100 m = 1.82 s.g. The prognosed fracture gradient at the casing shoe is 1.57 s.g. Clearly, if such a well control event took place, the shoe would not withstand the pressure, and an underground blowout would result. However, introducing a kick margin, this problem can be avoided. Assuming that we take in a volume of formation fluid at the bottom of the well, a height h from bottom of next hole section, the prognosed fracture gradient is reached at:

$$0.098\times1.57\times1100 = 250\text{ bar} - 0.098\times0.76\text{h} - 0.098 \times1.50(1820-1100-h)$$

The maximum height of formation fluid influx is h = 346 m, which in a 16 in. open hole amounts to 44.9 m^3 of influx.

The last point to check is to ensure that the weak point in the system is at the casing shoe, and not below the wellhead. The X-70 casing selected has a burst resistance of 197 bar. The operator uses a minimum factor of safety of 1.18, such that maximum allowable pressure is 197 bar/1.18 = 167 bar.

Now assume that the well is filled with formation fluid and shut in, and the pressure at the wellhead is the maximum allowable, 167 bar. The pressure at the shoe would then be:

$$167\text{ bar} + 0.098\times0.76(1100-225) = 232\text{ bar}$$

This is equivalent to a gradient at the casing shoe of: 232 bar/0.098 × 1100 m = 2.15 s.g. The prognosed fracture gradient is 1.57 s.g. We observe that the pressure gradient that would result in a bursted casing below the wellhead is much higher than the prognosed fracture gradient. The casing shoe therefore represents the weak point in the system. According to the definitions of Section 4.1, the 18-5/8 in. casing represents a reduced well integrity case.

The 13-3/8 in. Intermediate Casing

The design parameters are as follows:

Depth of casing:	1820 m
Depth to seabed:	225 m
Depth to sea level:	25 m
Depth to top of cement:	1685 m
Depth of next open hole section:	2365 m
Pore pressure gradient, next section:	1.55 s.g.
Fracture design gradient at shoe:	1.77 s.g.
Formation fluid density:	0.76 s.g.
Mud density:	1.50 s.g.
Cement density:	1.45 s.g.

The casing data are:

13-3/8 in. grade X-70, 72 lbs/ft. casing	
Weight:	107.1 kg/m
Cross-sectional inner area:	772 cm^2
Burst strength:	510 bar
Collapse resistance:	199 bar
Pipe body yield strength:	1016 × 10^3 daN

First, we will calculate the *collapse loading during installation.* If losses occur to a thief zone, the mud level may stabilise at a depth equal to a hydrostatic sea water pressure, or:

$$0.098 \times 1.03(1820 - 25) = 0.098 \times 1.50(1820 - h)$$

The height of mud from the drill floor is 587 m.

The external pressure at the casing shoe is: 0.098 × 1.50 × 1820 = 268 bar

The internal pressure at the casing shoe: 0.098 × 1.03(1820-25) = 181 bar

Figure 5.13a shows these pressures. The collapse loading is constant from 587 to 1820 m, and is 268–181 = 87 bar.

The collapse strength is derated due to biaxial forces. As shown later, the axial load becomes 244 × 10^3 daN, and the axial load/strength ratio is: 244/1016 = 0.24. From Figure 5.5b, this results in a collapse reduction of 0.87, and the derated collapse pressure is 0.87 × 199 = 173 bar. The collapse strength is also shown in Figure 5.13a. Below the wellhead the derated collapse strength applies. At the casing shoe, there is no axial load, so no correction applies here. Since the axial load in the casing string is a linear function with depth, a straight line connects the two extreme collapse points. The most severe loading is at 587 m depth. Interpolation of the collapse resistance results in 179 bar, and the strength/load ratio is 179 bar/87 bar = 2.06.

For burst design, we will use an oil-filled casing criterion, and we assume a production casing scenario with sea water behind the casing and obtain the following pressures:

External pressure at wellhead:	0.098 × 1.03 × 225 = 23 bar
at the casing shoe:	0.098 × 1.03 × 1820 = 184 bar
Pressure at bottom of next open hole section:	0.098 × 1.55 × 2365 = 359 bar

Assuming oil-filled casing, the internal pressures are:

at casing shoe:	359–0.098 × 0.76(2365–1820) = 318 ba
at wellhead:	318–0.098 × 0.76(1820–225) = 199 bar

The pressure profiles are shown in Figure 5.13b. We have also shown the burst strength, which is corrected for 10% wear by using Equation (5.4). Since all elements of this equation are constant except the burst pressure and the wall thickness, a proportionality relation gives us: P_{burst} = 0.9 × 510 = 459 bar, which also is shown in Figure 5.13b. We observe that the highest burst load occurs below the wellhead, with a burst load of 199–23 = 176 bar. This furthermore results in a strength/load ratio of 459/176 = 2.61.

For the *tensional load design*, the weight of the string is:

$$107.1\,\text{kg/m}(1820 - 225)\,\text{m} \times 0.981\,\text{daN/kg} = 168 \times 10^3\,\text{daN}$$

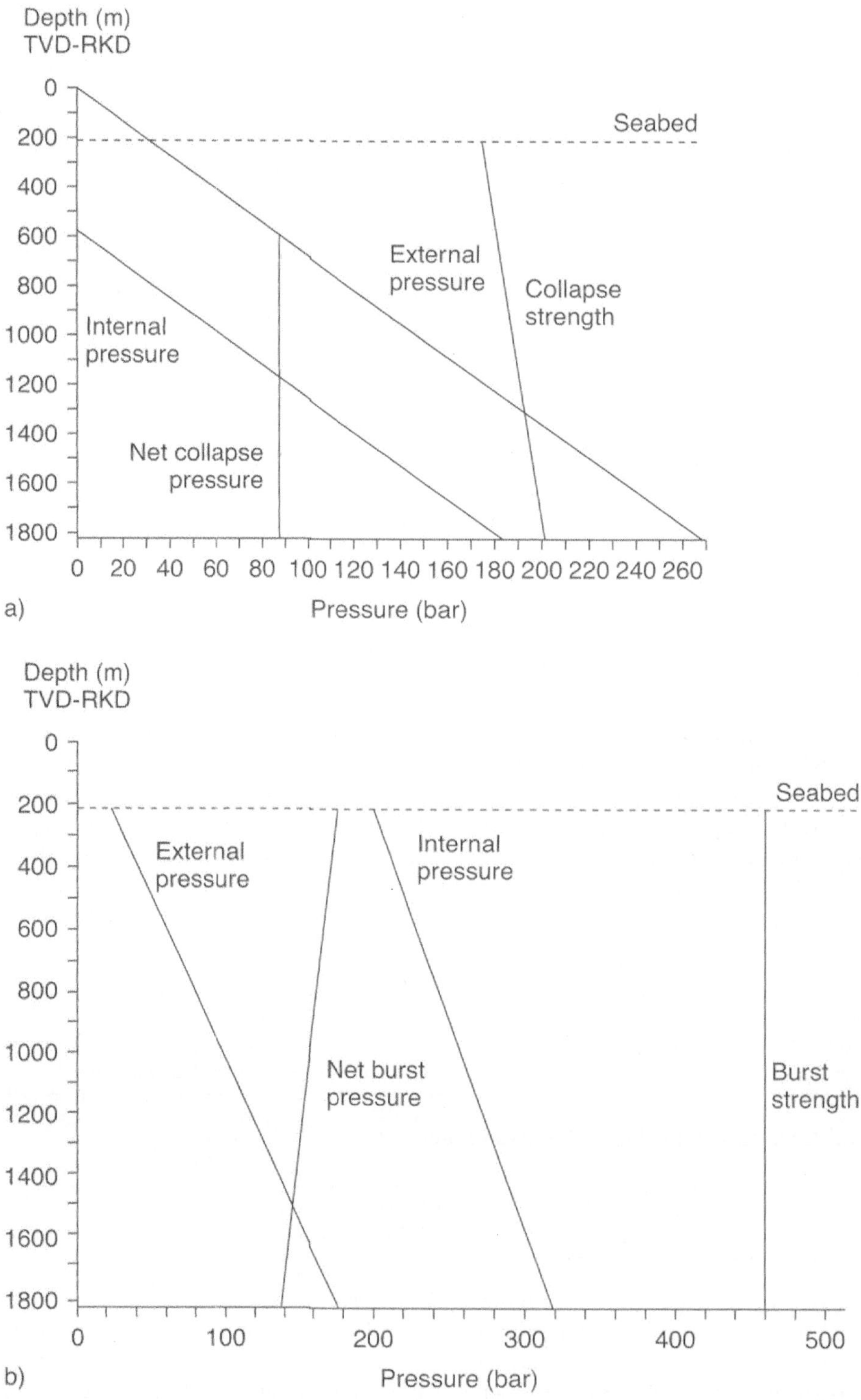

FIGURE 5.13 (a) 13-3/8 in. casing collapse design. (b) 13-3/8 in. casing burst design.

The casing is landed in a 1.50 s.g. mud, providing buoyancy. The net weight in mud is:

$$\left(1-\frac{1.5\,\text{s.g.}}{7.8\,\text{s.g.}}\right)168\times10^3\,\text{daN}=136\times10^3\,\text{daN}$$

Additional tension due to bending is estimated to 108 × 10^3 daN.

Based on the definitions in Section 5.2 the casing test pressure for both the bumping plug case and the cement set case is 176 bar.

This test pressure causes tension in the casing, which is a closed chamber during testing. The axial force is:

$$176\,\text{bar}\times772\,\text{cm}^2=136\times10^3\,\text{daN}$$

The design tensional load is the sum of the weight in mud, the bending, and the pressure testing, or:

$$(136+108+136)10^3=380\times10^3\,\text{daN}$$

The casing strength is derated 10% due to wear during continued drilling, which yields a design strength of 0.9 × 1016 × 10^3 daN = 914 × 10^3 daN.

The strength/load ratio in tension is then: 914 × 10^3 daN/380 × 10^3 daN = 2.41

For the biaxial collapse correction, only the buoyed weight and bending tension applies, giving a design load of:

$$(136+108)10^3=244\times10^3\,\text{daN}$$

Finally, an evaluation of well integrity is required. Again following Section 4.1 we will consider the integrity of this section of the well.

If the well is filled with oil and shut in, the pressure at the 13-3/8 in. casing shoe will be equal to 318 bar, as calculated above. This pressure is equal to a gradient of 318 bar/0.098 × 1820 m = 1.78 s.g. The prognosed fracture gradient at the casing shoe is 1.77 s.g. If the fracture gradient obtained during drilling exceeds 1.78 s.g. we actually have a full well integrity case. If it is lower than 1.78 s.g. we have a reduced well integrity case. Since the allowable kick volume will nearly fill the well also for this case, we will define this casing as a full well integrity case.

The last point to check is to ensure that the weak point in the system is at the casing shoe, and not below the wellhead. The X-70 casing selected has a burst resistance of 510 bar. The operator uses a minimum factor of safety of 1.18, such that maximum allowable pressure is 510 bar/1.18 = 432 bar.

Now assume that the well is filled with formation fluid and shut in, and the pressure at the wellhead is maximum allowable, 432 bar. The pressure at the shoe would then be:

$$432\text{ bar}+0.098\times0.76(1820-225)=551\text{ bar}$$

This is equivalent to a gradient at the casing shoe of 551 bar/0.098 × 1100 m = 3.09 s.g. The prognosed fracture gradient is 1.57 s.g. We observe that the pressure gradient that would result in a bursted casing below the wellhead is much higher than the prognosed fracture gradient. The casing shoe therefore represents the weak point in the system. According to the definitions of Section 4.1, the 13-3/8 in. casing represents a reduced well integrity case.

The 9-5/8 in. Production Casing

The design parameters are as follows:

Depth of casing:	2365 m
Depth to production packer:	2200 m
Depth to seabed:	225 m
Depth to sea level:	25 m
Depth to top of cement:	2060 m
Pore pressure gradient:	1.55 s.g.
Formation fluid density:	0.76 s.g.
Mud density:	1.70 s.g.
Cement density:	1.90 s.g.
Completion fluid density:	1.10 s.g.

The casing data are:

9-5/8 in. grade L-80, 47 lbs/ft. casing	
Weight:	69.9 kg/m
Cross-sectional inner area:	382 cm^2
Wall thickness:	0.472 in
Burst strength:	473 bar
Collapse resistance:	328 bar
Pipe body yield strength:	483×10^3 daN

First, we will calculate the *collapse loading during installation*. If losses occur to a thief zone, the mud level may stabilise at a depth equal to a hydrostatic sea water pressure, or:

$$0.098 \times 1.03(2365 - 25) = 0.098 \times 1.70(2365 - h)$$

The height of mud from the drill floor is 947 m.
The external pressure at the casing shoe is: 0.098 × 1.70 × 2365 = 394 bar
The internal pressure at the casing shoe: 0.098 × 1.03(2365-25) = 236 bar

Figure 5.14a shows these pressures. The collapse loading is constant from 947 to 2365 m, and is 394–236 = 158 bar.

The collapse strength is derated due to biaxial forces. As shown later, the axial design load during installation is 165×10^3 daN, and the axial load/ strength ratio is 165/483 = 0.34. From Fig. 5.15b, this results in a collapse reduction of 0.77, and the derated collapse pressure is 0.77 × 328 = 252 bar. The collapse strength is also shown in Fig. 5.14a. Below the wellhead the derated collapse strength applies. At the casing shoe, there is no axial load, so no correction applies here. Since the axial load in the casing string is a linear function with depth, a straight line connects the two extreme collapse points. The most severe loading is below 947 m, where the strength/load ratio is 278 bar/ 158 bar = 1.76.

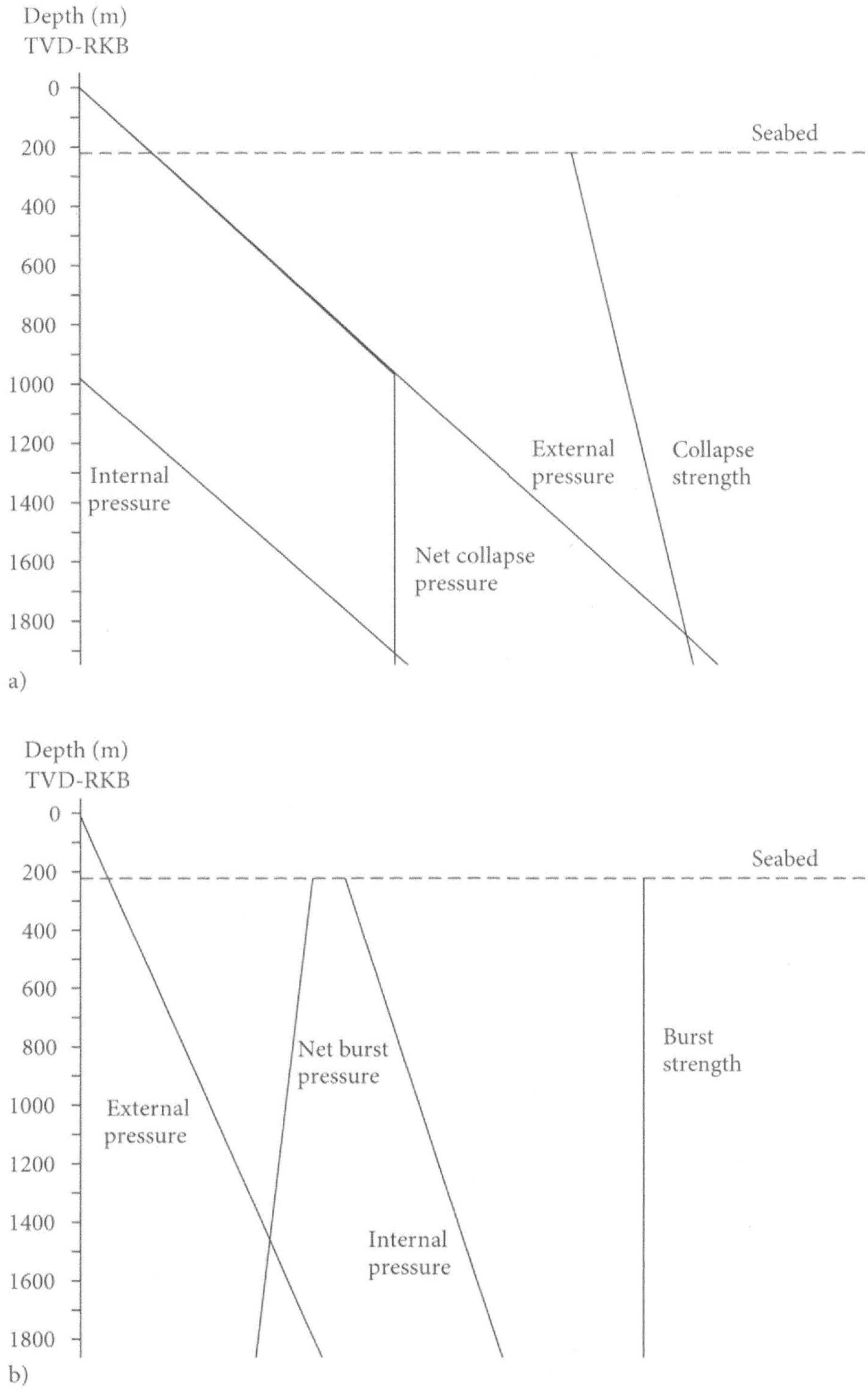

FIGURE 5.14 (a) 9-5/8 in. casing collapse design. (b) 9-5/8 in. casing burst design. Oil filled casing scenario. (c) 9-5/8 in. casing burst design. Leaking tubing scenario.

(Continued)

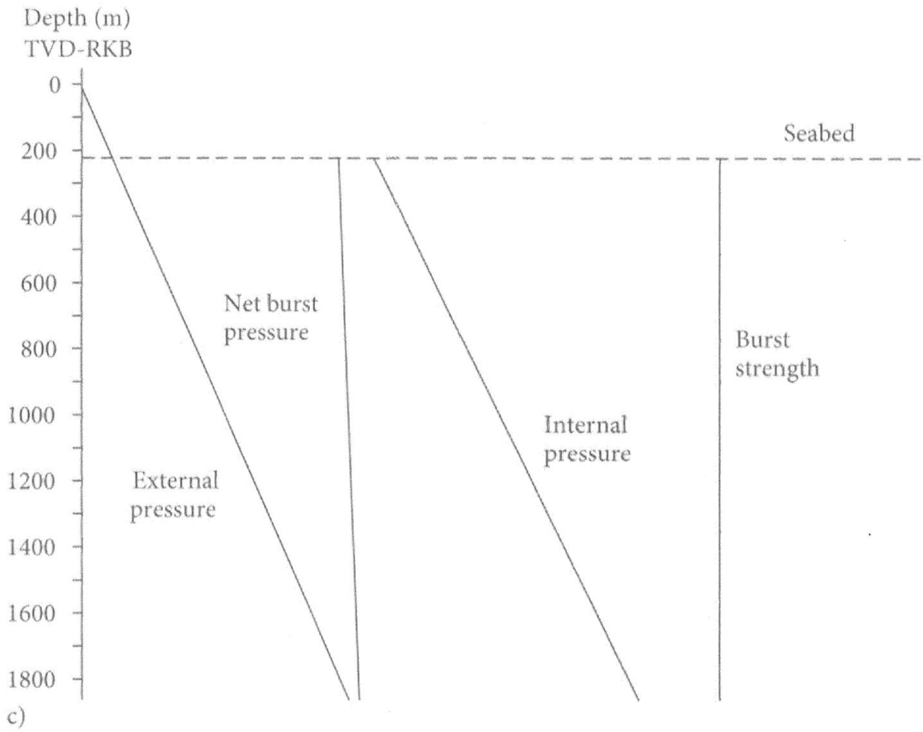

FIGURE 5.14 (Continued)

We will also calculate a collapse loading in the event of plugged perforations during production. This post-drilling scenario assumes the formation pressure as external loading in the reservoir interval, and formation fluid density inside.

The external pressure (pore pressure) at the casing shoe is 0.098 × 1.55 × 2365 = 359 bar
The internal pressure is a column of formation fluid is 0.098 × 0.76 × 2365 = 176 bar
The resulting load on the casing at the casing shoe is 359–176 = 183 bar
This calculation applies only at reservoir level. Below the packer we have to consider 10% corrosion allowance over time. Using Equation (5.22a), we see that the only parameter which changes with corrosion is the wall thickness, which initially was 0.472 in., but reduces to 0.9 × 0.472 = 0.425 in. Using Equation (5.22a) to set up a proportionality relation, we obtain:

$$\frac{2CE}{1-\nu^2} = P_c\left(\frac{D_o}{t}-1\right)^2\frac{D_o}{t}$$

$$328\text{ bar}\left(\frac{9.625\text{ in}}{0.472\text{ in}}-1\right)^2\frac{9.625\text{ in}}{0.472\text{ in}} = \text{P}_\text{c}\left(\frac{9.625\text{in}}{0.425\text{in}}-1\right)^2\frac{9.625}{0.425}$$

which gives a derated collapse pressure of 238 bar. The strength/load ratio for this case becomes 238/183 = 1.30.

For *burst design, we will first calculate the oil-filled casing scenario*, and we assume a production casing scenario (see Section 5.1.7) with sea water behind the casing and obtain the following pressures:

External pressure at wellhead:	0.098 × 1.03 × 225 = 23 bar
at the casing shoe:	0.098 × 1.03 × 2365 = 239 bar

Assuming oil-filled casing, the internal pressures are:

at casing shoe:	0.098 × 1.55 × 2365 = 359 bar
at wellhead:	359–0.098 × 0.76(2365–225) = 200 bar

The pressure profiles are shown in Figure 5.14b. We have also shown the burst strength, which is corrected for 10% wear by using Equation (5.4). Since all elements of this equation are constant except the burst pressure and the wall thickness, a proportionality relation gives us: P_{burst} = 0.9 × 473 = 426 bar, which is also shown in Figure 5.14b. We observe that the highest burst load occurs below the wellhead, with a burst load of 200–23 = 177 bar. This furthermore results in a strength/load ratio of 426/177 = 2.41.

The tubing leak criterion will be evaluated next. Assume that the production tubing is filled with formation oil and then shut in. The pressures will be as shown in Figure 5.14c. Now assume that a tubing leak occurs in the production tubing just below the wellhead. The pressure inside the production tubing will now also act on the outside, which is the annulus for the production casing. This annulus is filled with completion fluid.

Pressure in casing annulus below wellhead:	200 bar
on top of packer:	200 + 0.098 × 1.1(2200–225) = 413 bar
Pressure outside casing at packer level:	0.098 × 1.03 × 2200 = 222 bar

The burst load above the packer will be 413–222 = 191 bar. In this case, we observe that the leaking tubing gives a higher loading than the oil-filled casing criterion. Figure 5.14c therefore shows the final burst design. The strength/load ratio is 426/191 = 2.23.

For the *tensional load design*, the weight of the string is:

$$69.9\,\text{kg/m}\left(2365-225\right)\text{m}\times 0.981\,\text{daN/kg} = 147\times 10^{3}\,\text{daN}$$

The casing is landed in a 1.70 s.g. mud, providing buoyancy. The net weight in mud is:

$$\left(1-\frac{1.7\text{ s.g.}}{7.8\text{ s.g.}}\right)147\times 10^{3}\text{ daN} = 115\times 10^{3}\text{ daN}$$

Additional tension due to bending is estimated to 50 × 10^3 daN.

Based on the definitions in Section 5.2 the casing test pressure for both the bumping plug case and the cement set case is 192 bar. For the previous surface and intermediate casing strings, the test pressures from the burst evaluation were applied. The production casing, on the other hand, will be subjected to extensive testing and loading during completion and production. To allow for excessive pressures during workover operations, the production casing will be tested to full wellhead design pressure. For this case a 5000 psi wellhead is installed. The casing will therefore be tested to 5000 psi or 345 bar.

This test pressure causes tension in the casing, which is a closed chamber during testing. The axial force is:

$$345\text{ bar}\times 382\text{ cm}^{2} = 132\times 10^{3}\text{daN}$$

The design tensional load is the sum of the weight in mud, the bending, and the pressure testing, or:

$$(115+50+132)10^3 = 297\times10^3 \text{ daN}$$

Due to wear during continued drilling for the wells which will have liner installed, the tensional strength has to be derated 10% or, $0.9 \times 483 \times 10^3 = 435 \times 10^3$ daN

The strength/load ratio in tension is then: 435×10^3 daN/297×10^3 daN = 1.46

For the biaxial collapse correction, only the buoyed weight and the bending tension applies, giving a load of:

$$(115+50)10^3 = 165\times10^3 \text{daN}$$

The 7 in. Injection/Production Liner

The design parameters are as follows:

Depth of casing:	2365 m
Depth of liner:	2625 m
Depth to production packer:	2580 m
Depth to seabed:	225 m
Depth to sea level:	25 m
Depth to top of cement:	2300 m
Pore pressure gradient at 2365 m:	1.55 s.g.
at 2640 m:	1.50 s.g.
Formation fluid density:	0.76 s.g.
Design fracture gradient:	1.87 s.g.
Mud density:	1.70 s.g.
Cement density:	1.90 s.g.
Completion fluid density:	1.10 s.g.

The casing data are:

7 in. grade 13Cr, 29 lbs/ft. casing	
Weight:	43.2 kg/m
Cross-sectional inner area:	194 cm^2
Wall thickness:	0.407 in.
Burst strength:	562 bar
Collapse resistance:	484 bar
Pipe body yield strength:	301×10^3 daN

We will calculate a collapse loading occurring if the perforations get plugged during production. This post-drilling scenario assumes the formation pressure as external loading in the reservoir interval, and formation fluid density inside.

The external pressure at the casing shoe is:	0.098 × 1.55 × 2365 = 359 bar
at the liner shoe:	0.098 × 1.50 × 2625 = 386 bar
The internal pressure is a column of formation fluid:	0.098 × 0.76 × 2365 = 176 bar
	0.098 × 0.76 × 2625 = 196 bar
The resulting load on the casing at the casing shoe is	359 – 176 = 183 bar
at the liner shoe	386 – 196 = 190 bar

This calculation applies only at reservoir level. There is no need to derate the collapse rating for biaxial stresses, as the axial load is small, and the liner is cemented along its full length providing mechanical support. Also, there will be no drilling below the liner, so it is not necessary to adjust the collapse pressure for wear.

However, below the packer we have to consider a 10% corrosion allowance over time. Using Equation (5.5), we see that the only parameter which changes is the wall thickness, which initially was 0.407 in., but reduces to 0.9 × 0.407 = 0.367 in. Using Equation (5.5) to set up a proportionality relation, we obtain:

$$\frac{2CE}{1-v^2} = P_c\left(\frac{D_o}{t}-1\right)^2\frac{D_o}{t}$$

$$484\,bar\left(\frac{7\,\text{in}}{0.407\,\text{in}}-1\right)^2\frac{7\,\text{in}}{0.407\,\text{in}} = P_c\left(\frac{7\,\text{in}}{0.367\,\text{in}}-1\right)^2\frac{7\,\text{in}}{0.367}$$

which gives a derated collapse pressure of 351 bar. This scenario is shown in Fig. 5.15a. The strength/load ratio for this case becomes 351/190 = 1.85.

Burst during bullheading will be evaluated next. While pumping fluids into the formation, the perforations may plug off, resulting in a build-up of pressure across the casing wall. We assume that the production packer is set above the liner. Furthermore, we assume that the bullheading fluid is formation fluid. The maximum pressure inside the liner is defined equal to the fracturing pressure of the formation behind the casing. The internal pressure then becomes:

$$0.098\times1.87\times2365 \quad = 433\text{ bar}$$
$$0.098\times1.92\times2625 \quad = 494\text{ bar}$$

On the outside, we use the very conservative assumption that a sea water gradient exists behind the casing, giving a pressure of:

$$0.098\times1.03\times2365 \quad = 239\text{ bar}$$
$$0.098\times1.03\times2625 \quad = 265\text{ bar}$$

The net burst pressure is: 433 – 239 = 194 bar at 2365 m, and 494 – 265 = 229 bar at 2625 m.

The burst load has to be derated for corrosion. Assuming a 10% allowance, the derated burst strength is 0.9 × 562 = 506 bar, and the strength/load ratio becomes 506/194 = 2.61.

For the *tensional load design*, the weight of the string is:

$$43.2\,\text{kg}/\text{m}\times365\,\text{m}\times0.981\,\text{daN}/\text{kg} = 15\times10^3\,\text{daN}$$

The casing is landed in a 1.70 s.g. mud, providing buoyancy. The net weight in mud is:

$$\left(1-\frac{1.7\text{ s.g.}}{7.8\text{ s.g.}}\right)15\times10^3\,\text{daN} = 12\times10^3\,\text{daN}$$

Additional tension due to bending is estimated to be 23 × 10^3 daN.

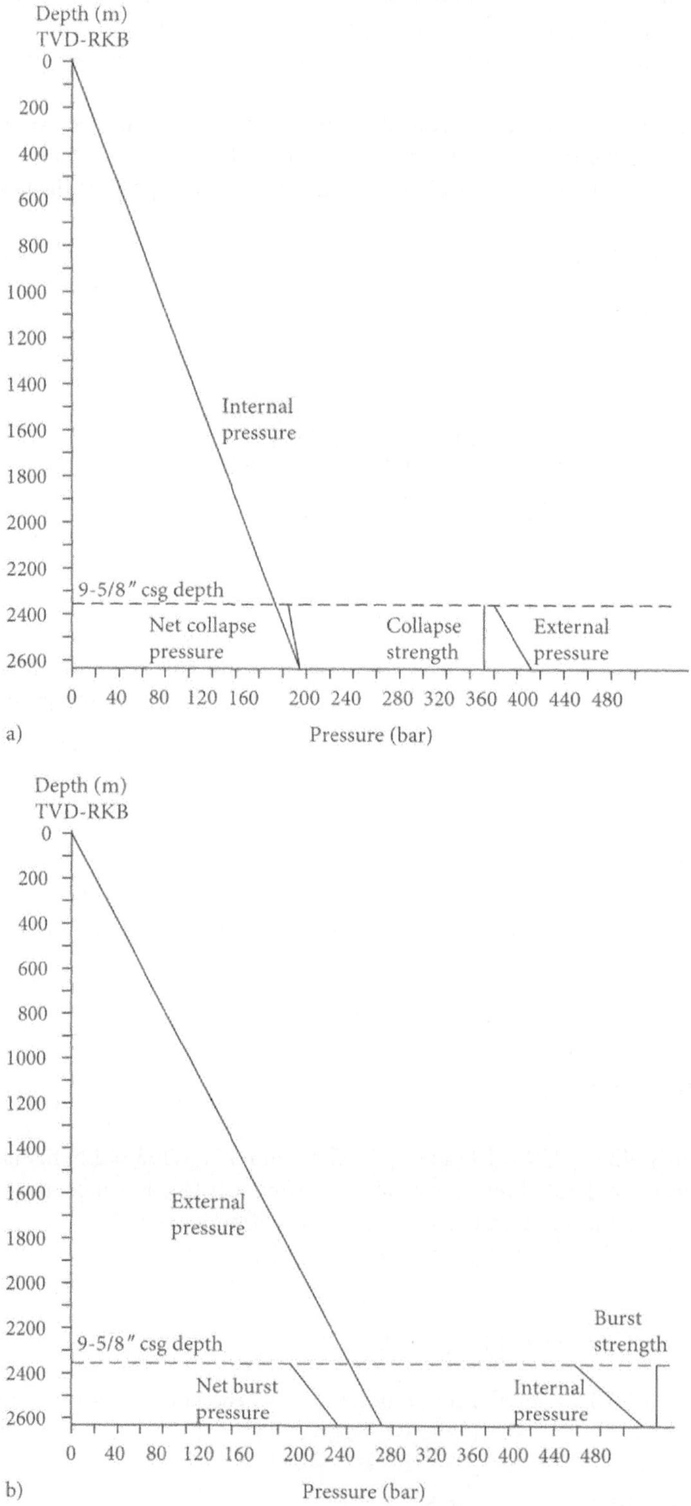

FIGURE 5.15 (a) 7 in. liner collapse design. (b) 7 in. liner burst design.

Based on the definitions of the 9-5/8 in. casing string, the casing test pressure for both the bumping plug case and the cement set case is 345 bar.

This test pressure causes tension in the casing, which is a closed chamber during testing. The axial force is:

$$345 \text{ bar} \times 194 \text{ cm}^2 = 67 \times 10^3 \text{ daN}$$

The design tensional load is the sum of the weight in mud, the bending, and the pressure testing, or:

$$(12+23+67)10^3 = 102 \times 10^3 \text{daN}$$

The strength/load ratio in tension is then: 301 × 10^3 daN/102 × 10^3 daN = 2.95.

5.3.5 Design Factors

The design factors from the foregoing analysis are shown in Table 5.2. For this particular field, the operator has the following minimum requirements for design factors:

Burst design factor > 1.18
Collapse design factor > 1.10
Tension design factor > 1.10

This design is therefore acceptable based on the operator's requirements.

5.3.6 Connectors

The actual connectors selected depends on the choice of manufacturer. The tensile requirement of all connectors is that the tensile strength must exceed that of the casing. Other requirements are:

- The 30 in. connectors should be of snap-on type.
- The 18-5/8 in. connectors should be of easy make-up type.
- The 13-3/8 in. connectors should be of standard buttress type.
- The 9-5/8 in. casing connectors and the 7 in. liner connectors should be of premium type, gas tight with metal to metal seal.

5.3.7 Discussion

In the following a complete casing design is carried out for a well. This chapter is intended to provide a basis for well design of other wells. Now we will briefly discuss the results and the implications.

From Table 5.2 we see that only the 18-5/8 in. hole section had reduced well integrity. This could have been shown in a simpler way. In Figure 5.16 we have shown the reservoir pressure as a

TABLE 5.2 Design factors from the design

CASING (INCH)	*BURST*	*COLLAPSE*	*TENSION*	*MAX. KICK SIZE (M3)*	*MAX. FRAC. GRAD (S.G.)*
18-5/8	1.84	1.13	1.44	44.9	2.15
13-3/8	2.61	2.06	2.41	–	3.09
9-5/8	2.41	1.30	1.46	–	–
7	2.19	1.92	2.95	–	–

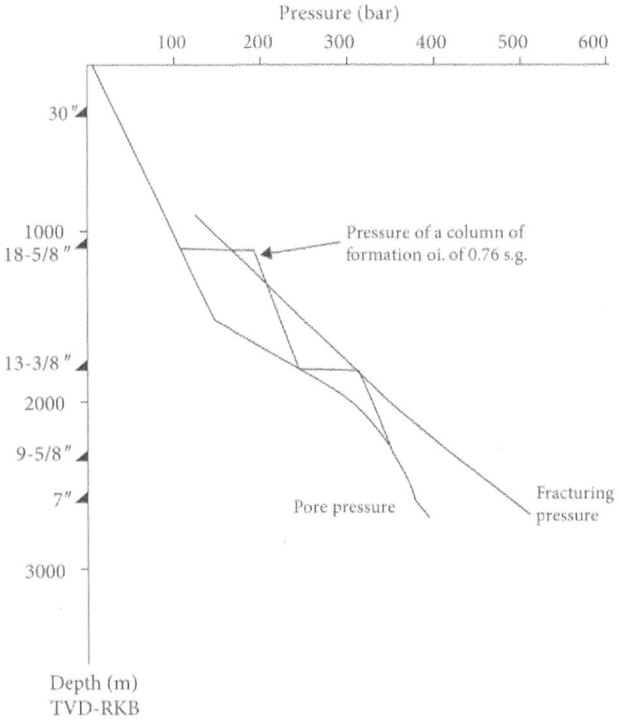

FIGURE 5.16 Pressures during oil-filled casing shut in.

function of well depth. From each casing point we plot the gradient of the formation fluid. This is the pressure that would arise if the well was oil-filled and shut in. Clearly we observe the conclusion from Table 5.2. Figure 5.16 is also useful if we are designing wells for specific conditions, i.e. flow testing of shallower secondary objectives. We can select casing setting depths to obtain full well integrity using this type of figure.

Inspection of Table 5.2 also reveals that we have the potential of using weaker tubing types, since we far exceed the minimum requirements. However, in the North Sea region one attempts to standardise on a few casing types for reasons of cost, storage, and transportation. Based on this condition we will expect to see some over-design.

We see clearly from the above example that the design factors are not the only governing parameters. As important are the design input data, such as density of formation fluid and density of fluids behind the casing. In traditional casing design one often used methane gas as formation fluid as it was the lightest gas type encountered. However, if geological knowledge can rule out this assumption one should use other fluid or gas densities as demonstrated in this chapter. The effective pressure in a cemented interval is often assumed to be equal to sea water pressure, as water is the only mobile phase. This assumption totally neglects mechanical support from the cement. As this is often a constraint, this issue should also be further addressed in well design.

A related issue is the effective density of the drilling mud behind the casings. In a short time perspective, the actual mud density can be used. If it is a production well with a long time frame, particle settling may be assumed, resulting in water being the only mobile phase providing pressure behind the casings.

Bending resistance during landing of the casing is included in the calculations. This is another issue that should be further addressed. Actually this is coupled to the drag problem in wellbores.

In the example presented, the intermediate casing string could handle a gas-filled situation. The reason is that we assumed oil to be the reservoir fluid instead of gas. Often, surface and intermediate casing strings cannot handle a gas-filled case. In these instances one must introduce a maximum allowable kick volume in the design. Examples of these are presented in Sections 6.6.4 and 6.7.2.

We now understand that when performing optimal well design, we need to work on the design premises, which actually govern the quality of the design. Therefore, post analysis of previous failures is a key element. Also, it is important to establish realistic design scenarios.

Problems

Problem 1

A 24 in. intermediate surface casing is planned to be installed in a well. Please perform burst and collapse calculations, and determine the design factors. No corrections for biaxial forces and derating of casing strength are required.

The data are:

Casing depth:	580 m
Depth to seabed:	110 m
Water depth:	68 m
Pore pressure gradient at casing shoe:	1.03 s.g.
next hole section at 1000 m:	1.03 s.g.
Fracture gradient at shoe:	1.39 s.g.
Mud density:	1.03 s.g.
Mud weight next section:	1.15 s.g.
Cement density:	1.70 s.g.

The casing data are:

24 in. grade X-52	
Outer diameter:	610 mm
Wall thickness:	15.88 mm
Burst strength:	168 bar
Collapse resistance:	60 bar

Problem 2

A 14 in. intermediate casing should be designed. Please perform burst, collapse, and tension calculations, and calculate the design factors. Perform biaxial derating of the collapse strength, but no derating otherwise. If the well cannot handle a gas-filled situation, use kick margins in the design.

The data are:

Casing depth:	3100 m
Frac. gradient at 3100 m:	2.06 s.g.
Pore press. gradient at next hole, 4730 m:	1.84 s.g.
Density of formation fluids:	0.547 s.g.
Mud density:	1.65 s.g.
Mud density next section:	1.90 s.g.
Depth to top cement:	2100 m
Cement density:	1.90 s.g.

The casing data are:

14 in. Grade P-100, 86 lbs/ft	
Weight:	127.4 kg/m
Outer diameter:	355.6 mm
Inner diameter:	325.1 mm
Burst strength:	569 bar
Collapse strength:	268 bar
Tension strength:	1235×10^3 daN

Other data needed can be taken from problem 5.1.

Problem 3

You are asked to design a production casing string. This string is intended for drilling only. However, also check if the casing is capable of handling flow testing, i.e. by evaluating the leaking tubing design scenario. Perform burst, collapse, and tension calculations and calculate the design factors. Perform corrections for biaxial loading on the collapse resistance, and a linear temperature degrading by applying a factor of 1.0 at surface and 0.9 at the casing shoe.

The data are:

Casing depth:	4730 m
Frac. gradient at 4730 m:	2.24 s.g.
Pore press. gradient at next hole, 5100 m:	2.16 s.g.
Density of formation fluids:	0.547 s.g.
Mud density:	1.90 s.g.
Mud density next section:	2.20 s.g.
Depth to top cement:	4000 m
Cement density:	1.90 s.g.

The casing data are:

9-5/8 in. Grade SMC-110, 66.4 lbs/ft	
Weight:	102.4 kg/m
Outer diameter:	250.8 mm
Inner diameter:	218 mm
Burst strength:	889 bar
Collapse strength:	797 bar
Tension strength:	936×10^3 daN

5.4 THREE-DIMENSIONAL TUBULAR DESIGN

5.4.1 Purpose

In the previous derivations the burst pressure model is one-dimensional, whereas the collapse pressure calculations are one- or two-dimensional (biaxial correction). This implies that the other dimensions are neglected. This raises several questions such as what is the error and why not always perform three-dimensional tubular designs? This is often performed today, but there are issues that need to be discussed.

One- and two-dimensional burst and collapse designs clearly identify the failure modes. In burst most often an axial rupture occurs but with high axial load the pipe can also part in two. In collapse the pipe deforms into a non-circular shape. Because the failure is so clearly defined these models serve well as educational tools.

A three-dimensional design can in simple terms be described as a vector summation of various loads. The resultant stress is then compared to the yield strength of the material. The ellipse shown in Figure 5.17 actually represents the yield strength of the material. When the applied load is inside the ellipse, failure has not occurred. Once the load exceeds the ellipse, failure occurs. In a three-dimensional load case it can be difficult to determine the actual failure mode. This is the problem with three-dimensional tubular design. To better understand the failure, we therefore recommend that all tubular design is performed in one or two dimensions, followed by a three-dimensional design as outlined in this section.

Nomenclature

DF	Design factor
F	Axial pipe force
P_i	Fluid pressure inside the pipe
P_o	Fluid pressure outside the pipe
D_i	Inside pipe diameter
D_o	Outside pipe diameter
t	Pipe wall thickness

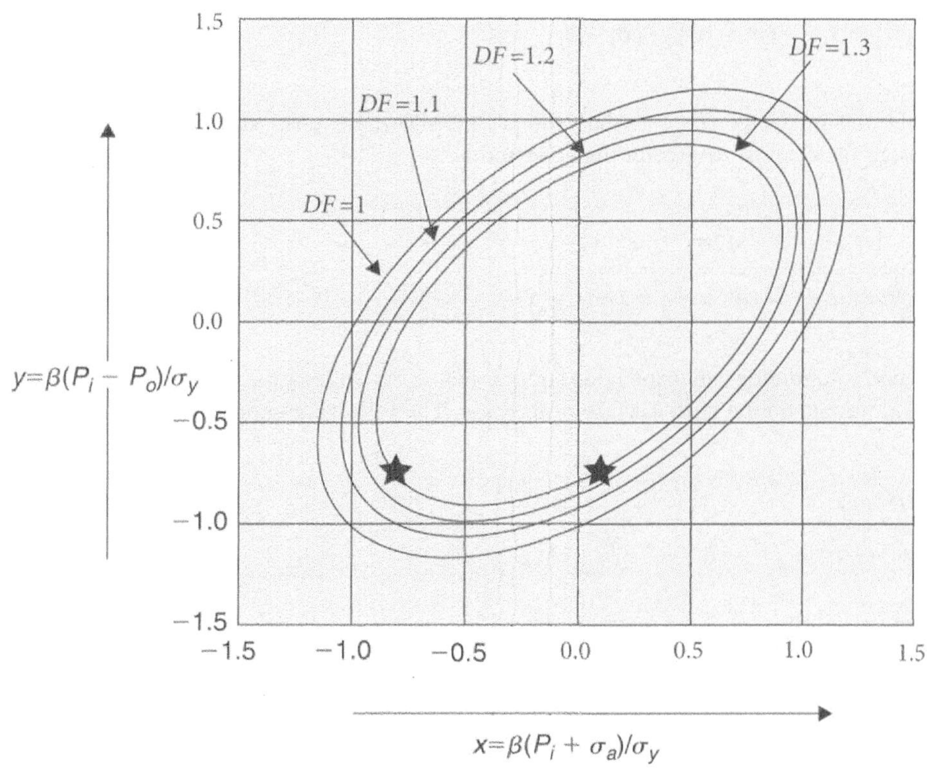

FIGURE 5.17 Three-dimensional design factors projected onto a zero axial stress plane.

x, y Dimensionless parameters
β Geometry factor
σ_a Axial stress
σ_t Hoop (tangential) stress
σ_r Radial stress
σ_y Yield strength
σ_{VME} Equivalent stress

5.4.2 Triaxial Design

Aasen and Aadnoy (2007) derived a dimensionless model for tubular stresses in three dimensions. This model can be used for all pipe materials and pipe dimensions. The model is also valid for cases with applied torque, bending, or buckling. It is based on incipient yielding at the inside pipe wall, and the collapse predictions from the model for diameter-to-thickness ratios less than 14 are in agreement with experimental evidence. For large casing strings, the model should be used with caution for collapse predictions. The model is valid for burst calculations for all pipe sizes.

Reference is made to Equation (5.9) which is the Hencky-von Mises yield theory. For the biaxial corrections of Section 5.1.5 this failure model is applied in a two-dimensional problem.

Three-dimensional analysis includes all three dimensions, i.e. the axial stress σ_a, the radial stress σ_r, and the hoop stress σ_t. These stresses may be combined to a single equivalent stress (σ_{VME}) using the Henky-von Mises distortion energy theorem:

$$2\sigma_y^2 = (\sigma_a - \sigma_r)^2 + (\sigma_a - \sigma_t)^2 + (\sigma_r - \sigma_t)^2 \tag{5.19}$$

We define the design factor (*DF*) as the ratio of the allowable yield stress to the applied stress. The resulting design factor must be greater than one and is:

$$DF = \frac{\sqrt{2}\sigma_y}{\sqrt{(\sigma_a - \sigma_r)^2 + (\sigma_a - \sigma_t)^2 + (\sigma_r - \sigma_t)^2}} \tag{5.20}$$

Using the Lamé solution for radial and hoop stresses in a thick-walled pipe, it is found that burst or collapse failures would initiate on the inner surface of the pipe. The following geometry factor is introduced:

$$\beta = \frac{(D_o / t)^2}{2(D_o / t - 1)} \tag{5.21}$$

The radial and hoop stresses at the inside pipe wall become:

$$\sigma_r = -P_i \tag{5.22a}$$

$$\sigma_t = \beta(P_i - P_o) - P_i \tag{5.22b}$$

In the absence of bending, the axial stress is calculated as:

$$\sigma_a = \frac{F}{A} \tag{5.23}$$

Introduce the following dimensionless variables:

$$x = \frac{P_i + \sigma_a}{\sigma_y} \tag{5.24a}$$

$$y = \beta \frac{P_i - P_o}{\sigma_y} \tag{5.24b}$$

Inserting these equations into Equation (5.20) the design factor becomes:

$$\begin{aligned} DF &= \frac{1}{\sqrt{x^2 - xy + y^2}} \quad \text{or} \\ y &= \frac{x}{2} \pm \sqrt{\frac{1}{DF^2} - \frac{3}{4}x^2} \end{aligned} \tag{5.25}$$

Equation (5.25) defines an ellipse as seen in Figure 5.17. The plus sign is selected for burst calculations, while the negative sign is used for calculating collapse pressure. Substituting for x and y in this equation gives the following expression for the collapse pressure for (DF=1):

$$P_{\text{collapse}} = \frac{P_i(2\beta - 1) - \sigma_a + \sqrt{4\sigma_y^2 - 3(P_i + \sigma_a)^2}}{2\beta} \tag{5.26}$$

The burst pressure is:

$$P_{\text{burst}} = \frac{\beta P_o(2\beta - 1) + \sigma_a(\beta - 2) + \sqrt{4\sigma_y^2(\beta^2 - \beta + 1) - 3\beta^2(P_o + \sigma_a)^2}}{2(\beta^2 - \beta + 1)} \tag{5.27}$$

These two equations define the collapse and burst pressure of a tubular loaded in three dimensions. The presented model is valid for collapse of drill pipe, snubbing pipe, production tubing, and small casing strings. For large pipe sizes (diameter-to-thickness ratios greater than 14), other collapse formulas should be employed. The burst theory is valid regardless of pipe size. Aasen and Aadnoy (2007) also included torsion and bending.

5.4.3 Comparing 1D, 2D, and 3D Models Using HP/HT Well Cases

We consider three design examples from HPHT wells. In the two first cases, the concern is the pressure integrity of the 10-3/4 in. × 9-7/8 in. production casing. The third example is collapse pressure during a snubbing operation.

Input data for the design examples are summarised in Table 5.3. Reservoir temperature is 170°C and wellhead temperature at full production is 155°C. Yield strength values are derated according to prevailing temperature and are named σ_y in the table.

In the examples, we calculate the design factors based on the three-dimensional model. These results are then compared to the design factors obtained using biaxial and uniaxial design. This comparison is shown in Table 5.4.

5.4.3.1 Design 1: Burst of Gas-Filled Casing

We consider burst of the production casing just below the wellhead. This scenario could occur during both drilling and production. During the drilling phase, the worst case is that the entire casing fills up with produced natural gas. After the well has been completed with the production tubing, a leak in the tubing below the wellhead could cause the same burst pressure. In both cases, the greatest differential between inside and outside pressures are located in the very top of the casing. Inside the casing at this depth, the pressure is equal to the reservoir pressure minus the hydrostatic head of the gas column. Outside the production casing the hydrostatic pressure is equal to 360 m of sea water in this subsea well. We consider in the following a leak during production, and derate the yield strength of the C110 casing to 6750 bar because of elevated temperature. Thermal expansion of the casing changes the axial load in the casing at the wellhead from tension to 104 tonnes of compression. The resulting axial stress is –634 bar as shown in Table 5.3.

From Equation (5.25), the general 3D solution, we obtain $DF = 1.37$. The uniaxial model calculates an optimistic $DF = 1.44$ in burst since the effect of axial compression is neglected.

5.4.3.2 Design 2: Burst of Casing Caused by Leaking Tubing

In this scenario we look at burst at the top of the production packer during production. The production tubing leaks just below the wellhead, and wellhead pressure enters the production casing on top of the packer fluid. This is the same situation described in the previous case, the difference being that we now look at burst at the bottom of the well. The proposed design calls for a 1.18 s.g. packer fluid. There is a significant compressive load on the production casing caused by thermal expansion during production. Once the wellhead pressure enters the tubing/casing annulus, this compression is relieved somewhat by wellhead growth. The resulting axial stress at the bottom of the casing is –2140 bar as shown in Table 5.3. Outside the production casing, we have a sea-water gradient. Inside the casing, the pressure is equal to wellhead pressure plus the hydrostatic head of the packer fluid column.

The three-dimensional model gives $DF = 1.10$. If this design is not considered acceptable, one solution is to select a lighter packer fluid. Uniaxial burst pressure gives $DF = 1.17$.

5.4.3.3 Design 3: Live Well Entry With Snubbing

A 460k snubbing jack is used to deploy a 5 in. pipe in this well with 683 bar wellhead pressure. The maximum snub force is experienced while pushing the first joint into the well. The concern is that the wellhead pressure will collapse the pipe, which is run closed-ended and empty in the well. The compressive force working on the pipe below the stripper rams is equal to the wellhead pressure multiplied by the cross-sectional area of the closed pipe. The resulting compressive stress on the pipe is 1970 bar as shown in Table 5.3.

TABLE 5.3 Data for HPHT well

CASE	PIPE	DO/T	β	PI (BAR)	PO (BAR)	σA (BAR)	σY (BAR)
1	10¾", 85.3 lb/ft, C110	13.51	7.292	717	36	-634	6750
2	9⅞", 62.8 lb/ft, Q125	15.80	8.434	1230	451	-2140	7850
3	5", 23.2 lb7ft, C95	10.46	5.783	0	683	-1970	5770

TABLE 5.4 Comparison of design factors

CASE	*3D MODEL*	*UNIAXIAL BURST*	*BIAXIAL COLLAPSE*
1	1.37	1.44	
2	1.10	1.17	
3	1.69		1.64

For a design factor equal to 1.25, calculate the collapse pressure $P_o = 911$ bar from Equation (5.26). The biaxial design factor based on collapse pressure is equal to 1.64.

These examples show that the three-dimensional analysis gives lower design factors in burst and higher in collapse. This difference is load dependent. Considering uncertainties and conservative assumptions, we advocate using the classical one- and two-dimensional design methods for ordinary wells. For critical well designs it is recommended to perform three-dimensional analysis in addition.

The three-dimensional model presented is simple and easy to use. Aasen and Aadnoy (2007) have also included torsion and bending which makes the model generally applicable.

Design of a HPHT Well 6

6.1 INTRODUCTION

In this chapter we will discuss the design of a high-pressure high temperature (HPHT) well. It is actually a summary of the design and decision-making process. First, we will define this particular type of well. The Norwegian Petroleum Directorate defines a well as a HPHT well if:

- it is deeper than 4000 m, or
- its reservoir pressure exceeds 10 000 psi, or
- the temperature exceeds 150°C.

It is readily observed that the conditions are more harsh in such a well compared to a standard shallower well. A HPHT well is a critical well, where there are small design margins, and where a well control problem is difficult to handle. This chapter will identify the most important design premises. Usually, the definition of the design premises and performing the well design requires several iterations because of the narrow margins.

For readers interested in more operational details, the publications by French and McLean (1993), Krus and Prieur (1991), and Seymour and MacAndrew (1994) are recommended.

Specific nomenclature for this chapter is as follows:

D_{sf}	Depth from drill floor to sea floor
h_f	Distance drill floor – sea level
δ_d	Difference frac. gradient – normal pore pressure gradient
d_p	Normal pore pressure gradient referred to drill floor
d frac.	Gradient referred to drill floor
d_{mw}	Mud density (s.g.)
h	Height of gas (m)
d_{wf}	Fracture gradient
P_{wf}	Fracturing pressure
$P_{wf\text{-}normalised}$	Normalised fracturing pressure
P_o	Pore pressure
σ_o	Overburden stress
NFG	Normalised frac. gradient
NLG	Normalised loss gradient

DOI: 10.1201/9781003618171-6

6.2 DESIGN PREMISES

6.2.1 Design Premises

The very first task to undertake in a well design process is to establish the objectives for the well. Furthermore, alternative approaches may be derived to reach these objectives. For the particular well in question the most important license requirements were as follows:

- to perform a flow test in a Jurassic sandstone reservoir,
- to drill through expected reservoir layers in upper or middle Jurassic,
- or, to drill to 5000 m,
- to core or collect other data in all prospective intervals according to governmental regulations.

The objectives can briefly be condensed to:

- Perform a flow test in upper Jurassic.
- Test hydrocarbon bearing intervals in lower Cretaceous.

According to the definitions above, the depth of the well will be 5000 m deep, but with a formation top uncertainty of ±100 m, the design depth is 5100 m. Evaluation of possible flow test intervals limits the possible reservoir interval from 4700 m to 5011 m.

In the geological evaluation it was determined that the well would likely produce small amounts of hydrogen sulphide. As this may cause hydrogen embrittlement in the casing string, one decided to use sour service steel in the production casing and liner, which is H_2S resistant.

The cost of the well was also evaluated. Two different scenarios were compared:

1. *Integrated design.* The well is designed to handle both the drilling phase and the testing phase. This will require expensive and strong production casing which can withstand both the "gas-filled casing" scenario and the "leaking tubing" scenario.
2. *Separate drilling and flow testing.* The production casing is for this alternative designed to withstand a "gas-filled casing" scenario during drilling only. This string will not be sufficiently strong for flow testing. If flow testing is performed, a tie-back string must be installed, which can handle the pressures if a "leaking tubing" scenario arises.

It was decided to go for option 2. If the well was not tested after all, one would save some of the cost of the production casing. If, on the other hand, flow testing was decided to be performed, one had to take into account the installation of an extra tie-back string.

The weight of the casing was another element that led to the conclusion of a relatively light production casing with a light tie-back string. A single casing which would handle both scenarios above would be considerably heavier, approaching the practical weight limit of the drilling rig.

Figure 6.1 shows the various scenarios the well is designed for. These can be briefly summarised as follows:

Alternative 1: Drilling through reservoir. If no hydrocarbons are observed, the well will be plugged and abandoned.

Alternative 2: Installation of a liner. The casing is not sufficiently strong to handle a flow test.

Alternative 3: If a well test is to be performed, a tie-back string must be installed from the liner to the wellhead.

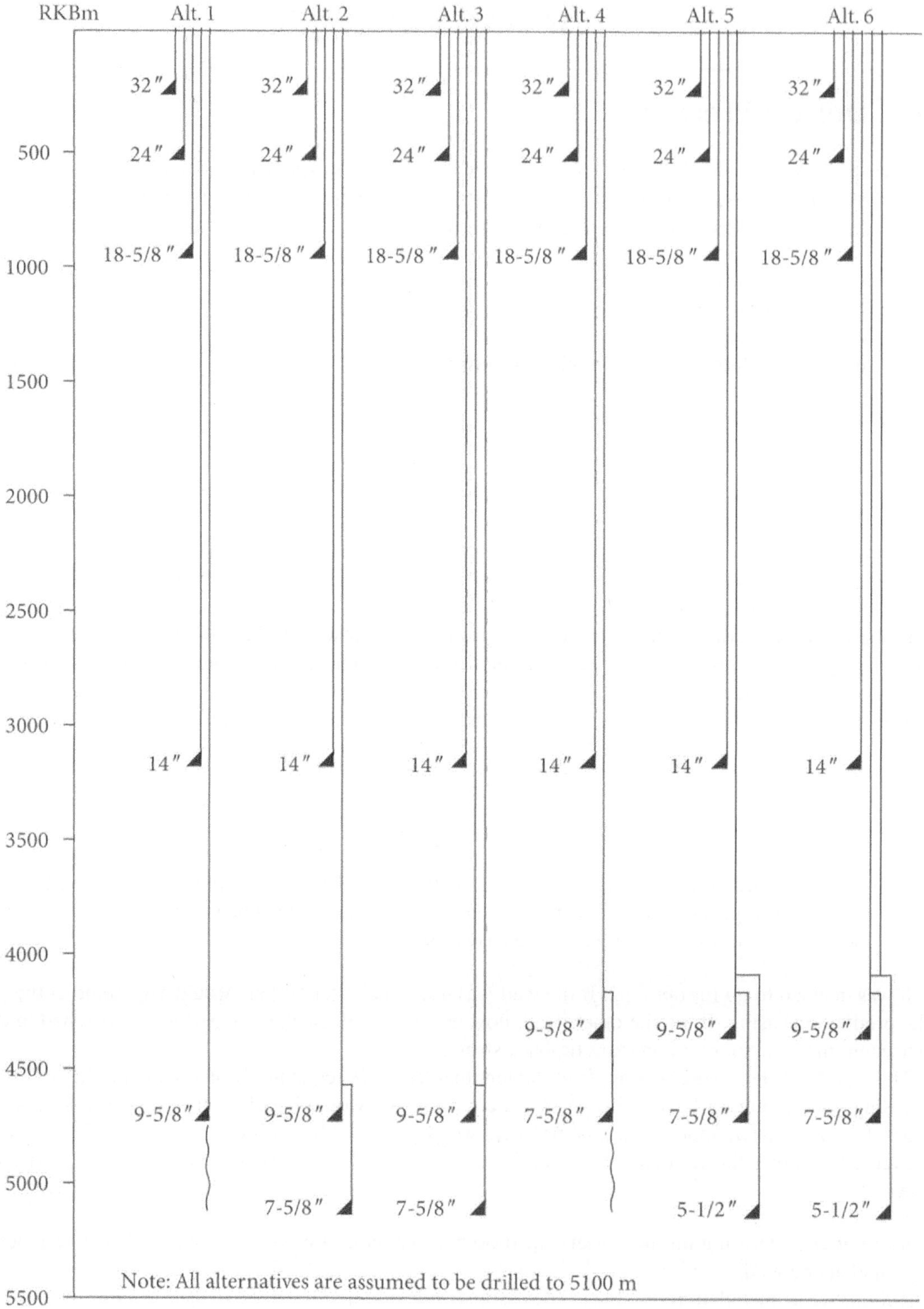

FIGURE 6.1 Casing string alternatives.

Alternative 4: If unexpected borehole problems like significant circulation losses arise, the production casing may have to be set shallower. In this case, the liner may not be set through the reservoir.

Alternative 5: If alternative 4 is used, a 5-1/2 in. contingency liner must be used in the reservoir interval.

Alternative 6: If the contingency liner is used, the tie-back string must also be installed when a flow test is performed.

Although contingency strings may be needed at shallower depths, they are not included in the analysis. The reason is that the most difficult and most critical interval is covered by the alternatives above. The setting depth of the shallower casing strings including the 14 in. intermediate casing were considered realistic, with no need for contingency solutions.

6.2.2 Setting Depth

Evaluation of three reference wells resulted in a preliminary setting depth evaluation. When the design was carried out, some of the depths were slightly adjusted, with the final results as follows:

- The 30 in. conductor casing was set at 210 m depth, that is a casing length of 100 m.
- The 24 in. casing was designed to be set at 580 m. This actually represents an extra surface casing designed to isolate several possible shallow gas intervals.
- The 18-5/8 in. surface casing was set at 1000 m, mainly to isolate shallow gas and swelling clays.
- The 14 in. intermediate casing was designed to be set at 3100 m, to isolate away all clay before entering the chalk.
- The 9-5/8 in., 9-7/8 in. production casing was designed to be set at approximately 4700 m depth, to provide well integrity before entering the reservoir, and to isolate possible lost circulation zones. This casing must be set as deep as possible before entering the reservoir because the operational margins between kick and losses is usually very small in wells of this type.

6.2.3 Pressure Prognosis

Figure 6.2 shows the resulting pressure gradient prognosis for the well. The various elements of the prognosis will be modelled and derived in the following.

6.3 GEOMECHANICAL DESIGN

6.3.1 Shallow Fracturing Model

In this evaluation we will simply apply the model derived in Section 3.3: Fracturing pressures for shallow penetration. In this chapter leak-off data from a number of soil drilling and shallow casing strings were compiled, and normalised with respect to water depth. Then a model was defined on the low side of these data. Using this model, there is very little likelihood that the actual leak-off obtained in this well obtains lower values. In other words, the model is conservative. One problem with shallow leak-off data is the

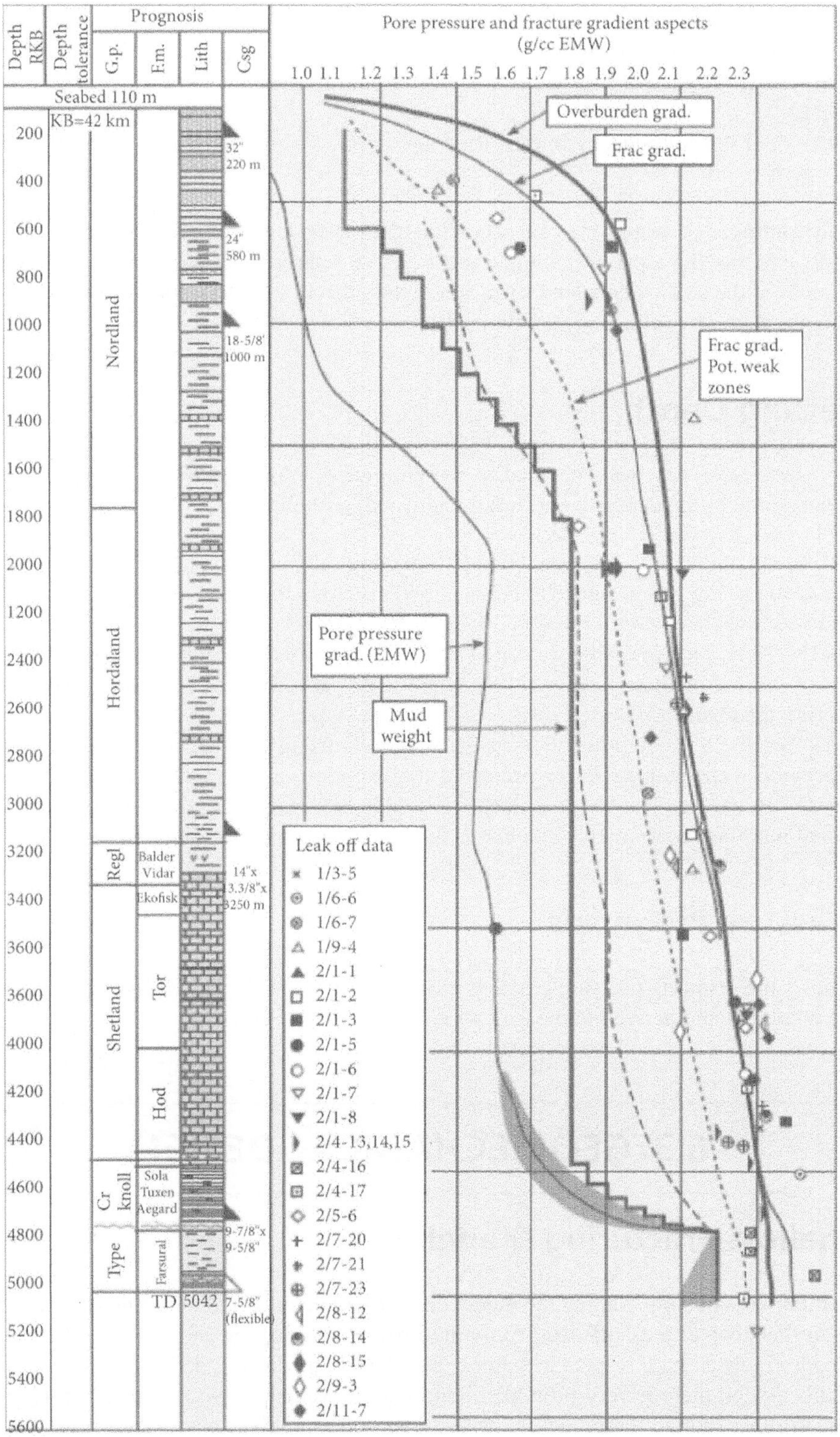

FIGURE 6.2 Pressure gradient design prognosis.

large spread. It has so far been impossible to model this spread, giving one of the reasons for using the low side of the data only.

The model from the chapter referred to above is defined by the following equations:

$$d = 1.03\left(D_{sf} - h_f\right) + 1.276\frac{D - D_{sf}}{D} \quad \text{for } 120\text{ m} > D - D_{sf} > 0 \tag{6.1a}$$

$$d = 1.03\left(D_{sf} - h_f\right) + 1.541\frac{D - D_{sf}}{D} - \frac{33.16}{D} \quad \text{for } 600\text{ m} > D - D_{sf} > 120\text{m} \tag{6.1b}$$

For our particular case, we have 68 m water depth, and the drill floor of the jack-up rig is located 42 m above sea level, resulting in a total distance of 110 m from the drill floor to the sea floor. The frac. gradient from Equation 6.1 is adjusted for the drill floor elevation. The pore pressure gradient, which is normal hydrostatic sea water, can be expressed from the drill floor as:

$$d_p = 1.03\frac{D - h_f}{D} \tag{6.2}$$

Inserting the depth numbers into the equations above, the difference between the leak-off pressure and the pore pressure can be expressed as:

$$\delta d = d - d_o = 0.246\frac{D - 110}{D} \quad \text{for: } 230\text{m} > D > 110\text{m} \tag{6.3a}$$

$$\delta d = 0.511\frac{D - 174.9}{D} \quad \text{for: } 710\text{ m} > D > 230\text{ m} \tag{6.3b}$$

Equation (6.3) gives the shallow frac. gradient or leak-off prognosis for depths down to 710 m penetration. This model will be plotted and shown on the gradient plot for the complete well. A point to note is that we have chosen to plot the difference between the frac. gradient and the normal pore pressure gradient. Since the drill floor air gap introduces normal pressure gradients less than one (from Equation (6.2)), it is observed that field personnel have difficulty handling this. By using the difference as shown in Equation (6.3), we do not have to relate to gradients less than one. Figure 6.3 shows this equation plotted and also a normal pore pressure gradient referred to sea level. By using this plot we actually obtain a higher frac. gradient because the pore pressure gradient is not adjusted for air gap, but we chose to do this in order to reduce some conservatism in the frac. gradient estimation process.

Figure 6.3 shows the resulting frac. gradient for this particular well. This will serve as a design parameter for the shallowest casing strings.

6.3.2 Shallow Kick Scenario

During the well design process several kick scenarios were investigated. One reason was that several possible shallow gas zones were identified from seismic surveys, specifically at 600 m and at 820 m depth. At an early stage, a floating drilling rig was also considered, and for this case one could drill the top sections with or without riser and with or without a seabed diverting system. Obviously a number of pressure control scenarios were investigated.

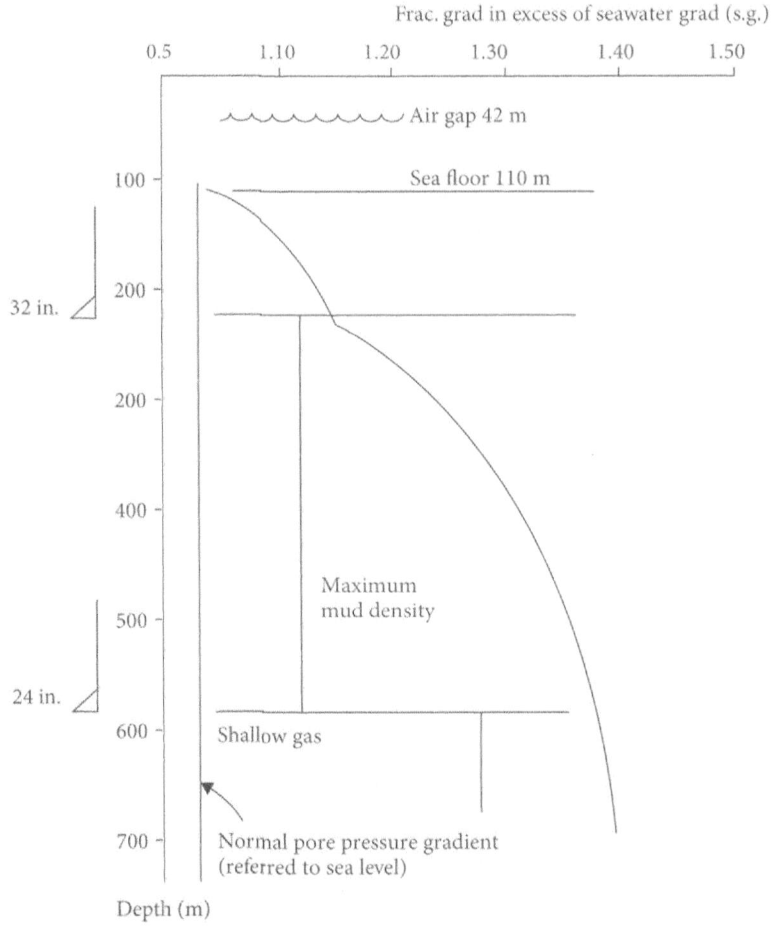

FIGURE 6.3 Design frac. gradient for shallow sections.

It was decided to use a jack-up rig. This rig type has the pressure control equipment at drill floor level, rather than at sea floor as floating rigs do. We will therefore limit our discussion to scenarios relating to the jack-up rig type.

The 32 in. conductor casing is drilled out for and set with no fluid return to surface. When drilling out the next hole section, a riser is installed providing mud return. On top of this riser is a low-pressure diverting system. After setting the 24 in. casing, a low-pressure blow-out-preventer is installed, making the unit capable of normal kick control.

We will here focus on the two phases when drilling out below the 32 in. and the 24 in. casing. Assume the situation shown in Figure 6.4a, a normally pressured volume of gas is entering the borehole, followed by a shut-in of the well. Since the bottom of the gas has normal formation pressure, and the gas can be considered weightless, this pressure also exists at the top of the gas.

Pressure on top and bottom of gas:

$$P = 0.098 \times 1.03D$$

The pressure above the gas will be:

$$P_1 = 0.098 \times 1.03D - 0.098\ d_{mw}\left(D - D_1 - h\right)$$

The previous casing shoe is considered the most critical point, and the fracturing pressure at any depth is:

$$P_{wf} = 0.098\,d_{wf}D_1$$

Now assume that the well is shut in, with a gas volume that just fractures the formation. Equating the two previous expressions ($P_1 = P_{fr}$) results in an expression for the gas height h:

$$h = \left(\frac{d_{wf}}{d_{mw}} - 1\right)D_1 + \left(1 - \frac{1.03}{d_{wf}}\right)D \tag{6.4}$$

Equation (6.4) is plotted in Figure 6.4b for two scenarios; at a well depth of 600 m, and at a well depth of 1000 m.

First, we will investigate the situation when drilling out below the 24 in. casing shoe. Here we assume that the depth of next bottom hole section is 600 m, that is just below the predicted shallow gas zone. Furthermore, we assume that the drilling fluid is viscosified sea water with drilled solids that has a resulting density of maximum 1.12 s.g. Equation (6.4) becomes:

$$h = \left(\frac{d_{wf}}{1.12} - 1\right)D_1 + \left(1 - \frac{1.03}{1.12}\right)600$$

Using the design frac. curve from Figure 6.3, the above equation is plotted in Figure 6.4b. Shown is the maximum gas height as a function of setting depth for the 32 in. casing. We observe that the setting depth of the casing is critical with respect to gas influx. If we accept a maximum influx height of 60 m, we place the 32 in. casing at 220 m depth.

A similar evaluation was performed for the next hole section. Here the open hole section goes to 1000 m, and the maximum mud density is 1.25 s.g. For this case we have decided to place the 24 in. casing at 580 m to have a strong shoe before penetrating the possible shallow gas zone at about 600 m. In this case we may safely handle a gas height of 240 m.

The previous discussion focused on the maximum allowable gas height one can handle and still maintain well integrity. If such a gas volume has entered the well during the circulation out phase, higher loading can arise due to gas expansion. We will not pursue this further here.

Before leaving the topic of kicks, we will briefly investigate the effect of inadvertent loss of circulation. During drilling one may suddenly encounter circulation losses. The fluid level in the annulus may drop. In shallow layers, we assume hydrostatic pore pressure. To the knowledge of the author, there is no reported lower pore pressure than normal sea water pressure in the North Sea. If this can be assumed to be a lower limit, one may say that the mud loss stops when the pressure in the bottom of the borehole is equal to the normal pore pressure. This is defined by the following equation:

$$P = 0.098d_{mw}\left(D - h'\right) = 0.098 \times 1.03\left(D - h_f\right)$$

$$\text{or: } h' = \left(1 - \frac{1.03}{d_{mw}}\right)D + \frac{1.03}{d_{mw}}h_f \tag{6.5}$$

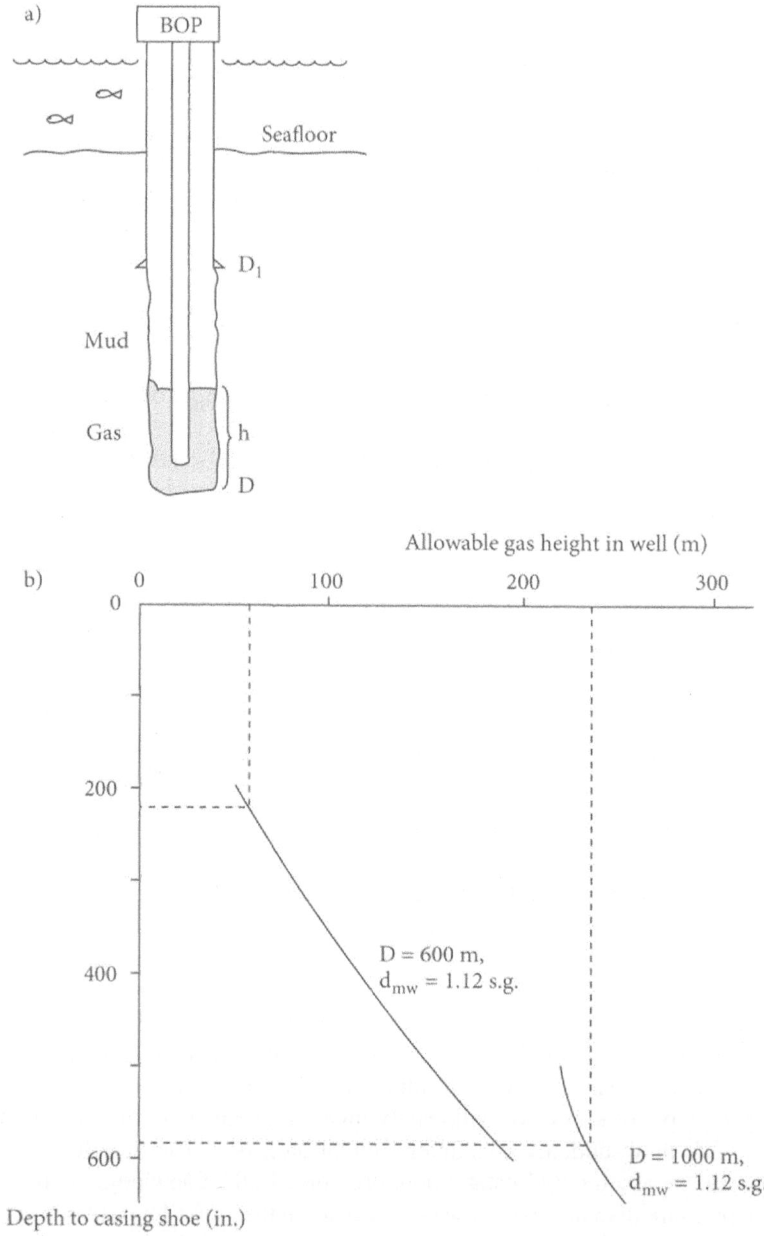

FIGURE 6.4 Shallow gas kick height versus casing shoe depth.

Figure 6.5 shows the critical annulus fluid level drop for various mud densities. If mud is suddenly lost at for example 400 m using a 1.15 s.g. drilling fluid, the maximum annular level drop is 64 m. If the mud density is 1.25 s.g., the fluid level drop increases to 105 m. The equation above can be used to estimate annulus level drops. If losses actually occur during the operation, one should attempt to check the validity of the equation above. If significant deviations occur, this should be tested against the assumption of a minimum pore pressure equal to the hydrostatic head of sea water.

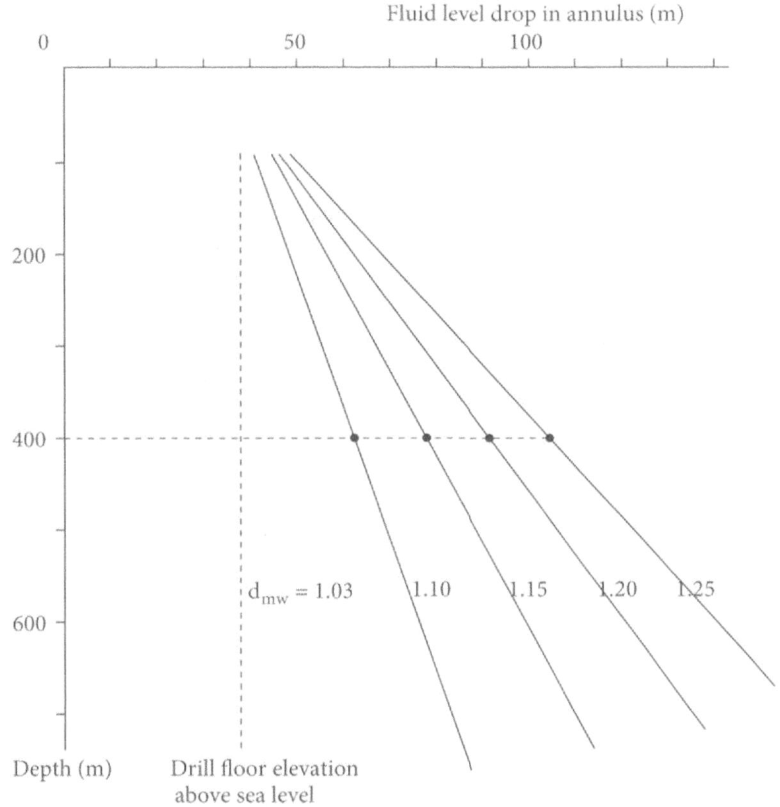

FIGURE 6.5 Maximum annulus level drop versus mud weight and depth.

Another problem arises if the annular fluid level drops, and several shallow gas zones or loss zones are exposed simultaneously. This is illustrated in Figure 6.6. Imagine that the upper zone was drilled, but no problems encountered because a pressure overbalance is retained inside the borehole. Drilling resumes, and at a deeper depth a loss zone is exposed, causing a drop in the annulus level. The fluid level stabilises when the bottom of the well reaches a pressure equal to the head of sea water. The mud weight is somewhat more dense, which results in the fluid level drop. However, now we observe that the borehole pressure at the shallow gas zone is lower than the formation pressure. An inward flow potential is created, which may result in a shallow gas kick.

Multiple shallow gas zones and loss zones must be evaluated in a common context in casing seat selection. In the particular design at hand, it was decided that the two shallow gas zones were close, and that the well integrity had to be ensured before penetrating these. Therefore the 24 in. casing was set at 580 m.

Before closing this section, a general comment will be given.

If drilling without a marine riser, multiple shallow gas or shallow loss zones can be exposed simultaneously with little risk of influx. The fluid level will never go below sea level. However, when drilling with marine riser and weighed mud, annulus fluid level may drop causing a shallow gas influx if multiple zones are exposed. For the latter case, operational procedures should ensure that the fluid level never drops below a certain limit.

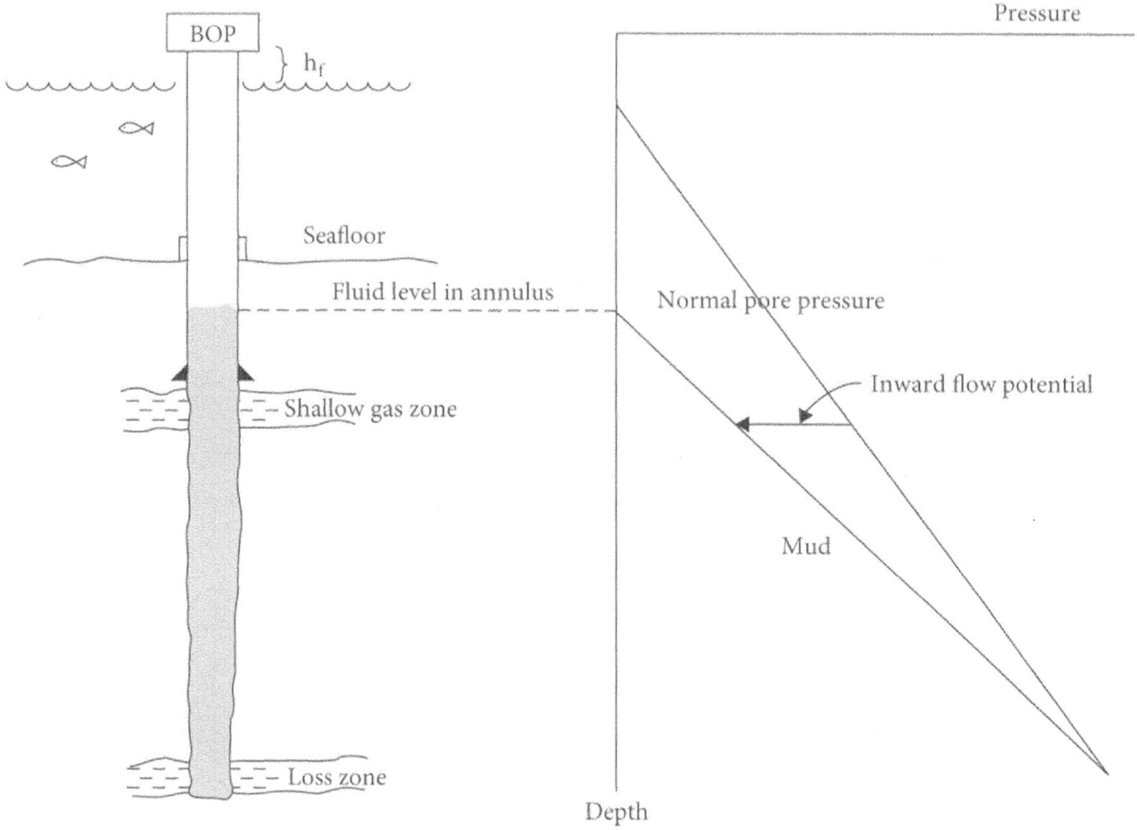

FIGURE 6.6 Flow potential caused by mud losses and multiple zone exposure.

6.3.3 Deep Fracturing Model for Central Graben

A number of HPHT wells have been drilled in the area for this new well. Despite this, there was considerable difficulty in establishing a good fracture prognosis. The main reason is that the allowable mud weight window is very small, and that the spread in leak-off data from reference wells are larger. Therefore, an innovative approach was chosen to improve the frac. prognosis.

The operator had previously performed a study over 70 HPHT wells in the southern North Sea. Also, it was decided that the best placement of the production casing was in the Cromer Knoll formation. It was further found that 36 of the reference wells had fracturing information from the Cromer Knoll. These data were then used for modelling.

Figure 6.7 shows the relevant leak-off data. We observe that the spread is in the order of 0.24 s.g. Even evaluating the proximity between the wells, the spread is too large to define a rational model.

From rock mechanics considerations it is well known that the fracturing pressure is a function of the pore pressure and the in-situ stress state. These two parameters were therefore added to the analysis. The pore pressure gradients at each leak-off point were found from final well reports. There is some uncertainty here, because the formation is basically impermeable, with no direct pressure measurements. At these depths the horizontal stress state is approaching the overburden stress. The overburden stress was therefore used to define the in-situ stress state.

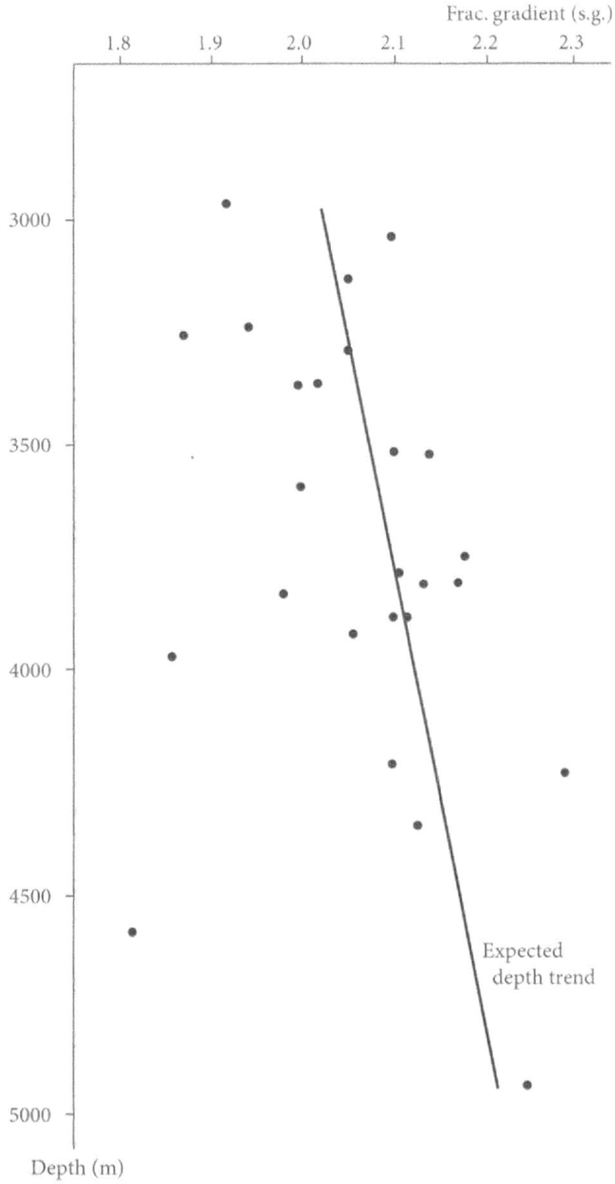

FIGURE 6.7 Leak-off reference data for Cromer Knoll in Central Graben.

Figure 6.8 shows the overburden stress gradient data at the depth of the leak-off tests. There is a considerable spread. Earlier overburden data were of varying quality. In particular, at shallow depth poor correlations have often been used in the absence of well logs. In the analysis, we used the data from Figure 6.8 to define an overburden trend line which was used in the analysis.

Finally, Figure 6.9 shows the leak-off data plotted against the pore pressure gradients for each point. We observe a considerable variation in pore pressure, and suspect that this is an important parameter in the magnitude of the leak-off data.

At this stage we have the following data: leak-off gradients, pore pressure gradients, and the stress state (overburden stress gradients) for the same rock type. The problem is to use these data constructively to derive a frac. model. Direct correlation between these parameters was attempted, but did not yield improvements. Therefore another approach was pursued.

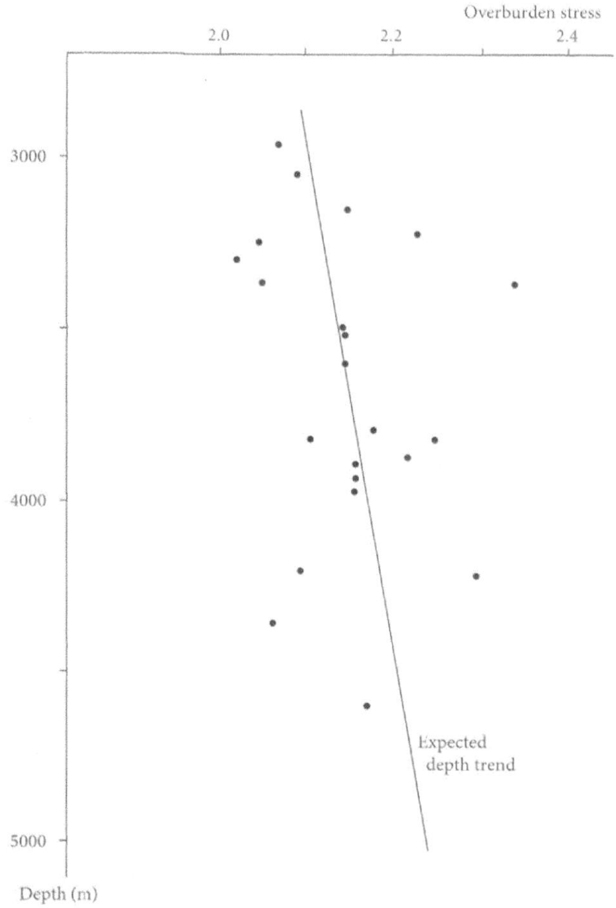

FIGURE 6.8 Overburden stress gradients for the leak-off data.

Aadnoy (1991) derived a simple compaction model, which yielded the modified leak-off pressure when the pore pressure was varied. Specifically, the model was initially derived to estimate the reduction in frac. pressure in depleted reservoirs. Since the present leak-off data were clearly pore pressure dependent, and since the pore pressure varied, all leak-off data were normalised to the same pore pressure gradient. The compaction model is presented in the Section 3.2.4.

The data presented in Figures 6.7, 6.8, and 6.9 will be normalised in the following way:

- The leak-off data will be divided by the overburden stress. It is assumed that the horizontal stress is governed by the overburden stress, and that the depth dependence is removed in this process. The leak-off/overburden stress ratio should be a depth-independent number with a vertical trend. This is done to develop a model suitable for predictions at other depths.
- The leak-off gradients are normalised to a common pore pressure gradient by the compaction equation. Since the leak-off data clearly are pore pressure dependent, this process should make the data more comparable. For predictions, the expected pore pressure gradient must be inserted into the model.

The average overburden stress trend from Figure 6.8 was found to be:

$$\sigma_o = 1.916 + 0.00006611D \quad (6.6)$$

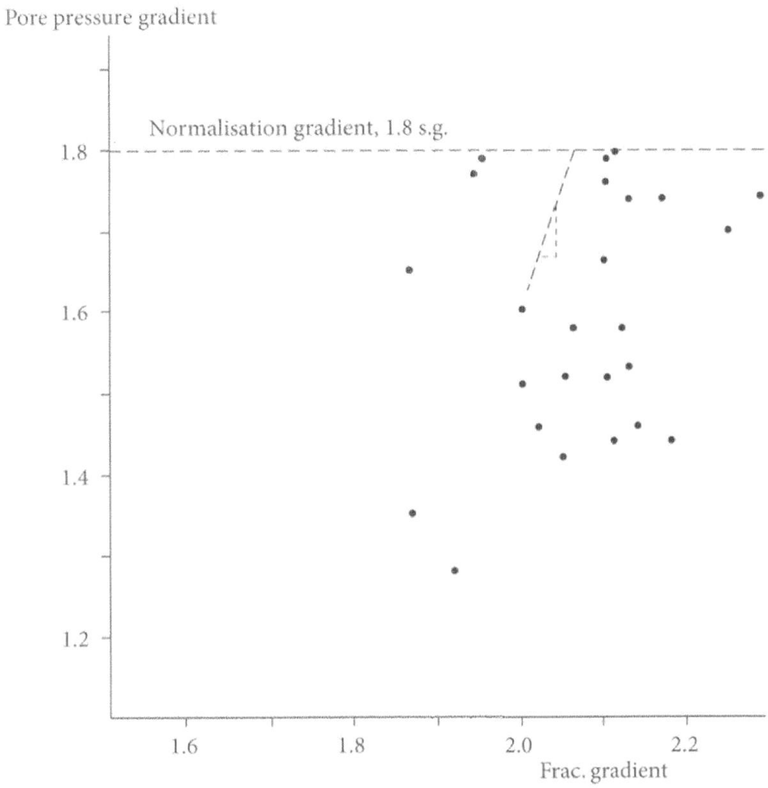

FIGURE 6.9 Pore pressure gradient versus leak-off pressure.

and the compaction equation for an arbitrarily chosen pore pressure reference gradient of 1.8 s.g. (see Section 3.2.4):

$$\frac{P_{wf-normalized} - P_{wf}}{\sigma_o} = \frac{1.8 - P_o}{7\sigma_o} \tag{6.7}$$

We have in the derivation of Equation (6.7) assumed a Poisson's ratio of 0.30, which gave the least spread in data.

The data from Figures 6.7, 6.8, and 6.9 were normalised using Equations (6.6) and (6.7) above. The result is shown in Figure 6.10.

In Figure 6.10a the initial leak-off/overburden stress data are plotted. The spread is large, again pointing to the difficulty of using these for rational prognosis. Since we have divided each of the leak-off data with the overburden stress from Equation (6.6), we expect a vertical trend.

In Figure 6.10b the same data are in addition normalised to a pore pressure gradient of 1.8 s.g. by Equation (6.7). The result is that the spread in the data is reduced from 0.27 s.g. to 0.11 s.g., if the extreme data points are neglected. Comparison of the two plots shows a considerable improvement. However, a spread of 0.11 s.g. is still too wide a margin for direct practical applications.

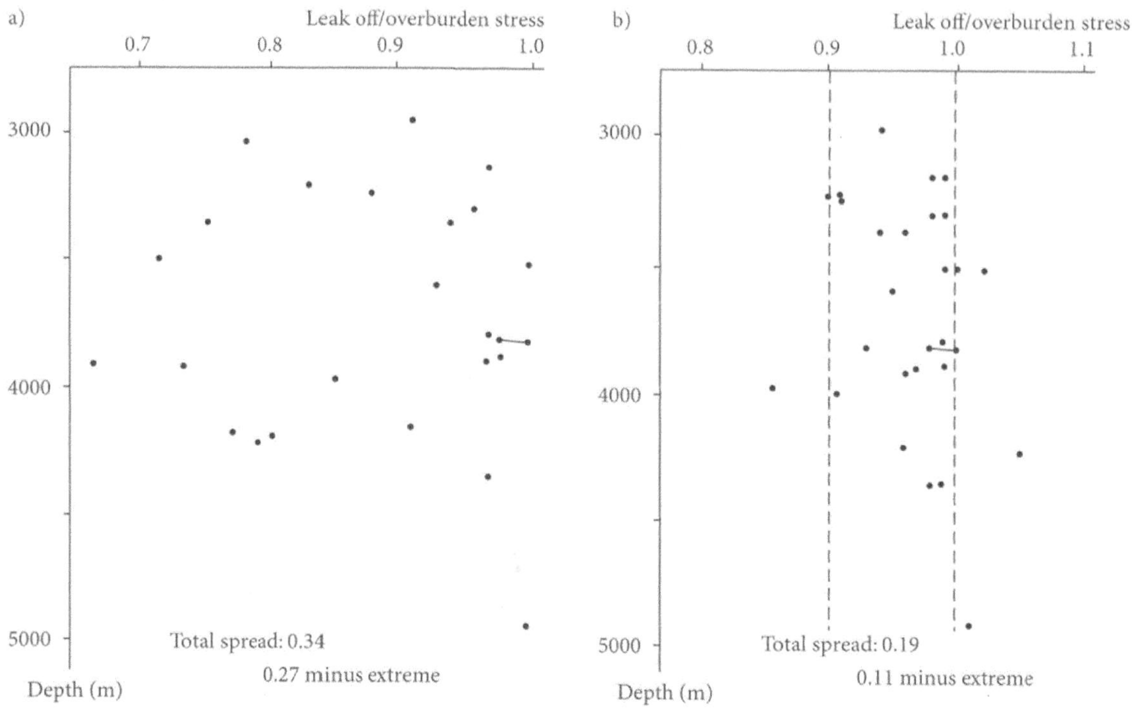

FIGURE 6.10 Comparison between depth normalised and depth-pore pressure normalised leak-off data. a) Leak-off data normalised with overburden stress. b) Leak-off data normalised with overburden stress, compaction, pore pressure (1.8 s.g.).

The data of Figure 6.10b were split into three groups, one for low frac. gradient, one for medium and one for high. These were mapped onto the map shown in Figure 6.11. The data from the three groups were not randomly scattered on the map, but seemed to concentrate in certain regions. The physical reason for this grouping is not fully understood, but one may relate it to the fault pattern and numerous salt domes in the area.

The well we are planning in this chapter is located in the area with the group for high frac. gradients (group III, Figure 6.10b).

The task that remains is to evaluate the fracture level of the well. The nearest reference well is shown in Figure 6.10b. Due to proximity, but also due to geological similarity, this is considered the best reference well. We therefore chose from Figure 6.10a a normalised frac. level of NFG = 1.01. The resulting fracture prediction equation then becomes:

$$P_{wf} = NFG\sigma_o - \frac{1.8 - P_o}{7} \quad \text{or:}$$
$$P_{wf} = 1.01(1.916 + 0.00006611D) - \frac{1.8 - P_o}{7} \tag{6.8}$$

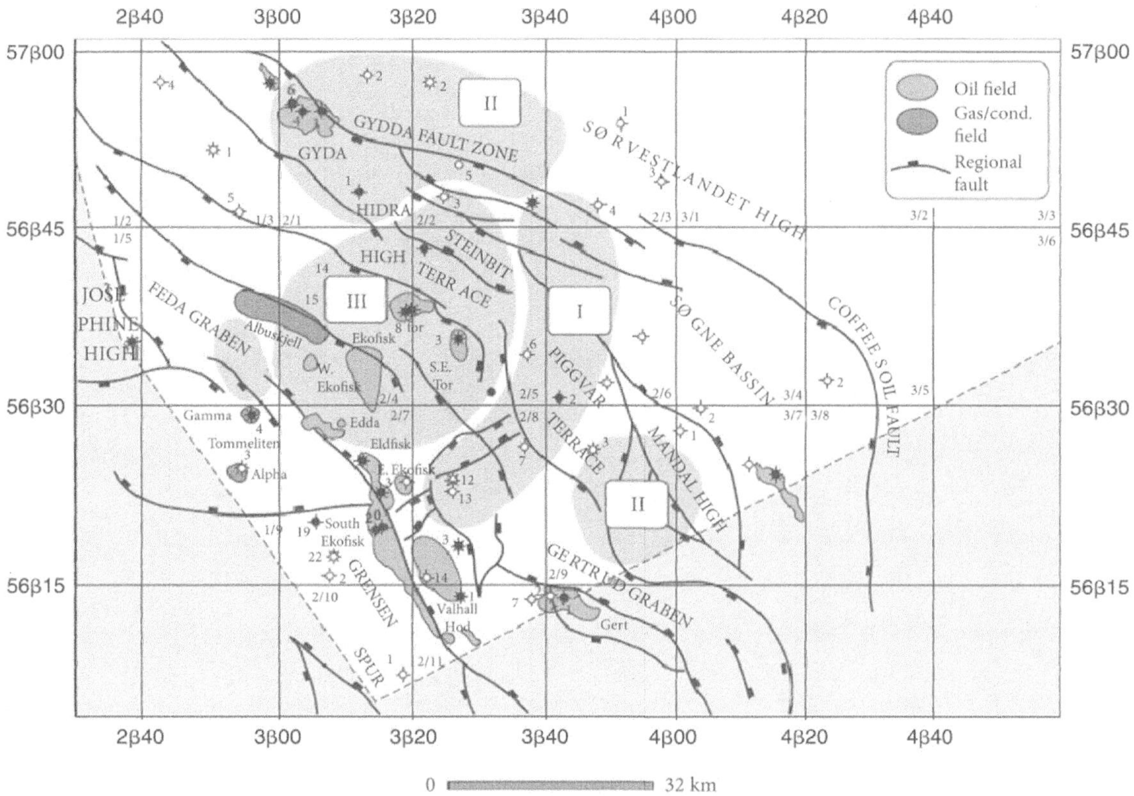

FIGURE 6.11 Geographical grouping of normalised leak-off data.

Figure 6.12 shows the predicted fracture gradient for our well, calculated from Equation (6.8). We observe that the pore pressure dependence is clearly reflected in the plot.

6.3.4 Deep Lost Circulation Model

It is well known in the field of drilling engineering that mud losses often occur. Unexpected losses may create hazards with respect to well control, and should be minimised. Since most casing string shoes are placed in competent rock, these often represent the stronger parts of the system. Losses often occur during continued drilling, and at lower levels than the leak-off positions.

We will adopt the philosophy that two fracture gradient plots are valid for the well, one for strong shales, and one for loss zones, like sand stringers. We also compiled loss data for the wells of the Central Graben area, and found twelve data sets that were relevant. These were normalised and analysed the same way as for the leak-off data.

Figure 6.13a shows the depth and pore pressure normalised loss data. We have arbitrarily made two groups, one high and one low. Figure 6.13b shows the area distribution of these data. Two nearby reference

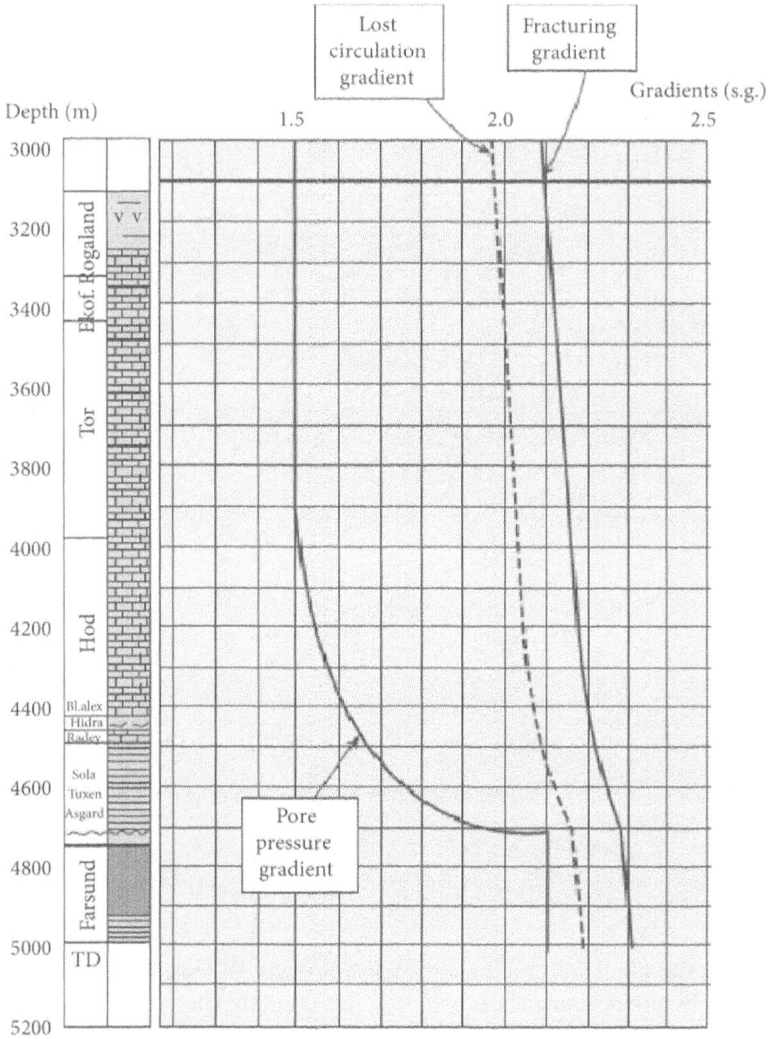

FIGURE 6.12 Prognosed fracture gradient at reservoir level.

wells have experienced losses. The best estimate of this normalised loss gradient is NLG = 0.95, which yields the following loss prediction equation:

$$P_{wf} = 0.95(1.916 + 0.00006611D) - \frac{1.8 - P_o}{7} \tag{6.9}$$

This equation is shown in broken line in Figure 6.12. This figure now defines the frac.-loss window for which we want to design our well. The difference between the frac. curve and the loss curve is about 0.11 s.g.

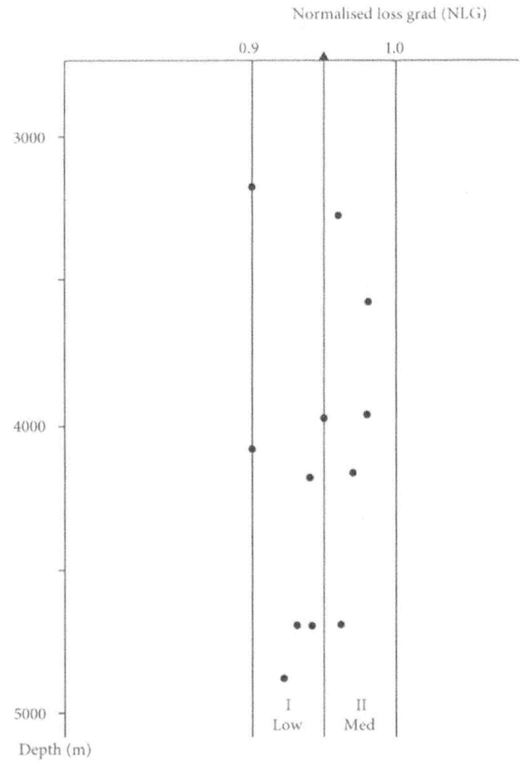

a) Normalised loss gradients

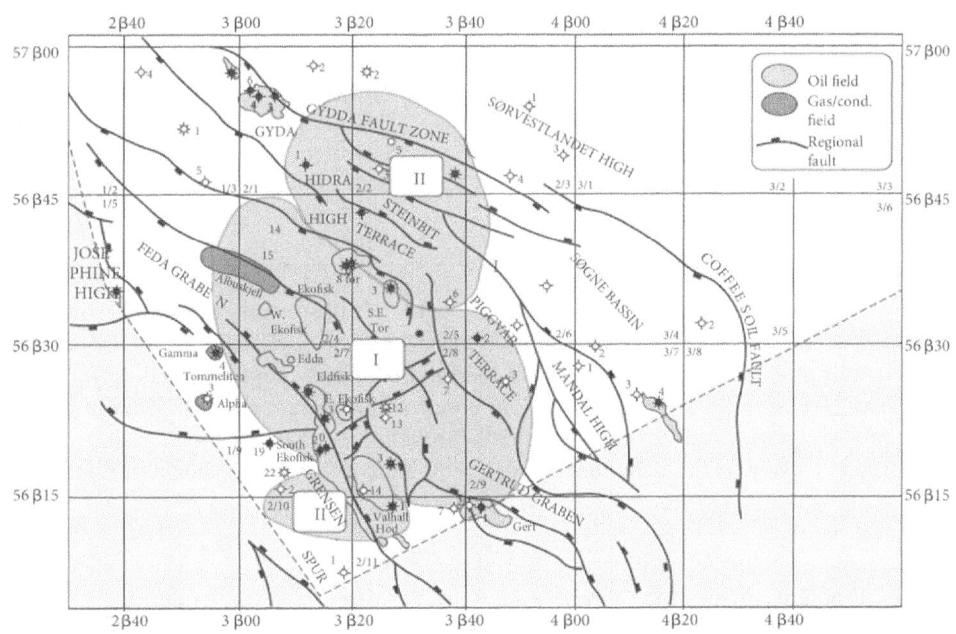

b) Geographical grouping of normalised loss data

FIGURE 6.13 Analysis of loss data.

6.4 DESIGN OF THE MUD WEIGHT SCHEDULE

The mud density is a critical parameter to minimise borehole problems. In this chapter we will determine the mud weight schedule for the whole well. There are two basic elements considered: evaluation of problems in reference wells and the application of the median line principle. The latter will be applied, but for background information see Section 2.2.

The operator has previously drilled three HPHT wells in the same area. These are all drilled in a similar geological setting. Due to the similarities both in geology and in the wells, we will in the following evaluate the problems encountered during drilling of these wells, in order to minimise borehole problems for the well under planning. We will call the three reference wells A, B, and C, respectively.

Rock mechanical collapse analysis is not performed, but this chapter serves to illustrate an alternative, but practical approach, which is very useful in a practical well design.

Evaluation of Well A

In the following is a short list of reported drilling problems from well A. The information is taken from the daily drilling reports:

	Depth (m):	Problem:
The 17-1/2 in. section:	931–1221	Tight hole, max. overpull 45 tons
	1600–1700	Tight hole
	1640	Tight spot, max. overpull 45 tons
	2130	Max. overpull 85 tons
	2125–2135	Tight hole, max. overpull 55 tons
	2295–2334	Reaming, max. overpull 55 tons
The 12-1/4 in. section:		Max. overpull 30 tons
	3538	Stuck
	3335–3388	Reaming
	3388–3487	Reaming
	3423	Kick-off for side-track
	4038	Tight hole
	4023–4048	Reaming

The mud weight schedule used is shown in Figure 6.14. We observe that relatively low mud weights have been used, and that the problems reported are typical for these sections in many wells. This is a typical exploration well mud weight schedule, where the mud density is kept low, mainly for pore pressure mapping purposes. The low mud weight gives low borehole pressure, resulting in an inward creep. By increasing the mud weight towards the median line, the tight hole conditions should reduce. Since the new well will be drilled in the same area, the most recent geological prognosis is shown in Figure 6.14.

Evaluation of Well B

This well is actually a relief well drilled to kill a blowout in well A. This well is deviated, contrary to the other wells. Problems arose in the top hole sections, resulting in several side-tracks. The following problems were reported from the daily drilling reports:

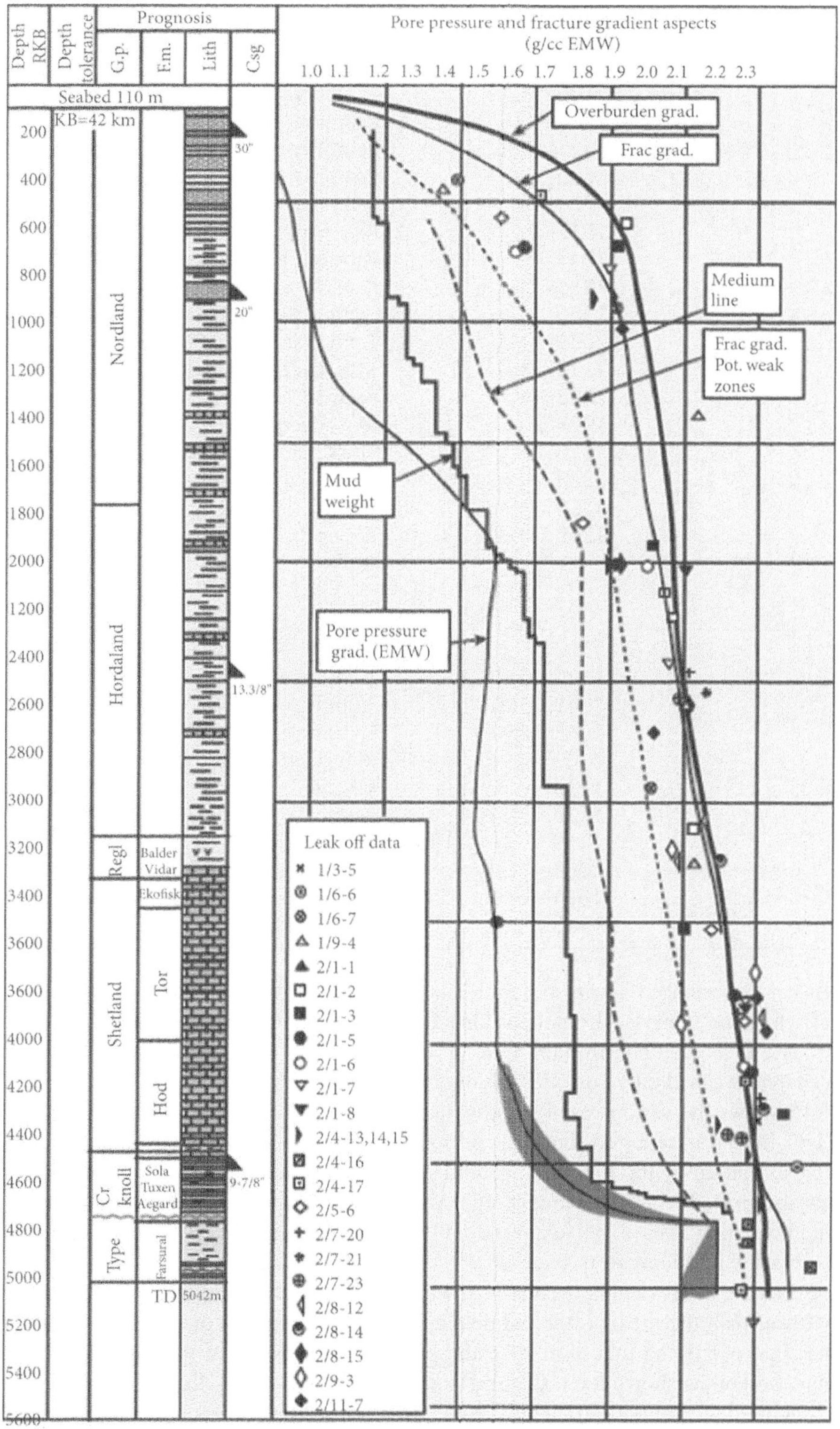

FIGURE 6.14 Mud weight schedule used in well A.

The 17-1/2 in. hole section:	Depth (m):	Problems:
	916–931	Reaming
	1250–1280	Tight hole
	1406–1426	Reaming
	1483–1507	Reaming
	1521–1536	Reaming, max. overpull 75 tons
	1609–1807	Reaming, cavings observed
	1798–1970	Reaming, cavings
	1915–2077	Reaming, stuck
	1943	Pack off, lost circulation, max. overpull 100 tons
Side-track no. 1:	1283	Kick off
	1980	Tight hole, reaming, max. overpull 77 tons
	2031–2110	Tight hole, ream, 68 tons overpull
	2110–2197	Max. overpull 77 tons
	2363	Pack off, circulation loss
	1963–2040	Max. overpull 113 tons
	1616	Stuck, circulation loss
	2328	Stuck with underreamer
	1350	Stuck
	1262	Back off

Side-track no. 2:	1262	Kick off
	1120–1160	Tight hole
	1945–2100	Tight hole
	1643–1850	Tight hole
	2100	Max. overpull 68 tons, 170 tons, 200 tons
Side-track no. 3:	1730	Kick off
	1739	Pack off
	1755	Pack off
	2068	Pack off, max. overpull 123 tons
	1014–1040	Underreamer, overpull 114 tons
The 12-1/4 in. section:	3703	Max. overpull 120 tons

This well has experienced severe borehole stability problems. The deeper sections were relatively problem free, while the upper sections had considerable problems. This can relate to the mud weight, reactive clays, and hole cleaning, in particular. Of course each of these issues must be addressed separately. In this chapter we will only consider the mud weight.

Figure 6.15 shows the various mud weights used. In the 17-1/2 in. section the mud weight was initially low. Tight hole was reported, but also presence of cavings, which is an indicator of mechanical borehole collapse, caused by too low mud weight. Also, hole cleaning problems were observed. The first side-track was also drilled with a low mud weight, again going stuck. In the second side-track, a considerably higher mud weight was used, but again the drill string went stuck. The third side-track was successful, and the well gave few problems from there on.

In Figure 6.15 the mud weights used are shown. We observe that side-track no. 1 used a mud weight considerably lower than the median line, while side-track no. 2 used a considerably higher mud weight. This illustrates that overdoing may not solve the problem. It is believed that the median line gives the most ideal mud weight, as shown later. Comparison between side-tracks 1 and 2 is interesting. By going directly to a high mud weight (side-track no. 2) one does not get the advantage of successive "jacking" of the borehole wall.

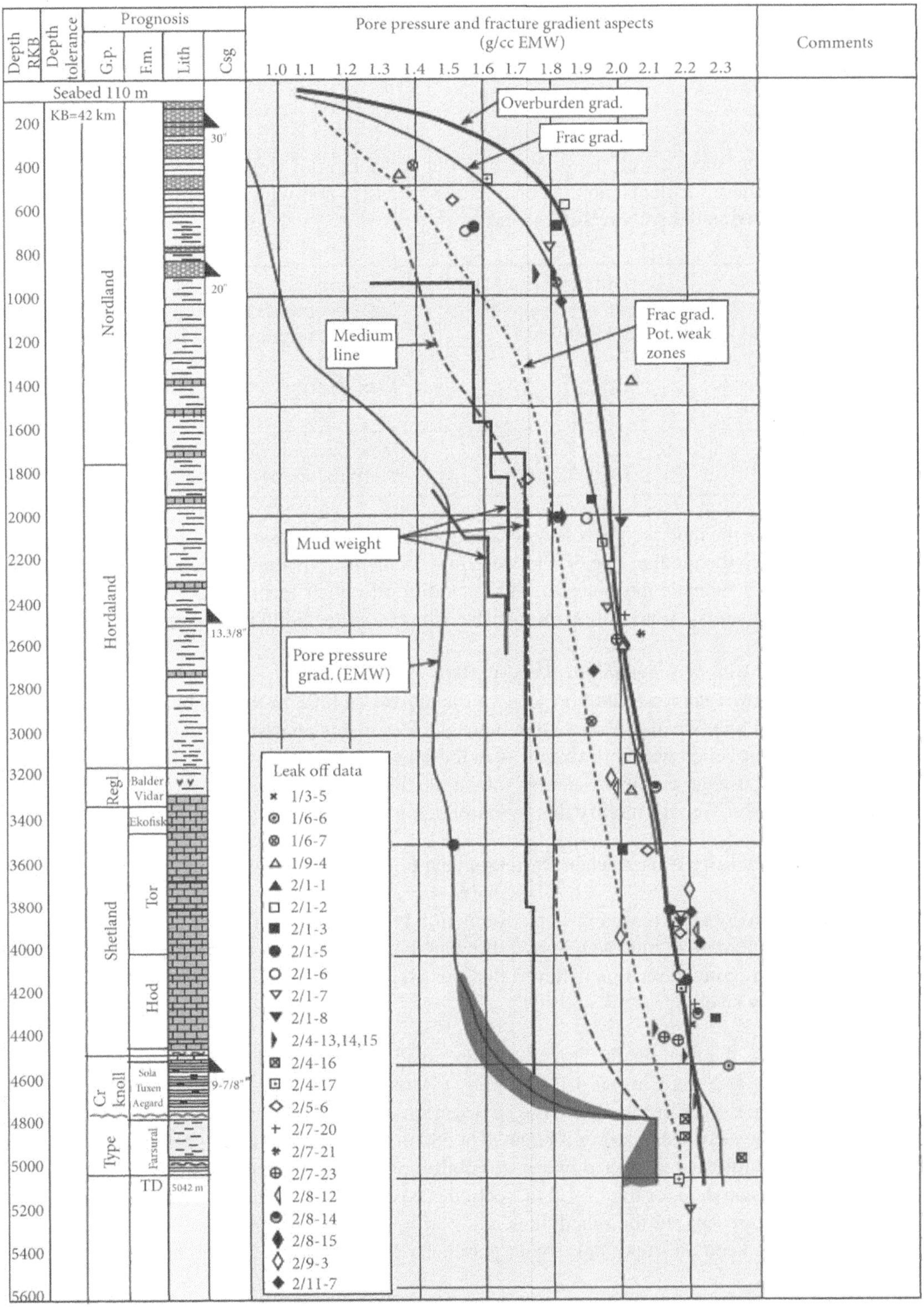

FIGURE 6.15 Mud weights used during drilling of well B.

Side-track no. 3 had mud weights close to the median line initially. The deeper portion of side-track no. 3 had mud densities lower than the median line. This is mainly due to a wish to tag and map the pore pressure increase above the reservoir. At these depths, borehole stability problems have been relatively small.

Evaluation of Well C

The two previous cases show a well that took a blowout and the relief well used to kill it. The next well was not involved in these incidents, and is considered successful and with few problems. The following problems were reported in the daily drilling reports:

The 17-1/2 in. section:	Depth (m):	Problem:
	1269	Max. overpull 55 tons
	1269–973	Stuck, jarring, overpull 140 tons
	937–1335	Reaming
	1380	Max. overpull 140 tons, cavings

The 12-1/4 in. section:	3254	Max. overpull 60 tons
	3370–3520	Washing and reaming

Figure 6.16 shows the mud weight schedule used for this well. We observe that it is no longer extreme, but reasonably close to the median line. There were few problems. Of course the mud weight schedule is not the only parameter which contributes to this. In addition to other technical issues like hole cleaning and mud inhibition, there is a practical learning curve which is reflected in the operational practices.

Establishing the Mud Weight Schedule for the New Well

The previous three cases demonstrate how one can analyse borehole problems by comparing the mud weights to the median line. We also observe that there is a reasonable consistency in the approach. The last well had a mud weight which was not extreme as in the previous cases.

Actually, we will use the previous discussion to argue that the median line is a reasonable design concept for the mud weight. We will modify this by introducing the following guidelines:

- A low mud density is used when starting a new hole section because this may reduce later loss problems.
- The mud density is increased in steps. Reduction is avoided as it may lead to tight hole.
- A stepwise increase in mud weight by following the median line principle is preferred rather than an instant large increase. It is believed that this stepwise "jacking" of the hole counteracts tight hole and clay swelling.

Figure 6.17 shows the final result. The following considerations are taken into account:

The 38 in. section (32 in. casing) is drilled with sea water without return to the rig. After installing the 32 in. casing, a marine riser is installed, and the next section will be drilled with weighed mud. Because of fear of shallow gas, the next casing string is planned at 580 m. In all these shallow sections the mud density is kept as high as possible, but restricted by the possibility of circulation losses.

The mud weight schedule for the 17-1/2 in. section is based on the information for the reference wells. Well C is the best reference, but the schedule is adjusted towards the median line. Also here, the fear of circulation losses has been an important design parameter for both the mud weight and the casing seat selection.

The 12-1/4 in. section is being drilled in chalk, and swelling problems in the formation are not expected here. Experience from well C showed few problems here. We therefore decide to use a mud weight lower than the median line, primarily to be able to establish the formation pressure at the bottom of

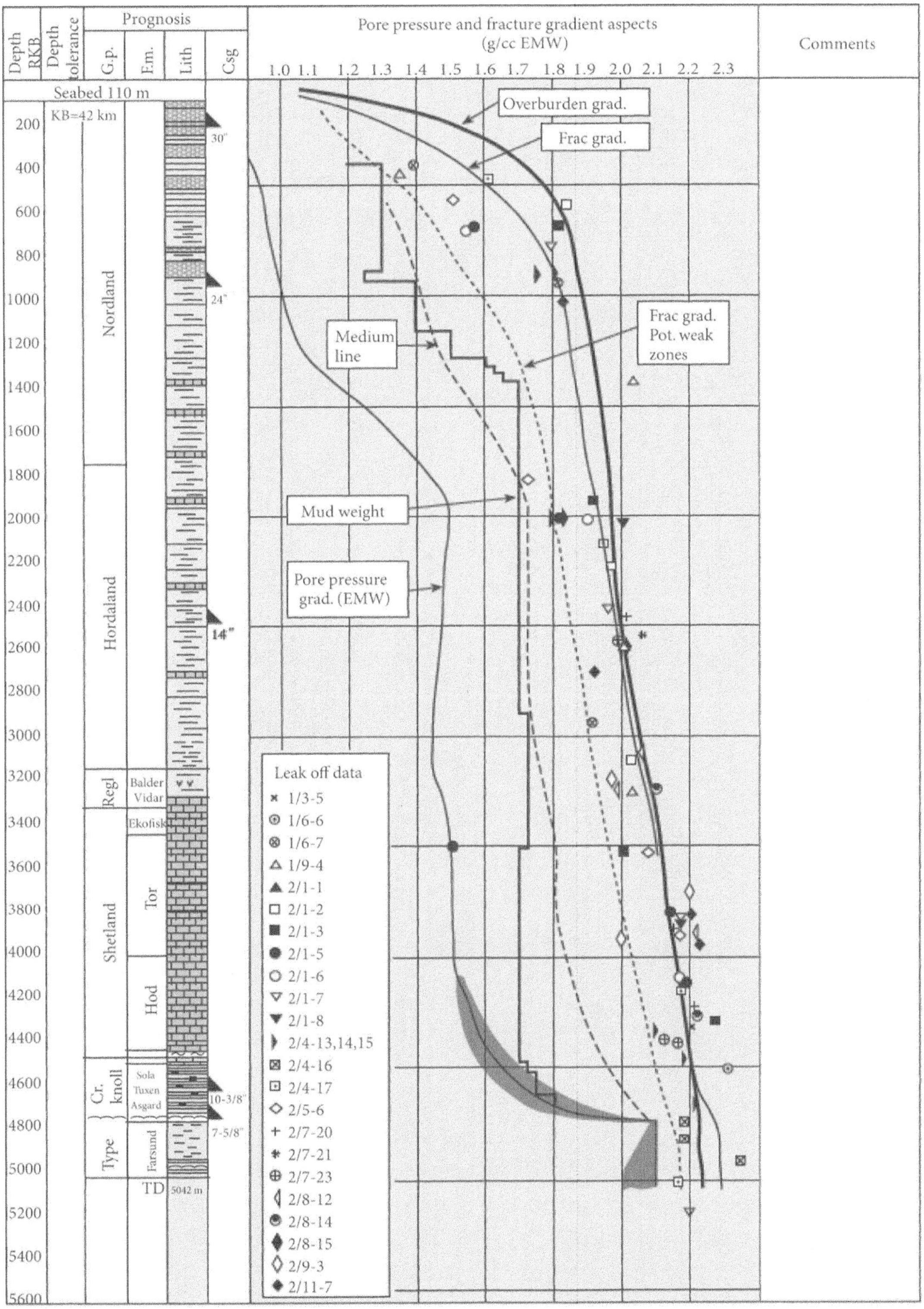

FIGURE 6.16 Mud weight schedule for well C.

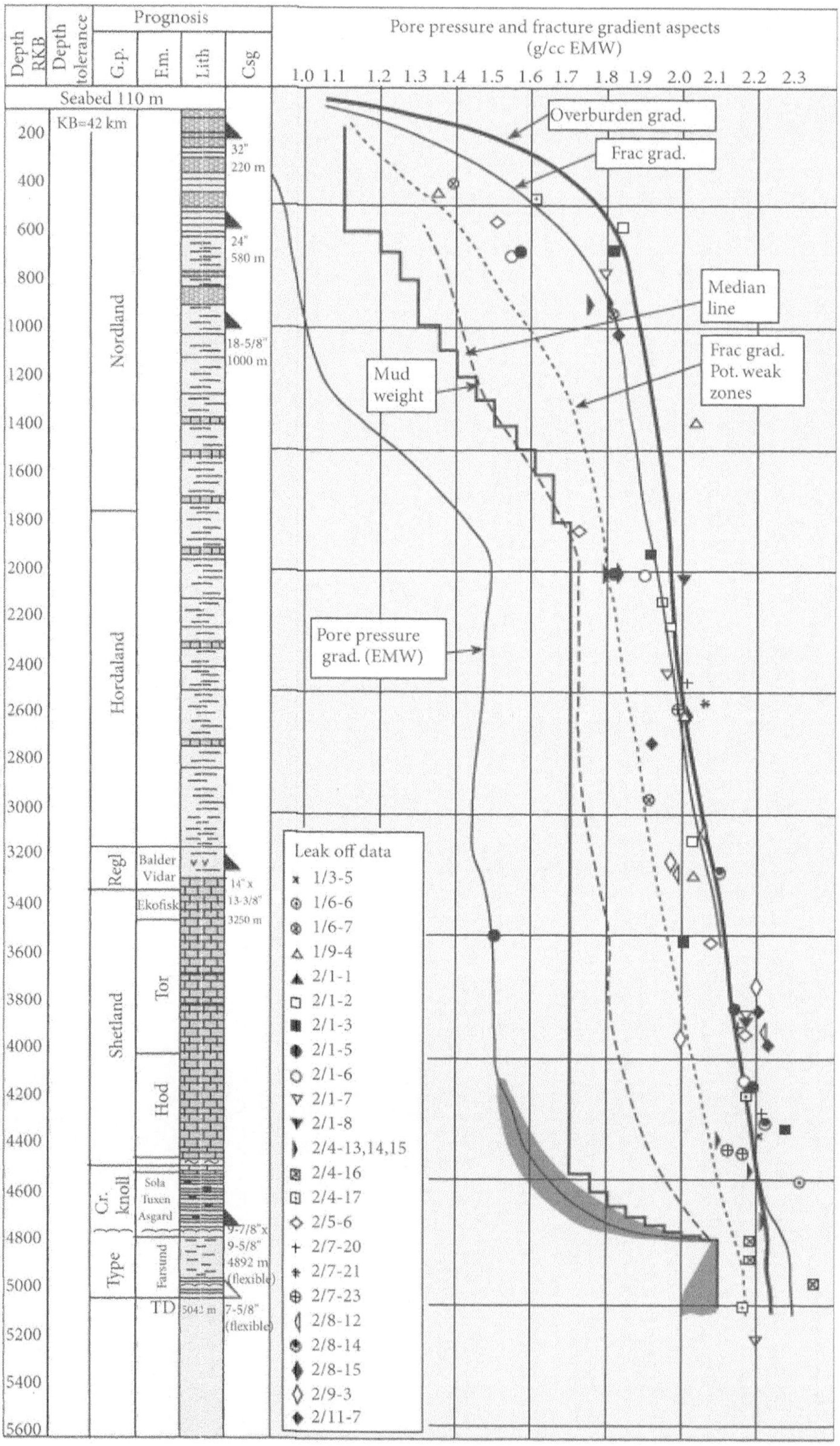

FIGURE 6.17 Proposed mud weight schedule for the new well.

the section. After the pore pressure has been tagged, a stepwise increase is called for until the production casing seat has been reached.

6.5 PRODUCTION CASING CONSIDERATIONS

6.5.1 Setting Depth Based on Mud Weight and Kick/Loss

In the previous sections the frac. prognosis for the well has been derived. This is used as a key input parameter in the design of the well. Most casing strings will be set according to the prognosis, with only minor modifications. The reason is that they are not especially critical.

The production casing is the most critical casing string. Due to the rapid pore pressure build-up at about 4700 m and the reduced difference between the frac. and the pore pressure gradient curve, this casing location is very critical. The following general criteria were identified:

- The production casing should be set as deep as possible before drilling through the reservoir, to provide sufficient fracturing integrity.
- The Cromer Knoll group in the lower Cretaceous therefore gives the setting interval.
- By setting the production casing in the deeper part of the Cromer Knoll group, a good integrity should result, also reducing the likelihood of possible weak zones or hydrocarbon-bearing intervals.
- Rapid pore pressure increase with depth is experienced in most Cromer Knoll zones in the area. Therefore, the maximum pore pressure gradient of the well should nearly be obtained before drilling out the reservoir.

Based on the above discussion, the production casing should be set as deep as possible into the lower part of the Cromer Knoll group.

6.5.1.1 Setting Depth Versus Mud Weight Used During Drilling

The prognosis in Figure 6.2 will be used to evaluate the production casing setting depth. The 14 in. casing is planned to be set at 3250 m. When drilling out below for the production casing, the 14 in. shoe location is probably the weakest point. At this depth the prognosed frac. gradient is 2.08 s.g., and there is an assumed lost circulation gradient of about 1.92 s.g. From these considerations we may use a mud weight range of 1.8–1.9 s.g. when drilling out for the production casing. Even if a higher frac. gradient is obtained at 3250 m, losses may occur in the Shetland formation below.

To study the tentative production casing setting depth further, a pressure-depth plot is made and shown in Figure 6.18. The figure clearly shows that the production casing must be set near the bottom of the Cromer Knoll if the reservoir is to be drilled in one run. Here the prognosis uncertainty comes into play. If we obtain full integrity in the whole interval below the production casing, it can be placed anywhere inside the Cromer Knoll. However, if we encounter a lost circulation zone immediately below the production casing, it should be placed below 4670 m if we proceed with a 2.14 s.g. mud into the reservoir.

We will carry the mud density evaluation a step further. In Figure 6.18 the final mud density before setting production casing determines the maximum depth. If we record the depths at which the mud pressure curve matches the pore pressure curve, we can plot the setting depth versus mud density. This is shown in Figure 6.19. This figure mainly predicts the minimum mud weight required to reach a given depth. In the following we will determine this depth from a pressure control point of view.

6.5.1.2 Setting Depth Based on a Kick/Loss Scenario

Now we will assume that the production casing has been set, and we are concerned with the drilling of the reservoir section. This is the most critical phase of drilling as the margin between mud loss and pore

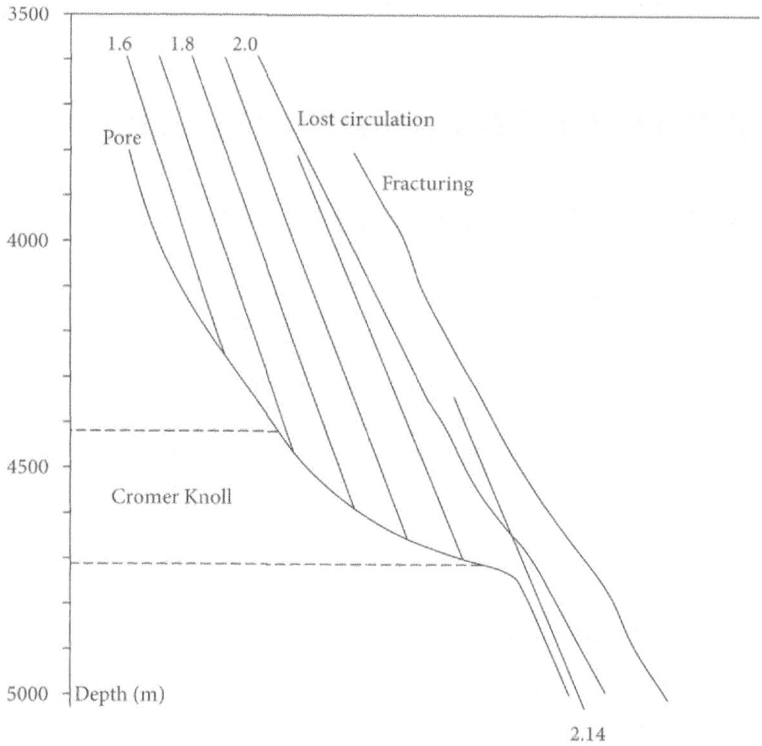

FIGURE 6.18 Mud weight versus setting depth.

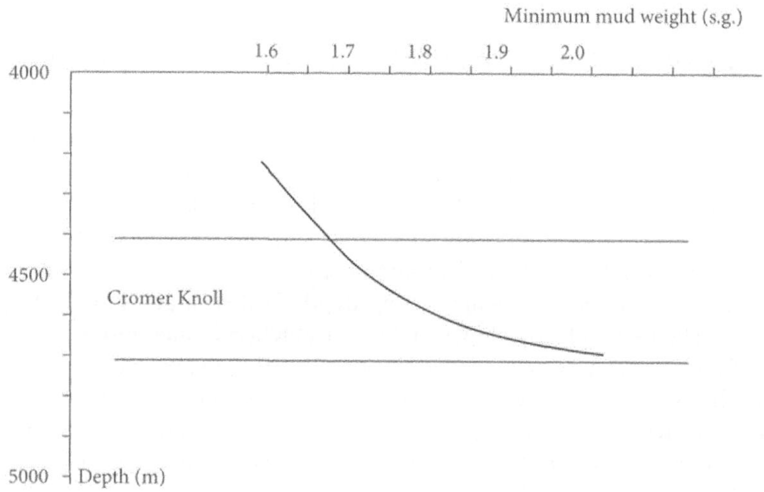

FIGURE 6.19 Setting depth versus mud density.

fluid influx is very small. The most likely scenario is that formation fluids enter the wellbore at a given depth, the well is shut in, and the well fractures at a weak zone below the casing shoe. This is illustrated in Figure 6.20, where the density of the formation fluids is assumed equal to 0.547 s.g. From this figure we observe that an open hole length of a few hundred meters can at most be tolerated if a weak zone is encountered. The most significant element is the existence of a loss zone. If, however, no loss zone is encountered, there is a sufficient margin between the pore pressure curve and the fracturing curve to allow for drilling of the whole reservoir section and maintaining well integrity even if a kick is taken.

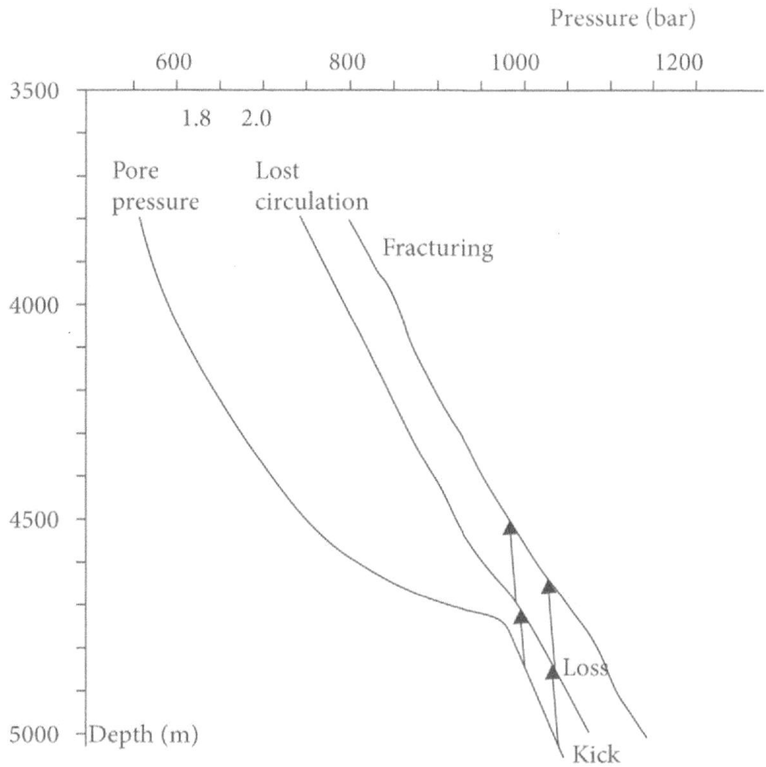

FIGURE 6.20 Kick/loss scenario. (Formation fluid density assumed equal to 0.547 s.g.)

The most important reason for setting the production casing as deep as possible is illustrated in Figure 6.20. A deep casing will isolate possible loss zones. Also, the critical loss pressure is expected to increase with depth.

If we assume a kick is taken at 5000 m and the well is shut in, the formation fluid pressure approaches the loss curve at about 4850 m and the fracturing curve at 4630 m. The maximum permissible open hole length is therefore between 150–370 m, depending on which curve applies. In Figure 6.21 we have performed this evaluation throughout the reservoir interval. The area between the two curves represents the depths at which full well integrity can be maintained. As an example, setting the production casing at 4620 m, we may drill to 5000 m if the fracture prediction is encountered, but only to 4730 m if a loss zone is encountered below the shoe.

6.5.2 Operational Strategy

In the previous discussion we evaluated the setting depth of the production casing mainly from the point of view that the well should be capable of handling a kick without significant mud losses. The obvious question is: What if reality deviates from the prognosis? It is the belief of the author that the analysis still applies, but the prognosis must be corrected when new information becomes available. In other words, a dynamic strategy must be implemented. This will be outlined in the following.

6.5.2.1 Continuous Integrity Evaluation

The models presented are depth and pore pressure normalised. This means that if new information becomes available, this may change the reference level, but the same equations apply. After each leak-off or integrity test is performed, the fracturing model should be adjusted. Likewise, if losses occur at any time, this information should be used to modify the loss model. Adjustment of the pore pressure curve should also be used for corrections. The main idea is that once the reference levels have been established, the model

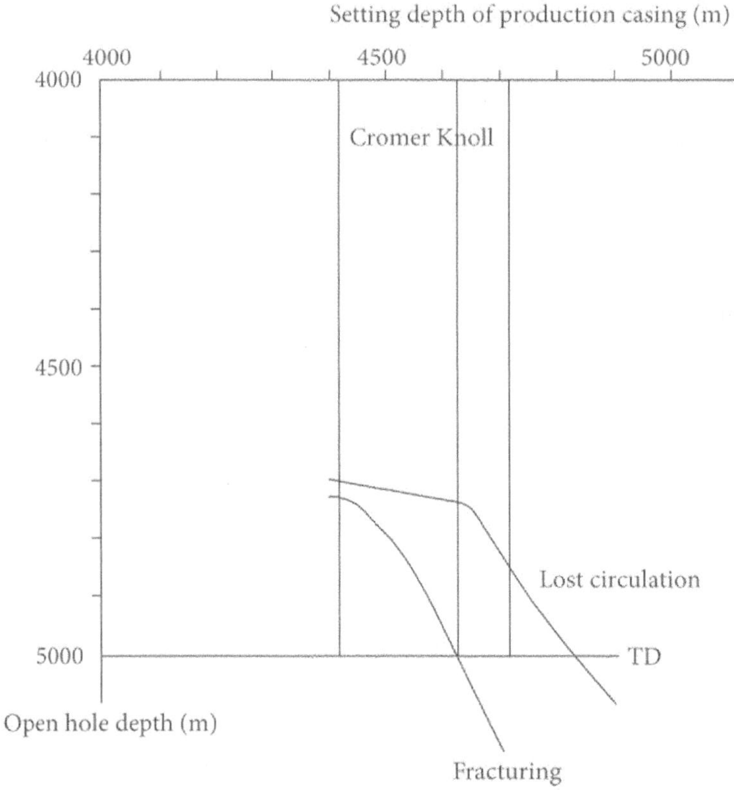

FIGURE 6.21 Permissible open hole depth to avoid losses after taking a kick.

provides a tool for the depth trend prediction. In addition to the maximum open hole length, the maximum kick size should also be continuously evaluated.

6.5.2.2 Kick Simulation

A kick will initiate very close to a potential loss zone in a HPHT well. From a kick simulation point of view this may simplify the evaluation since the casing shoe is covered by formation fluids very early. However, since the kick/loss margins are very close, it is of utmost importance to maintain a constant bottom hole pressure. Therefore, a kick simulator should be dedicated to the drilling phase. Remember also that there are significant frictional pressure drops in the annulus.

Another related element is the mud ballooning, which is caused by the pressure drop and by the transient temperature history of the mud. This is an effect which may cause problems during drilling of the reservoir, and should therefore be evaluated continuously. This is addressed in Appendix B.

6.5.2.3 Drillability Analysis

During drilling below the production casing, a drillability evaluation should be performed. This can either be done with the d-exponent method, or preferably, the mud logger unit should compute a modified algorithm. The main purpose of the drillability analysis is to identify weak zones. Comparative data from reference wells should be established (see Section 3.5).

6.5.2.4 Hydraulic Monitoring

A post analysis of reference well C, discussed earlier, revealed that the pumping pressure increased 50–90 bars above normal before taking the kick. This is believed to be a key contributor to the well control situation that arose, but which was handled well. In this particular case, mud gelling in the annulus was given the blame. Remember that we are in these wells finding extreme temperature and pressures, which may lead to unexpected events.

A simple hydraulic friction model can give valuable information about the state of the drilling fluid and the hole, and abnormalities can be corrected at an early stage. Furthermore, this can be a valuable tool to establish the bottom hole pressure if a loss situation occurs. Another related case is pack-off of drilled cuttings, which may also lead to losses.

In the previous discussion we have focused on obtaining good predictions, and on an active follow-up during drilling. For the particular well under planning, we will recommend the following operational procedures:

The 12-1/4 in. section:

- Drill the interval 2400–4400 m with 1.7–1.8 s.g. mud.
- Perform a leak-off test.
- Increase the mud weight towards leak-off value.
- Drill to the production casing setting depth using geological markers, vertical seismic profiling, and pore pressure indicators as casing seat selection criteria. Increase the mud weight according to the pore pressure indicators, attempting to set the casing as deep as possible without penetrating weak zones below.
- Set the production casing, cement, and perform a leak-off test.

The 8-1/2 in. reservoir section:

- Increase the mud weight to just exceed the last registered pore pressure. Increase the mud weight with increase in pore pressure.
- If potential loss zones are exposed, and the open hole length exceeds that allowed in the model, intermediate integrity tests may be required before continuing drilling. This to ensure that the well can handle a shut-in kick situation.

6.6 DESIGN OF SHALLOW CASING STRINGS

6.6.1 Design Basis

Collapse, burst and compression, and bending forces are usually not considered a problem for the conductor casing. Therefore, no design calculations will be performed.

For the two surface casing strings, an evaluation of several collapse criteria was performed. Fluid level drop assuming a thief zone was found to give lower loading than the collapse during cementation criterion, which was therefore used for both strings.

Since the interval down to 1000 m may contain several shallow gas pockets, this assumption is used in the burst design. The assumption is that a shallow gas pocket of normal pressure is penetrated at the bottom of the next hole section for the 24 in. casing string, and that the gas-filled well is shut in. For the 18-5/8 in. casing string the burst scenario is a gas-filled, shut-in well with the formation pressure acting at the bottom of the next hole section. The burst is a post installation scenario, and it is assumed that the cement is set and that the only mobile phase is water. Reference is given to Section 5.1.7, where the effective pressure behind the casing is discussed. We assume hydrostatic weight of saltwater behind the whole casing strings.

The fracturing prognosis shown in Figure 6.2 consists of two curves, a high fracture gradient and a gradient for weak zones. Normally the fracture gradient applies, but if weak zones are encountered mud losses may occur. For the design of the 24 in. surface casing, the low fracture gradient is used due to the unpredictability at shallow depth. For all other casing strings, the ordinary fracture prognosis is used.

Tension design is only performed for the 18-5/8 in. surface casing. The tension design is based on the weight of the casing in mud, bending effects plus the forces when pressure testing the casing. Maximum tensile loading occurs during installation, so the calculations have not been adjusted for later wear.

6.6.2 The 32 in. Conductor Casing Design

As discussed in the design basis, no calculations will be performed for the 32 in. casing, which has the main function of sealing off the soft material of the sea bottom.

6.6.3 The 24 in. Surface Casing Design

The most important design parameters are as follows:

Depth of casing:	580 m
Depth to seabed:	110 m
Depth to sea level:	42 m
Depth to top of tail cement:	480 m
Depth to next open hole section:	1000 m
Maximum density of drilling fluid:	1.12 s.g.
Lead cement density:	1.50 s.g.
Tail cement density:	1.90 s.g.
Minimum fracture pressure at 580 m:	1.39 s.g.
Mud weight in next hole section:	1.30 s.g.
Pore pressure at 580 and 1000 m:	1.03 s.g.
Formation gas density:	0.085 s.g.

The casing data are:

24 in., grade X-52, 0.625 in. wall thickness	
Burst strength	168 bar
Collapse strength	60 bar

First, we will calculate the *collapse loading during cementation*. The external load is cement on the outside of the casing and sea water as a displacing fluid inside. The pressures are:

External casing pressure at wellhead: = 0 bar

- at seabed: 0.098 × 1.03 (110-42) = 7 bar
- at top of tail cement: 7 + 0.098 × 1.50(480–110) = 61 bar
- at casing shoe: 61 + 0.098 × 1.90(580–480) = 80 bar

Internal pressure at wellhead: = 0 bar

- at seabed: 0.098 × 1.03 × 110 = 11 bar
- at top of tail cement: 0.098 × 1.03 × 480 =48 bar
- at casing shoe: 0.098 × 1.03 × 580 =59 bar

Since Chapter 5 on casing design is graphically showing these pressures, we will shorten this process by using the computational results. The collapse load is the difference between the outside pressure and the inside pressure:

Collapse load at wellhead: 0–0 = 0 bar

- at seabed: 7–11 = 0 bar
- at top of tail cement: 61–48 = 13 bar
- at casing shoe: 80–59 = 21 bar

The maximum collapse load will arise at the casing shoe, with a strength/load ratio of 60 bar/21 bar = 2.86, which is the design factor.

For *burst design*, we assume sea water behind the casing and obtain the following pressures:

External casing pressure at wellhead: = 0 bar

- at casing shoe: 0.098 × 1.03 × 580 = 59 bar

Since the hole below the intermediate surface casing will be exposed to a potential shallow gas zone, a realistic burst criterion is that gas fills the hole at the bottom of next hole section. The formation pressure at this depth is:

$$0.098 \times 1.03 \times 1000 = 101 \text{ bar}$$

During this condition, the pressures further up the hole are

- at 580 m: 101 bar – 0.098 × 0.085(1000 – 580) = 97 bar
- at surface: 97 bar – 0.098 × 0.085 × 580 = 92 bar

The burst load at the shoe is: 97 – 59 = 38 bar

- at wellhead: 92 – 0 = 92 bar

The assumption is that the complete casing is gas-filled and the well shut in.

The burst design load is largest in the top of the casing string. Here we obtain a strength/load ratio of 168 bar/92 bar = 1.83.

6.6.3.1 Integrity Evaluation

The gas pressure at the shoe is calculated to 97 bar. This corresponds to a gradient of 97 bar/(0.098 × 580 m) = 1.71 s.g. This is significantly higher than the frac. prognosis. Therefore, we actually have a reduced integrity case; the casing is sufficiently strong but the open hole below cannot withstand a shut-in situation with a completely gas-filled casing and hole. Therefore, we have to establish the critical kick margin.

The upper pressure limit at the shoe is (assuming a frac. gradient of 1.39 s.g.): 0.098 × 1.39 × 580 = 79 bar

Let us now evaluate the maximum gas height that can be allowed to enter the bottom of the open hole by establishing a pressure balance between the fracturing pressure and the formation pressure minus the hydrostatic weight in the open hole.

$$79 \text{ bar} = 101 \text{ bar} - 0.098 \times 0.085 \times h - 0.098 \times 1.30(1000 - 580 - h)$$

This yields a maximum height of $h = 265$ m, which amounts to a maximum allowable kick size of 65 m^3 if the open hole below the casing has a diameter of 22 in.

The X-52 casing has a burst rating of 168 bar. Assuming a minimum factor of safety of 1.18, the maximum allowable pressure is 142 bar. If this pressure exists at the wellhead, the casing shoe would see a pressure of:

$$142\,\text{bar} + 0.098 \times 0.085 \times 580\,\text{m} = 147\,\text{bar}$$

This corresponds to a pressure gradient at the casing shoe of 147 bar/(0.098 × 580 m) = 2.58 s.g. This means that the wellhead area will be the weak point in the system only if the frac. gradient exceeds this value. Since the frac. prognosis is considerably lower (1.39 s.g.), the casing shoe will represent the weakest point in the well.

Since the casing string is very short, no tension design is performed. The test pressure for the casing string is 92 bar.

Before closing the design on this casing string, a comment on the cementation operation is required. We observe that the density of the cement far exceeds the design fracturing pressure below the casing shoe. However, the design fracture gradient is a minimum value, such that a larger value is expected during the actual operation. Therefore, the cement operation may be performed without losses. Should losses occur, then remedial cementation may be required.

6.6.4 The 18-5/8 in. Surface Casing Design

The design parameters are:

Depth of casing:	1000 m
Depth to seabed:	110 m
Depth to sea level:	42 m
Depth to top of tail cement:	900 m
Depth of next open hole section:	3100 m
Maximum density of drilling fluid:	1.30 s.g.
Lead cement density:	1.50 s.g.
Tail cement density:	1.90 s.g.
Fracture pressure at 1000 m:	1.83 s.g.
Mud weight in next hole section:	1.70 s.g.
Pore pressure at 3100 m:	1.45 s.g.
Density of gas:	0.176 s.g.

The casing data are:

18-5/8 in., grade X-70, 84.5 lbs/ft	
Weight:	
1235 N/m	
Burst strength	197 bar
Collapse strength	43 bar
Tension rating:	8000 kN
Internal cross-sectional area:	0.16 m^2

A pre-evaluation concluded that the worst collapse loading arose during cementation. The cement is filling the outside of the casing up to the seabed, while the inside is drilling mud used as displacing fluid.

It was found that sea water gave too high a collapse load, therefore the cement has to be displaced with mud of density 1.15 s.g.

The pressures are:

External casing pressure at wellhead: = 0 bar

- at seabed: 0.098 × 1.30 × 110 = 14 bar
- at top of tail cement: 14 + 0.098 × 1.50(900–110) = 130 bar
- at casing shoe: 130 + 0.098 × 1.90(1000–900) = 149 bar

Internal pressure at wellhead: = 0 bar

- at seabed: 0.098 × 1.15 × 110 = 12 bar
- at top of tail cement: 0.098 × 1.15 × 900 = 102 bar
- at casing shoe: 0.098 × 1.15 × 1000 = 113 bar

Since the previous chapter on casing design graphically shows all these pressures in detail, we will only present the results in Figure 6.22. The collapse load is the difference between the outside pressure and the inside pressure:

Collapse load at wellhead: 0 – 0 = 0 bar

- at seabed: 14 – 12 = 2 bar
- at top of tail cement: 130 – 102 = 28 bar
- at casing shoe: 149 – 113 = 36 bar

The maximum collapse load will arise at the casing shoe, with a strength/load ratio of 43 bar/36 bar = 1.19.

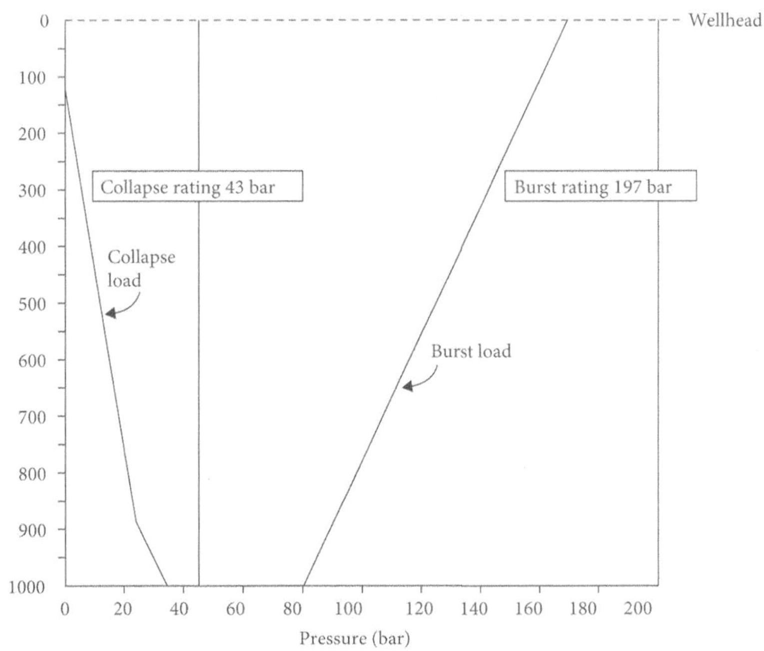

FIGURE 6.22 Burst and collapse design for the 18-5/8 in. surface casing.

For *burst design*, we have assumed a post installation scenario where the only mobile phase behind the casing is sea water. Therefore, the external pressure is assumed equal to a sea water gradient. The internal pressure is a gas column with a base pressure equal to the maximum pore pressure at the end of the next open hole section. Since we are considering a long open hole section, we will take into consideration the weight of the gas column. First, we will evaluate the internal pressure at the casing shoe.

Pressure at bottom of next open hole section: 0.098 × 1.45 × 3100 = 440 bar

Weight of gas in open hole section: 0.098 × 0.176(3100 – 1000) = 36 bar

Internal pressure at casing shoe: 440 – 36 = 404 bar

External pressure at shoe: 0.098 × 1.03 × 1000 = 101 bar

Burst pressure at shoe: 404 – 101 = 303 bar

The burst rating of this casing string is only 197 bar, and we observe that the load is significantly higher if the well is allowed to be filled with gas. This casing can therefore not handle a full well integrity case (see Section 4.1). A maximum pit gain must be defined. If an actual pit gain is less, full integrity is obtained, but if this critical pit gain is exceeded during operation, the well may not withstand the load.

This casing will only be subjected to a partially gas-filled condition, and no flow tests will be performed. For this case we know that the maximum load will arise just below the wellhead. The burst strength here is 197 bar. Regulations require that the actual load is less than 85% of nominal value, which gives an allowable burst pressure of 197 bar × 0.85 = 167 bar. The operator also uses a minimum design factor of 1/0.85 = 1.18. A surface pressure of 167 bar will result in a pressure at the shoe of (adding the weight of the gas):

$$167 + 0.098 \times 0.176 \times 1000 = 184 \text{ bar}$$

The pressure gradient at the shoe then becomes: 184 bar/(0.098 × 1000) = 1.88 s.g.

This means that if the well is taking a severe kick and shut in, the shoe is the weakest point if the frac. gradient is less than 1.88 s.g. A higher frac. gradient moves the weakest point in the system up to the top of the casing, which is unacceptable. The burst load becomes:

Burst loading at wellhead: 167 – 0 = 167 bar

– at the casing shoe: 184 – 101 = 83 bar

This load curve is also shown in Figure 6.22.

The kick margin must be determined next. Assume that the bottom of the next open hole section is filled with gas, with mud on top, and a shut-in well. For this case the frac. gradient at the shoe will represent the weak point, and we will use the prognosis of 1.83 s.g. If this gas height is called h, there is pressure balance between the formation pressure and the frac gradient if:

$$\begin{aligned}0.098 \times 1.83 \times 1000 = {} & 0.098 \times 1.45 \times 3100 - 0.098 \times 0.176h \\ & - 0.098 \times 1.70(3100 - 1000 - h)\end{aligned}$$

or: $h = 595$ m

The maximum allowable kick size for a 17-1/2 in. hole is then about 92 m^3. Please note that this kick limit is based on shut-in conditions. The loading when circulating out the kick may be larger, and should be investigated further. However, it is established that the allowable kick is easily detectable with the current gain/loss measurement accuracy on the rigs.

The test pressure must be established. Since the casing is landed in a 1.30 s.g. drilling mud, the highest burst loading during testing will be at the shoe. If we assume the casing tested to 85% of its nominal burst strength, according to regulations, the surface pressure required is:

167 bar = P + 0.098(1.30–1.03)1000

or P = 141 bar

or 167 bar if the casing is tested during cementing (bumping plug).

For tension design, the weight of the casing is:

$$1235\,\text{N}/\text{m} \times 1000\,\text{m} = 1235\,\text{kN}$$

Since the casing is landed in 1.30 s.g. drilling fluid, buoyancy reduces the effective weight to:

$$1235\,\text{kN}\left(1 - 1.3/7.85\right) = 1030\,\text{kN}$$

Additional tension load from bending effects, if the casing must be lifted during installation. This is estimated to be 850 kN. If the casing is tested during installation (cement plug bumping), the internal pressure will impose an additional axial load of

$$167\ \text{bar} \times \left(100\,\text{kN}/\text{m}^2\right)/\ \text{bar} \times 0.16\,\text{m}^2 = 2672\,\text{kN}$$

All these loads will not arise simultaneously. The largest loading at one point in time would be the buoyed weight of the casing plus the loading due to pressure testing, giving a total of 1030 + 2672 kN = 3702 kN. The strength/load ratio in tension is therefore 8000 kN/3702 kN = 2.16.

6.7 DESIGN OF THE 14 IN. INTERMEDIATE CASING STRING

6.7.1 Design Basis

The collapse design for this casing string is based on a fluid loss scenario. Assuming that drilling mud is lost at the end of the next open hole section, it is assumed that the fluid level stabilises when a bottom hole pressure equals that of normally pressured sea water. Because we are now considering a very long casing string with a considerable axial load, biaxial force correction will be used in the design for casing collapse.

Since this is an intermediate casing string, the burst design is only considering loading arising when a kick is taken. This casing string is therefore not designed for a flow test. The kick scenario assumes the maximum pore pressure in the next open hole section, and assumes gas condensate as formation fluid. This is based on geological evaluation of the area, and on fluid samples taken from reference wells. The annulus behind the casing is assumed to be loaded with mud down to the top of the cement, and sea water as a mobile phase in the cemented interval. The well is defined as a short time frame exploration well, as discussed in Section 5.1.7. Support from cement is not included because it may be difficult to verify.

The tensile load design will comprise the buoyed weight, bending loads and loads arising during pressure testing.

6.7.2 Casing Design

The design parameters are:

Depth of casing:	3100 m
Depth to seabed:	110 m
Depth to sea level:	42 m
Depth to top of cement:	2310 m
Depth of next open hole section:	4700 m
Maximum density of drilling fluid:	1.70 s.g.
Cement slurry density:	1.90 s.g.
Fracture pressure at 3100 m:	2.07 s.g.
Max. mud weight in next hole section:	2.00 s.g.
Pore pressure at next casing point (4700 m):	2.00 s.g.
Density of gas/condensate next section:	0.547 s.g.

The casing data are:

14 in., grade P-110, 86 lbs/ft	
Weight:	1257 N/m
Burst strength	569 bar
Collapse strength	268 bar
Tension rating	12352 kN
Internal cross-sectional area	0.078 m^2

First, we will evaluate the *collapse design load* for the casing string. We assume that this is a post-installation scenario, that is the collapse may take place when drilling the next hole section. If mud losses occur at the bottom of the next hole section, and the mud levels stabilise when a normal pressure is obtained at the bottom, the top of the fluid column is:

$$0.098 \times 1.03 \times 4700 = 0.098 \times 2.0(4700 - h)$$
$$\text{or: } h = 2280\,\text{m}$$

If this scenario arises, we have an air-filled annulus down to 2280 m, and mud inside the remainder of the casing string. The outside pressure of the casing is:

External pressure at 2280 m: 0.098 × 1.03 × 2280 = 230 bar

– at the casing shoe: 0.098 × 1.03 × 3100 = 313 bar

The inside pressure at the shoe is: 0.098 × 2.0(3100 – 2280) = 161 bar

The collapse load is then at surface: 0 bar

– at 2280 m: 230 – 0 = 230 bar
– at 3100 m: 313 – 161 = 152bar

Figure 6.23 shows the collapse design load. The maximum collapse load is found at 2280 m depth. The collapse strength must be derated however, because of a significant axial load. Later in this chapter it is found that the axial design load at the surface is 3053 kN after installation. The axial load/strength ratio

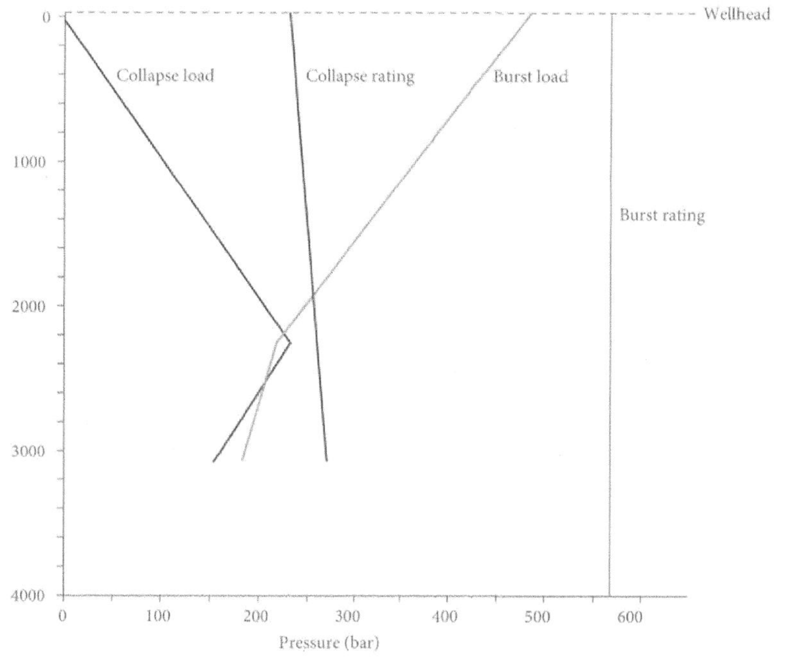

FIGURE 6.23 Burst and collapse design for the 14 in. intermediate casing.

is then 3053kN/12,352 kN = 0.25. Referring to Section 5.1.5, the collapse strength reduction is found to equal 86%. Therefore, the collapse strength at the surface is 268 bar × 0.86 = 230 bar. At the bottom of the casing string no correction applies since here there are no axial loads. A linear collapse strength curve is obtained from these two end points, as shown in Figure 6.23. We observe that the most critical point still is at 2280 m, and here the collapse strength/load ratio is 258 bar/230 bar = 1.12.

For *burst design* we assume that a kick is taken at the bottom of the next open hole section. The pressures are as follows:

Pore pressure of next section: $0.098 \times 2.00 \times 4700 = 921$ bar

Pressure at shoe (3100 m): $921 \text{ bar} - 0.098 \times 0.547(4700 - 3100) = 835$ bar

Pressure below wellhead: $921 \text{ bar} - 0.098 \times 0.547 \times 4700 = 669$ bar

The equivalent density gradient at the shoe is 835 bar/(0.098 × 3100) = 2.75 s.g. Since this is significantly larger than the prognosed fracture gradient of the shoe (2.07 s.g.) the section will not have full well integrity, and we will have to design the well for reduced well integrity which implies introducing a kick margin, and ensuring that the weak point of the well is in the formation below the casing shoe. Also, both the casing and the open hole cannot withstand the loading during a fully gas-filled condition.

The burst strength of the casing string is 569 bar, and we assume again that we will use only 85% of this value according to regulations. Therefore, maximum allowable casing loading is 0.85 × 569 = 484 bar.

We will now establish the condition that ensures that the weak point in the well stays below the casing shoe and not below the wellhead. A surface pressure of 484 bar corresponds to a pressure in the well of (if gas-filled):

- at 2310 m: $484 + 0.098 \times 0.547 \times 2310 = 608$ bar
- at casing shoe: $484 + 0.098 \times 0.547 \times 3100 = 650$ bar

A pressure below the shoe of 650 bar corresponds to a gradient of 650 bar/(0.098 × 3100 m) = 2.14 s.g. This means that if the weak point is to stay below the casing shoe, the maximum acceptable leak-off value is 2.14 s.g. The prognosis gave 2.07 s.g., so this condition is accepted.

In order to establish the burst pressure profile, we need to calculate the pressure behind the casing string. We assume mud density down to the top of the cement and sea water pressure gradient in the cemented interval.

The external pressure at top of cement is: 0.098 × 1.70 × 2310 = 385 bar

– at casing shoe: 385 bar + 0.098 × 1.03(3100 – 2310) = 465 bar

The burst loads can now be established as follows:

Burst load at wellhead: 484 bar

- at 2310 m: 608 – 385 = 223 bar
- at the casing shoe: 650 – 465 = 185 bar

These pressures are shown in Figure 6.23. The highest loading arises just below the wellhead, where we obtain the following strength/load ratio: 569 bar/484 bar = 1.18.

The kick margin must also be determined next. Assume that the bottom of the next open hole section is filled with gas, with mud on top, and a shut-in well. For this case the frac. gradient at the shoe will represent the weak point, and we will use the prognosis of 2.07 s.g. If this gas height is called h, there is pressure balance between the formation pressure and the fracture gradient if:

$$0.098 \times 2.07 \times 3100 = 0.098 \times 2.0 \times 4700 - 0.098 \times 0.547h - 0.098 \times 2.0(4700 - 3100 - h)$$

or: h = 149 m

The maximum allowable kick size for a 12-1/4 in. hole is then about 36 m^3. Please note that this kick limit is based on shut-in conditions. The loading when circulating out the kick may be larger, and should be investigated further. However, it is established that the allowable kick is easily detectable with the current gain/loss measurement accuracy on the rigs.

The test pressure must be established. The casing is landed in a 1.70 s.g. drilling mud, which implies that the test pressure will first be reached at the casing shoe. Considering the density difference on the inside and outside of the casing in the cemented interval, the following surface test pressure then results:

$$484 \text{ bar} = P + 0.098(1.70 - 1.03)(3100 - 2310) \quad \text{or: } P = 432 \text{ bar}$$

or the test pressure is 484 bar if tested during cementing (plug bumping).

For tension design, the weight of the casing is:

$$1257\,\text{N/m} \times 3100\,\text{m} = 3897\,\text{kN}$$

Since the casing is landed in 1.70 s.g. drilling fluid, buoyancy reduces the effective weight to:

$$3897\,\text{kN}(1 - 1.7/7.85) = 3053\,\text{kN}$$

Additional tension load from bending effects, if the casing must be lifted during installation, is estimated to be 670 kN. If the casing is tested during installation (cement plug bumping), the internal pressure will impose an additional load of

$$484 \text{ bar} \times \left(100\,\text{kN}/\text{m}^2\right)/\ \text{bar} \times 0.076\,\text{m}^2 = 3678\,\text{kN}$$

All these loads will not arise simultaneously. The largest loading at one point in time would be the buoyed weight of the casing plus the loading due to pressure testing, giving a total of 3053 + 3678 kN = 6731 kN. The strength/load ratio in tension is therefore 12352 kN/6731 kN = 1.84.

6.8 DESIGN OF THE PRODUCTION CASING STRING

6.8.1 Design Basis

The production casing has been subject to a number of constraints. The most important will be outlined below.

A conventional 9-5/8 in. production casing would not handle the pressures that could arise during production testing. It was investigated to see if a thick-walled 10-3/4 in. casing could be used instead. The advantage was that only one casing string was needed above the reservoir both for the drilling and the testing phases, but the weight was so large that it was on the limit of the capacities of present drilling rigs. This concept was therefore abandoned.

Instead, it was decided to separate the drilling and testing phases. A lighter 9-7/8 in. × 9-5/8 in. casing was designed for the drilling phase only, but if a flow test should be conducted, a tieback string would have to be installed on top of the liner to serve as a production casing. Figure 6.1 illustrates the various combinations which are covered in this design, and are meant to be contingency solutions to reach the well objectives if problems arise.

The geological design premises defined possibilities for gases such as H_2S. It was decided that all production casing strings should be capable of resisting hydrogen embrittlement. Therefore, the upper production casing was of a sour service quality. Because this problem is pronounced only at lower temperatures, ordinary casing qualities were used in the deeper sections where the temperature makes the casing material more ductile. However, because of the high temperatures, all casing strength data were derated for temperature.

The collapse design assumes that drilling fluid is lost to a thief zone; the burst design assumes that a kick is taken and the well is shut in. Only the shut-in phase is covered in this design. Loading due to circulating out the kick should be evaluated separately. The tension design consists of evaluation of the buoyed weight, bending forces, and forces that arise during pressure testing. The well is defined as an exploration well, as discussed in Section 5.1.7. External loading for burst calculations is therefore mud density to the cemented interval, and sea water density inside the cemented interval.

6.8.2 The 9-7/8 in. × 9-5/8 in. Production Casing Design

This casing is designed to handle the drilling phase only. No provisions for flow testing are made. The most important design data are:

Casing depth:	4700 m
Pore pressure at shoe:	1.90 s.g.
Frac. gradient at shoe:	2.24 s.g.

(continued)

Maximum mud weight:	2.00 s.g.
Maximum pore pressure at 5100 m:	2.11 s.g.
Minimum pore pressure at 5100 m:	2.01 s.g.
Maximum mud weight in next section:	2.20 s.g.
H_2S service required above 85°C or above:	2400 m
Depth to top of cement:	4000 m

The casing data are as follows:

9-7/8 in., SMC 110, 66.4 lbs/ft	
Depth interval:	0–2400 m
Weight:	970 N/m
Collapse rating:	797 bar
Burst rating	914 bar
Tension rating:	9356 kN
Drift diameter:	21.59 cm
9-5/8 in. Q 125, 53.5 lbs/ft	
Depth interval:	2400–4700 m
Weight:	781 N/m
Collapse rating:	581 bar
Burst rating	854 bar
Tension rating:	8640 kN
Drift diameter:	21.59 cm
Inner area, both casings:	0.037 m^2

First, we will design the casing for *collapse*. A realistic scenario is that drilling fluid losses occur during drilling of the reservoir section. We will for this case use the minimum reservoir pressure as the lowest pressure that the fluid level can stabilise at. The fluid level will for this case drop to:

$$0.098 \times 2.01 \times 5100 = 0.098 \times 2.20(5000 - h)$$

or $h = 440$ m

The internal pressure at the casing shoe is then: $0.098 \times 2.20(4700-440) = 918$ bar

- and at top of cement: $0.098 \times 2.20(4000-440) = 767$ bar

The external pressure is:

- at 440 m: $0.098 \times 2.0 \times 440 = 86$ bar
- at 4000 m: $0.098 \times 2.0 \times 4000 = 784$ bar
- at 4700 m: 784 bar + $0.098 \times 1.03(4700 - 4000) = 855$ bar

The collapse load is then as follows:

- at surface: 0 bar
- at 440 m: $86 - 0 = 86$ bar
- at 4000 m: $784 - 767 = 17$ bar
- at 4700 m: $855 - 918 = 0$ bar

We observe that the collapse load is very small, and that the nominal collapse resistance/load ratio is 797 bar/86 bar = 9.27. Since this casing string is exposed to a considerable temperature, and high tensional loads, a derating of collapse strength for temperature and bi-axial stresses is required. However, these two effects arise at opposite places in the string; the tension effect is largest at the top, while the temperature derating is largest at bottom. The collapse load is largest near the surface, where only axial effects apply. The maximum tension is 3074 kN. The axial load/strength ratio is 3074 kN/9356 kN = 0.33, which gives a bi-axial correction factor of 0.79. The derated collapse resistance at the wellhead is therefore 0.79 × 797 bar = 630 bar. The corrected collapse resistance/load ratio is therefore approximately 630 bar/86 bar = 7.32.

The *burst* evaluation assumes that a kick is taken at the very bottom of the well, that is at 5100 m. A geological evaluation concluded that the expected formation fluid in the reservoir zone would be a gas/condensate of 0.547 s.g. density.

The pore pressure at the bottom of the well is: 0.098 × 2.11 × 5100 = 1055 bar

The pressure at the shoe is: 1055 bar – 0.098 × 0.547(5100 – 4700) = 1033 bar

- at 4000 m: 1055 bar – 0.098 × 0.547(5100 – 4000) = 996 bar
- at 2400 m: 1055 bar – 0.098 × 0.547(5100 – 2400) = 910 bar
- at wellhead: 1055 bar – 0.098 × 0.547 × 5100 = 782 bar

This corresponds to a pressure gradient at the shoe of 1033/(0.098 × 4700) = 2.24 s.g.

We observe that the pressure gradient at the shoe is equal to the prognosed fracture gradient. With reference to the prognosis evaluating loss zones, we therefore conclude that this casing string has full well integrity.

The external pressures acting on the casing are caused by the density of the drilling mud above the cemented interval, and by sea water pressure in the cemented interval, or more specifically:

External pressure at wellhead: 0 bar

- at 2400 m: 0.098 × 2.0 × 2400 = 470 bar
- at 4000 m: 0.098 × 2.0 × 4000 = 784 bar
- at 4700 m: 784 bar + 0.098 × 1.03(4700 – 4000) = 855 bar

The burst loads for this casing string are:

- at wellhead: 782 – 0 = 782 bar
- at 2400 m: 910 – 470 = 440 bar
- at 4000 m: 996 – 784 = 212 bar
- at 4700 m: 1033 – 855 = 178 bar

These loads are shown in Figure 6.24. This well will approach bottom hole temperatures of 180°C. Therefore, the burst strength derated for temperature must be used. Table 6.1 shows the results. Since both burst and tension depend directly on the yield strength of the steel, a tension derating is calculated simultaneously.

These strength data are also shown in Figure 6.24. We observe that the largest loading arises at the top of the casing string, yielding a strength/load ratio of 914 bar/782 bar = 1.17, which was accepted by the operator.

This casing string has full well integrity because the loading from the kick is lower than the burst resistance. However, the maximum fracture gradient to ensure a weak point below the shoe is equal to the prognosed fracture gradient used in the design. If the actual frac. gradient is higher, full well integrity is

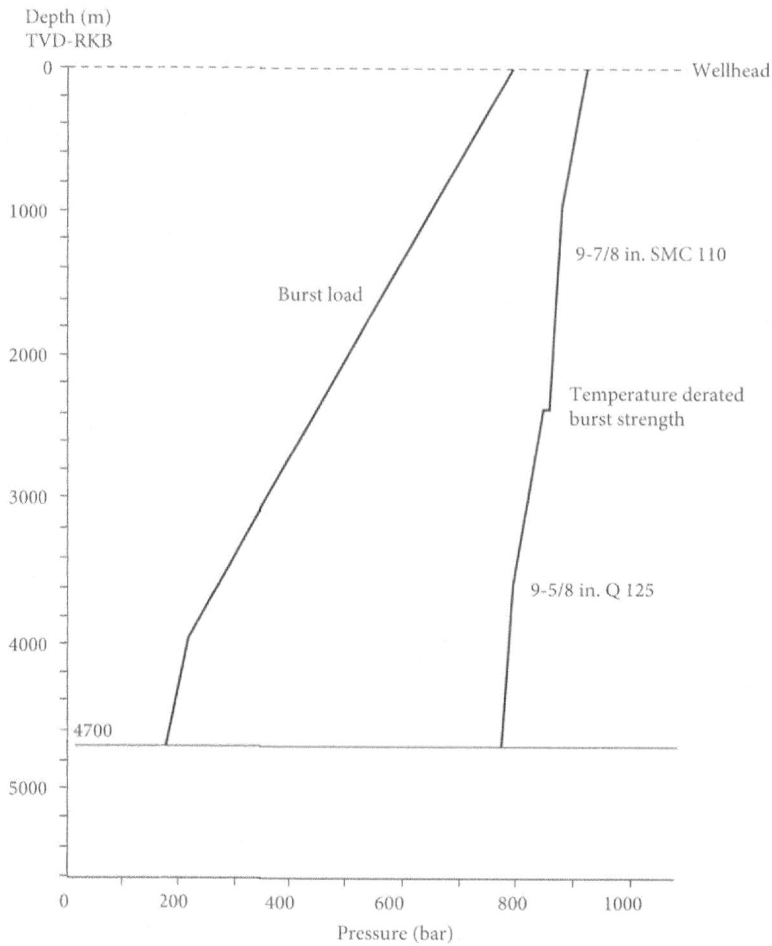

FIGURE 6.24 Burst design for the 9-7/8 in. × 9-5/8 in. production casing.

TABLE 6.1 Derated burst and tension strengths for the production casing

CASING STRING	*DEPTH (M)*	*TEMP. (°C)*	*DERATING FACTOR*	*DERATED BURST (BAR)*	*DERATED TENSION (KN)*
9-7/8 inch SMC 110	0	–	1.0	914	9356
	700	49	0.975	891	9122
	1000	58	0.966	883	9038
	2400	85	0.945	864	8841
9-5/8 inch Q125	0	–	1.0	854	8640
	2400	85	0.952	813	8225
	3600	118	0.931	795	8044
	4200	145	0.920	783	7923
	4700	164	0.908	775	7845

obtained. If the actual frac. gradient is lower, the well may fracture below the casing shoe if a serious kick is taken.

For tension design, the weight of the casing is:

$$970\,\mathrm{N/m} \times 2400\,\mathrm{m} + 781\,\mathrm{N/m}\,(4700 - 2400)\,\mathrm{m} = 4124\,\mathrm{kN}$$

Since the casing is landed in 2.00 s.g. drilling fluid, buoyancy reduces the effective weight to:

$$4124\,\text{kN}\left(1-2.00/7.85\right)=3074\,\text{kN}$$

Additional tension load from bending effects, if the casing must be lifted during installation, is estimated to 400 kN. If the casing is tested during installation (cement plug bumping), the internal pressure will impose an additional load of:

$$782\,\text{bar}\times\left(100\,\text{kN}/\text{m}^2\right)/\,\text{bar}\times 0.037\,\text{m}^2 = 2893\,\text{kN}$$

All these loads will not arise simultaneously. The largest loading at one point in time would be the buoyed weight of the casing plus the loading due to pressure testing, giving a total of 3074 + 2893 kN = 5967 kN. The strength/load ratio in tension is therefore 9356 kN/5967 kN = 1.57.

6.8.3 The 7-5/8 in. Liner and Tieback Casing Design

The liner and the tieback casing must be designed to handle the flow test phase. This implies that, in addition to being able to handle a kick, the so-called leaking tubing criterion must also be satisfied. Figure 6.25 illustrates this.

Assume that a well flow test is performed. We know that the highest burst loading arises just below the wellhead, where there is least back-pressure. The test tubing may therefore burst in this location. If this happens, then the inside of the test tubing is exposed to a gas-filled tubing loading. The outside of the test tubing is actually the inside of the tieback string. This annulus is filled with mud. Adding a leaking tubing pressure on top of this mud column creates a large pressure on top of the DST packer, which is the highest loaded position. This is the scenario we will use for the designs to follow. See also the chapter on well integrity.

The design parameters are as follows:

Liner:	
Liner interval:	4650–5100 m
Depth to top of cement:	4650 m
Weight of mud ahead of cement:	2.20 s.g.
Assumed depth of DST (drill stem test) packer:	4950 m

The tieback string:	
Tieback string interval:	0–4650 m
Weight of fluid behind string:	2.20 s.g.
Weight of fluid inside during test:	2.20 s.g.
The liner/tieback ratings are:	
7-5/8 in. SMC 110, 42.8 lbs/ft.	
Weight:	625 N/m
Burst rating:	1005 bar
Collapse rating:	959 bar
Tension rating:	6826 kN

First, we will evaluate the *collapse* loading of the liner. The most realistic scenario is collapse loading due to plugging of the perforations during well testing, combined with venting of the production tubing. The deepest point in the well, relevant to this design is considered to be at 5000 m depth.

The external pressure on the liner is: 0.098 × 2.11 × 5000 = 1034 bar

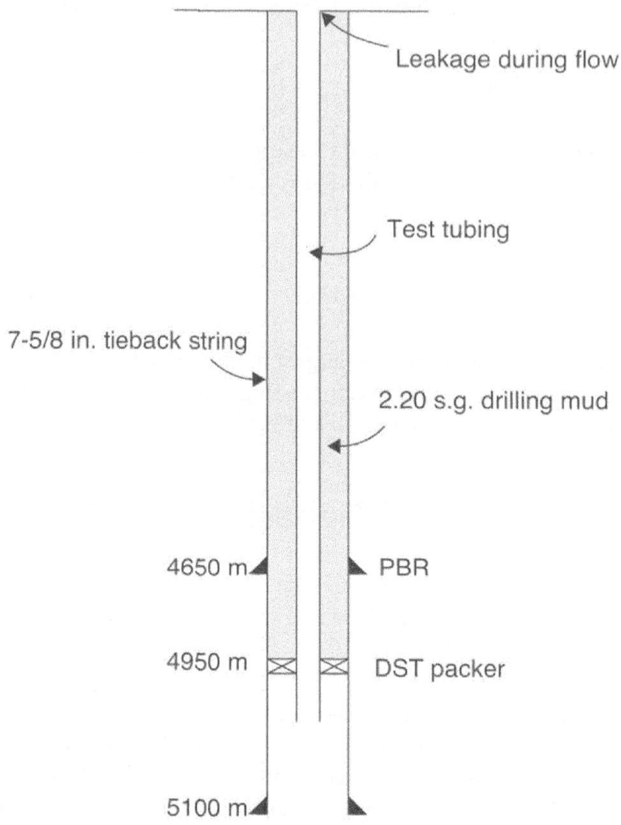

FIGURE 6.25 Physical situation during flow testing.

The internal pressure, if filled with condensate is: 0.098 × 0.547 × 5000 = 268 bar
The collapse loading is: 1034 – 268 = 766 bar

Assuming a temperature derating of 88%, the collapse resistance is 0.88 × 959 bar = 844 bar. The strength/load ratio becomes 844 bar/766 bar = 1.10.

Next the *burst* design will be performed. The tieback string from wellhead to the DST packer, which also includes the upper part of the liner, will be designed for a leaking tubing criterion. The part of the liner which is below the packer is not designed for burst since no realistic burst criteria were established. The deepest point in the well relevant for burst is 5000 m.

The external pressure at the wellhead is: 0 bar

- at the liner hanger: 0.098 × 2.20 × 4650 = 1003 bar
- at the DST packer: 1004 bar + 0.098 × 1.03(4950 – 4650) = 1033 bar

The internal pressure at 5000 m is: 0.098 × 2.11 × 5000 = 1034 bar

- at wellhead: 1033 bar – 0.098 × 0.547 × 5000 = 765 bar
- at liner hanger: 765 + 0.098 × 2.20 × 4650 = 1768 bar
- at packer: 765 + 0.098 × 2.20 × 4950 = 1832 bar

TABLE 6.2 Derated burst and tension for the tieback string

CASING STRING	*DEPTH (M)*	*TEMP. (°C)*	*DERATING FACTOR*	*DERATED BURST (BAR)*	*DERATED TENSION (KN)*
		20	1.0	1005	6826
	0	130	0.900	905	6143
	1000	144	0.887	891	6055
7-5/8 inch SMC 110	2400	150	0.885	889	6041
	4650	170	0.880	884	6007
	4950	180	0.880	884	6007

The burst load at the wellhead is: 765 – 0 = 765 bar

- at the liner hanger: 1768 – 1003 = 765 bar
- at top of packer: 1832 – 1033 = 799 bar

During well testing, a significant amount of heat is transported up the well. The design premises set an upper limit of 130°C at the wellhead. The test should be stopped if this temperature limit is exceeded, and the well should be allowed to cool before continuing flow testing. The whole tieback string must be temperature derated due to this dynamic temperature profile. Table 6.2 shows the results.

Both the burst loading and the derated burst strength data are shown in Figure 6.26. Clearly, the highest loading will arise on the top of the packer at 4950 m. The burst strength/load ratio is here: 884 bar/799 bar = 1.11. Although this is lower than the minimum requirement of 1.18, it was accepted since the casing strength has been derated for temperature.

The *tension load* on the liner is negligible. The tieback string will hang freely from the wellhead, and no load is transmitted to the liner because of a telescopic joint (polished bore receptacle). For simplicity, we neglect buoyancy on the tieback string, of which the weight is: 625 N/m × 4650 m = 2906 kN. The axial strength/load ratio becomes 6826 kN/2906 kN = 2.35.

The liner will be integrity tested prior to tieback preparations. The liner lap should be tested to a pressure equivalent to the integrity of the casing shoe of the production casing.

The liner and the tieback string should be simultaneously pressure tested prior to flow testing of the well. The test pressure should be equal to or exceed the design burst pressure from Figure 6.26 in the whole string interval. To obtain this test pressure scenario, a procedure outlined in Section 5.2.2 should be employed. The wanted pressure profile may be obtained by displacing the upper part of the well with sea water before pressure testing.

6.8.4 The 5-1/2 in. Contingency Liner and Tieback Design

In the event that the 7-5/8 in. liner is used to handle borehole problems (alternatives 4, 5, and 6 in Figure 6.1), the 5-1/2 in. liner must be used to reach the objectives of the well. The design loads for the contingency liner will basically be the same as for the 5-1/2 in. liner. We will here briefly check the design factors. The same tieback string will be used, with only the bottom portion replaced by a 5-1/2 in. liner. The data are as follows:

5-1/2 in. Q 125, 20 lbs/ft.	
Design interval:	4630 – 5100 m
Burst strength:	989 bar
Collapse strength:	834 bar

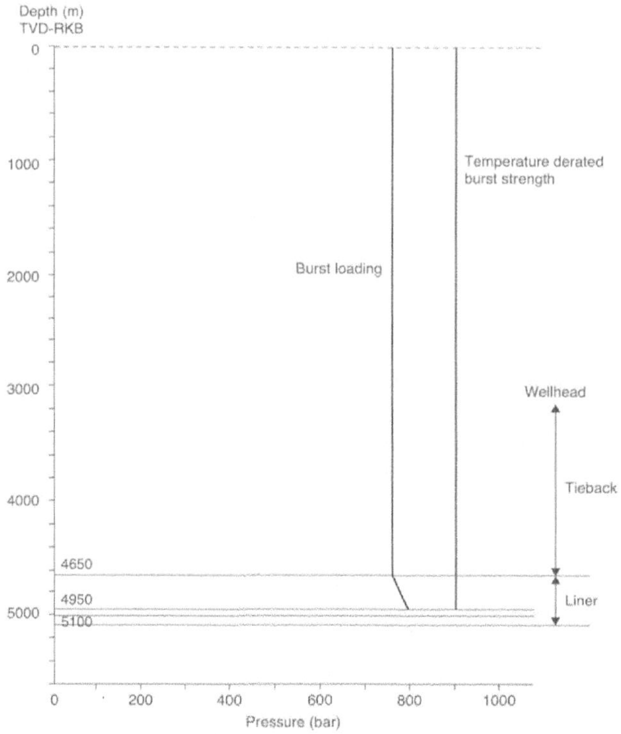

FIGURE 6.26 Burst design for the liner and the tieback string based on a leaking tubing scenario.

The burst derating for this casing quality is 0.90 at 170°C, which yields a strength of 0.90 × 989 bar = 890 bar. With reference to the previous chapter the strength load ratio becomes 890 bar/799 = 1.11.

Assuming the same derating on the collapse resistance, the collapse strength becomes: 0.90 × 834 bar = 751 bar. The strength/load ratio is 751 bar/766 bar = 0.98. Clearly, this is not acceptable, so a minimum wellhead pressure condition must be imposed. Assuming that the minimum design factor is 1.10, the critical collapse load is 766 bar × 1.10 = 843 bar. In other words, the minimum acceptable wellhead pressure during a flow test operation is 843 – 751 = 92 bar. A lower pressure may result in a collapsed liner if the perforations become plugged.

6.8.5 Design Summary

In the following, various factors will be listed for the design. Please note that these factors are valid only for the conditions they were derived for. In case of modifications during the actual operations, a check of the design factors must be made. The criteria for acceptance are for this case that the design factors for collapse and tension exceed 1.10, and for burst 1.18. If derated strength data are used, lower values can be accepted. Various operators may choose other design factors. However, the design factors for various operators are usually not directly comparable, as they may be based on different assumptions.

Various operators use different design factors. However, when studying this example well design one realises that the design assumptions and conditions are as important as the design factor itself. Of particular mention are the fluid density behind the casing and the assumed density of the formation fluid.

Often we assume mud density in open intervals behind the casing strings. In the cemented interval a conservative assumption is that the only mobile phase after the cement is set is water. Typically, support from the hardened cement is neglected because it is difficult to verify.

TABLE 6.3 Summary of design factors

CASING SIZE (INCH)	*COLLAPSE*	*BURST*	*TENSION*	*MAX. KICK SIZE (M3)*	*MAX. FRAC. (S.G.)*
32	–	–	–	–	–
24	2.86	1.83	–	65	2.58
18-5/8	1.19	1.18	2.16	92	1.88
14	1.12	1.18	1.84	36	2.14
9-7/8 × 9-5/8	7.32	1.17	1.57	–	–
7-5/8	1.10	1.11	2.35	–	–
5-1/2	1.10	1.11	–	–	–

In exploration drilling one often assumes that methane gas is entering the borehole during kicks. This is considered the lightest gas occurring, and will provide a safe design. In critical HPHT wells, the design is difficult using this assumption, and one often assumes a heavier reservoir fluid. In our case, the geological evaluation concluded that there is a high likelihood for a gas condensate of heavier density.

One other assumption is the lowest thief zone pressure as used in collapse design. Common for all these design premises is the necessity of continuously improving the predictions. Therefore, geological evaluation is one of the key areas for further improvements in well design, in addition to improved post analysis of wells already drilled.

Problem

An HPHT well is planned in an unknown area. The following leak-off data are available from adjacent wells.

DEPTH (M)	*LOT (S.G.)*	*P_o (S.G.)*	*σ_o (S.G.)*
4000			2.18
4200	2.10	1.80	
4600	2.11	1.70	
4750	2.17	2.06	
5000			2.25

The new well is planning to set the production casing in the bottom Cretaceous at 4400 m. The estimated pore pressure is 1.6 s.g. Please perform analysis as follows.

a) Plot the reference data.
b) Normalise the data to a common pore pressure of 1.8 s.g. Is there a better trend now?
c) Find the normalised leak-off point for the new well from the figure. Calculate the predicted leak-off value for the expected pore pressure gradient of 1.6 s.g.

Drilling Operations and Well Issues

7

7.1 PLATFORM AND WELL TYPES

There exist a variety of platform types such as jackets, concrete platforms and subsea facilities. We will not list every possible design but from a well design point of view the wellheads have two locations, either at the surface or at the seabed. A brief summary is as follows:

Surface wellheads:

- Land wells
- Offshore production platforms
- Jack-up drilling rigs
- Deep water semisubmersible (rare)

Subsea wellheads

- Semi-submersible drilling rigs

The wellhead here means the wellhead equipment plus either the Blow-Out-Preventer during drilling or the Xmas tree during production. The placement of the wellhead area makes a difference during the design of a well.

Surface wellheads have the pressure control point (BOP/Xmas tree) at the platform level. This means that the well pressure is controlled from this location. Annuli behind the production casing are also accessible and the pressure in each annulus is controlled.

Subsea wellheads are located at the seabottom. Typically, there is no access to the closed annuli behind each casing. These may therefore be subjected to pressures caused by thermal expansion.

During drilling of the conductor and surface casing holes, the return is to seabed, where a pump typically moves drilled cuttings away from the well area. After setting the conductor casing the BOP and the marine riser is installed, and the drilled cuttings are brought to the drill rig.

From a well design perspective there are some small differences. When computing the pressure acting behind each casing string, the pressure is computed from sea level before the riser is installed. After the riser is installed the reference level becomes the drill floor. On a typical semisubmersible rig this difference may be in the order of 25 m. In the worked example of Chapter 5 we have assumed a semisubmersible rig, whereas in Chapter 6 we have assumed a jack-up rig.

There are other differences as well. For the subsea well the entire weight is carried by the template and the cemented casings. For a jack-up rig often a mud line suspension system is used at seabed to carry the weight of the well.

DOI: 10.1201/9781003618171-7

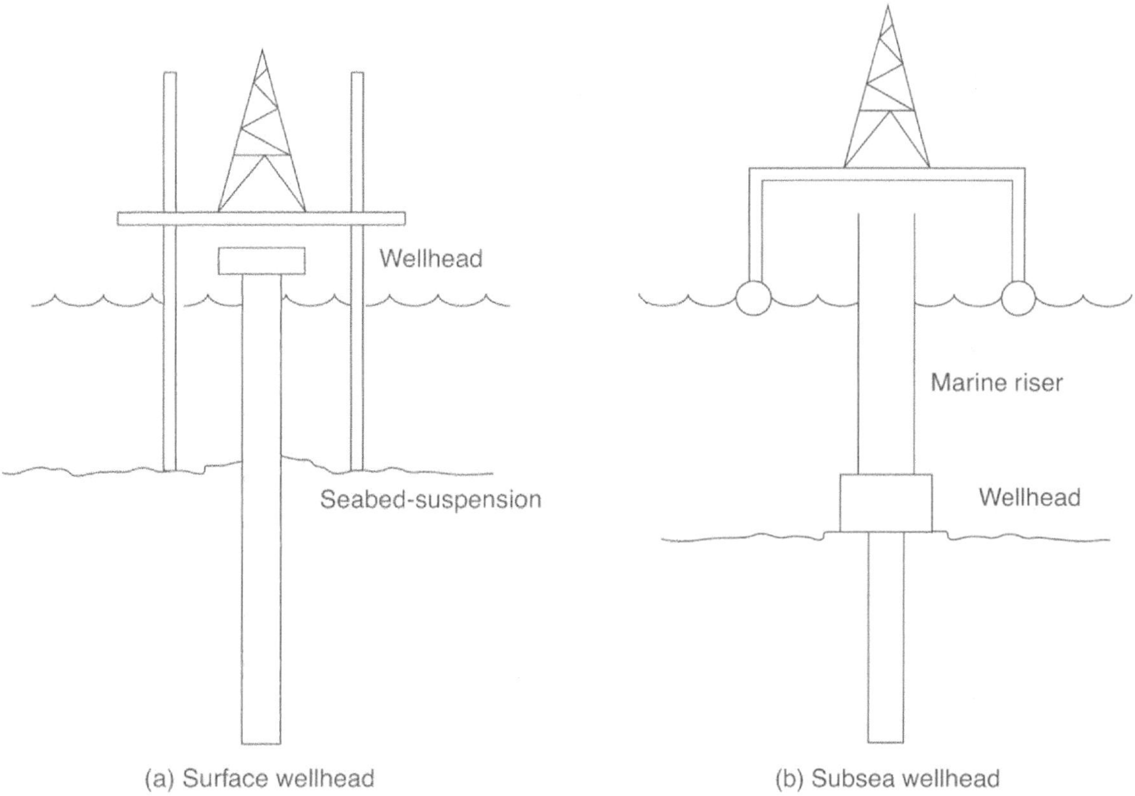

FIGURE 7.1 Surface and subsea wellheads.

These issues above are presented to show that there are many differences between wells. We will not exhaust this discussion, just point out the importance of identifying the important design parameters before carrying out the actual well design.

7.2 SEQUENCES OF A DRILLING OPERATION

Following the design of the well the actual drilling operation can start. The drilling rig and materials must be acquired, and various service companies are hired to deliver specific services for the construction phase of the well. These are large processes involving many people. We will not show all details here. In the following a typical well construction operation will be described including the time of each operation. The times given are obtained from typical daily drilling reports. Future work will be to reduce cost by reducing time or optimizing operations.

The well used as an example is a subsea production well drilled in a water depth of 350 m. The well is a horizontal. The following analysis covers the entire well construction from the start to when the rig leaves the well producing.

The first two holes, the 36 in. and the 26 in. holes, are drilled with return to the seabed. At the seabed a pump is installed which transports drilled cuttings away from the wellhead area.

The 36 in. hole and the 30 in. conductor casing:

OPERATION:	HOURS USED:
Set anchors and pretension	24
Ballast rig down. Make up BHA no. 1 and 1300 m drill pipe	20
Stab into well	1.3
Drill 50 m of 36 in. hole	9
Circulate hole clean	2
Perform wiper trip	1
Displace to hivis mud, pull back to mudline	2
Position rig using thrusters	
Rig up for running 30 in. conductor	0.5
Make up 3.5 in. cement stinger and run in hole	2
Run in 30 in. conductor	6
Circulate and cement 30 in., wait on cement	4
Pull out cement tools and drill pipe	4
Position rig using thrusters	0
Total hours	32

The 26 in. hole and the 20 in. surface casing:

OPERATION:	HOURS USED:
Pull out and rack 36 in. BHA	1
Break down 36 in. BHA	1
Pull out and rack cement head in derrick	1
Make up 26 in. BHA and run to template	1.5
Position rig w/thrusters, stab in and run to shoe	1
Drill 150 m of 26 in. hole	12
Circulate hole clean	3
Wiper trip, run to bottom, and displace to hivis mud	4
Pull out to seabed	2
Pull to surface and rack pipes	4
Position rig using thrusters	
Rig up and run 145 m 20 in. casing	12
Circulate and cement 20 in. csg.	4
Release running tool and pull out of hole	2
Total hours:	49

After the 30 in. and the 20 in. casing strings are installed and cemented all the way to the seabed, the blowout preventer (BOP) is installed on the well. This operation is described below.

Running BOP

OPERATION:	HOURS USED:
Position rig using thrusters	
Rig up for running BOP stack	4
Run marine riser and BOP	16
Land BOP	4
Rig down riser handling equipment	4
Prepare test plug for running in the hole	0.5
Run in hole with test plug	3
Test connector	1
Test BOP	3
Pull out BOP test tool	3
Lay down test tool	0.5
Total hours:	39

Drilling can now continue. Because the marine riser is installed, mud return goes up to the drilling rig, goes through the mud cleaning equipment and back to the active mud pit.

The 17.5 in. hole and the 13-3/8 in. intermediate casing:

OPERATION:	HOURS USED:
Make up 17.5 in. BHA	3
Run in hole with BHA	3
Drill 20 in. cement, plugs, and shoetrack	3
Circulate contaminated mud and perform LOT	2
Pick up drill pipe	1
Drill 700 m of directional 17.5 in. hole.	24
Circulate hole clean, wiper trip, pull out of hole	10
Run log	8
Rig up for running of 13-3/8 in. casing	3
Set back cementing head, retrieve seat protector	6
Run 500 m of 13-3/8 in. casing	10
Circulate and cement 13-3/8 in. casing	3
Set seal assembly and pressure test	1
Perform BOP test and pull out	6
Run and set 13-3/8 in. wear bushing	6
Lay down 17-1/2 in. BHA	3
Service cement head	1
Total hours:	93

The 12.25 in. hole and the 9-5/8 in. production casing

OPERATION:	HOURS USED:
Make up 12.25 in. BHA	4
Run in hole to 900 m	6
Drill 1100 m of 12.25 in. hole	86
Circulate, wiper trip, circulate	8
Pull out and rack 12.25 in. BHA	5
Lay out 12.25 in. BHA	2
Retrieve wear bushing	6
Rig up for running 9-5/8 in. casing	3
Run in 680 m of 9-5/8 in. casing	14
Circulate and cement casing, pressure test	6
Set and test seal assembly	1
Pull out of well	3
Run in wear bushing. Temporary P/A. Establish barrier	4
Disconnect and secure BOP	8
Total hours:	156

In this well we will install a horizontal wellhead before drilling the reservoir and completing the well for production. To ensure well control the BOP will be landed on top of the Xmas tree.

Wellhead

OPERATION:	HOURS USED:
Pull marine riser and BOP	8
Prepare for X-mas tree running	8
Run and install X-mas tree	24
Pull running tool	4

OPERATION:	HOURS USED:
Run marine riser and BOP	9
Position and latch	2
Total hours:	55

Now the reservoir section will be drilled.

The 8.5 in. hole through the reservoir

OPERATION:	HOURS USED:
Make up drill collars and heavy weight drill pipe	4
Pick up and make 8.5 in. BHA	6
Run in hole to 800 m	7
Drill composite bridge plug	2
Run in hole to 810 m	6
Drill shoe, circulate, perform FIT	12
Drill 1000 m horizontal section	160
Total hours:	197

The last phase is to complete the well first by installing the lower completion, which includes everything below the production packer.

Install lower completion:

OPERATION:	HOURS USED:
Clear rig floor	0.8
Make up scraper and magnet assy	1.6
Pick up heavy weight drill pipe	12
Make scraper run	18.9
Rig up tubing tongs	3.1
Make up wirewrap screens, blanks, and swellpackers	27.2
Install screen packer/hanger and running tool	1.6
Rig down tubing tongs	0.8
Run screens on drill pipe	15.7
Drop ball, pump down, and set packer	3.1
Test packer from above	1.6
Release running tool and flowcheck	1.6
Displace well above packer to brine	0
Pull out and lay down running tool	7.9
Make up anchor packer assy	4.7
Run in assembly on drill pipe	15.7
Orient and set anchor packer	1.6
Pull out and lay down running tool	7.9
Total hours:	126

The upper completion is the tubing and everything above the production packer.

Install upper completion:

OPERATION:	HOURS USED:
Pull bore protector	3.1
Install lateral zone assemblies	7.9
Splice and terminate el. cable	15.7
Make up main completion assembly	6.3

OPERATION:	HOURS USED:
Run in 5.5 in. tubing	29.9
Install TRSCSSV, splice cables	6.3
Run in 5.5 in. tubing	1.9
Make up tubing hanger	6.3
Make up THRT and STT??	6.3
Run in on WOR??	18.9
Install lifting frame, flow head	9.4
Land and lock tubing hanger, test seals	1.6
Drop ball, press. Tubing, set and test packer	3.1
Close TRSCSSV, inflow test, equalize, and open	1.6
Rig up wirelin, pull packer setting plug	12.6
Pull plug	3.1
Press. Up annulus and test packer from above	1.6
Total hours:	136

Finally, the well must be started to produce oil and gas. The drilling rig will be disconnected and moved away from the well.

Start well for production:

OPERATION:	HOURS USED:
Prepare rig for well flow	3.1
Displace WOR to nitrogen	9.4
Flow out brine and mud and oil	12.6
Bullhead tubing with diesel and wax inhibitor	3.1
Displace WOR to brine	3.1
Close TRSCSSV	0.8
Install tubing head crown plug and test on wireline	12.6
Close HXT valves and test	3.1
Remove and lay-out WOR, SST etc.	12.6
Install tree cap, test, pull, and lay down running tool	9.4
Pull BOP and marine riser	15.7
Install corrosion cap	6
Deballast and prepare to move rig	6.3
Total hours:	98

The table above shows the time distribution for the various phases of the well construction. It is evident that the completion phases are very time-consuming amounting to nearly half of the total well construction time.

For deeper waters, there is additional time spent on running/retrieving riser and tripping in/out of the well.

Summary of well construction time:

OPERATION:	HOURS USED:	PERCENT TIME:
36 in. hole	32	3.3
26 in. hole	49	5
Running BOP	39	4
17.5 in hole	93	9.5
12.25 in. hole	156	15.9
Wellhead	55	5.6
8.5 in. hole	197	20
Lower completion	126	12.8

(continued)

OPERATION:	HOURS USED:	PERCENT TIME:
Upper completion	136	13.9
Start well	98	10
Total hours:	981	
	40.9 days	100%

7.3 TORQUE AND DRAG IN WELLBORES

7.3.1 Introduction

This section presents analytic friction models for application in petroleum wells. Although very simple, they apply for all wellbore shapes such as straight sections, dropoff bends, build-up bends, side bends or a combination of these.

The entire well can be modelled by two sets of equations, one for straight wellbore sections and one for curved wellbores. The latter is based on the absolute directional change, or the dogleg of the wellbore.

Three worked examples are given: a two-dimensional well, a three-dimensional well, and combined hoisting and rotation in the three-dimensional well.

One main purpose is to provide a simple explicit tool to model and to study friction throughout the well by separating gravitational and tensional friction effects. For more details, please see Aadnoy, Fazaelizade, and Hareland (2010) and Fazaelizade, Hareland, and Aadnoy (2010).

Symbols

A	Wellbore inclination
B	Buoyancy factor
Φ	Wellbore azimuth
Θ	Absolute change in direction
μ	Coefficient of friction
Ψ	Angle between axial and tangential pipe velocities
P	Density
A_r, A_o	Inner, outer cross-sectional pipe areas
DL	Dog-leg
DLS	Dog-leg severity
N_r	Rotary pipe speed
V	Velocity
r	Pipe/connection radius
F	Force in string
T	Torque in string
N	Normal force
L	Pipe length
w	Unit pipe weight

7.3.2 The General Wellbore Orientation

Standard surveying techniques measure wellbore inclination and azimuth. These are used to determine vertical depth and geographic reach. Furthermore, two slope parameters called dogleg and dogleg severity are computed. The dogleg is the absolute change of direction, and the dogleg severity is the derivative of the dogleg. The dogleg is determined by the following equations:

$$\begin{aligned} \cos\theta &= \sin\alpha_1 \sin\alpha_2 \cos(\phi_1 - \phi_2) + \cos\alpha_1 \cos\alpha_2 \\ DL(°) &= \frac{180}{\pi}\left|\theta(\text{rad})\right| \end{aligned} \quad (7.1)$$

where indexes 1 and 2 refer to two consecutive survey measurements, or to the start and end of a longer wellbore section. Defining the distance between these measurements as ΔL, the derivative (called dog-leg-severity) is:

$$DLS = \frac{DL}{\Delta L(\text{m})}(\text{degrees}/\text{m}) \quad (7.2)$$

It is customary in the oil industry to present the DLS as degrees per 30 m, i.e. multiplying Equation (7.2) by 30.

Although the inclination α is measured in a vertical projection and the azimuth ϕ in a horizontal projection, the dogleg θ is measured in an arbitrary plane, as shown in Figure 7.2. Inspection of Equation 7.1 reveals that it depends on both inclination and azimuth. We will utilize these properties in the following when we present general friction models that are not restricted to a plane.

7.3.3 The Static Pipe Weight

7.3.3.1 The Buoyancy Factor

The effective string weight or the string tension in a fluid-filled well is the unit pipe weight w multiplying by the buoyancy factor β. The buoyancy factor is defined as:

$$\beta = 1 - \frac{\rho_o A_o - \rho_i A_i}{\rho_{pipe}(A_o - A_i)} \quad (7.3)$$

where subscript o refers to the outside pipe area and subscript i to the inside. If there is equal fluid density inside/outside the pipe, the buoyancy equation becomes:

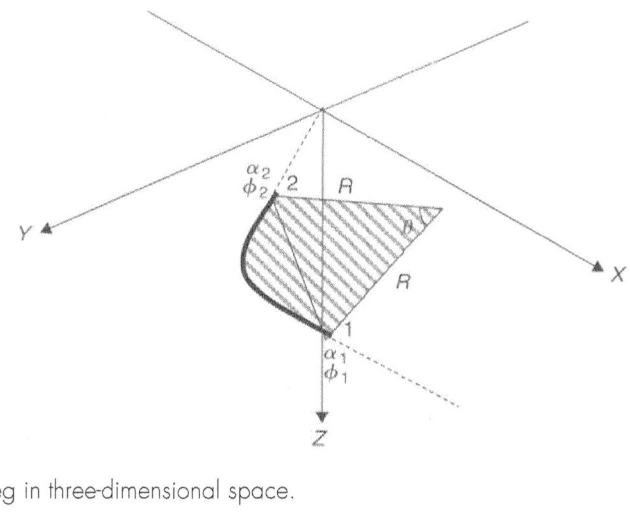

FIGURE 7.2 The dogleg in three-dimensional space.

$$\beta = 1 - \frac{\rho_o}{\rho_{pipe}} \tag{7.4}$$

Equation (7.4) is most commonly used during drilling operation, whereas Equation (7.3) is used in cases where there is a density difference between the inside of the string and the annulus, such as during cementing operations. During well intervention operations, the wellhead may be shut˙in and an annular pressure applied. The same buoyancy equation applies, but one must add end reactions caused by the annular pressure.

7.3.3.2 Static Hook Load

The static hook load is equal to the buoyed pipe weight multiplied by the projected vertical height of the well, regardless of wellbore orientation. Thus, a vertical depth D_1 has equal static hookload as a deviated well with the same projected height D_1 (Aadnoy, Larsen, and Berg, 1999). Another way of computing the static string weight is to set the coefficient of friction equal to zero for the friction equation defined in the section below.

7.3.4 The Three-Dimensional Friction Model

The analysis to follow assumes a soft string model. This implies that pipe bending is so small that bending stiffness can be neglected.

The equations derived here define the hook loads for hoisting and lowering operations and also torque for a string in a wellbore. There are two sets of equations, one for straight well sections and another for arbitrary well orientation.

7.3.4.1 Drag for Straight Inclined Wellbore Sections Without Pipe Rotation

A characteristic of a straight wellbore is that pipe tension is not contributing to the normal pipe force, and hence not affecting friction. Straight sections are *weight-dominated* as only the normal weight component gives friction. The top force F_2 of an inclined pipe is given by:

$$F_2 = F_1 + \beta \Delta L w \{\cos\alpha \pm \mu \sin\alpha\} \tag{7.5}$$

where + means hoisting and – means lowering of the pipe.

7.3.4.2 Torque for Straight Inclined Wellbore Sections Without Axial Pipe Motion

The torque is defined as the normal weight component multiplied by the coefficient of friction and the pipe tool joint radius. The result is:

$$T = \mu r \beta w \Delta L \sin\alpha \tag{7.6}$$

7.3.4.3 Drag for Curved Wellbore Sections Without Pipe Rotation

For curved borehole sections, the normal contact force between string and hole is strongly dependent on the axial pipe loading. This is therefore a *tension-dominated* process. In, for example, a short bend, the tension may be much larger than the weight of the pipe inside the bend. In the following derivation we will assume that the pipe is weightless when we compute the friction, but add the weight at the end of the bend.

Furthermore, the dogleg angle θ depends both on the wellbore inclination and the azimuth. Because the pipe will contact either the high side or the low side of the wellbore, its contact surface is given by the dogleg plane.

For build-up, dropoff, sidebends, or a combination of these, the axial force becomes:

$$F_2 = F_1 e^{\pm\mu|\theta_2-\theta_1|} + \beta w \Delta L \left\{ \frac{\sin\alpha_2 - \sin\alpha_1}{\alpha_2 - \alpha_1} \right\} \tag{7.7a}$$

where + means hoisting and – means lowering of the pipe.

For a build-up section, using the circle segment $\Delta L = R\alpha$, the above equation can be expressed as:

$$F_2 = F_1 e^{\pm\mu|\theta_2-\theta_1|} + \beta w R \sin\alpha_1 \tag{7.7b}$$

Torque for Curved Wellbore Sections Without Axial Motion

The torque for the bend is:

$$T = \mu r N = \mu r F_1 \left|\theta_2 - \theta_1\right| \tag{7.8}$$

Friction for any wellbore shape can thus be computed by dividing the well into straight and curved elements. The forces and torques are summed up starting from bottom of the well. Equations (7.5) and (7.7) give the drag, whereas Equations (7.6) and (7.8) give the torque.

7.3.4.4 Combined Axial Motion and Rotation

The solutions above must be modified if a combined motion takes place. Aadnoy and Andersen (2001) showed how the frictional capacity is decomposed into the two directions: axial motion and rotation. The effect of combined motion is well known, for example when rotating a liner. A high rotational speed reduces axial drag.

During combined motion, the axial velocity is V_h, and the tangential pipe speed is V_r. These give a resultant velocity V. The angle between the two velocities is defined in Figure 7.3.

The angle between the axial and tangential velocity is:

$$\psi = \tan^{-1}\left(\frac{V_h}{V_r}\right) = \tan^{-1}\left(\frac{60V_h\,(\mathrm{m/s})}{2\pi N_r\,(\mathrm{rpm})\,r\,(\mathrm{m})}\right) \tag{7.9}$$

The torque and drag for combined motion are:

For straight pipe sections:

$$\begin{aligned} F_2 &= F_1 + \beta w \Delta L \cos\alpha \pm \mu\beta w \Delta L \sin\alpha \sin\psi \\ T &= r\mu\beta w \Delta L \sin\alpha \cos\psi \end{aligned} \tag{7.10}$$

For curved pipe sections:

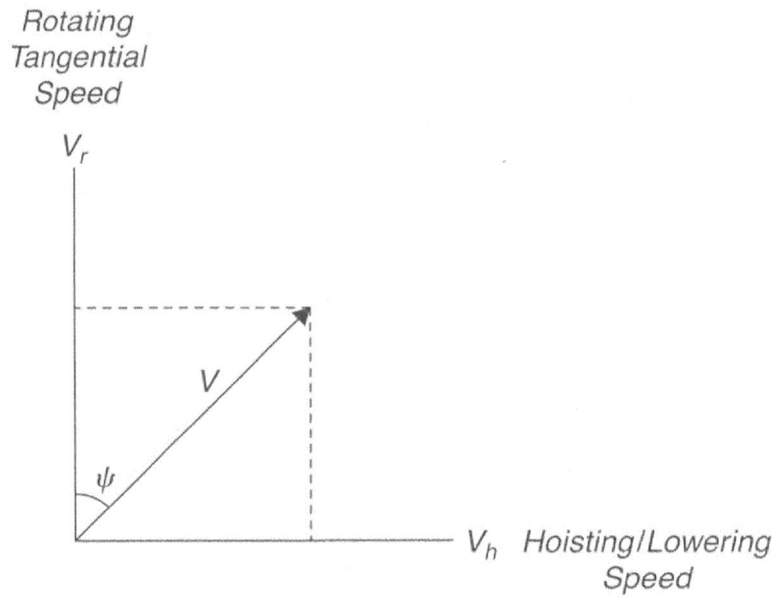

FIGURE 7.3 Resultant velocity of axial and tangential velocity.

$$F_2 = F_1 + F_1\left(e^{\pm\mu|\theta_2 - \vartheta_1|} - 1\right)\sin\psi + \beta w \Delta L \left\{\frac{\sin\alpha_2 - \sin\alpha_1}{\alpha_2 - \alpha_1}\right\} \tag{7.11}$$

$$T = \mu r N = \mu r F_1 |\theta_2 - \theta_1| \cos\psi$$

TABLE 7.1 Forces in the drill string during hoisting and lowering

POSITION	*STATIC WEIGHT (KN)*	*HOISTING (KN)*	*LOWERING (KN)*
Well Bottom	0	0	0
Bottom dropoff section	286	286	286
Bottom sail section	286 + 0.237 × 120 = 286 + 28.4 = 314.4	286 × 1.17 + 28.4 = 363	286 × 0.855 + 28.4 = 272.9
Top sail section	314.4 + 0.237 × 925 = 314.4 + 219.2 = 533.6	363 + 0.237 × 1308 (cos 45° + 0.20 sin 45°) = 626	272.9 + 0.237 × 1308 (cos 45° - 0.20 sin 45°) = 448.3
Top build-up section	533.6 + 0.237 × 120 = 533.6 + 28.4 = 562	626 × 1.17 + 28.4 = 760.9	448.3 × 0.855 + 28.4 = 411.7
Top well	562 + 0.237 × 335 = 562 + 79.4 = 641.4	760.9 + 79.4 = 840.3	411.7 + 79.4 = 491.1

7.3.4.5 Bottom End Condition

At the very bottom of the string, tension is small and the weight dominates friction also for curved bends. During tripping in and out of the well $F_1 = 0$ is used as an end condition. This will be shown in Case B to follow.

7.3.5 Application of the New Model

The models presented can be applied during drilling by computing the friction for each survey point. Alternatively, the well can be divided into a number of segments. Three cases will be presented where entire wells are analyzed. These are based on wells where the drill string is in tension.

7.3.5.1 Case A: Analysis of a Two-Dimensional S-Shaped Well

This example will demonstrate the application of the model, and also identify some frictional effects.

Figure 7.4 shows an S-shaped well that is drilled in a vertical plane. The total length is 2111 m, and the drill string consists of 161 m of 8" × 3" drill collars (2.13 kN/m) and 1950 m of 5"-19.5 lbs/ft drill pipe (0.285 kN/m). The drill collar radius is 0.1 m, and the drill string connection radius is 0.09 m. The well is filled with 1.3 s.g. drilling mud and the coefficient of friction is estimated to be 0.2. The bottom-hole-assembly starts out just below the drop-off section, and is vertical. For this case there is no change in azimuth, and the dogleg of Equation (7.1) becomes equal to the change in inclination.

The buoyancy factor is from Equation (7.4) (pipe density 7.8 s.g.: $\beta = 1 - 1.3/7.8 = 0.833$).

Assuming that the drill bit is off bottom, we will compute the forces starting from the bottom of the well. For simplicity, the frictional factors of the bends are:

$$e^{\pm\mu\theta} = e^{\pm 0.2\left(45\frac{\pi}{180}\right)} = e^{\pm 0.157} = \begin{cases} 1.17 \\ 0.855 \end{cases}$$

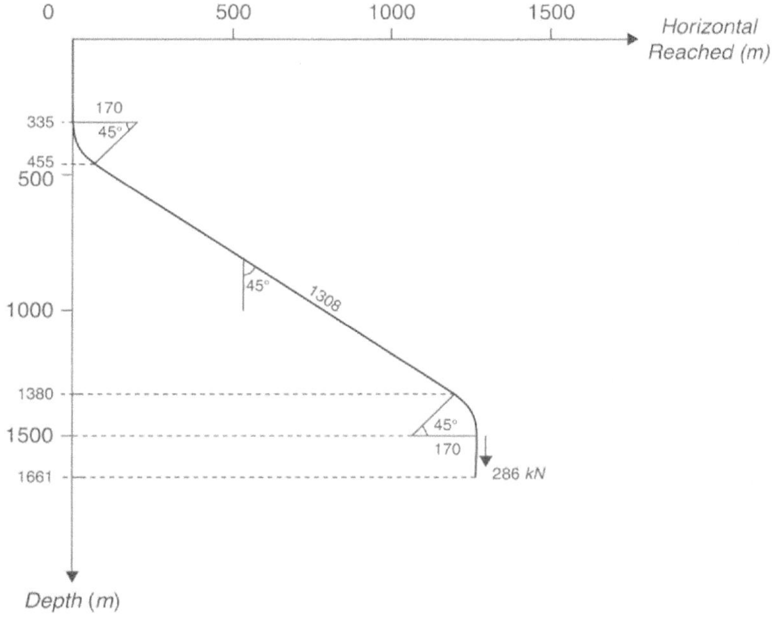

FIGURE 7.4 Geometry of S-shaped well.

The net weight of the bottom-hole-assembly: 0.833 × 2.13 (kN/m) × 161 (m) = 286 kN

The buoyed pipe weight: 0.833 × 0.285 kN/m = 0.237 kN/m

Figure 7.4 shows the geometry of the well.

When calculating torque, we will use two scenarios: 1) with bit off bottom, and 2) with a bit force of 90 kN. The static weight for the last example is simply obtained by subtracting the bit force throughout the string.

From Figure 7.5 it is obvious that the build-up and dropoff bends have a dominating effect on well friction. This is further seen in Figure 7.6, which shows the torque. When the bit force is applied, the tension in the string decreases, leading to less string torque.

Figure 7.6 shows the reduction in string tension that leads to a reduced string torque. The numerical values are given in Table 7.2. When the bit force is applied, the driller observes an increase in torque of 6.46 kNm, to a total value of 22 kNm. However, due to reduction in string torque, the bit torque is actually 9 kNm. From this example it is obvious that the bit torque is always higher than the torque increase at surface for a deviated well.

The simplicity of the solutions presented comes from neglecting the friction caused by pipe weight throughout the bend. As shown in Appendix E the simple equations can be used because the BHA is in a straight section. Comparing the exact solution with our simplified model, an error of about 1% for hoisting and nearly 0% for lowering was found.

7.3.5.2 Case B: Analysis of a Three-Dimensional Well

Figure 7.7 shows this well. It is complex as its direction changes in three-dimensional space. The analysis is similar to the analysis of Case A, except that the bends are not restricted to a vertical plane, but in a three-dimensional plane.

TABLE 7.2 Torque in drill string during drilling and with bit off bottom

POSITION	*STATIC WEIGHT, BIT OFF BOTTOM (KN)*	*TORQUE, OFF BOTTOM (KNM)*	*STATIC WEIGHT, 90 KN BIT FORCE*	*TORQUE, IN STRING (KNM)*	*TORQUE IN WELL (KNM)*
Well Bottom	0	0	−90	0	22 − 13.0 = 9
Bottom dropoff section	286	0	286 − 90 = 196	0	9
Bottom sail section	314.4	0.2 × 0.09 × 286 × π/4 = 4.04	314.4 − 90 = 224.4	0.2 × 0.09 × 196 × π/4 = 2.77	9 + 0.2 × 0.09 × 196 × π/4 = 9 + 2.77 = 11.77
Top sail section	533.6	4.04 + 0.2 × 0.09 × 0.237 × 1308 × sin45 = 4.04 + 3.95 = 8.0	533.6 − 90 = 443.6	2.77 + 3.95 = 6.72	11.77 + 3.95 = 15.72
Top build-up section	562	8.0 + 0.2 × 0.09 × 533.6 × π/4 = 8.0+ 7.54 = 15.54	562 − 90 = 472	6.72 + 0.2 × 0.09 × 443.6 × π/4 = 6.72 + 6.27 = 13.0	22
Top well	641.4	15.54	641.4 − 90 = 551.4	13.0	22

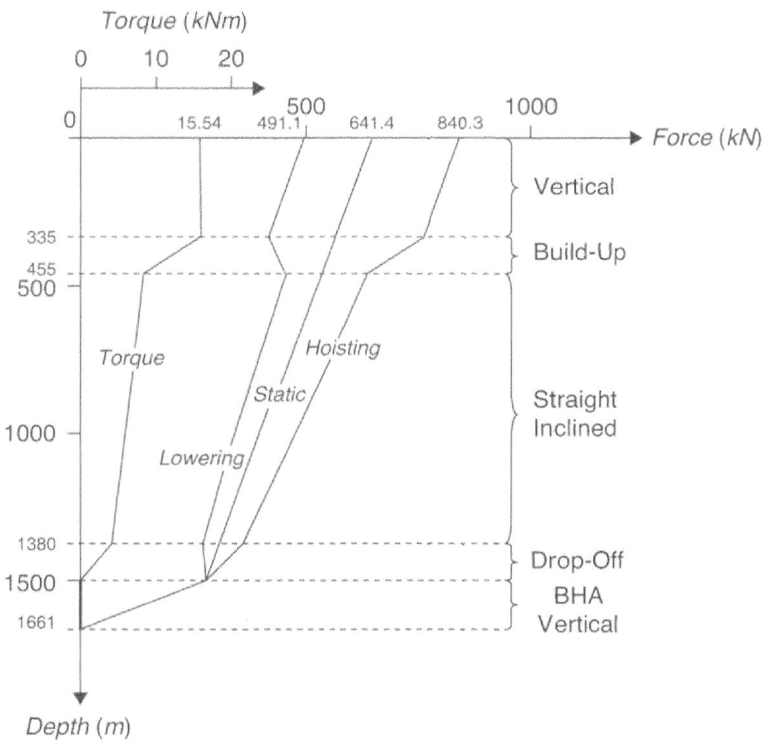

FIGURE 7.5 Torque and drag for the S-shaped well.

The pipe data are the same as for Case A. Performing the analysis, the results are shown in Fig. 7.8. Clearly, a different friction picture is seen in this well which has three bends. By increasing the total angle change of a well it increases friction significantly.

7.3.5.3 Case C: Combined Motion in Three-Dimensional Well

Some well operations are performed with combined motion. For high inclination wells it is common to rotate the liner in order to bring it to the desired depth. In the following we will investigate the frictional picture during such events.

Again assuming the same pipe data, we will investigate the effect of combined motion in the well of Case B. The pipe is rotated at 100 RPM, and it is hoisted or lowered at 0.27 m/sec. From Equation 7.9, the angle between the axial and tangential pipe velocities becomes 30 degrees.

Using Equations (7.10) and (7.11), the friction for combined motion is computed. The results are shown in Figure 7.9, which compares pure hoisting/lowering with combined motion. A clear reduction in drag is observed.

Both when hoisting and lowering the string, a reduced axial friction is seen. It is seen that both the hoisting and the lowering forces approached the static pipe weight, with the limiting value $\psi \rightarrow 0°$, which is obtained with low hoisting speed or high rotational speed. In other words, high rotation removes the axial friction.

Using Equations (7.10) and (7.11) the torque during hoisting and lowering is compared to the static (bit off bottom) torque. See Figure 7.10. During hoisting while back reaming the torque decreases slightly, whereas during lowering the torque decreases more.

The examples above serve to show the potential of torque and drag analysis using the simple three-dimensional models presented in this section. To determine the coefficients of friction from field data, not only can the hoisting-lowering-rotation models be used, but also the models for combined motion.

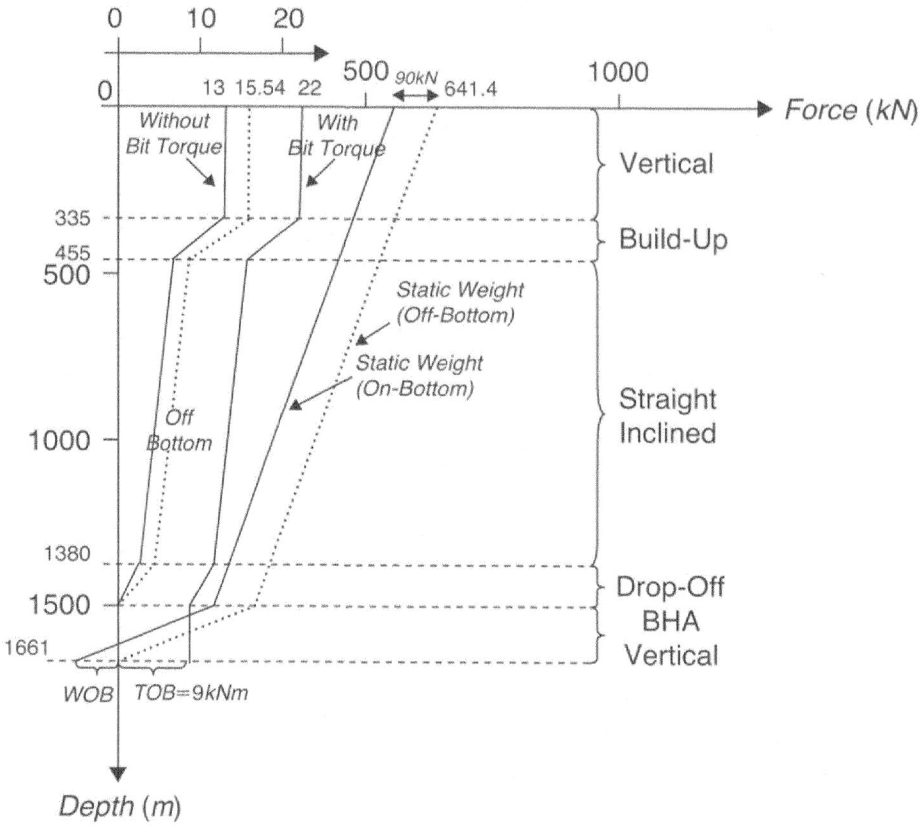

FIGURE 7.6 Torque during drilling.

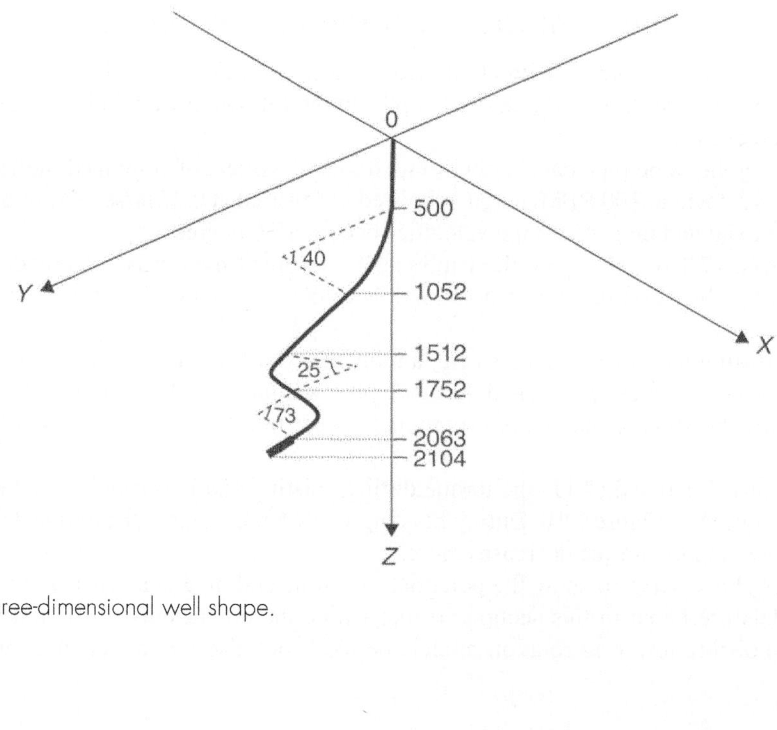

FIGURE 7.7 Three-dimensional well shape.

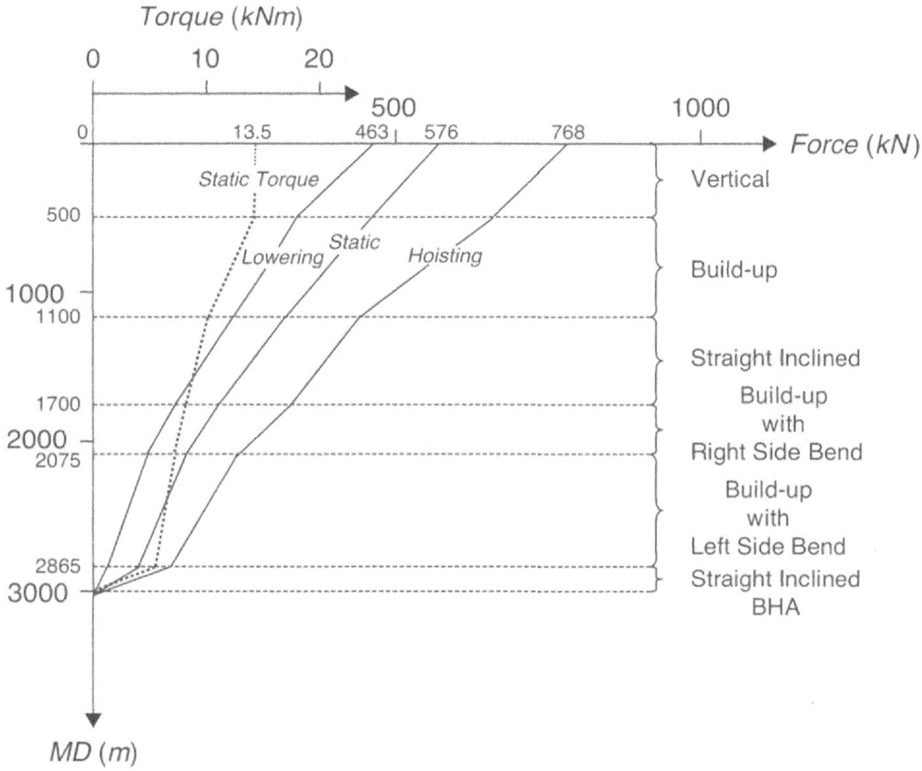

FIGURE 7.8 Well friction for three-dimensional well.

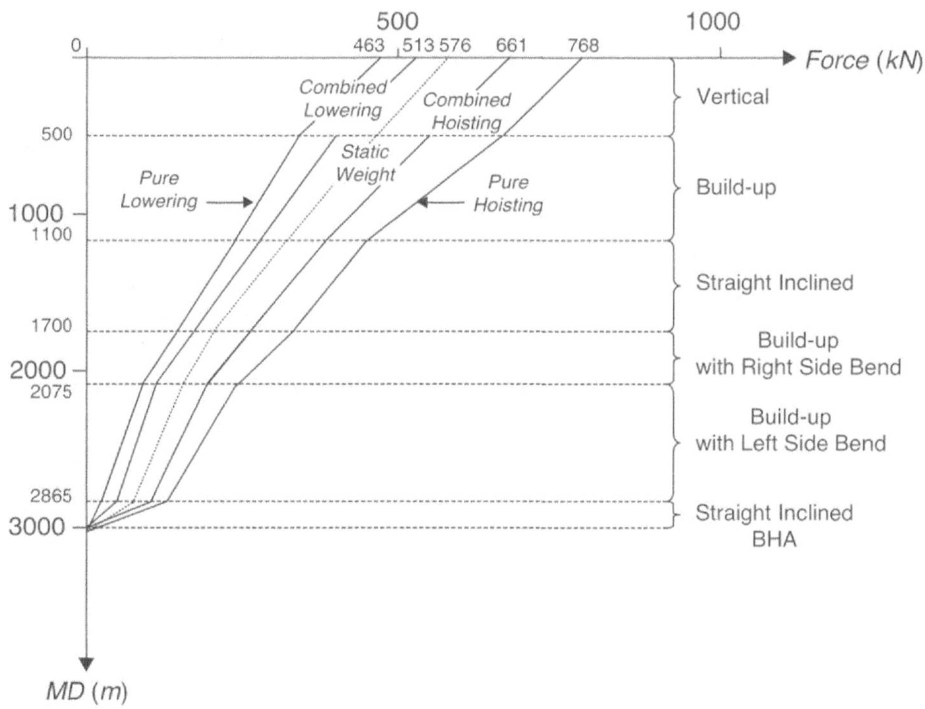

FIGURE 7.9 Comparison between pure hoisting/lowering and combined motion.

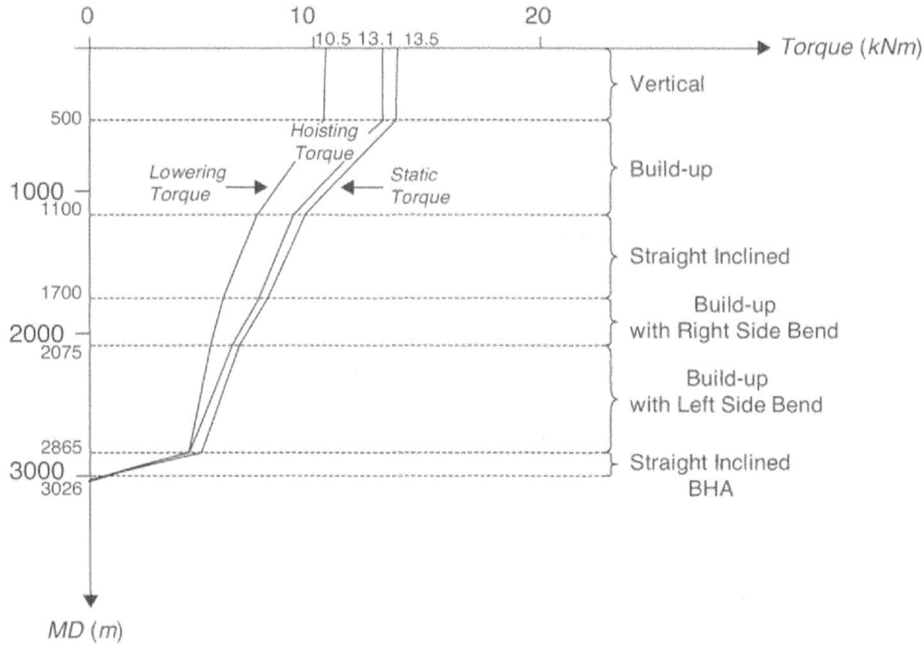

FIGURE 7.10 Torque with combined motion.

This should improve the determination of the frictional forces in a well, and is also applicable to any well operation.

The examples above show how bends in the wellbore affect friction. For more details and derivation of the models please see Aadnoy, Fazaelizade, and Hareland (2010). Extended applications of these models are found in Fazaelizade, Hareland, and Aadnoy (2010).

7.4 ANALYSIS OF STUCK PIPE IN DEVIATED WELLBORES

7.4.1 Introduction

This section presents equations to determine depth to the stuck point in deviated wellbores, based on pulling tests and torsion tests. Also, methods to free a stuck pipe are given. In particular it is shown that bends in the wellbore lead to more friction, which results in a deeper stuck point in a deviated well compared to a similar vertical well. The section is based on a method derived by Aadnoy, Larsen, and Berg (2003). The analysis is limited to differential sticking. Mechanical stuck pipe can also be analyzed by replacing the term for the differential stuck force with a mechanical resistance.

Three methods will be presented to free the pipe: 1) maximum mechanical force method, 2) minimum density method, and 3) maximum buoyancy method. A detailed field case study demonstrates these applications.

When a drill string gets stuck during drilling, operational procedures are applied to release the drill string. These procedures include working the drill string up or down, attempting to rotate the string, and pumping mud through the drill bit to aid pipe release. One may also place different fluids around the stuck area, and use a drilling jar. Current stuck point formulas used by the oil industry neglect well friction and deviation, and are therefore strictly valid for vertical wells only. A free-point indicator can be run to determine the stuck point, but this operation takes considerable time.

7.4.2 Forces in the Drill String

Axial Weight and Buoyancy
The buoyant weight of a string is defined in Section 7.3. Also defined in this section is the axial weight of a string in a deviated well. This is called "the projected height principle".

Pipe Strength
We are only considering axial pipe failure in this section. This case only considers the axial stretching of the drill pipe at the rig floor, and is expressed as:

$$\sigma_a \leq \sigma_{yield} \tag{7.12}$$

7.4.3 Differential Sticking

The two most common reasons for a stuck drill string are mechanical sticking due to jamming in the hole and differential sticking. It is often difficult to determine which type has occurred.

For a deviated well, the force needed to pull the drill string can be expressed as the sum of the pipe weight, the drag force and the differential sticking force, the latter being:

$$F = \mu dh\Delta P \tag{7.13}$$

Figure 7.11 illustrates a differential sticking. Assume that the stuck-point is in a straight hole section. The force normal to the stuck point is the component of the weight of the pipe plus the hydraulic force that causes differential sticking.

From Equation (7.13), the stuck force is equal to the normal force multiplied by a coefficient of friction. The pressure differential, ΔP, is the difference between the outside mud pressure and the pore pressure in the rock. The borehole pressure may be the static mud pressure plus the surface pressure, if the well is pressurized in the annulus.

If the string is stuck for mechanical reasons, for example if solids have packed off the annulus around the stabilizers, Equation (7.13) must be replaced with a mechanical resisting force. It is difficult to measure this. One can however assume values and perform simulations to investigate the forces throughout the wellbore. The two stuck scenarios, differential sticking and mechanical, are very different. In the former it is important to keep the pressures low, whereas this has no impact on mechanical sticking.

7.4.4 Rig Tests

7.4.4.1 Pull Test

When a drill string is stuck it is important to determine the depth to the stuck point. Often the string has to be cut just above the stuck point before a sidetrack operation can be initiated. To estimate the depth to the stuck point, the string is pulled with an additional force dF, and the elongation dL is measured on the surface. Since steel behaves linearly elastic, Hooke's law describes the relationship between force and elongation, $dF = AEdL/L$. The reference is the string hanging on its own weight. If the well is vertical, and

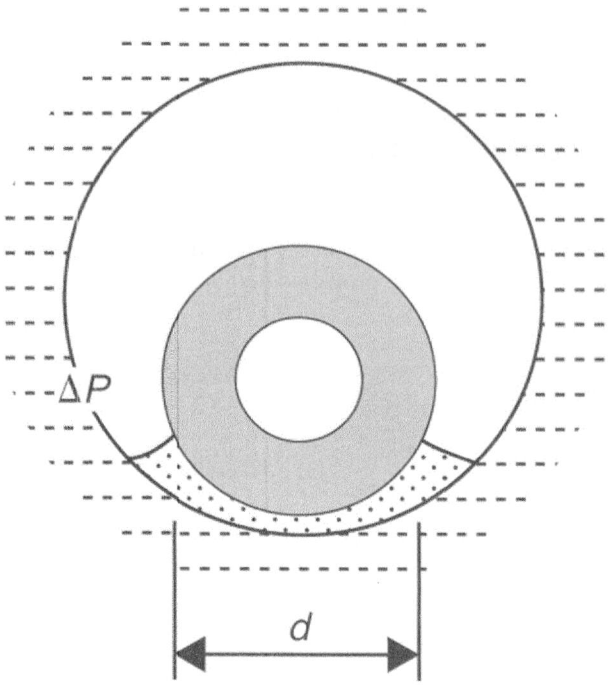

FIGURE 7.11 Differential sticking in a wellbore.

if friction is neglected, the depth to the stuck point becomes (if the drill string is composed of i different elements):

$$dL = \sum_{n=0}^{i} dL_n = \frac{1}{E}\sum_{n=0}^{i}\frac{L_n}{A_n}dF_i \tag{7.14}$$

Assuming that the bottom-hole-assembly is negligible because it is much stiffer than the drill pipe, and that one drill pipe size is used, the depth to the stuck point becomes (for a vertical well):

$$L = EA\frac{dL}{dF} \tag{7.15}$$

This is the equation currently used in industry. It is here assumed that there is no friction in the well. This means that when a pull force dF is applied at surface, this force is reflected all the way to the stuck point.

If we introduce drag, some of the pulling force goes into friction. The pull-rate is slow (nearly static), and it can be a reasonable assumption to neglect drag in a vertical well.

Aadnoy, Larsen, and Berg (2003), showed that curved sections of the borehole amplify the friction. The vertical well shows the pull force all the way down to the stuck point in Figure 7.12a. In Figure 7.12b (the deviated well), the pull force remains constant all the way to the build-up section. Through the build-up section some of the pull force is taken up as friction. Below the build-up section, a smaller pull force occurs.

Assume that the well consists of a vertical section to the kick-off point, L_1.A build-up section of radius R builds up to an angle α. A sail section maintaining this inclination continues to the stuck point.

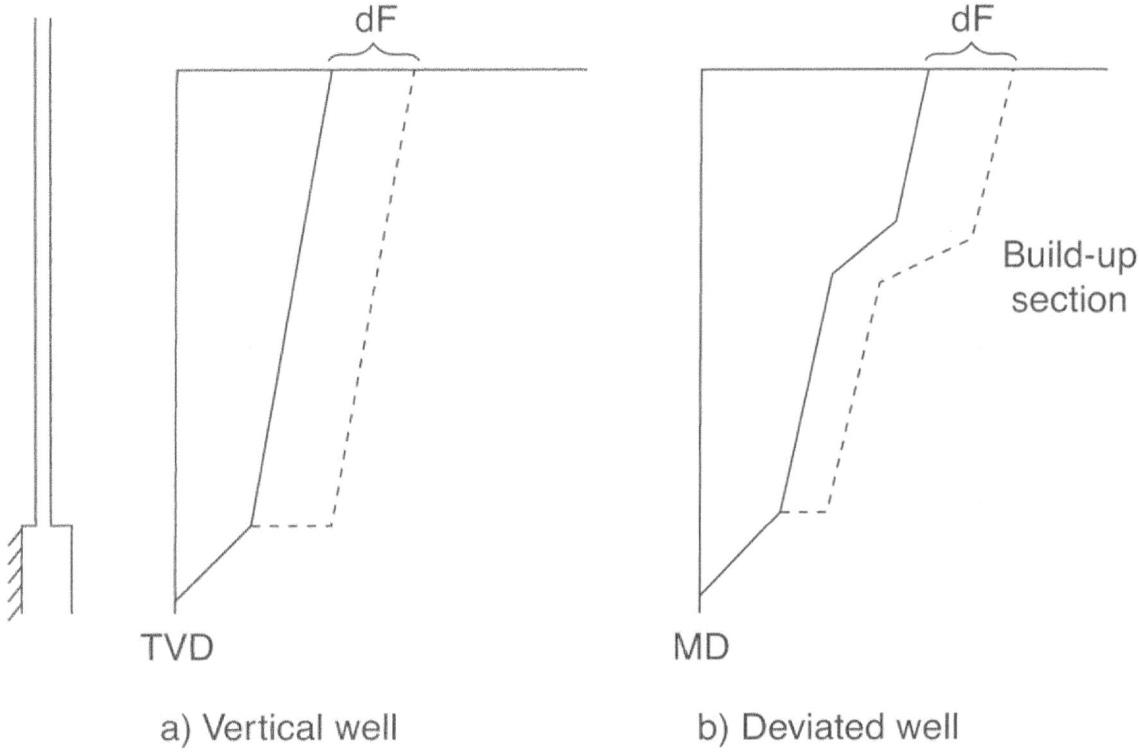

FIGURE 7.12 Pull force at the stuck point for vertical and deviated wells.

We assume that the pipe is pulled slowly with a force dF a distance dL. This force dF is reflected down to the kick-off point. The estimated length to the stuck point for this case is:

$$L = AEe^{\mu\alpha}\frac{dL}{dF} - \left(e^{\mu\alpha} - 1\right)\left(L_1 + \frac{1}{2}\alpha R\right) \tag{7.16}$$

Equation (7.16) is valid if one drill pipe size is used, and for a well consisting of one build-up section and a constant sail angle in the section below. For two drill pipe sizes the equation can be modified. Referring to the top pipe size with index 1, and the bottom pipe size with index 2, the depth to the stuck point is now:

$$L = A_2Ee^{\mu\alpha}\frac{dL}{dF} - \frac{A_2}{A_1}\left(e^{\mu\alpha} - 1\right)\left(L_1 + \frac{1}{2}\alpha R\right) - L_2\left(\frac{A_2}{A_1} - 1\right) \tag{7.17}$$

7.4.4.2 Torsion Test

When the drill string is rotated at the surface, an applied moment dM leads to a twist angle $d\theta$. The corresponding expression for the angle of rotation as a function of depth is similar to the pulling case:

$$d\theta = \frac{1}{G}\sum_{n=0}^{i}\frac{L_i}{J_{pi}}dM \tag{7.18}$$

The shear modulus is $G = \frac{E}{2(1+v)}$ and the polar moment is $J_p = \frac{\pi}{32}(D^4 - d^4)$. The elastic modulus is 215 kN/mm^2, and the Poisson's ratio is 0.25 for steel.

If only one drill pipe type is used, the stuck point can be estimated from the following equation:

$$L = \frac{JE}{2(1+v)} \frac{d\theta}{dM} \tag{7.19}$$

The equation may be expanded to the case of using two drill pipe sizes. The depth to the stuck point now becomes:

$$L = \frac{J_2 E}{2(1+v)} \frac{d\theta}{dM} - L_1 \left(\frac{J_2}{J_1} - 1 \right) \tag{7.20}$$

The equations above are valid for well paths consisting of a vertical section to the kick-off point, a constant build section, and a constant inclination sail section to the stuck point. For more complex geometries similar equations can be derived, but these will become more cumbersome. A simple numerical routine may be used for these cases.

7.4.5 Field Case Study

During drilling of a long deviated well several stuck pipe incidents occurred. Two of these stuck pipe incidents resulted in the bottom-hole-assembly (BHA) being left in the hole and the well had to be sidetracked twice in the 12-1/4 in. section. The second incident, which resulted in the BHA being shot off, will be analyzed, due to better data from this incident. Figure 7.13 shows the well path for this case.

The friction factor of 0.12 (which was measured during drilling), will be used in the analysis below. The same coefficient of friction is assumed to be also valid after the pipe was stuck.

7.4.5.1 Description of Case

After sidetracking the hole (following the first stuck pipe incident), a sand reservoir was entered. The washpipe started leaking, and it was decided to pull out to above the reservoir for replacement. The first stand was pulled without excessive drag. Then pipe rotation was stopped for 7 minutes to set back the stand and an attempt was made to pull the second stand. It was not possible to pull the drill string using 260 tons hook load. The full string weight also was applied, without any effect. Combined torque and pull were also applied without success. The jar did not go off in any of the attempts to free the string.

Full circulation was maintained during these attempts, leading to the conclusion of differential sticking. The 13-3/8 in. section (shoe at 2281 m) was drilled with a mud weight of 1.65 s.g. for borehole stability reasons, and the pore pressure specific gravity was determined to be 1.24 s.g.

First, we will estimate the depth to the stuck point in the well. Pipe data are shown in Table 7.3. The following additional data from the stretch test on the rig are given in Table 7.4.

Inserting these data into Equation (7.16), the depth to the stuck point is as follows:

First, $\mu\alpha = 0.12 \frac{68\pi}{180} = 0.142$ and: $e^{\mu\alpha} = 1.153$ and: $\frac{1}{2}\alpha R = \frac{68\pi}{2 \times 180} 840 = 498\,\text{m}$

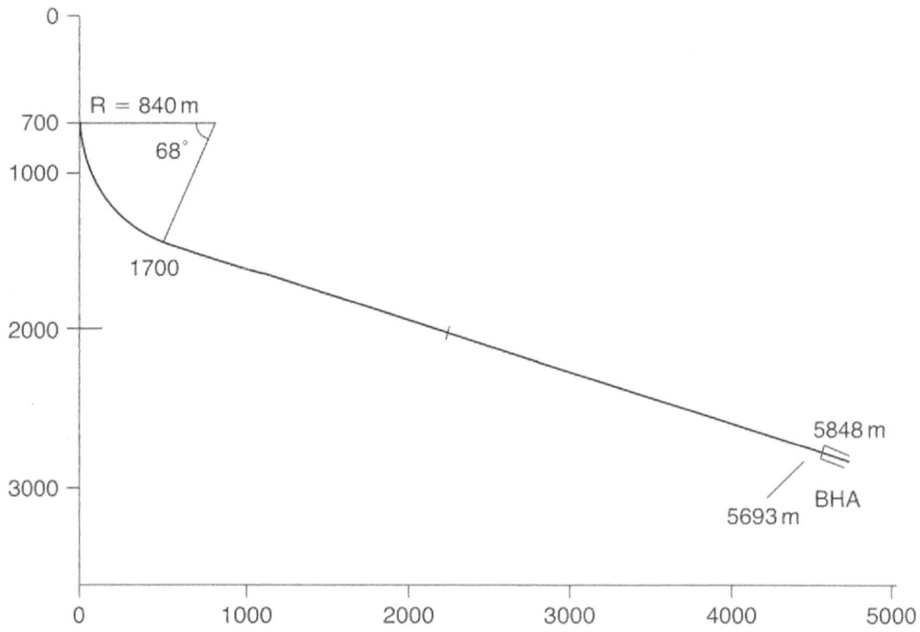

FIGURE 7.13 Well path of the well.

TABLE 7.3 Drill string data for the field case

PIPE SIZE (M)	*5 IN*	*BHA*
Quality	S135	
Length (m)	5693	155
Weight (kN/m)	0.336	1.344
OD (mm)	127.00	
ID (mm)	108.6	
t (mm)	9.19	
A (mm²)	3401	
J (mm⁴)	11.88 × 10⁶	
Tensile strength (kN)	2495	
Torsional strength (kNm)	78.8	
E (kN/mm²)	215	
Friction factor:	0.12	

$$L = 3401\left(\text{mm}^2\right)215\left(\text{kN}/\text{mm}^2\right)1.153\frac{2.60(\text{m})}{373(\text{kN})} - (1.153-1)(700+498) = 5693\text{ m}$$

For comparison, the estimated depth for a vertical well is 5097 m. This is found by setting $\mu = 0$ in Equation (7.16). The standard industry model which neglects friction would give this value. The calculation above is repeated with different friction coefficients. Figure 7.14 shows the depth to the stuck point for various friction coefficients.

TABLE 7.4 Data for the pull and rotation tests

BUILD RADIUS (M)	*840*
Inclination (°)	68
Friction coefficient	0.12
Pull force (kN)	373
Pull length (m)	2.60
Torque (kNm)	9.05
Pipe rotation (turns)	8

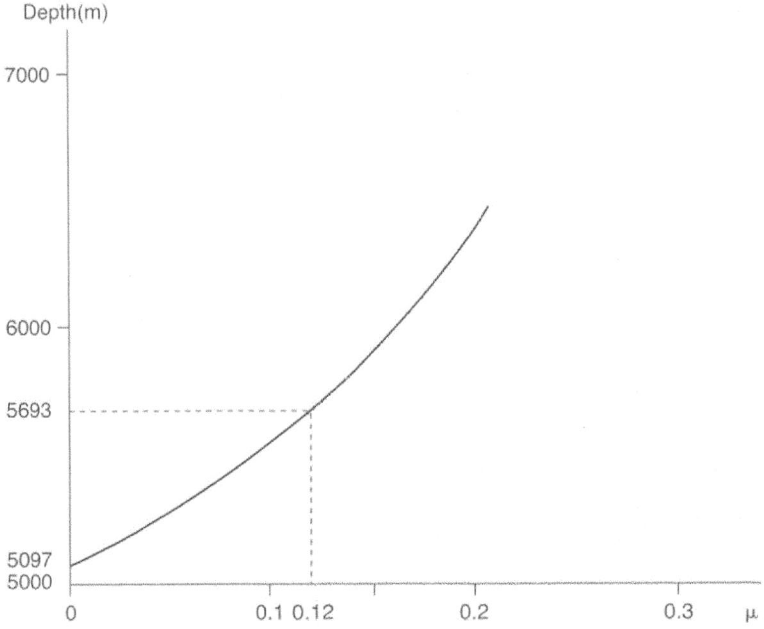

FIGURE 7.14 Depth to stuck point versus friction coefficient.

The torsion test also provides important information. The drill string was rotated eight turns, with an applied moment of 9.05 kNm. Using data from Tables 7.3, 7.4 and Equation (7.19), the estimated length of the drill string is:

$$L = \frac{11.88 \times 10^6 \left(\text{mm}^4\right) 215 \left(\text{kN}/\text{mm}^2\right)}{2(1+0.25)} 10^{-6} \left(\text{m}^2/\text{mm}^2\right) \frac{8 \times 2\pi}{9.05} = 5675\,\text{m}$$

Thus two independent tests can be performed to determine the depth to the stuck point, both pull test and torsion test. The top of the bottom-hole-assembly was located at 5691 m. From the two tests above, we concluded that the drill string was most likely stuck in the top of the bottom-hole-assembly. There is at that depth a 22-m-thick sand stringer, and we assume that the string is stuck along the entire stringer.

7.4.6 Methods to Free the Pipe

The analysis below will demonstrate the effects of the various ways a stuck pipe problem can be handled. Forces in the drill string are summed up starting from the bottom to the stuck point, and from the stuck point to the surface. In this particular case the drill string is stuck in the bottom part of the drill string above the drill collar.

The force below the drill collar is zero.

Force at top of drill collar: (Equation (7.5))

$$F_1 = \beta w_{DC} \Delta L_{DC} (\cos\alpha + \mu \sin\alpha)$$

Additional pull force at stuck point: (Equation (7.13))

$$F_2 = \mu d h \Delta P$$

Force above stuck point:

$$\mathrm{F}_3 = \mathrm{F}_1 + \mathrm{F}_2$$

Force at lower end of build (top of 5 in. drill pipe straight section): (Equation (7.5))

$$F_4 = F_3 + \beta w_{DP1} \Delta L_1 (\cos\alpha + \mu \sin\alpha)$$

Force at top of build section: (Equation (7.7b))

$$F_5 = F_4 e^{\mu\alpha} + \beta w_{DP1} R \sin\alpha$$

Pull force at surface:

$$F_6 = F_5 + \beta w_{DP1} \Delta L_{KOP}$$

7.4.6.1 Maximum Mechanical Force

This simplest approach is to pull or rotate the drill string towards the strength limit of the drill pipe. For the present field case, the mud specific gravity used was 1.65 s.g. and the pore pressure specific gravity in the permeable interval was estimated to be 1.24 s.g. The projected area at the stuck position was estimated to be 0.05 m times 22 m. The differential pressure and the buoyancy are:

$$\Delta P = 0.098(1.65 - 1.24)2936 = 11.8 \text{ MPa} = 11800 \text{ kN/m}^2$$
$$\beta = 1 - \frac{1.65}{7.85} = 0.79$$

Inserting these data, the data from Tables 7.3 and 7.4 and a stuck point of 5691 m we get the following forces during pipe release:

$$F_1 = 0.79 \times 1.344 \left(\frac{\text{kN}}{\text{m}}\right) \times 155(\text{m})(\cos 68^\circ + 0.12 \sin 68^\circ) = 80\,\text{kN}$$

$$F_2 = 0.12 \times 0.05(\text{m}) \times 22(\text{m}) \times \frac{11800\,\text{kN}}{\text{m}^2} = 1558\,\text{kN}$$

$$F_3 = 80 + 1558 = 1638\,\text{kN}$$

$$F_4 = 1638\,\text{kN} + 0.79 \times 0.336 \times 3991(\text{m})(\cos 68^\circ + 0.12 \sin 68^\circ)$$

$$= 1638\,\text{kN} + 514\,\text{kN} = 2153\,\text{kN}$$

$$F_5 = 2153(\text{kN}) e^{0.12 \times 68^\circ \frac{\pi}{180^\circ}} + 0.79 \times 0.336(\text{kN/m}) \times 840(\text{m}) \times \sin 68^\circ$$

$$= 2482\,\text{kN} + 207\,\text{kN} = 2689\,\text{kN}$$

$$F_6 = 2689\,\text{kN} + 0.79 \times 0.336\left(\frac{\text{kN}}{\text{m}}\right) \times 700(\text{m}) = 2875\,\text{kN}$$

The axial strength of the drill pipe is 2495 kN. Therefore, this is not sufficient to free the pipe.

7.4.6.2 Applying Pull Force

Aadnøy and Andersen (1998) presented a comprehensive set of equations for calculating string weight and well friction across various well geometries. In this section, only the equations relevant to the current case are applied; for other geometries, the reader is referred to the original reference.

During pipe freeing, the maximum force in the drill string typically occurs at the surface. For the Yme well geometry shown in Figure 7.13, the pipe is stuck at the bottom-hole assembly.

By inserting the well data into the appropriate equations for each section, and applying the last derived expression for the total system, the required pull force at the surface can be expressed as:

$$F_{surface}(kN) = 1507\beta + 0.152 + P(kN/m^2) \tag{7.21}$$

7.4.6.3 Applying Torsion

In addition to axial pull, a torsional load is applied to the drill string to help free it. Aadnøy and Andersen (1998) derived the torque equation for string rotation, where torque is defined as the product of the normal force (as presented earlier), the friction coefficient, and the contact radius. An average contact radius of 0.1 m is used for both the bottom-hole assembly and the drill pipe connections. By combining the relevant equations from the bottom-hole assembly to the surface, the total required torque can be expressed as:

$$M_{surface}(kN\,m) = 32.4\beta + 0.013 + P(kN/m^2) \tag{7.22}$$

Inserting the same data into Equation 7.22, the torque required is 181.3 kNm, which also far exceeds the torque limit of the drill pipe of 78.8 kNm. The conclusion is that by applying maximum mechanical force only, the string will remain stuck, and a parted drill pipe may result.

7.4.6.4 Minimum Density Method

It is obvious from the example above that the dominating parameter is the differential pressure that is the main cause of differential sticking. This can be reduced by displacing the well with a lighter mud. We will repeat the analysis assuming that the well has been displaced to a 1.3 s.g. mud. The differential pressure and the buoyancy now become:

$$\Delta P = 0.098(1.30 - 1.24))2936 = 1.736 \text{ MPa} = 1730 \text{ kN} / \text{m}^2$$

$$\beta = 1 - \frac{1.30}{7.85} = 0.83$$

$$F_1 = 0.83 \times 1.344(\text{kN}/\text{m}) \times 155(\text{m}) \times (\cos 68^\circ + 0.12 \times \sin 68^\circ) = 84\,\text{kN}$$

$$F_2 = 0.12 \times 0.05(\text{m}) \times 22(\text{m}) \times 1730\text{kN}/\text{m}^2 = 228\text{kN}$$

$$F_3 = 84 + 228 = 312\,\text{kN}$$

$$F_4 = 312 \text{ kN} + 0.83 \times 0.336 \times 3991(\text{m})(\cos 68^\circ + 0.12 \sin 68^\circ)$$

$$= 312 \text{ kN} + 541 \text{ kN} = 853 \text{ kN}$$

$$F_5 = 853(\text{kN}) e^{0.12 \times 68^\circ \frac{\pi}{180^\circ}} + 0.83 \times 0.336(\text{kN}/\text{m}) \times 840(m) \times \sin 68^\circ$$

$$= 983 \text{ kN} + 217 \text{ kN} = 1200 \text{ kN}$$

$$F_6 = 1200 \text{ kN} + 0.83 \times 0.336(\text{kN}/\text{m}) \times 700(\text{m}) = 1200 \text{ kN} + 195 \text{ kN}$$

$$= 1395 \text{ kN}$$

Now the required load is below the pipe strength. In other words, displacing the well with a lighter mud is the best measure for freeing the drill string. The buoyancy decreases, but this effect is negligible to the effect of reducing the bottomhole pressure.

7.4.6.5 Maximum Buoyancy Method

The last method analyzed is displacement of the inside of the drill pipe with seawater to increase buoyancy. This is permissible providing that well control is maintained as the seawater is later displaced up the annulus. For this case we assume the initial mud of 1.65 s.g. in the annulus which results in the initial differential pressure of 11.8 MPa. The buoyancy increases and is given by:

$$\beta = 1 - \frac{1.65 \times 127^2 - 1.03 \times 108.6^2}{7.85(127^2 - 108.6^2)} = 0.57$$

Inserting these numbers into Equation (7.21) leads to a required pull force of 2652 kN and Equation (7.22) leads to a required torque of 174 kNm. Table 7.5 summarizes the results.

This example demonstrates that there is some potential in increasing buoyancy by displacing the inside of the drill string with seawater. The most important measure, however, is to reduce the annulus pressure by displacing the well with a lighter mud.

$$F_1 = 0.57 \times 1.344(\text{kN}/\text{m}) \times 155(\text{m}) \times (\cos 68^\circ + 0.12 \times \sin 68^\circ) = 58\,\text{kN}$$

TABLE 7.5 Summary of analysis of the field case

METHOD	*REQUIRED PULL FORCE (KN)*
Maximum mechanical force	2875
Minimum density	1395
Maximum buoyancy	2574
Pipe strength	2495

$$F_2 = 0.12 \times 0.05(\mathrm{m}) \times 22(\mathrm{m}) \times 11800\,\mathrm{kN/m^2} = 1558\,\mathrm{kN}$$

$$F_3 = 58 + 1558 = 1616\ \mathrm{kN}$$

$$F_4 = 1616\,\mathrm{kN} + 0.57 \times 0.336 \times 3991(\mathrm{m})(\cos 68^\circ + 0.12\sin 68^\circ)$$

$$= 1616\,\mathrm{kN} + 371\,\mathrm{kN} = 1987\,\mathrm{kN}$$

$$F_5 = 1987(\mathrm{kN})e^{0.12\times 68^\circ \frac{\pi}{180^\circ}} + 0.57 \times 0.336(\mathrm{kN/m}) \times 840(\mathrm{m}) \times \sin 68^\circ$$

$$= 2291\,\mathrm{kN} + 149\,\mathrm{kN} = 2440\,\mathrm{kN}$$

$$F_6 = 2440\,\mathrm{kN} + 0.57 \times 0.336(\mathrm{kN/m}) \times 700(\mathrm{m}) = 2440\,\mathrm{kN} + 134\,\mathrm{kN}$$

$$= 2574\,\mathrm{kN}$$

This section presents a mechanistic analysis of differentially stuck pipe in a deviated well. The following mechanisms are investigated: buoyancy effects, pipe strength, differential sticking, and well friction.

Equations are presented to estimate the depth to the stuck point in deviated wellbores based on pull and rotation tests. Due to friction in bends, the stuck point appears deeper in a deviated well compared to a vertical well.

For derivation of equations and more details, the readers are referred to Aadnoy, Larsen, and Berg (2003). Here other scenarios are also developed such as combined loading by both pulling, applying pressure and rotating at the same time.

7.5 WELL INTEGRITY

7.5.1 Introduction

In recent years well integrity has become an important subject. It has become clear that many wells suffer from leaks of various types. It is also evident that these problems are not only related to old wells, but fairly new wells also. With the advent of the increasing CO_2 injection in the future, well integrity will be a challenge. Improvements in both technology and understanding of subsurface processes is required. This section is based on an early study by the Petroleum Safety Authority in Norway published by Vignes and Aadnoy (2008).

A number of serious well failures in recent years led to investigations of well integrity issues. The Petroleum Safety Authority Norway (PSA) performed a "pilot well integrity survey" based on supervisory audits and requested input from seven operating companies, 12 pre-selected offshore facilities and 406 wells. The wells were a representative selection of production and injection wells with variation both in age and development categories.

The pilot project indicates that 18% of the wells in the survey have integrity failure, issue, or uncertainties and 7% of these are shut in because of well integrity issues. The selection of wells and the companies indicate that the statistics are representative. A later study suggested that as many as each fifth production well and each third injection well may suffer from well integrity issues.

The well incidents in the past and the results of "pilot well integrity survey" revealed that the industry needs to increase focus on the barrier philosophy. Control of barrier status is an important HSE factor to avoid major incidents caused by, e.g., unintentional leaks and well control situations. Knowledge of well integrity status at all times enables the companies to take the right actions in a proactive manner and thereby prevent incidents.

Nomenclature

ASV	Annulus safety valve
DHSV	Downhole safety valve
GLV	Gas lift valve
HSE	Health safety environment
NCS	Norwegian continental shelf
P&A	Plug and abandonment
PBR	Polished bore receptacle
PSA	Petroleum Safety Authority Norway
SD&P	Simultaneous drilling and production
TOC	Top of cement
WAG	Water and gas injector

7.5.2 Status of Wells on the Norwegian Continental Shelf

Active development wells are production and injection wells. The production wells include water production, gas production, and oil production. The injector wells include gas injection, water injection, and WAG. P&A wells are not included in this survey.

Table 7.6 shows that the total number of active wells on the NCS at 1.1.2006 is 1908 wells and the number of active wells in the pilot well integrity survey is 406 wells. The pilot well integrity survey includes 21% of active development wells at the NCS.

Table 7.7 furthermore shows the split between the injection and the production wells on platforms, whereas Table 7.8 shows the split on subsea wells.

The pilot well integrity survey includes 24% of the platform wells and 16% of the subsea wells with production and injection on the NCS.

TABLE 7.6 Development wells in the pilot survey per 2006

	ACTIVE DEVELOPMENT WELLS IN THE "PILOT SURVEY"	*TOTAL ACTIVE DEVELOPMENT WELLS*
Production	*323 wells*	*1539 wells*
Injection	*83 wells*	*369 wells*
Total	***406 wells***	***1908 wells***

TABLE 7.7 Active platform wells in the survey

	ACTIVE PLATFORM WELLS IN THE SURVEY	*TOTAL ACTIVE PLATFORM WELLS*
Injection	*51 wells*	*244 wells*
Production	*249 wells*	*1011 wells*
Total	***300 wells***	***1255 wells***

TABLE 7.8 Active subsea wells in the survey

	ACTIVE SUBSEA WELLS IN THE SURVEY	*TOTAL ACTIVE SUBSEA WELLS*
Injection	*32 wells*	*125 wells*
Production	*74 wells*	*528 wells*
Total	***106 wells***	***653 wells***

7.5.3 Results of the Well Integrity Survey

Figure 7.15 shows the number of production and injection wells with well integrity issues. These are 48 production and 27 injection wells, and the total is 75 wells.

Most of the integrity problems are within barrier elements such as tubing, annular safety valve (ASV), casing, cement, and wellhead. Figure 7.16 shows a breakdown into technical categories.

- Reported tubing problems were leakage in prod. tubing above down hole safety valve (DHSV), tubing to annulus leakage or internal leak in tubing hanger neck seal.
- Problems with annular safety valve (ASV) are ASV leakage or that the ASV failed.
- Casing problems like casing leakage (non-gastight connections) or collapsed casing were reported in the pilot survey.
- Cement problems such as no cement behind casing and above production packer, leaks likely along cement bonds or leak through cement micro-annulus.
- Reported wellhead problems are leakage in wellhead from annulus A to B, because of the wrong seal type in the wellhead.

Figure 7.17 shows the age of the reported wells. The majority of the well integrity problems occur in wells during the period from the early 1990s. The frequency of wells with integrity issues in age group 0–14 years is twice as high as for well group 15–29 years.

The figure illustrates that tubing leak is one dominant failure factor (39%) in wells from 0 to 19 years old. Wells from 0–14 years old have barrier element issues like tubing, ASV, and cement. Wells from 15–29 years old have barrier element issues like tubing, casing, and pack offs. One interpretation of the figure is that the recent technological advancement may not have resolved old integrity issues.

A relatively low number of subsea wells have reported well integrity failure/issues or uncertainty. This can be explained by the limited possibility of monitoring these wells.

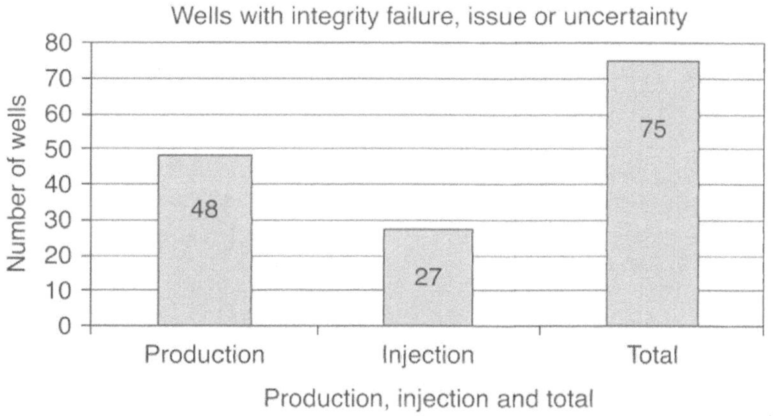

FIGURE 7.15 Production wells, injection wells, and total wells reported with integrity failure, issue, or uncertainty in this survey.

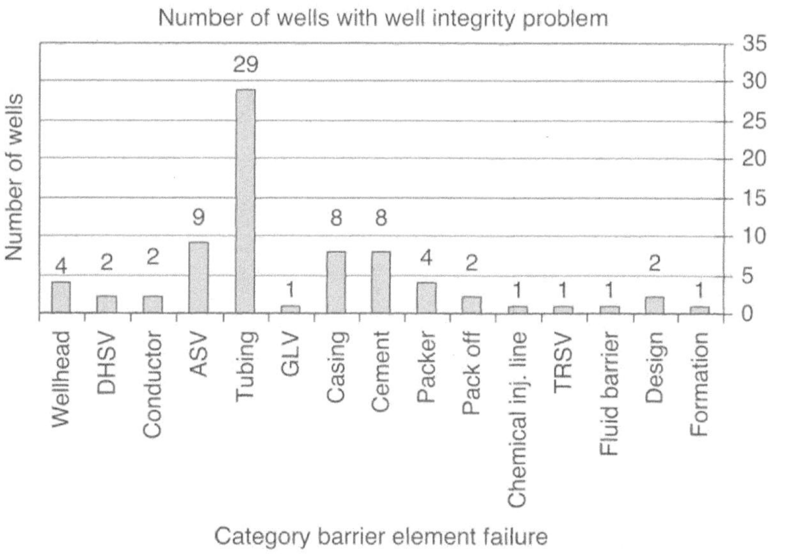

FIGURE 7.16 Number of wells with integrity failure, issues or uncertainty, and category of barrier element failure.

7.5.3.1 Examples of Well Failures

The PSA has also conducted a number of technical audits in the period from 2003 to gain an understanding of the technical aspects of well failures. In addition to the technical and operational investigations, one also evaluated the design manuals, methodology for data collection, and the actual well design. These were compared to the actual well construction and later problems.

A number of interesting observations came out of this work. In the following we will discuss five of these well incidents to gain further insight.

Well Failure: Example I

During workover of a well on a production platform, the load-bearing surface casing collapsed, resulting in a wellhead that dropped onto the platform structure. The root cause was severe corrosion near the top of the surface casing annulus. Corrosive seawater got access through a cement port that was left open during the drilling phase. Temperature and tidal variations through platform shaft leaks were accompanying factors. This particular well was shut in for approximately one year before production commenced.

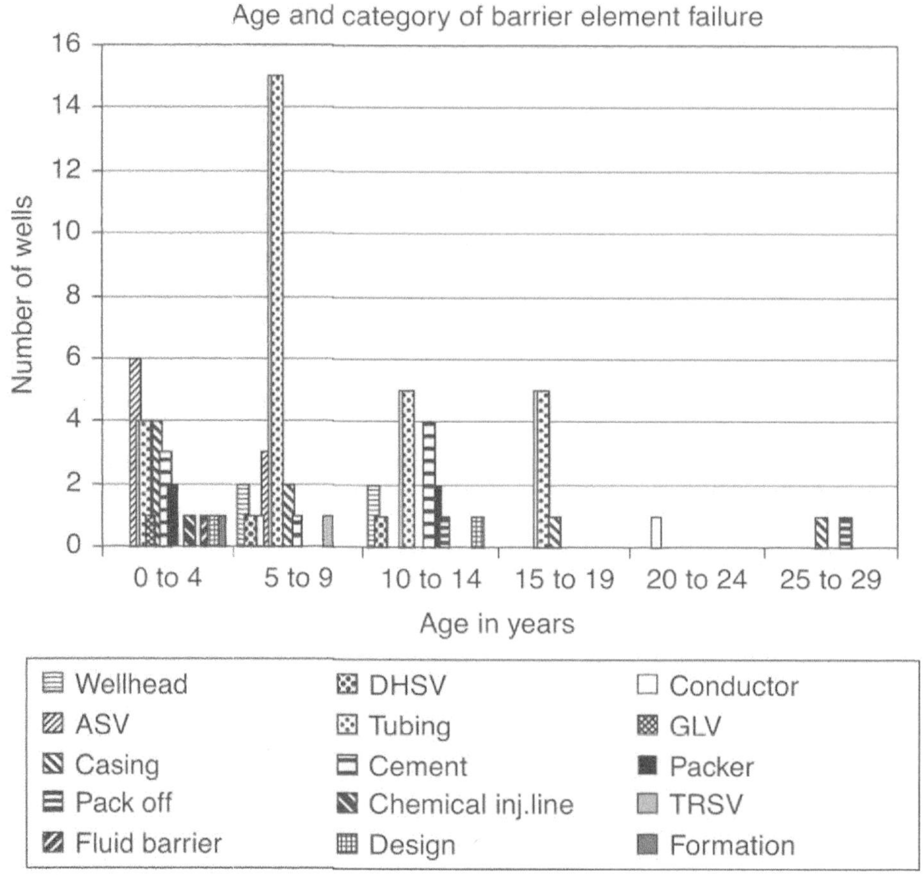

FIGURE 7.17 Age and category of barrier element failure per 1.3.2006.

Well Failure: Example II

The operator had installed a specially designed "slim" wellhead in a field. The casing hanger had only 8 degrees taper as opposed to the usual 40 degrees. During pressure testing the casing hanger was pulled through the wellhead. Later, the operating company experienced a similar incident with a tubing hanger. The root cause of the failures was axial overload as the "slim" wellhead tapers had limited capacity. The wellhead capacity had been uprated from 350 tons to 600 tons. The investigation report, however, showed that the manufacturer's test had failed, leaving it incorrect to allow such up-rating. The operating company have after these incidents limited the load to the initial design value.

Another finding in this case was that the operator did not have in place an updated well design manual. This was corrected after the audit.

Well Failure: Example III

During drilling of a depleted reservoir severe circulation losses arose, followed by a well control problem. During pumping of a "gunk" plug, the drill string became plugged. This complicated the well control situation. The problem was resolved. However, during the well control phase the well was left open, contrary to regulations. The operating company revised their well-killing procedures including field procedures for lost circulation pills after the incident.

Another issue during the well control incident was annulus pressure build-up exceeding MAASP. This downgrading was introduced after some years with production falling due to depletion. After several years with injection and subsidence, the pressure measured during the well control situation exceeded the new established/down-graded pressure limit. This caused serious problems and a need to determine absolute maximum allowable pressure. The well-killing operation was at the end successfully executed just below the new pressure limit. The operating company rechecked their MAASP-limits in adjacent wells.

This is an example of the importance of lifecycle evaluations and continuous monitoring and evaluation of well integrity rather than just at the well design stage.

Well Failure: Example IV

Tubing leaks were found in 14 production wells on an offshore platform. As the leak rates exceeded the acceptance criteria, an investigation was initiated. It was difficult to establish a root cause. However, the conclusion was that it was likely to be leaks through the PBRs (telescopic expansion joints in the tubing). These were considered weak points in the production tubing. A possible explanation was wear damage when the production tubing was tripped into the well. The operating company decided that the PBR will be shielded during installation in future wells.

Well Failure: Example V

An offshore production well had tubular collapse in both the production casing and the production tubing at the same depth. These had to be removed in one operation, a task that was complicated and time consuming. During this operation a well control incident happened during perforation of the production casing. The investigation revealed that the well control procedures were not applied. The root causes were leaks from the reservoir to the outside of the production casing and installation of one single casing joint (grade 47 lbs/ft) in a 53.5 lb/ft casing string. This wrong casing joint collapsed.

7.5.3.1.1 Summary of Well Failure I–V

There is a common denominator between these five examples.

- Operational decisions during abnormal situations often lead to well failures.
- Design issues such as long-term effects were not sufficiently considered.
- The challenge is to account for rare events that may lead to major incidents. The normal approach is to focus on frequent and low-consequence incidents.

7.5.3.2 Improvement Potential

Figure 7.18 shows the issues for improvement from A–L and the number of companies. The figure shows that B: well documentation, C: handover documentation, D: regular condition monitoring, E: Comments related to NORSOK.

D010 compliance and H: competence and training are issues for improvement for almost all of the companies in the pilot well integrity survey.

B: Well documentation. Different approaches for transfer of well barrier responsibilities, and insufficient transfer of critical information during operations and license acquisitions, may result in reduced barrier control.

There is a need for improved and user-friendly access and visualization of key well information.

C: Handover documentation. Several companies include pressure test verification charts in their handover documentation, without interpretation. These data could be several years old and do not necessarily represent the present status. Well integrity or well barrier schematic illustrations, verifications, and guidance on how to monitor status, were not easily accessible.

While completion schematics are readily available, included in the handover documents and kept updated at all times, the barrier schematic illustrations with descriptions were often inadequate.

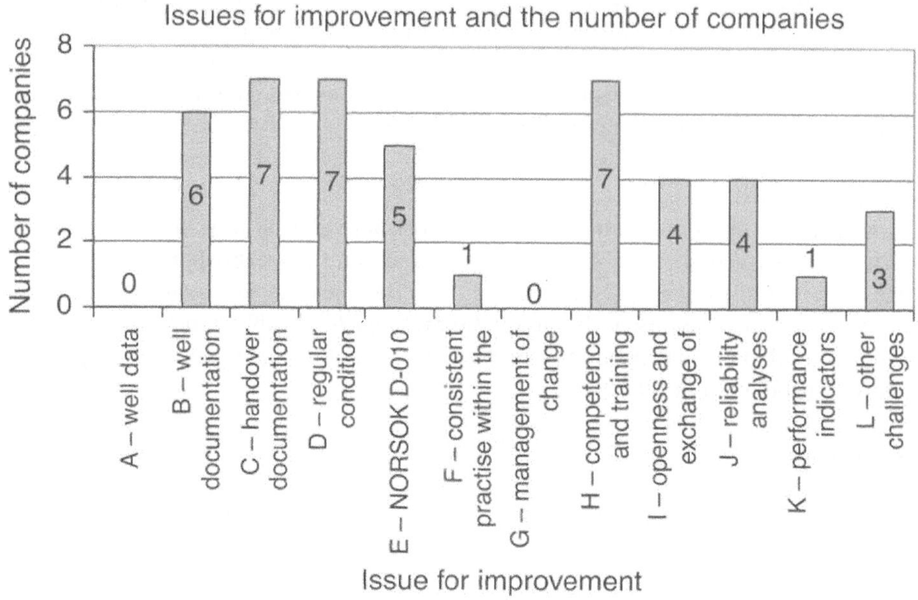

FIGURE 7.18 The issues for improvement and the number of companies.

D: Condition monitoring. Consideration of and evaluations of initial casing design for the lifetime of the well, and possible changes to the well usage (example, from production to injection), are also difficult to obtain.

E: Comments related to NORSOK D010 compliance. The NORSOK D-010 defines requirements and guidelines, including the established well barrier philosophy. It appears to be less known and adhered to than PSA expected at the time when the audits were conducted.

The lack of information on well names, dates, and revision numbers complicates easy overview of the status on well barriers. Colour coding and descriptions of the barrier envelopes varies from company to company, but also within the same company.

The barrier illustrations in several cases do not contain details of perforations, Xmas tree, or cement presence, and in some cases are lacking barrier elements. Such well details, with possible inclusion of dimensions and depth data, should be included for updated versions.

Verification details of the casing cement barriers are often lacking. Hence, proper planned top of cement (TOC), packer setting depth, and later proper zonal isolation for later plugging and abandonment (P&A) could imply uncertainties.

H: Competence and training. The questionnaire revealed a need for strengthening knowledge, communication, and requirements to well status. The companies generally expressed high interest in sharing and exchanging training materials for practical use.

7.5.4 Summary

The results of the well integrity survey revealed that the industry needs to have an increased focus on their barrier issues. Control of barrier status is considered an important HSE factor that could limit unintentional leaks, well control situations, and accidents. Known well status enables the companies to take the right actions in a proactive manner and thereby prevent potential losses and expenses.

- 7% of the wells were shut in because of integrity failures/issues or uncertainty.
- 9% of the wells were working under conditions/exemptions.
- 2% of the wells had insignificant deviations for current operation.

The integrity problems exist within barrier elements such as tubing, ASV, casing, and cement.

The well integrity problems occur most frequently in wells from the early 1990s. This indicates that the integrity problems often occur in more recent wells.

The improvement potential for the industry:

- Improve systems of reliability and condition-based surveillance for well barriers and well integrity aspects, to improve well safety.
- Provide better visualisation of barriers for both onshore and offshore users to improve the safety level and user-friendliness.
- Agree on standard ways of visualisation, technical qualification, documentation, and abbreviations.
- Consider developing a standard handover package containing basic well engineering data required to give a full overview of the well barrier situation from spud to abandonment.
- The industry needs to further develop and acquire suitable technology for condition monitoring of wells, to improve on systematic and preventive maintenance, and to keep better control on degradation mechanisms.
- Improved attention to verification and condition monitoring of well barriers and well integrity is needed.
- Improve competence and training of involved personnel both from operators and contractors.
- Start the process of updating the NORSOK drilling standards to include recent experiences related to well barriers. This should include the qualification process for barrier elements.

There is a common denominator between the examples of well failures (I–V).

- Operational decisions during abnormal situations often lead to well failures.
- Design issues such as long-term effects are not sufficiently considered.
- The challenge is to account for rare events that may lead to major incidents. The normal approach is to focus on frequent and low-consequence incidents.

Plug and Abandonment of Wells

8

8.1 WELL BARRIER

8.1.1 Introduction

When well productivity declines to the point where continued production becomes unprofitable, the fate of the well is to be plugged and abandoned (P&A). Regulatory authorities require that the well be sealed to prevent potential leakage from the reservoir to the surface. For example, the Petroleum Safety Authority of Norway has developed specific guidelines for P&A operations, outlined in NORSOK D-010, which apply to the Norwegian Continental Shelf (NCS) (NORSOK D-010, 2013). These guidelines set minimum requirements and acceptance criteria that operators must adhere to. Similarly, the UK sector (O&G UK, 2015) has regulatory frameworks governing P&A operations.

P&A operations can be time-consuming and costly, with no immediate financial return for operators. For this reason, the oil and gas industry is increasingly focused on developing technologies, methodologies, and alternative materials to make these operations more efficient and cost-effective.

This section will outline the general technologies and materials used in P&A operations and potential future alternatives. The NORSOK D-010 guideline will also be discussed as an example of industry practices for P&A operation. NORSOK D-010 defines a well barrier as:

> [an] envelope of one or several well barrier elements (WBE) preventing fluids from flowing unintentionally from the formation into the wellbore, into another formation, or to the external environment.

The well barriers must be durable and effective enough to provide long-term isolation and secure the wellbore from leakage during the well's lifecycle. The barrier must prevent the migration of fluids both vertically (up and down the well) and horizontally (across the casing cement, the plug, and formations), as illustrated in Figure 8.1. Old logs must verify the barrier's quality, or new logs must be run. If the cement behind the casing is not in place, we need to remedy it, for instance, by perforating and using a cement injection method.

8.1.2 Approved Well Barrier Elements

According to the NORSOK D-010, the permanent well barrier elements can be used to establish a secure and effective seal in the wellbore, including (NORSOK D-010, 2013):

- *Cement plug*
- *Casing*

DOI: 10.1201/9781003618171-8

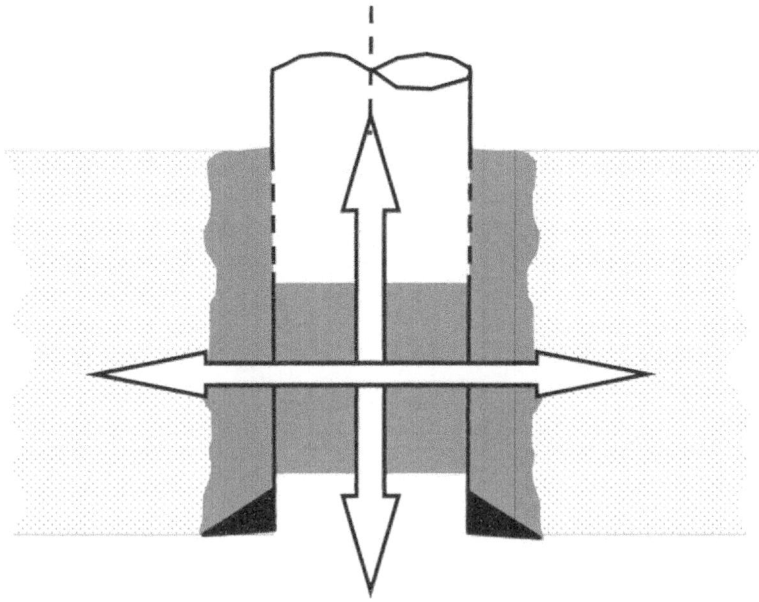

FIGURE 8.1 Horizontal and vertical seal for a permanent well barrier.

Source: NORSOK D-010 (2013).

- *Casing cement*
- *In-situ formation*

The well barrier system is a dual-layer system that includes both **primary** and **secondary barriers**:

1. **Primary barriers** (**blue elements**) are the first line of defence, preventing the flow of reservoir fluids into the wellbore or surrounding formations. Figure 8.2 shows that a well has six primary well barrier elements:
   - *Formation /cap rock above reservoir*
   - *Casing cement*
   - *Casing*
   - *Production packer*
   - *Completion string (below the DHSV)*
   - *Surface-controlled subsurface safety valve (DHSV)*

2. **Secondary barriers** (**red elements**) provide additional protection, ensuring that the well remains sealed if the primary barrier fails to prevent further fluid migration. (e.g., secondary well barrier elements):
   - *Formation above production packer*
   - *Casing cement*
   - *Casing with seal assembly*
   - *Wellhead*
   - *Tubing hanger with seals*
   - *Annulus access line and valve*
   - *Production tree (X-mas tree) with X-mas tree connection*

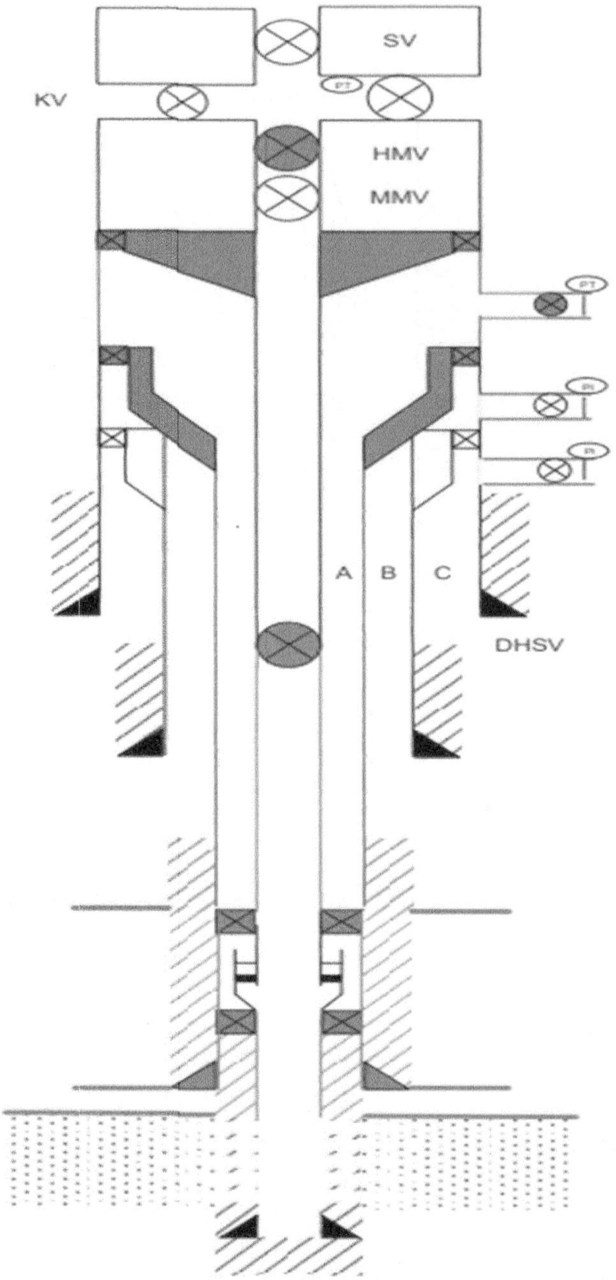

FIGURE 8.2 Well barrier elements (primary and secondary).

Source: Torbergsen et al. (2012).

8.1.3 Requirements for the Permanent Well Barrier

In addition to the *cement properties requirement* outlined in the NORSOK D-010 standard and Chapter 8.3 section § 8.3.2, the following *four key conditions* must be considered when designing and implementing well barriers for plug and abandonment (P&A) operations. These are:

1. *Length:*
The length of the well barrier (especially cement plugs) ensures proper isolation. Regional requirements vary. For instance, the NORSOK D10 typically requires cement plugs to be at least 100 meters MD in open-hole sections with a minimum of 50 m MD above any inflow/leakage point. The recommended guideline is local/national and designed based on the nation's environmental and safety policy.
2. *Position:*
Barriers should be placed above any known leakage or inflow zones to ensure full isolation. The formation integrity pressure must be carefully evaluated using tests like the Formation Integrity Test (FIT), Leak-Off Test (LOT), and Extended Leak-off test (XLOT) to ensure the selected depth for the plug is appropriate and can withstand the pressures from below.
3. *Verification:*
The barrier must be verified to guarantee the well integrity. The verification tests outlined in NORSOK D-010 are to ensure that well barriers perform their intended function of preventing fluid migration and protecting the environment.
By conducting *differential pressure tests, negative and positive pressure tests, tagging and load tests*, and utilising advanced tools like CBL and USIT, operators can ensure that the internal and external well barriers are intact and reliable.
4. *The Number of Well Barriers:*
The number of well barriers required for fluid containment ensures that fluids from the reservoir do not migrate to the surface or into other formations.
Each of these conditions plays a critical role in sealing the well effectively, preventing fluids (such as hydrocarbons, gas, or water) from migrating to the surface or other formations.
It must be noted that different regions have specific guidelines for well barrier design and Plug and Abandonment operations, but the principles of length, position, verification, and number of barriers remain universally important for ensuring wellbore integrity and environmental protection.

Summary
It is important to note that the information presented here covers the basic elements and requirements of the well barrier, as outlined in the NORSOK D10 guidelines (as per the published year). However, these guidelines are regularly updated, and readers should refer to the latest version for the latest requirements.

8.2 PLANNING FOR P&A IN THE DESIGN PHASE AND PROCEDURES

8.2.1 Introduction

P&A operations are the last part of a well's lifecycle, which seals the well permanently to prevent subsurface fluid leakage to the surface. Even though every well has unique characteristics and requirements, the general P&A process involves several common features. This section outlines the typical procedure for planning and executing P&A operations. Most of the workflow is based on the NORSOK D10 guidelines (Fjelde, 2012; Torbergsen et al., 2012; Oil & Gas UK, 2015; Fatemeh et al., 2015; Saasen et al., 2013). The typical P&A processes to be discussed are in the following three phases:

- Phase 1: Data gathering, planning, and submission for approval
- Phase 2: Preparatory—plug and abandonment execution phase
- Phase 3: Establish barrier—P&A phases

8.2.2 P&A Planning and Execution Processes

Abandoning a well is a complex process that requires detailed planning, the right personnel, and a coordinated approach to ensure safety, environmental protection, and compliance with regulatory requirements. These allow the P&A operation to be executed with minimal risk and maximum effectiveness, ensuring the well is safely isolated and abandoned with a reduced risk of future leaks.

8.2.2.1 Phase 1: Data Gathering and Planning Works

- Data Gathering and Planning

Incorporating this data into detailed planning helps minimise risks, enhance efficiency, and optimise the P&A operations. According to NORSOK D10 (§ 9.3.1), the data that should be gathered as a basis for the *well barrier design* and *abandonment* are:

- Well Configuration (Original and Present)
- Stratigraphic Sequence
- Formation Properties
- Specific Well Conditions
- Preliminary Well-Site Inspection and Work

Before initiating the physical P&A operations, preliminary work is crucial to assess the well's current condition and ensure everything is in place for the main operations. The main steps are:

- Well-head integrity check
  - Inspect wellhead and X-mas tree valves:
    - Verify functionality and integrity to ensure safe operation.
- Well integrity logging
  - Wellbore inspection with logging to check for blockades or issues with the tubing's internal diameter.
    - If blockages or constrictions are found, preparatory work is needed to clear the wellbore and allow access for P&A procedures.
- Planning Considerations and Submission for Regulatory Approval:

Effective planning is critical for a cost-effective, secure, and safe P&A operation. Proper planning minimises non-productive time (NPT) and operational costs. The following are typical planning activities that are related to the NORSOK D10 Well integrity solution, such as:

- Design the P&A program to meet regulatory recommended requirements.
- Outline the specific tasks, methodologies, e.g., abandonment techniques (i.e., placement methods), timelines, and resources necessary for executing the P&A operation.
- Provide HSE with a focus on safety concerns, environmental protection measures, and contingency plans in case of unforeseen issues.
- The organisation focuses on duties and equipment delivery with all parties, ensuring clear roles and responsibilities.
- Ensure that all personnel are trained and experienced for their roles in the P&A operation.

8.2.2.2 P&A Execution Phase

The preparatory stage deals with preparing the well for the main barrier establishment works.

- Mobilisation of the rig or vessel to the location

Rigs and vessels have different categories, and their selection depends on the P and A operation (Fjelde, 2012). It is important to select the right vessel, which is cost-effective and efficient.

- For instance, the Light Well Intervention vessels can perform initial P&A work (part of Phase 1), such as run logs/diagnostic runs, kill wells, punch tubing, set temporary plugs, and remove X-mas tree (XMT).
- Section milling is conducted with a rig to remove casing over a specified interval.
- Check the pressure in the annulus

In a closed annulus filled with fluid, heat transfer causes fluid expansion, leading to a significant annular pressure increase. Before tubing cutting and pulling operations (even well-kill operations), it is essential to:

- Check for significant annular pressure buildup (APB) in the A-annulus.
  - If detected, bleed off elevated pressure through the wellhead equipment to ensure safe operation
- X-mas tree removal and install BOP
- The vertical X-mas tree (VXT) is removed before the tubing is pulled, while horizontal X-mas trees (HXT) are typically removed after setting deep and shallow plugs.
- The blowout preventer (BOP) is installed to provide pressure integrity and safety during the remainder of the operation, as described in the following section.
- Kill and secure the well.

To prevent any uncontrolled flow of fluids or gas during the P&A process, the well must be "killed". This involves filling the wellbore with heavy fluids to create an overbalanced condition. The well-killing techniques and plug-setting process are:

- *Normal killing*: Displace the existing lighter fluid with a heavier fluid to ensure an overbalanced pressure condition.
- *Bull heading method*: When standard techniques are insufficient, this method involves pumping kill mud into the well and replacing it with a high-density fluid.
  - *A deep mechanical* plug is set after the bullheading.
  - After the plug setting, the ***tubing is punched*** above the deep mechanical plug set depth, and the tubing and annulus are displaced to kill fluid/brine.
- Downhole and annular safety valve removal.

Once the well is secured and the pressure is stabilised, the next step is to remove the equipment. The Equipment Removal Steps:

- Downhole safety valve (DHSV) removal
  - DHSV should be pulled using Wireline equipment.
- Annular safety valve (ASV) removal
  - If present, the valves may be wireline retrievable (easily removed without a rig) or tubing retrievable (removed along with the tubing).
- Cut tubing and control lines and tube removal
  - Cut the tubing above the production packer but below any downhole gauges.
  - If control lines are attached to the tubing, these should also be cut to ensure the tubing can be pulled out of the hole.
  - According to NORSOK D10, control lines and downhole equipment should not be part of the permanent barrier, as they can create leak paths.
  - Pulling the tubing hanger and tubing along with the control line, DSV/ASV.

- Run logging tools
    - Once the tubing is removed, logging tools such as cement bond logs (CBL) and ultrasonic imaging tool (USIT) are run to assess the cement quality behind the casing.
    - If the cement integrity is good, plugs can be set inside the casing.
    - If the formation has a creeping tendency, evaluate and use it to seal the annular spacing.

8.2.2.3 Establish Barrier—P&A Phases

P&A operations are carried out in three primary phases to ensure all zones are properly sealed and the well is safely abandoned, as illustrated in Figure 8.3 (NORSOK D10, 2013).

Phase 1: Primary Reservoir Abandonment

- The primary objective of this phase is to isolate the reservoir-producing or injection zones from the wellbore. Permanent barriers (usually cement plugs) are placed in the reservoir zones to prevent the migration of hydrocarbons or other fluids.

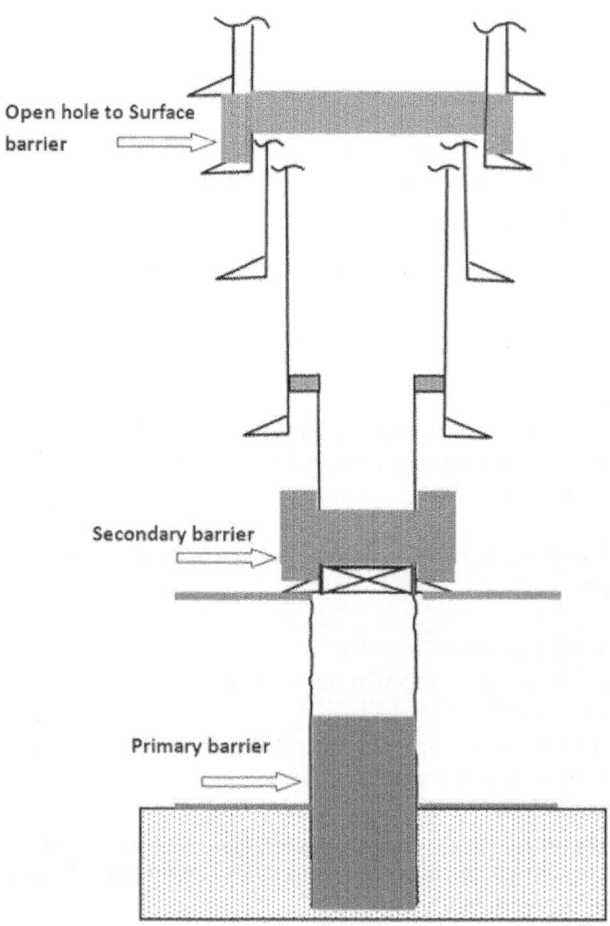

FIGURE 8.3 Example of established P and A phase barriers.

Source: NORSOK D10 (2013).

Phase 2: Secondary/Intermediate Abandonment

- In this phase, additional permanent barriers are set in the intermediate zones to seal any potential water-bearing or fluid-bearing zones. Barriers may also be placed between different zones to prevent communication between them.
- A surface barrier is typically placed at the shallower section of the well.

Phase 3: Wellhead and Conductor Removal

- The final phase of P&A operations involves removing the wellhead and conductor. This is done by cutting the equipment below the seabed, usually around 5 meters, to prevent future environmental impact or safety risks.

Summary

The P&A operation is a critical phase in a well's lifecycle, marking the end of production and ensuring the well is safely sealed. While each operation is unique, P&A follows common steps: *thorough planning*, *pressure management*, *well-killing techniques*, *equipment removal*, *logging*, *bonding evaluation*, and finally, *establishing permanent barriers*. These steps are vital for ensuring proper abandonment, minimising environmental risks, and maintaining long-term well integrity. Following the established procedures and prioritising safety, the petroleum industry can safely abandon wells and protect personnel and the environment. Employing efficient and cost-effective methods is crucial to maintaining operational safety and financial feasibility.

8.3 DEVELOPMENT WITHIN PLUGGING MATERIALS

8.3.1 Introduction

The primary objective of plugging materials in well abandonment is to provide a permanent, long-term seal that isolates the wellbore and ensures environmental protection. Although Portland cement remains the most common material for well plugging in the oil and gas industry, its long-term performance is influenced over time, among others, by temperature, in-situ stress, and other chemical interactions. In addition, even if high-quality cement slurry is used, poor execution of the cementing job can create channels or gaps that act as fluid leak paths.

The next section discusses the cement properties requirement as defined by NORSOK D10 standards, the challenges with conventional cement, the mechanisms that lead to potential failures, and the ongoing research to improve cement performance for long-term integrity and alternative materials to be used as barrier elements.

8.3.2 Standards Requirement and Issues with Cement

NORSOK D-010 provides specific requirements for the properties of cement used in well plugging to ensure long-term integrity. These requirements are critical for creating reliable seals that endure over time and withstand challenging subsurface conditions.

NORSOK D10 Requirements for Cement Properties

- Long-term integrity (eternal perspective)
- Impermeability
- Non-shrinking
- Mechanical load resistance

- Chemical resistance
- Bonding to steel
- Non-harmful to steel integrity

Challenges with Conventional Cement
Conventional cement has *several limitations* that can undermine its effectiveness in well abandonment operations. These challenges are primarily related to its physical properties and its performance over time:

- Cement shrinkage
- Cement micro-cracking
- Cement brittleness
- Casing corrosion
- Geological shifts due to pressure and tectonic stress
- Cement degradation due to chemical and temperature effects

Potential Leak Paths in Cement (Gasda et al., 2004)
Gasda et al. indicated that the possible potential leak paths exist in the cement plug, which can lead to well integrity issues:

1. Through cement plug
2. Between the cement plug and the interior of the casing
3. Between casing cement and the exterior of casing
4. Through casing cement
5. In cement fractures
6. Between cement and rock
7. Through the corroded casing

These leak paths are often associated with micro annuli, small gaps between the casing, cement, and formation, that provide the preferred fluid flow path. Figure 8.4 illustrates this.

Failure Rates in Well Integrity
Well integrity surveys conducted on 71 wells in the North Sea, including 31 production wells and 40 injection wells, revealed concerning failure rates for cement and casing. The results in Figure 8.5 indicate an 11% failure rate for cement and casing, highlighting that conventional cement materials often fail to meet the long-term integrity requirements set by industry standards. This failure suggests that conventional cement does not always perform as expected, and improvements are necessary to achieve reliable long-term sealing in abandoned wells.

8.3.3 Research into Improved Cementing Materials

To address conventional cement's limitations, ongoing research is focused on improving its mechanical, petrophysical, and elastic properties, making it more suitable for long-term well integrity.

8.3.3.1 Engineered Micro-Sized Cement Particles

Researchers have developed engineered micro-sized cement particles to improve the packing structure within the cement (Meek et al., 1993). By refining the internal structure, these particles reduce the *cement's permeability and porosity, increasing its resistance to fluid migration and enhancing* its mechanical properties. The field and laboratory studies have shown that cement with engineered micro-sized particles has a higher success rate in sealing damaged cement and casing.

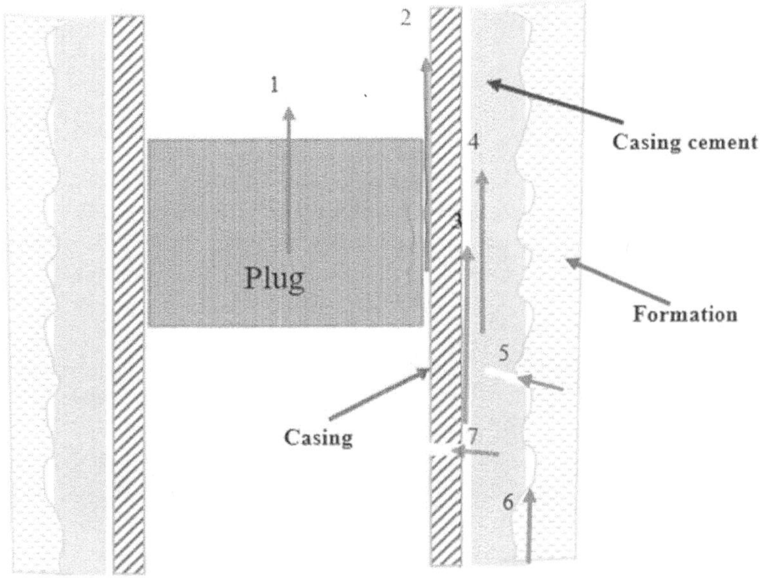

FIGURE 8.4 Possible leak paths.

Source: Resketched from Gasda et al. (2004).

8.3.3.2 Nanoparticle Additives

The effect of nanoparticles (1–100nm) on cement is another promising development. The laboratory-based research results indicate that nanoparticles improve conventional Portland cement properties, including *mechanical strength (Compression, tensile, flexural), rheology, setting time, and permeability* (Li et al., 2004; Li et al. 2006; Ershadi et al., 2011; Liu et al., 2012; Patil et al., 2015; Rahimirad et al., 2012; Roji et al., 2012; Shenghua et al., 2013; Zhang et al., 2015; Jafariesfad et al., 2016; Murtaza et al., 2016). These additives help create a denser cement matrix, reducing the likelihood of cracks or fractures that can form under pressure. Despite the nanoparticle's impressive results in enhancing the performance of cement, a further field-scale test needs to be conducted to validate the efficiency of the nanoparticles for long-term cement durability and capability to withstand harsh subsurface conditions.

8.3.4 Alternative Materials

The industry is also searching for alternative materials for P&A operations. Based on the material's activation, chemistry, and physical properties, the O&G UK Guidelines categorised materials qualified for the Abandonment of Wells (O&G UK, 2015; Rios et al., 2021). Among the list presented in Table 8.1, some potential barriers under the operation and research phases will be selected and summarised in the following. These materials are as an alternative to Portland cement.

8.3.4.1 ThermaSet

Table 8.1 shows that ThermaSet is listed under the type C category. ThermaSet is a particle-free polymer resin used for plugging, lost circulation, leak stopping, and sand consolidation in oil wells. It is a liquid material that solidifies through a thermally activated curing process, with customisable viscosities (10–2000 cP), and densities (0.7–2.5 SG) to match the formation's temperature (20–150°C) (Knudsen et al., 2015). ThermaSet exhibits faster setting times, minimises the "wait on cement" (WOC) time, and exhibits

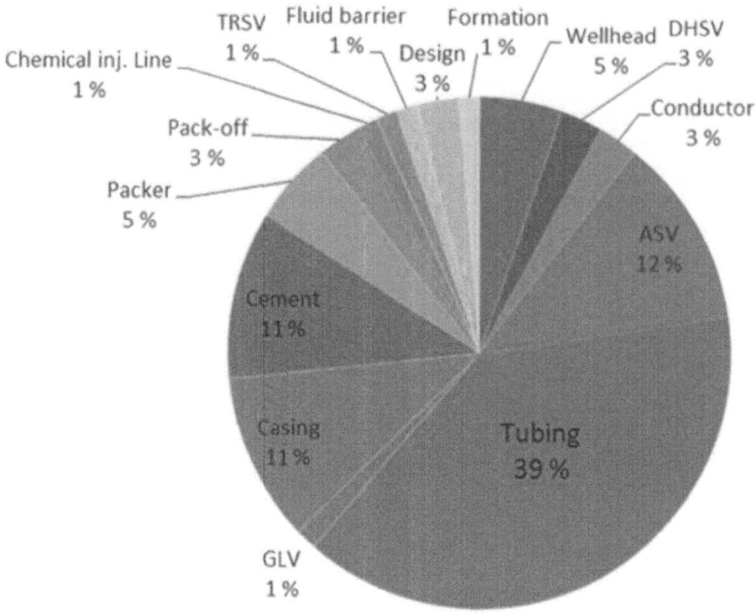

FIGURE 8.5 Well integrity survey.

Source: Vignes et al. (2010).

properties such as higher mechanical strength, lower elastic modulus, and lower permeability (Beharie et al., 2015). However, it is costly relative to cement.

8.3.4.2 Sandaband—Sand for Abandonment

Sandaband is a Bingham Plastic unconsolidated material that plugs and prevents lost circulation. It consists of Quartz, water, dispersant, and viscosifier, creating a gas-tight, non-segregating, non-shrinking, and non-fracturing compound. Quartz, a stable material, is resistant to degradation and unaffected by downhole fluids like CO_2, H_2S, and hydrocarbons (Rygg et al., 2010).

Unlike cement, Sandaband does not solidify through chemical reactions and requires no setting time. It deforms and reshapes under dynamic loading, avoiding fractures, and microannuli, which reduces leakage. However, it needs a solid base, such as a mechanical plug, to maintain its position; if the base fails, the Sandaband will sink, reducing its effectiveness. Despite this, Sandaband has been used since 2004 for temporary abandonment in the North Sea (Saasen et al., 2010). Sandaband is economically viable for remedial work and sealing annular spaces, making it more cost-effective.

8.3.4.3 Thermite

As shown in Table 8.1, Thermite material for P&A is listed under modified in-situ materials (i.e., the J category). The barrier materials are created by changing the casing, cement, and the formation by using thermal and chemical processes to alter the formation (e.g., thermite plug). Thermite is a mixture of aluminium (Al) and metal oxides (e.g., Fe_2O_3). When Thermite ignites, aluminium is oxidised, and metal oxides (Al_2O_3) are produced. Example: $Fe_2O_3 + 2\,Al \rightarrow 2\,Fe + Al_2O_3$, producing iron, aluminium oxide, and a large amount of heat through an exothermic reaction. During the reaction, a high heat emission will raise the temperature to over 2000°C, which exceeds the melting point of iron (about 1538°C), cement, and the formation. Under thermite reaction, the structure around the well (casing, cement, and formation)

TABLE 8.1 Well P&A material types

TYPE	*MATERIAL*	*EXAMPLES*
A	Cements/ceramics (setting)	Portland cement, pozzolanic cement, slag, phosphate cements, hardening cements, geopolymers
B	Grouts (non-setting)	Sand or clay mixtures, bentonite pellets, barite plugs, calcium carbonate, and other inner particle mixtures
C	Thermosetting polymers and composites	Resin, epoxy, polyester, vinylesters, including fibre reinforcements
D	Thermoplastic polymers and composites	Polyethylene, polypropylene, polyamide, PTFE, PEEK, PPS, PVDF, and polyurethane, including fibre reinforcement
E	Elastomeric polymers and composites	Natural rubber, neoprene, nitrile, EPDM, FKM, FFKM, silicon rubber, polyurethane, PUE, and swelling rubbers, including fibre reinforcement
F	Formation	Claystone. Shale, salt
G	Gels	Polymer gels, polysaccharides, starches, silicate-based gels, clay-based gels, diesel/ clay mixtures
H	Glass	
I	Metals	Steel, other alloys such as bismuth-based materials
J	Modified in-situ materials	Barrier materials formed from casing and/or formation through thermal or chemical modification

Sources: O&G UK (2015); Rios et al. (2021).

melts to form a magma plug. Then, the molten magma cools down, solidifies, and creates an impermeable artificial plug. The method allows the plug to seal the annular flow.

The thermite plugging process requires two trips. First, place the mechanical plug with an overlaying sand layer lower than the depth of the formulation to be plugged. The sand protects the mechanical plug from high heat. Then, run the thermite tool and ignite it through the electrical cable (Thorstensen et al., 2022).

8.3.4.4 Bismuth

Due to the growing environmental concerns and the limitations of cement, bridge plugs, and resins, the industry also uses bismuth and thermite as plugging materials. Bismuth is a post-transition metal. The material is one of the plug-in candidates listed under the I category of Table 8.1. Like ice, Bismuth has a property of expansion (more volume) when it solidifies. This property makes for good bonding with both casing and formation. The solidified and expanded bismuth then creates a gas-tight seal in the well. The bismuth needs to be melted at the plugging depth. Bismuth melts at 271.5°C. The downhole temperature is not sufficient to melt bismuth. Therefore, one can use a small amount of thermite to melt down bismuth. Bismuth sets rapidly, as opposed to hours of waiting for cement to set. The length of the bismuth determines the cost of the operation. The shorter plug is economical (Fulks et al., 2020).

8.3.4.5 Creeping Shale (Formation)—Green Shale

As shown in Table 8.1, formation (i.e., shale, clay) is also one of the P&A materials listed under categories F and J. Field observations have shown that when drilling operations are halted for any reason, the creep behaviour of shale formations can cause the annulus to close (Williams et al., 2009; Kristiansen et al.,

2018). Creep refers to the time-dependent plastic deformation under constant stress, typically at elevated temperatures. The extent and rate of creep deformation are influenced by various factors, including the mechanical properties of the shale, its clay composition (e.g., Kaolinite, Smectite), elasticity, porosity, the chemistry of the annular fluids, and the surrounding temperature and pressure. Based on the North Sea Lark-Horda shale, the candidate for the shale barrier should possess the following properties (Kristiansen et al., 2018; Oort et al., 2022).

- High clay content measured by CEC, especially high smectite content that provides:
    - High ductility
    - Low permeability
- High porosity (correlates with low rock strength properties)
- Low acoustic velocity (correlates with low rock strength properties)
- Low content of cementing minerals like quartz and carbonates will promote ductile deformations and creep
- Low stiffness
- Failure parameters (low cohesion and low friction angle)

The laboratory test results indicate that the primary driving forces for shale creep include a pressure drop in the annulus, the chemical effect of exposure to water, and the application of elevated temperatures (Kristiansen et al., 2018). In the presence of water alone, the activation of creep is relatively slow; however, combining water exposure with increased temperature and pressure accelerates the process significantly. Sodium silicate and lithium silicate exhibit the fastest barrier activation (Oort et al., 2022). With their high alkalinity, these brines promote ion exchange that weakens the shale structure, thereby accelerating the creep process.

While further field trials and laboratory research are needed to understand its capabilities fully, creeping shale has the potential to be cost-effective and sustainable for P&A operations.

8.3.4.6 Geopolymer

The cement industry is a major contributor to global CO_2 emissions, with 0.73 to 0.85 tonnes of CO_2 emitted per tonne of cement produced (He et al., 2019). Ordinary Portland Cement (OPC) alone accounts for approximately 5% of global CO_2 emissions (Lafarge et al., 2004). This environmental impact has stimulated interest in sustainable alternatives and durable cement materials, particularly for applications like oil and gas well abandonment (P&A). However, the selected material must meet technical requirements such as long-term integrity, pressure resistance, and durability.

Under category A, in Table 8.1, Geopolymer cement has emerged as a promising alternative to OPC due to its lower environmental impact and favourable performance characteristics. Geopolymer cement, primarily based on fly ash, offers a potential solution to mitigate the carbon footprint associated with cement production.

Fly ash is a by-product of coal combustion-driven power plants. It is the main precursor in geopolymer formulations. The concept of geopolymer cement was first introduced by Davidovits (1991), who developed the material by combining aluminosilicate-rich materials (e.g., fly ash and blast furnace slag, metakaolin, and certain clays) with alkaline activators (e.g., sodium hydroxide (NaOH) and sodium silicate (Na_2SiO_3)), resulting in a hardened material under the geopolymerisation process. Unlike OPC, which sets and hardens at ambient temperatures, geopolymer cement requires exposure to elevated curing temperatures to achieve comparable strength.

Geopolymers can exhibit a wide range of properties depending on the type of precursors and alkaline activators used, and the curing conditions (such as temperature and pressure). These properties include *high compressive strength*, *low shrinkage*, *acid resistance*, *fire resistance*, and *low thermal conductivity* (Duxson et al., 2007). Such attributes make geopolymer cement particularly suitable for use in demanding environments like oil and gas wells, where long-term integrity and resistance to harsh conditions are crucial.

For plug and abandonment (P&A) operations, geopolymer cement presents a promising alternative to OPC due to *its environmental benefits*, *low shrinkage*, *acid resistance*, and *low permeability*. These characteristics align well with the technical requirements outlined in standards such as NORSOK D-010, which stipulates the need for long-term well integrity after abandonment. However, despite its promising properties indicated by the laboratory test conditions, further field trials are necessary to assess and confirm its efficacy under realistic downhole conditions.

If successful in field applications, geopolymer cement could significantly reduce the carbon footprint of well abandonment operations while ensuring reliable long-term sealing and well integrity.

Summary

It is important to note that alternative materials are field-tested and are used in an oil well. Some of them are the ongoing research and development phases. Therefore, the information provided here is valid as per this book edition. More material could be developed, and the updated information will be included in the next edition.

8.4 P&A OPERATIONS

8.4.1 Introduction

The conventional methods used for P&A operations include *Section Milling*, *Cut and Pull*, and *Perforate Wash and Cement (PWC)*. These methods are designed to seal and isolate the well from surrounding formations, preventing any leakage of hydrocarbons or other fluids to the surface.

In this section, we will discuss the conventional P&A methods and their associated challenges and provide insights from field case studies.

8.4.2 Cut and Pull

Conventional cut and tube/casing removal is time-consuming, which is a cost factor. The industry developed a cut-and-pull method to perform P&A without removing the tubing.

The cut-and-pull process involved the following:

1. Cut the tubing above the packer and below the downhole gauge along with the control line (in red) (Figure 8.6, A).
2. Pull the tubing equal to the length required to log the casing cement (Figure 8.6, B).
3. Run the USIT-CBL tool string and record and verify the casing cement integrity (Figure 8.6, B).
4. Pump cement through the tubing to create a cement plug on the mechanical plug following the NORSK D10 length requirement (Figure 8.6, C).
5. Lower the tube on top of the cement after it has hardened (Figure 8.6, C).

This process saves time, which would otherwise be spent picking up the entire tubing from the well and disconnecting it on the surface. Field case studies in the North Sea have estimated that this method has saved an average of 12.8% operation time (Steinar, 2016).

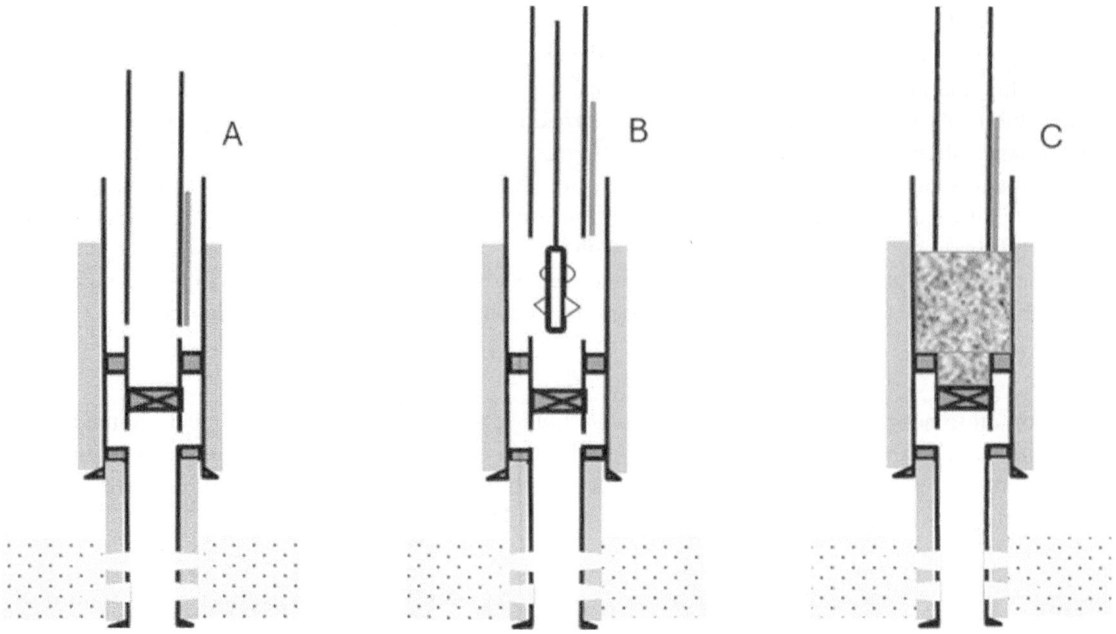

FIGURE 8.6 Step-by-step cut-and-pull and plugging process.

Source: Steinar (2016).

8.4.3 Section Milling

The section milling method is used if the casing cannot be removed by cut and pull. Figure 8.7 illustrates the plug placements. The milling tool removes the casing, cement, and then the formation, called underreaming. Milling through the formation ensures the cement plug will have a good bond against the fresh formation.

Overview of the section milling process:

1. Milling the casing and cement
2. Swarf removal
3. Installation of a mechanical plug
4. Cement plug installation as a secondary and surface barrier

Section milling is a time-consuming process. The main issues of section milling are:

- the worn-out milling knives, and
- the production of huge amounts of swarf.

Replacing milling knives and removing the swarf process increases the non-productive time. In general, the operation is a cost factor for the industry.

To address these challenges, the industry is focusing on developing more efficient and durable milling tools, better swarf removal techniques, and innovative milling strategies, all of which aim to improve the efficiency and cost-effectiveness of section milling operations.

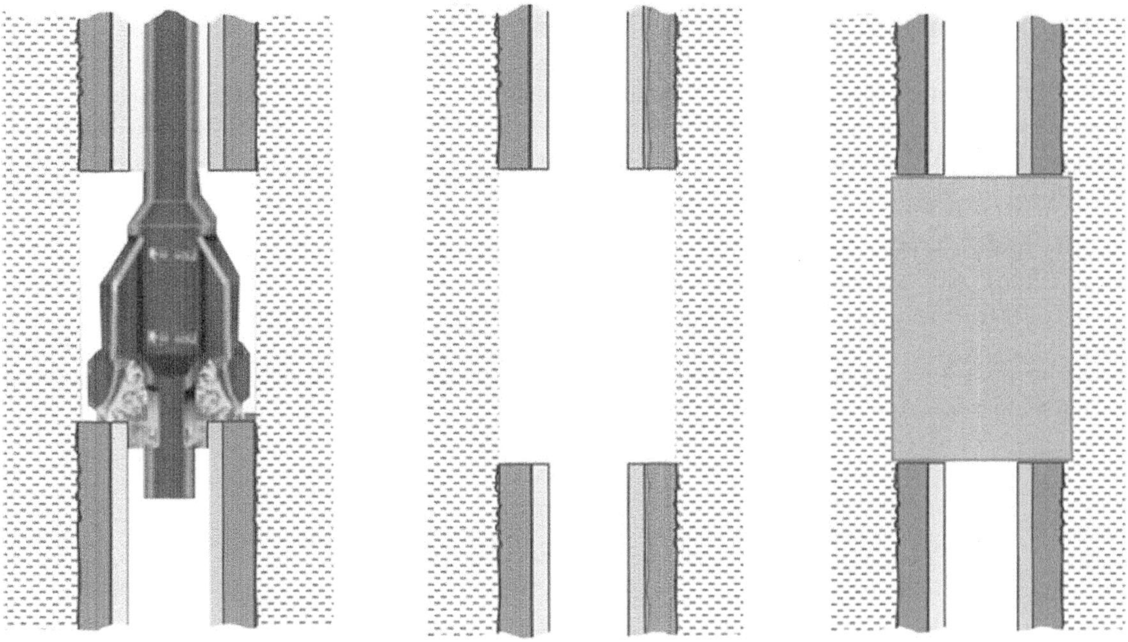

FIGURE 8.7 Permanent abandonment, section milling.

Source: NORSOK D10 (2013).

TABLE 8.2 Section milling time comparison: conventional vs. swarfless

CONVENTIONAL SECTION MILLING	*TIME (HR.)*	*"SWARFLESS" SECTION MILLING*	*TIME (HR.)*
Rig up Swarf Handling System	4		-
Pick up BHA	1	Pick up BHA	1
RIH to 10,000 ft.	3.5	RIH to 10,000 ft.	3.5
Displace to milling fluid	2		-
Section Mill 50 m. (165 ft.)	30	Section Mill 50 m. (165 ft.)	30
Circulate hole clean	6		-
POOH with section milling assembly	3.5	POOH with section milling assembly	3.5
BOP Jetting Run (allow add'l time for subsea)	3		-
Lay down BHA	1	Lay down BHA	1
Total Time	**54**		**39**

Source: Jonathan et al. (2018).

For instance, a "swarfless" section milling method has been introduced that leaves the swarf behind instead of transporting it to the surface (Jonathan et al., 2018). As shown in Table 8.2, the swarfless method eliminates five operations compared to the swarf removal-based method, resulting in a 15-hour reduction in operational time (from 54 to 39 hours).

8.4.4 Perforate, Wash, and Cement (PWC)

Pulling of the casing, tripping, and milling are the most time-consuming processes that account for a large percentage of the total P&A operation expenditure. The PWC technology has been developed to minimise the P&A cost. The method is a single-trip operation. After running the tool, the processes are (Larsen, 2016):

- *Perforate*: The method starts by perforating the chosen section of the casing.
- *Wash*: Followed by a thorough washing.
- *Cement*: The cement slurry is squeezed into the annulus through the perforations.

Regardless of the method's reliability, the operation time is reduced compared with the conventional section milling-based plug placement method. This is due to the operation being a single trip.

Digitalisation in Drilling

9

9.1 DIGITAL TWINS, AUTOMATION, AND REAL-TIME OPERATIONS CENTRES

Using a digital twin can significantly enhance wellbore design. The drilling industry is undergoing a transformative shift with the integration of digital twins, automation, and real-time operations centres. Digital twins, which are virtual replicas of physical wellbores and drilling equipment, allow operators to simulate and optimise drilling processes before actual implementation. This technology enhances decision-making by providing a comprehensive view of the well's lifecycle, predicting potential issues, and improving overall efficiency.

Automation further streamlines drilling operations by reducing manual interventions and increasing precision. Automated systems can control drilling parameters such as rate of penetration (ROP) and weight on bit (WOB), ensuring optimal performance and minimising risks. This leads to safer, faster, and more cost-effective drilling operations.

Digital Twin is a virtual and simulated model or a true replica of a physical asset. It is a computerised companion of the physical asset and can be used for various purposes, as depicted in Figure 9.1.

The model in Fig.9.1 specifically finds expression through five enabling components: sensors and actuators from the physical world, integration, data, analytics, and the continuously updated digital twin application. These constituent elements are explained at a high level below (Parrott and Warshaw, 2017):

***Sensors*:** Sensors distributed throughout the manufacturing process create signals that enable the twin to capture operational and environmental data pertaining to the physical process in the real world.

***Data*:** Real-world operational and environmental data from the sensors are aggregated and combined with data from the enterprise.

***Integration*:** Sensors communicate the data to the digital world through integration technology (which includes edge, communication interfaces, and security), creating a two-way link between the physical world and the digital world.

***Analytics*:** Analytics techniques are used to analyse the data through algorithmic simulations and visualisation routines that are used by the digital twin to produce insights.

***Digital twin*:** The "digital" side of Figure 9.1 is the digital twin itself, an application that combines the components above into a near-real-time digital model of the physical world and processes. The objective of a digital twin is to identify intolerable deviations from optimal conditions along any of the various dimensions. Such a deviation is a case for business optimisation; either the twin has an error in the logic (hopefully not), or it has identified an opportunity for saving costs, improving quality, or achieving greater efficiencies. The identification of an opportunity may result in an action in the physical world.

***Actuators*:** Should an action be warranted in the real world, the digital twin produces the action by way of actuators, subject to human intervention, which trigger the physical process.

DOI: 10.1201/9781003618171-9

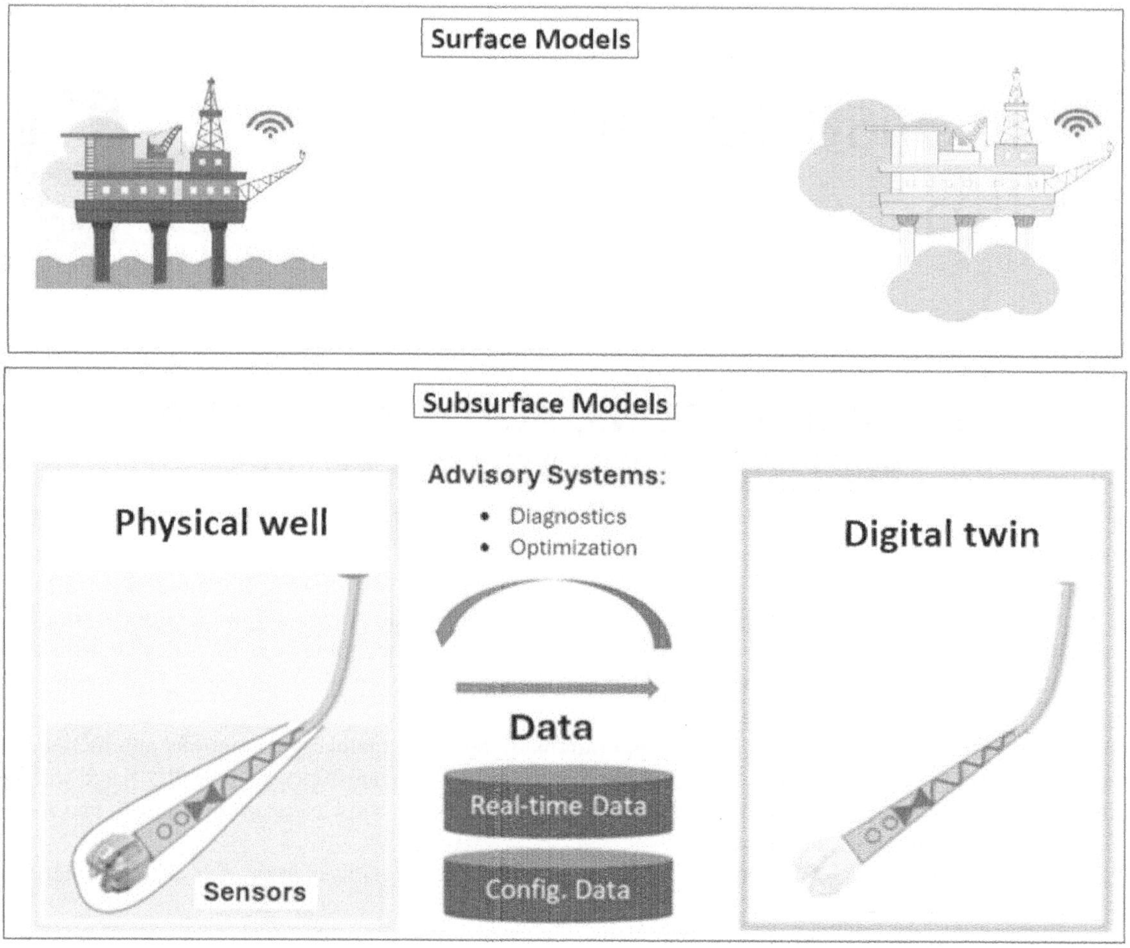

FIGURE 9.1 Visualisation of digital twin creation showing integration of sensors, data, and analytics for real-time modelling.

Source: Adapted from Khosravanian and Aadnoy (2021).

Surface and subsurface models:

- ***Surface Model*:** Represents the semisubmersible rig and surface operations, capturing data from sensors and actuators on the rig.
- ***Subsurface Model*:** Represents the downhole conditions and equipment, capturing data from sensors within the wellbore.

Advisory systems:

- ***Diagnostics*:** Utilise real-time data and analytics to identify issues and anomalies in the drilling process, providing actionable insights to improve operations.
- ***Optimisation*:** Apply advanced algorithms and simulations to enhance drilling performance, ensuring optimal conditions are maintained and efficiency is maximised.

9.1.1 Real-Time Operations Centres

Real-time operations centres play a crucial role in this digital transformation. These centres monitor drilling activities around the clock, providing real-time data and analytics to support decision-making. By integrating data from digital twins and automated systems, real-time operations centres enable proactive management of drilling operations, reducing nonproductive time (NPT) and enhancing safety (Figure 9.2).

Over the years, the role of the real-time operation centre has evolved beyond just facilitating fast decision-making and cost saving. As drilling engineering becomes more and more complicated, the real-time centre needs to be able to cater for the whole drilling cycle. To deliver a complete real-time communication solution, integrated real-time operation centre services were created with the intention of providing end-to-end solutions to the oil and gas operators, starting with the well planning phase, covering the well execution, and extending all the way to post-job analysis. The diagram below shows the integrated real-time operation services deployed in an operator's office, and the illustrations show the layout of the centre in the office space. This solution covers collaborative well planning, well engineering and planning, real-time data aggregation and visualisation, real-time monitoring and interventions, predictive modelling, drilling optimisation, training and mentoring, and finally data management and archiving (Iskandar et al., 2018).

In the section below, we explain in more detail the function of each of the services and how each one contributes to the success of an integrated real-time operation. (Figure 9.3)

9.1.1.1 Collaborative Well Planning

A complete real-time operation cycle starts with a collaborative well planning discussion, especially for challenging environment well cases. This discussion session involves all the stakeholders in the centre from different backgrounds. During the planning stage a three-dimensional (3D) integrated platform technology is used. This seamlessly integrates both geoscience and engineering data for detailed analysis.

Typically, drilling a well involves two fields of expertise: geoscience and drilling engineering. Historically, communication between the two has always been an issue because despite having a common goal, the two fields have different primary interests. The main concern for Geology & Geophysics (G&G) is to reach the depth of interest, but that is sometimes challenging to achieve according to the engineering principles. The drilling engineer's primary concern is to drill quickly and safely, which sometimes requires the G&G team to adjust its geological target.

Communication between different departments was usually done via emails or phone calls, and it could take several months to agree on a finalised well plan. The generally imperfect communication between geologists, geophysicists, reservoir engineers, and drilling engineers involved in well planning exacerbates this uncertainty. The end goal of a collaborative well planning session is to have a formal binding agreement between all parties, ratified by all stakeholders prior to the start of drilling. An effective collaborative well planning session is likely to result in the smooth running of the drilling process.

9.1.1.2 Well Engineering and Planning

Once the reservoir and drilling target has been finalised, the Drilling Application Engineer (DAE) will perform a study and conduct a consultation on the trajectory design, BHA design, torque and drag analysis, casing stress check, etc. At this stage, all the technical aspects of well planning are verified via simulation. The DAE will also provide a predictive torque and drag roadmap for various drilling parameters based on specified friction factors. Most importantly, the DAE is responsible for highlighting any risks associated with the well together with providing potential solutions.

Real-time data aggregation and visualisation depends on the operator, and the network architecture usually utilises the operator's existing infrastructure to preserve data confidentiality. A data aggregation server, commonly referred to as a rig server, is deployed on the rig site at the beginning of each job. The server is connected to the data acquisition systems of all the service companies such as mud loggers, MWD (measurement-while-drilling) providers, cementing operators, and well-test consultants, to centralise all drilling data. The data is stored locally on the rig before being transmitted to the town via the internet network.

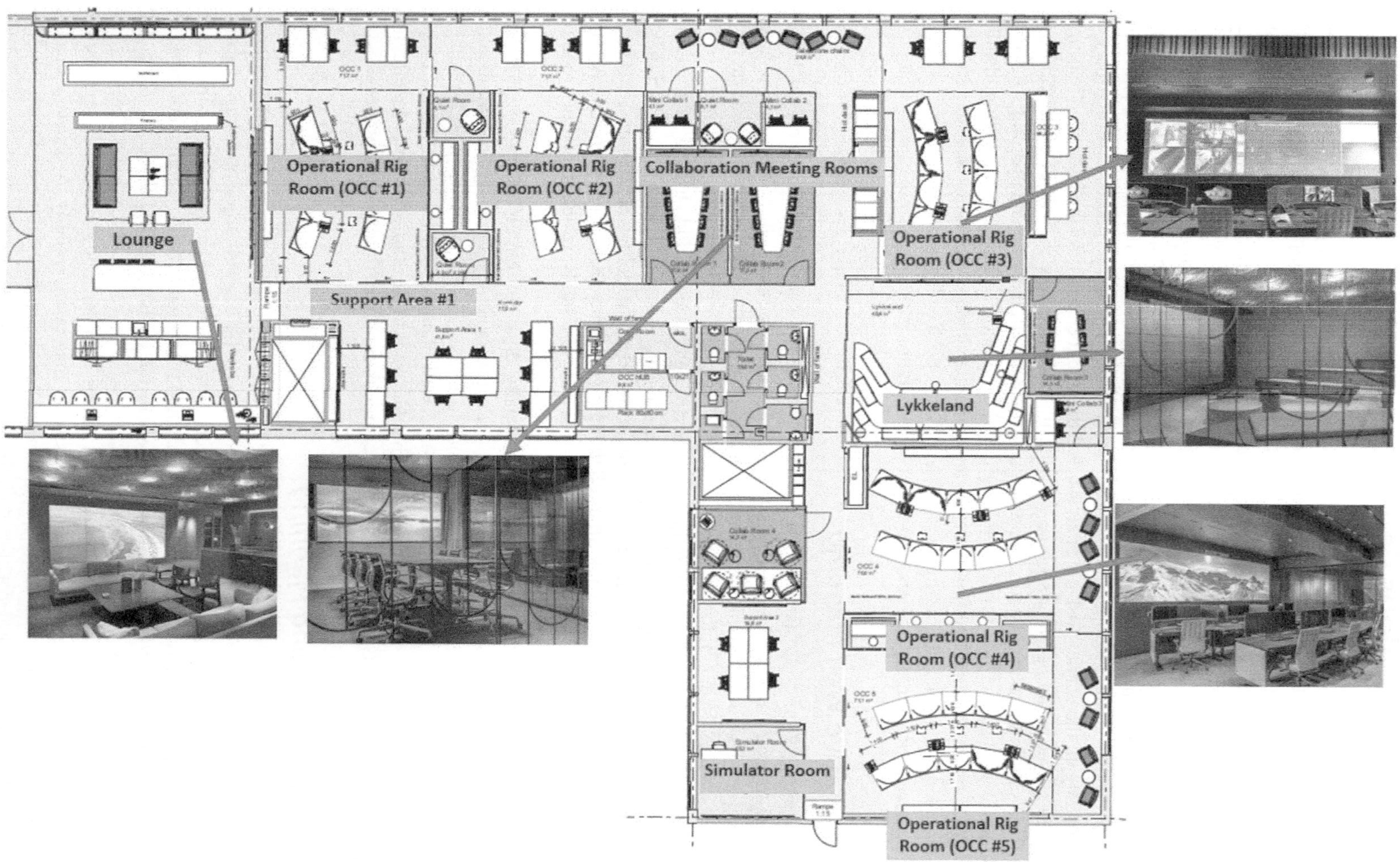

FIGURE 9.2 Aker BP's earlier Onshore Collaboration Centre used for real-time monitoring and collaborative decision-making.

Source: Khosravanian and Aadnoy (2021).

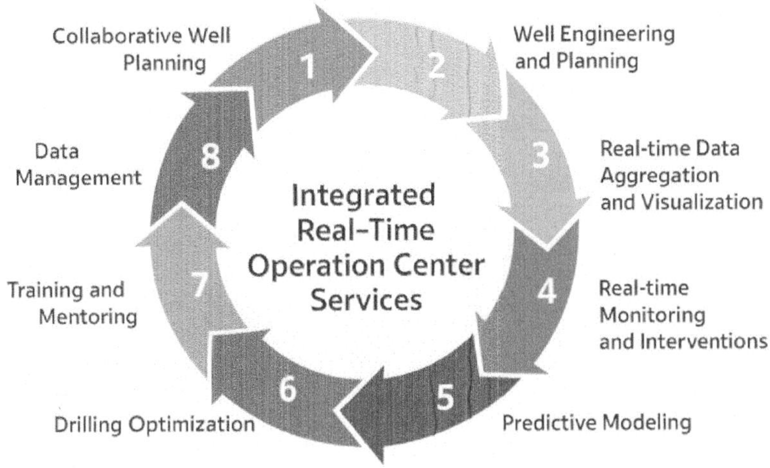

FIGURE 9.3 Integrated real-time operation centre services covering planning, monitoring, predictive modelling, and optimisation.

Source: Khosravanian and Aadnoy (2021).

End users, such as the drilling engineer and G&G team, are then able to visualise the data through a web-based application on a desktop or mobile device. The real-time operation (RTO) engineer's job is to monitor data streaming from the rig site through the network and to prevent data loss by acting immediately if any problems arise. At the same time, they are responsible for entertaining any issue faced by the end user regarding data visualisation (for solving any problems the end user has with the data visualisation).

9.1.1.3 Real-time Data Aggregation and Visualization

This section explains how drilling data from multiple rig sources is consolidated and securely transmitted to the operations center. Visualization tools display this data through dashboards and charts, enabling quick, informed decisions. Alerts highlight anomalies for proactive control.

9.1.1.4 Real-Time Monitoring and Interventions

A monitoring specialist is what makes the operation centre a real-time centre. Real-time monitoring (RTM) is a process through which operational personnel can review, evaluate, and adjust data on a database or a system (such as offshore drilling, well completions, or production). RTM allows operational personnel to review the overall processes and functions performed on the data in real time. Typically, RTM software or an RTM system provides visual insights into the data, which can be collected from multiple or various sources on the MODU. RTM can also provide instant notifications or alerts concerning specific data-driven or administrator-specified events, such as when a data value goes out of a specified range (Figure 9.4).

A team of monitoring specialists look at the data logs 24/7 to ensuring smooth drilling operations. The people who staff a real-time operation centre are typically highly experienced, with a minimum of 10 years' experience of working on a rig site. Based on their experience, monitoring specialists are expected to identify any potential hazards while drilling and to alert the people on the rig site and in the town using a traffic light system to highlight the incident severity. In many situations, the proactive measures taken by the oil workers in town have helped to reduce if not prevent any negative impact from an incident on the rig.

9.1.1.5 Predictive Modelling

One of the highlights of a real-time operation centre is that it is designed to allow communication between multiple software systems, especially those that support Wellsite Information Transfer Standard Markup Language (WITSML, the petroleum industry standard for sharing data). Typically, in an integrated

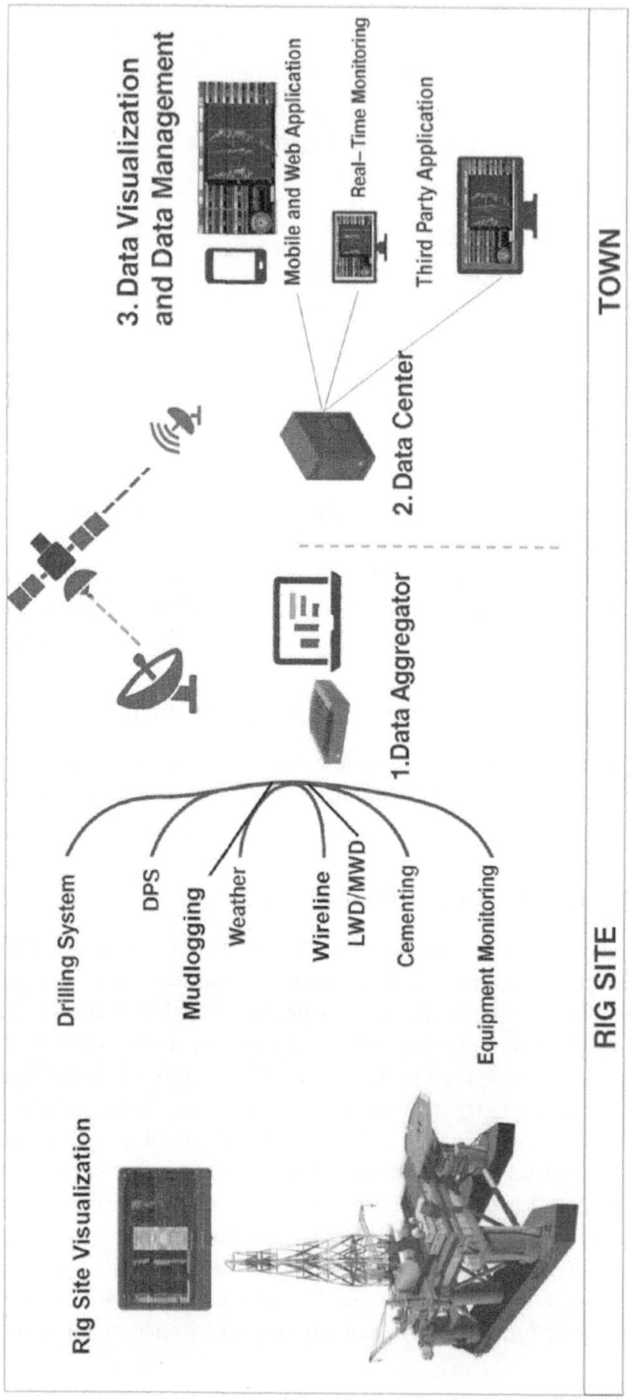

FIGURE 9.4 Real-time operation network architecture illustrating data flow between rig site and operations centre.

Source: Khosravanian and Aadnoy (2021).

real-time operation centre, there are several software systems running simultaneously to enhance the monitoring. The most recent technology to enhance monitoring is predictive modelling software which is able to predict the hole conditions in real time as well as in advance using the applied drilling parameters.

The RTO engineers are responsible for ensuring that the data from the rig site is fed into the predictive modelling application in real time. A dedicated calculation server for predictive modelling is installed to catch the real-time data from a WITSML store. The software calibrates the simulated model by applying the actual parameters used while drilling. By having the combination of hydraulic, thermodynamic, and mechanical inputs, the monitoring engineer is then able to predict the behaviour of cuttings bed height development according to the parameter being used. Inefficient drilling parameters applied while drilling, such as a slow revolutions per minutes (RPM) and flow rate, will result in additional trips such as wiper trips and backreaming activities to clean the hole. The development of the cutting proportion and cutting bed height generated in predictive modelling is based on the number of solids generated calculated on the ROP, mud carrying capacity, trajectory, flow rate, and RPM. These help to identify in which section of a well the cuttings are being accumulated at the low side, even with pumping or without pumping, if the applied drilling parameters (RPM and flow rate) are insufficient.

9.1.1.6 Drilling Optimisation and Detailed Technical Analysis

A dedicated drilling optimisation specialist (DOS) is responsible for finding the best possible solution to any technical problem for which there are usually several competing or conflicting considerations that might contribute to it (for which there are usually several possible contributing factors). The DOS is responsible for interpreting data from various down-hole and surface sensors and identifying the best possible solution for optimising the drilling operation. On top of that, DOS is responsible for performing post-run analysis, event reports (for example stuck pipe), identifying what was learned from the incident, and setting out the best practices that would prevent such issues from happening again.

9.1.1.7 Training and Mentoring

A highly sophisticated real-time centre does not necessarily translate into a collaborative working environment. Although the security level to get into this room is tight, there is usually a wide range of authorised personnel, especially from among those involved in the drilling operation. This includes the G&G team, drilling engineers, and the subject matter experts to encourage knowledge transfer from the experts to the junior engineers. The DAE and the monitoring specialist are responsible for managing the knowledge flow obtained throughout the drilling event and to document all lessons learned and best practices for future reference. Typically, during the drilling operation, the monitoring specialist will facilitate the discussion among the junior engineers. This is to bridge the knowledge gap and speed up the development of the junior engineers.

9.1.1.8 Data Management and Archiving

After the entire process from pre-planning to post analysis has been completed, the RTO Engineer is responsible for ensuring that all the data are being kept in a proper manner for future reference. This includes raw data from the rig site, logs, interventions, daily reports, failure reports, best practices, and lessons learned. The data are kept in a standard folder structure and all the files are named following a standard format. This is important to keep the data clean.

9.2 INTRODUCTION TO DIGITALISATION IN WELL DESIGN

Before, engineering software for drilling was offered as a monolithic software solution that combined several engineering modules into a single large product that was installed locally (on premises). The ability to break down large, complex goods into smaller, independent modules that can function on their own with the help of micro-services, which can be delivered on the cloud or on the edge, has been made possible by recent technological breakthroughs. The microservices only need to receive input protocols, and we may anticipate the desired results in the output protocol (Samuel). As seen in Figure 9.5 several microservices are inter-connected for the calculations. If one input variable is changed in one of the microservices, all the interlinked microservices get updated so that the system is always in stable condition. The microservices include well engineering calculations covering fluid mechanics, fluid dynamics, solid mechanics, string mechanics, string dynamics, thermal engineering, coupled engineering, etc. The input channel can also be dynamic and distributed (Samuel, 2023).

Digitalisation is transforming well design and planning, making the process faster and more efficient. Different companies are adopting various strategies for digitalisation. In the drilling industry, numerous software and microservices are being developed or are already in development. One of the key goals is to eliminate the concept of silos by integrating different microservices. However, a significant challenge is the data exchange between different microservices or applications. For example, one workflow related to well design being implemented in digital well planning (DWP) connects several concepts, such as collaborative well planning (CWP), drilling a well on a simulator (DWOS), and real-time well engineering. Therefore, we can say that defining well design without these concepts and their connections is not possible. Previously, we relied on Excel and multiple stand-alone applications, where each concept was handled manually, making the process very labour-intensive. Now, we can investigate historical data and lessons learned from other offset wells in real time, using various types of drilling and well simulation tools. This reflects a broader industry shift, and, over time, these technologies are expected to evolve into more mature and comprehensive systems.

Collaborative Well Planning (CWP): Collaborative well planning is an integrated way of working. In the session, the subsurface team collaborates with Drilling and Wells in real time with all subsurface data available in the software where the well trajectories are created. The goal is to find the optimal path for drilling through the complex overburden and into the reservoir, with as low risk and cost as possible. The new paradigm for automatic well path generation using multidisciplinary constraints significantly optimises well trajectory planning. This method allows for the generation of multiple well path alternatives based on experience-based constraints from various disciplines, leading to more efficient and high-quality well planning (Cayeux et al., 2023) (Figure 9.6).

The use of digital technologies and AI has completely transformed well trajectory optimisation. By adopting a full-stack AI-assisted solution, the well planning process has become much more efficient. This approach automates the creation of well trajectories, bringing together different disciplines to find the best well locations and paths. As a result, the planning cycle time has been significantly reduced, and overall performance has improved, showcasing the powerful impact of digital solutions in the drilling industry (Al Tamimi et al., 2024).

Digital Well Program (DWP): Once the team has identified optimal well placements and trajectories, the work is moved into the Digital Well Program. This is where wells are ultimately designed and planned. It provides a platform for integrated well planning and design, which promotes increased collaboration and connectivity across drilling activities. The Digital Well Program results in increased efficiency and extreme increase in quality of the data sets. This results in reduced risk and uncertainty in our well designs.

Drilling a Well on a Simulator (DWOS): The DWOS room concept has various functionalities in the drilling industry. Here, we want to focus on the aspects related to well design. The DWOS room allows us to test different risk scenarios. Before rigs and crews are moved into place to start drilling, the management and planning teams perform a "Drill Well on Simulator" under the risk scenarios in the well design, based on the offset well analysis.

Real-Time Well Engineering:

Connecting real-time engineering with the planning phase is essential for several reasons, especially when it comes to achieving more accurate simulations and effective decision-making. According to Santoro et al. (2023), here are some key points:

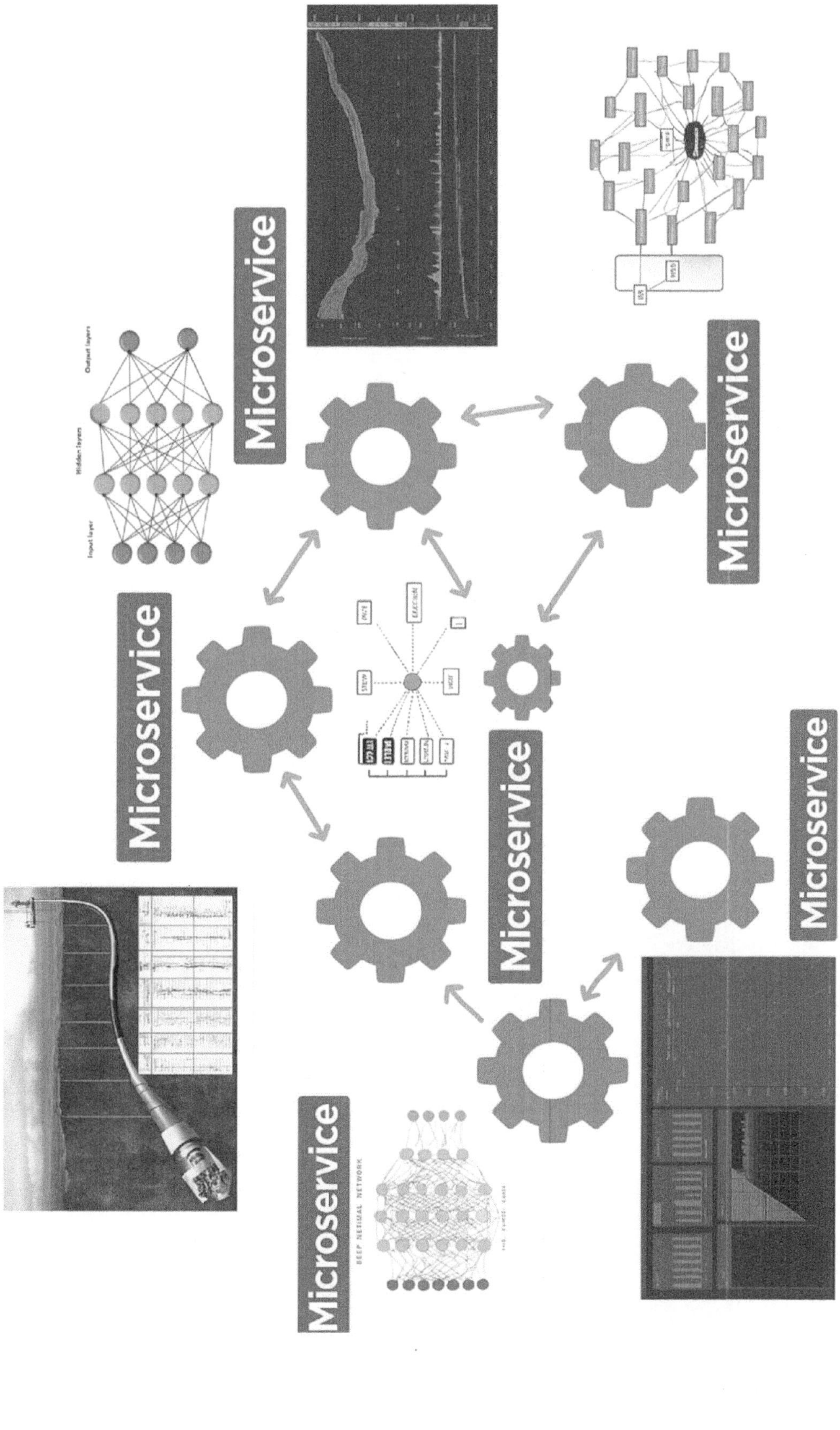

FIGURE 9.5 Connected microservice-based solution showing interlinked modules for well engineering calculations.

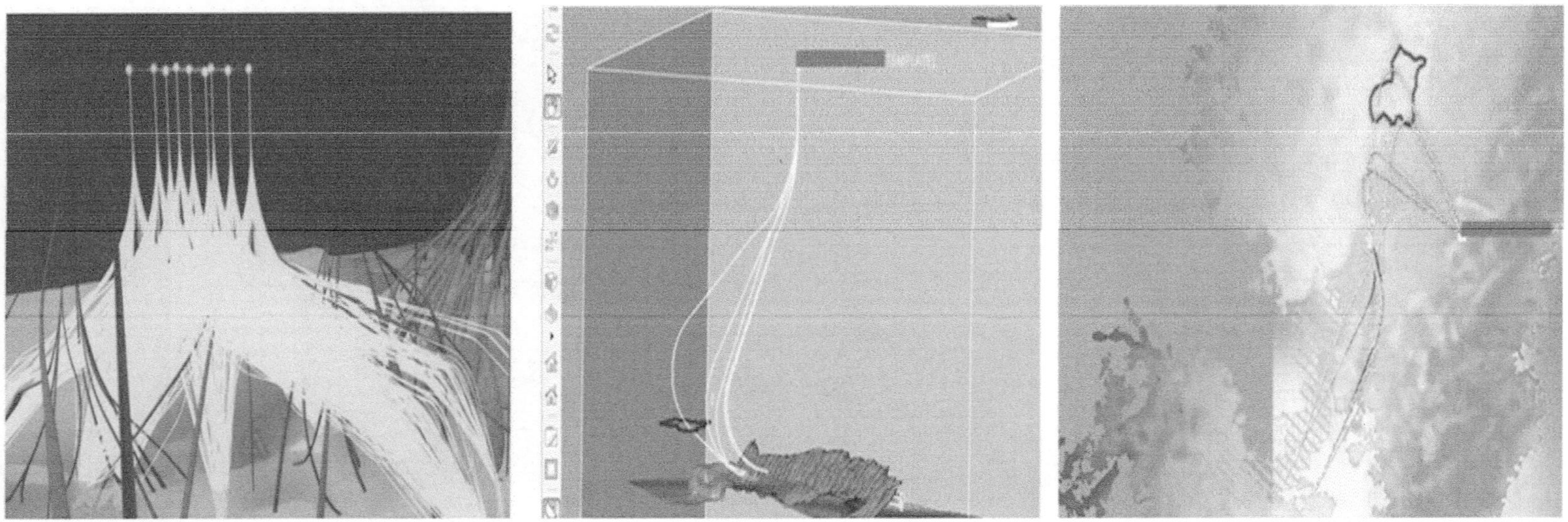

FIGURE 9.6 The left picture shows the auto-generation of well trajectories; the middle picture displays four selected possible trajectories, including the most extreme inclination and build path quickly discounted based on risk and cost; and the right picture provides a plan view of these trajectories, highlighting valid options that meet all requirements and represent contingency well paths.

Source: Revised from Cayeux et al. (2023).

1. Accurate Calibration: Dynamic models, embedded into the software, are calibrated to reach an overlap between simulated and measured drilling parameters. This ensures that simulations reflect actual drilling conditions.
2. Dynamic Decision-Making: Data collected on the field can provide real-time operation insights while feeding predictive analytics tools with the goal to improve safety and performance. This allows for quick updates to well designs to address new risks.
3. Enhanced Simulations: The tools can also provide support for advisory on drilling parameters, suggesting potential improvements. Drillers can request more accurate simulations to manage specific situations.
4. Operational Adaptability: Anomalous behaviour of the drilling parameters can be recognised, controlled, and adjusted. Real-time data allows for immediate adjustments to handle unexpected changes.
5. Improved Planning: These outputs can also be used as input for advanced analyses for both planning and post-job phases. Continuous feedback refines planning with the latest data.

By integrating real-time engineering with the planning phase, companies can enhance safety, optimise performance, and prevent costly delays, ultimately leading to more successful drilling operations.

9.3 BIG DATA ANALYTICS AND PREDICTIVE MODELLING

Big data predictive analysis can be implemented to detect downhole problems, helping companies identify trends and predict occurrences throughout their operations. This enables swift responses to disturbances and enhances operational efficiencies. By using big data techniques, the focus is on analysing existing data to make accurate forecasts of future events. Figures 9.7 and 9.8 illustrate the pipeline of an anomaly detection system. Here is a step-by-step overview (Altindal et al., 2024):

- **Data Acquisition**: The process begins with gathering raw drilling data from a database. This includes time-based logs/curves such as: Hook load, Block position, Time (Date Time), Flow rate, RPM (Rotary speed), Bit depth, Hole depth, Mud density in, ROP (m/hr), WOB, Torque and Standpipe pressure.
- **Data Pre-processing**: The next step involves cleaning and organising the data to ensure its quality and usability for analysis.
- **Feature Selection**: Relevant variables are selected to enhance model performance and reduce complexity.

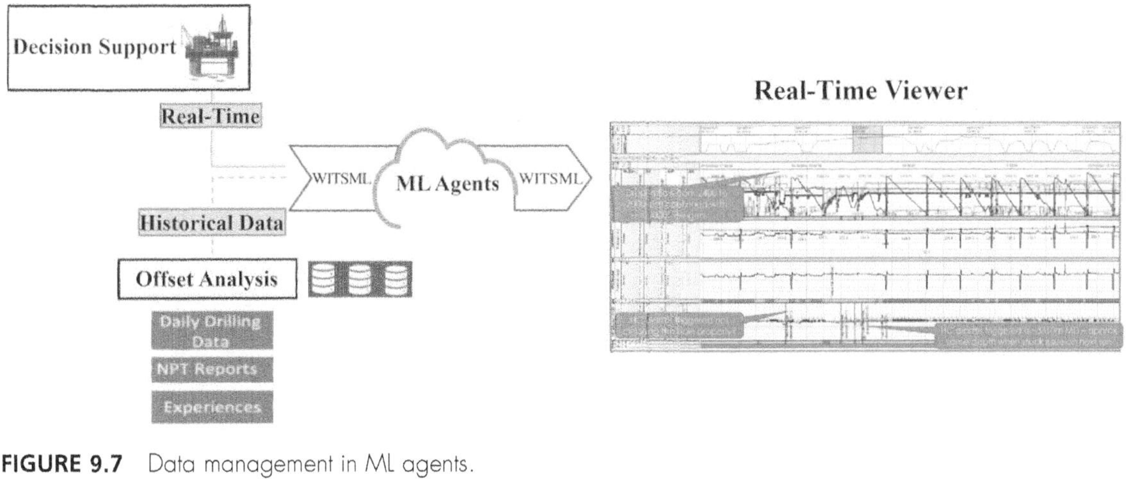

FIGURE 9.7 Data management in ML agents.

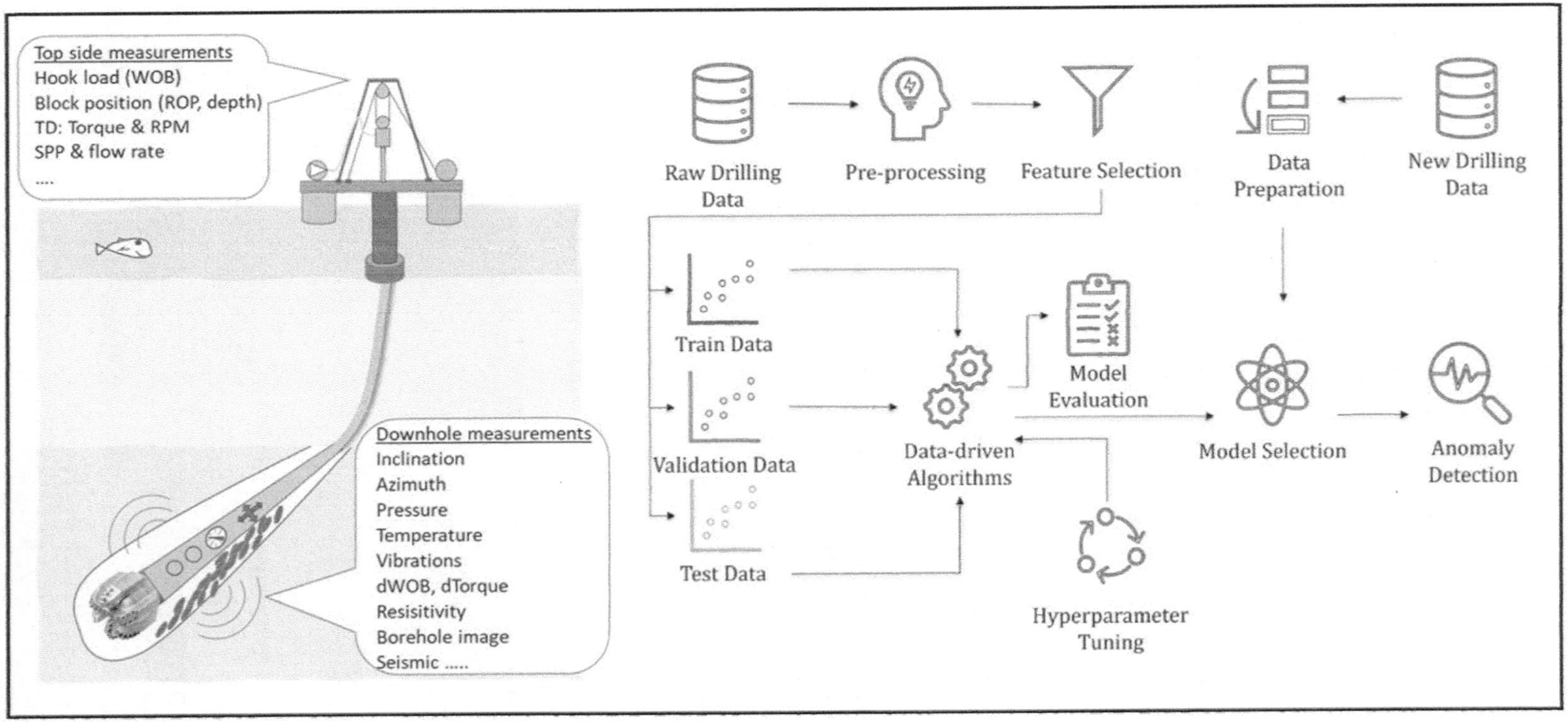

FIGURE 9.8 Predictive model (flowchart) to identify the downhole problems using Big Data.

Source: Combined and revised from Altindal et al. (2024).

- **Data Splitting**: The dataset is divided into training, validation, and test sets, which is crucial for evaluating and improving model accuracy.
- **Model Training**: Machine learning algorithms are applied, incorporating techniques such as K-fold cross-validation to fine-tune hyperparameters and optimise model performance.
- **Model Evaluation**: The most optimal model, determined through the cross-validation process, is used to predict outcomes on unseen test data, providing insights into the model's effectiveness and performance.

9.4 AI APPLICATIONS IN DRILLING OPERATIONS

AI and machine learning have become integral to various industries, including oil and gas. These technologies help drillers and operators derive powerful insights and predictions from the vast amounts of data gathered from rigs, wells, and facilities. These models are rapidly developing and being used in the drilling industry from the planning phase to plug and abandonment in the lifecycle of well delivery. Therefore, many different tools are developed or are developing very quickly in the drilling industry.

AI applications in this field exhibit different Technology Readiness Levels (TRL). For research and development (R&D) projects, AI models typically range from TRL 4 to TRL 5, indicating that they are in the validation and demonstration stages. On the other hand, commercialised AI tools used in the industry have achieved TRL 8 to TRL 9, signifying that they are fully operational and proven in real-world environments.

Here, we will highlight an example of stuck pipe incidents. These machine learning (ML) agents can be run in real-time operations to minimise the risk of stuck pipe incidents and manage the operation effectively by predicting potential issues before they escalate into severe stuck pipe situations. Additionally, these ML agents can be used based on offset well data as replay simulations to inform the planning phase. This example is an important sample, acknowledging that we cannot cover all AI models in detail. Figure 9.9 outlines the various AI applications across different phases of the well delivery lifecycle:

9.4.1 Stuck Pipe Prediction Using Machine Learning

Stuck pipe and other related drilling hazards are major causes of non-productive time while drilling. Detecting early indications of potential drilling risks manually by analysing drilling parameters in real-time has been a significant challenge for engineers. However, this task can be successfully executed by modern data analytics tools based on ML technologies. The objective of the presented study is to prove and demonstrate the ability of such ML algorithms to process and analyse simultaneously a variety of surface drilling data in real time in order to (Payrazyan and Robinson, 2023):

- Detect anomalies that are, in most cases, invisible to the human eye.
- Provide early warnings of possible upcoming drilling risks with sufficient time in advance, so that the rig crew can execute the appropriate mitigation actions.

In the realm of AI applications, various tools have been developed in the industry for different objectives. One notable example is the use of ML technologies with high Technology Readiness Levels (TRL) up to TRL 9 as commercialised tools to predict and manage stuck pipe incidents. To effectively implement these ML technologies, it is essential to understand the technical specifications and data requirements for different types of stuck pipe conditions.

The following table outlines a typical example of the key parameters and data inputs/outputs for differential sticking, mechanical sticking, and hole cleaning scenarios. It is important to note that different models may be used in the industry, and this technology is continuously evolving.

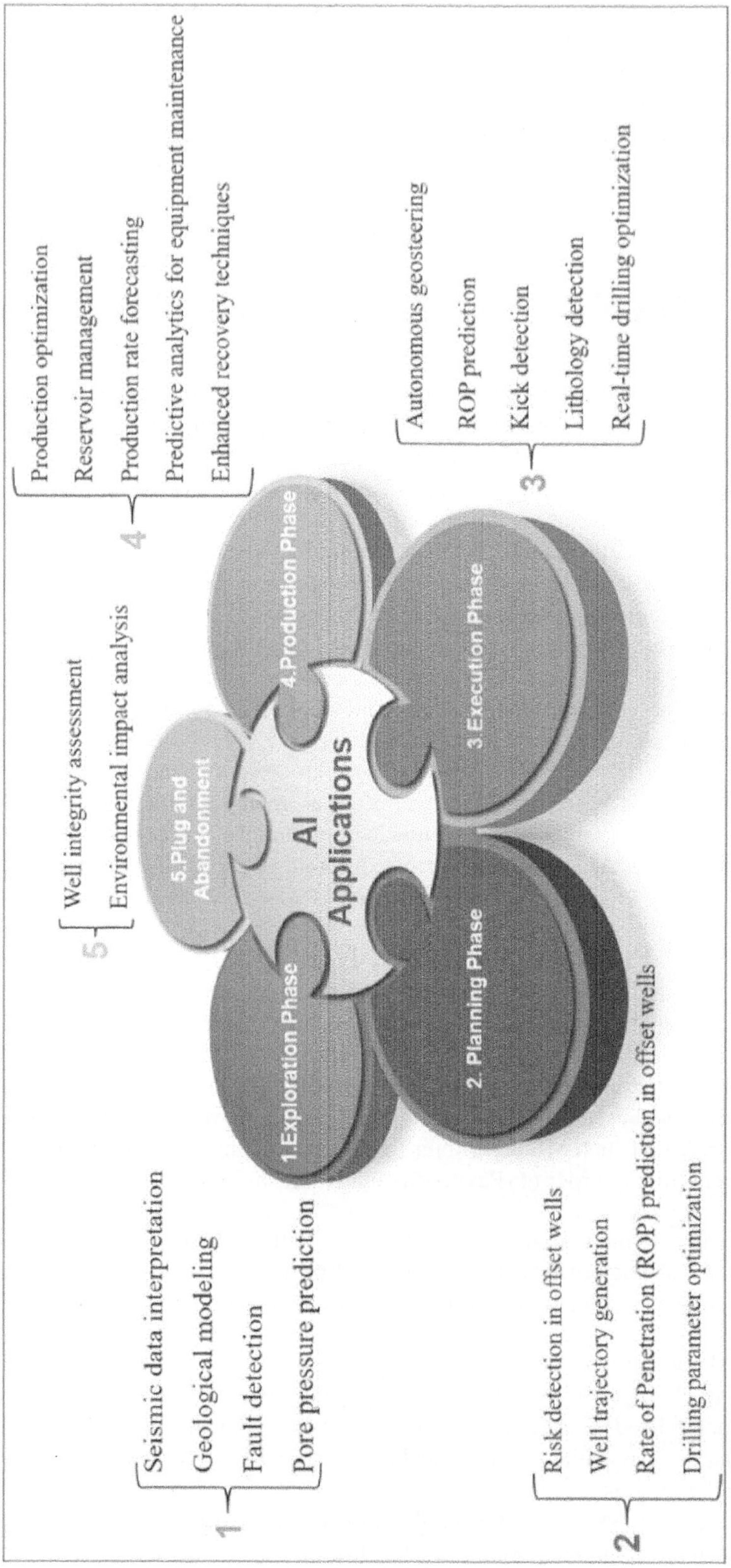

FIGURE 9.9 AI applications across the well delivery lifecycle.

TABLE 9.1 Stuck pipe agents: Technical specifications

CATEGORY	*DIFFERENTIAL STICKING*	*MECHANICAL STICKING*	*HOLE CLEANING*
Well Trajectory	0-90 degrees	0-90 degrees	0-90 degrees
Operation Types	Tripping, drilling, sliding, hole opening, reaming, circulation, and drill pipe, casing, and liner operations	Tripping	Drilling, sliding, hole opening, reaming, circulation, washing pipe up/down
Hole Sizes	30″ – 6″	30″ – 6″	30″ – 6″
BHA Size	26″ – 6″	26″ – 6″	26″ – 6″
Input Data	Time, bit depth, hole depth, hookload, flow rate, RPM, block position	Time, bit depth, hole depth, hookload, flow rate, RPM, block position, trajectory	Time, bit depth, hole depth, hookload, flow rate, RPM, trajectory, mud density in, ROP, standpipe pressure
Output Data	Hookload prediction, warning annotation	Hookload prediction, warning annotation	ECD prediction, warning annotation

TABLE 9.2 Stuck pipe agents: Technical specifications

CATEGORY	*DIFFERENTIAL STICKING*	*MECHANICAL STICKING*	*HOLE CLEANING*
ML Prediction Triggers	Yellow – lower risk Red – higher risk Message: "Static friction risk observed"	Yellow – lower risk Red – higher risk Message: "Dynamic friction risk observed"	Yellow – lower risk Red – higher risk Message: "Fluid friction risk observed"
Instant Triggers	Static friction, e.g., pumps are off, not rotating, not moving the pipe, for more than • Yellow: 8 min (configurable) • Red: 12 min (configurable) Message: "Long time with string stationary – risk of DS"	Overpull • Yellow – lower risk • Red – higher risk Taking weight • Yellow – lower risk • Red – higher risk Message: "Overpull" and "Taking weight"	Note! The alarms are configurable w.r.t. sensitivity and tolerance.

Table 9.2 outlines the warning and alarm types for different stuck pipe conditions, including differential sticking, mechanical sticking, and hole cleaning. It details the ML prediction triggers, such as risk levels and specific messages, as well as instant triggers that provide real-time alerts based on various operational parameters. These alarms are configurable in terms of sensitivity and tolerance, ensuring that the system can adapt to different drilling scenarios and provide timely warnings to mitigate potential risks.

9.4.2 Data Workflow

The algorithms developed have favourable characteristics, such as adaptiveness to real-time data and agnosticism to well types, BHAs, mud types, lithologies, or any other specific well characteristics. This supports out-of-the-box usage, which enables scalability to large numbers of wells. The system connects to WITSML data stores and outputs warnings with specific information regarding the identified symptom of the potential drilling incident (Payrazyan and Robinson, 2023).

In the realm of drilling operations, the use of ML agents for real-time predictions and offset analysis has become increasingly vital. These agents offer several key features that enhance their effectiveness and ease of use (Figure 9.10):

For real-time predictions and offset analysis, the system offers several key features:

- **Scalable for Any Well**: Designed to handle data from wells of any size and complexity, ensuring broad applicability across different drilling operations.
- **Plug & Play Functionality**: Easy integration with existing systems, requiring minimal setup for quick deployment and immediate use.
- **No Manual Configuration Required**: Operates with automated settings, eliminating the need for manual configuration and reducing the potential for human error.
- **Easy Interpretation within the Real-Time Platform**: Data and insights are presented in a user-friendly manner, making it simple for users to understand and act upon the information.
- **Support Tool for Engineers and Data Analysts**: Serves as a valuable resource for engineers and data analysts, providing the tools needed to monitor and optimise drilling operations effectively.

9.4.3 Results of ML Agents

Targeted sub-systems detect the current operation type (tripping, drilling, reaming, etc.) and identify symptoms related to differential sticking, hole cleaning issues, mechanical sticking, pack-offs, tight holes, obstructions, and other risks by analysing standard surface drilling time logs in real-time, such as hookload, weight on bit (WOB), revolutions per minute (RPM), bit depth, mud pressure, etc. The ML models and wider risk detection system have been demonstrated to generalise to new wells and consistently produce high performance across those tested (Payrazyan and Robinson, 2023).

9.4.4 Performance

The system provides drilling engineers with live warnings on average 1.5–4 hours prior to incidents, giving rig crews enough time to react. This also allows drilling engineers to know in advance a specific source of potential risk, which assists in selecting the right strategy for implementing corrective actions. The technology's performance has been successfully verified in live operations and post-drill studies on historical data on over 300 wells worldwide during 2.5 years, with mean recall and precision metrics of 0.986 ± 0.050 and 0.712 ± 0.181, respectively, across historical test wells, and significantly reduced occurrence rates of stuck pipe incidents in both onshore and offshore operations (Payrazyan and Robinson, 2023).

Here, we present another example where we tested the ML agent models on two wells as a blind test using offset well data, without sharing the Daily Drilling Reports (DDR) and information. The objective was to evaluate the models' ability to provide alarms or predictions ahead of stuck pipe incidents. The results demonstrated very good accuracy in predicting stuck pipe events in these two cases. As shown in Table 9.3, initial symptoms were detected well in advance for both wells. Figure 9.11 illustrates how real-time predictions based on offset well data can support planning efforts. The results shown in this figure were generated by running ML agents in shadow mode simulation using offset WITSML data. The two wells, SP03 and SP06, serve as examples of successfully commercialized ML agents in the drilling industry, currently operating at TRL 9. For more detailed information, please refer to the cited references. As the industry continues to evolve, we expect the emergence of more advanced ML agents built on diverse algorithms and logic frameworks.

9.5 OFFSET WELL REPLAY FOR REAL-TIME MONITORING

One aspect of real-time monitoring during the planning phase is the ability to replay operational parameters based on offset wells. By analysing offset well events, we can gain valuable insights into potential risks

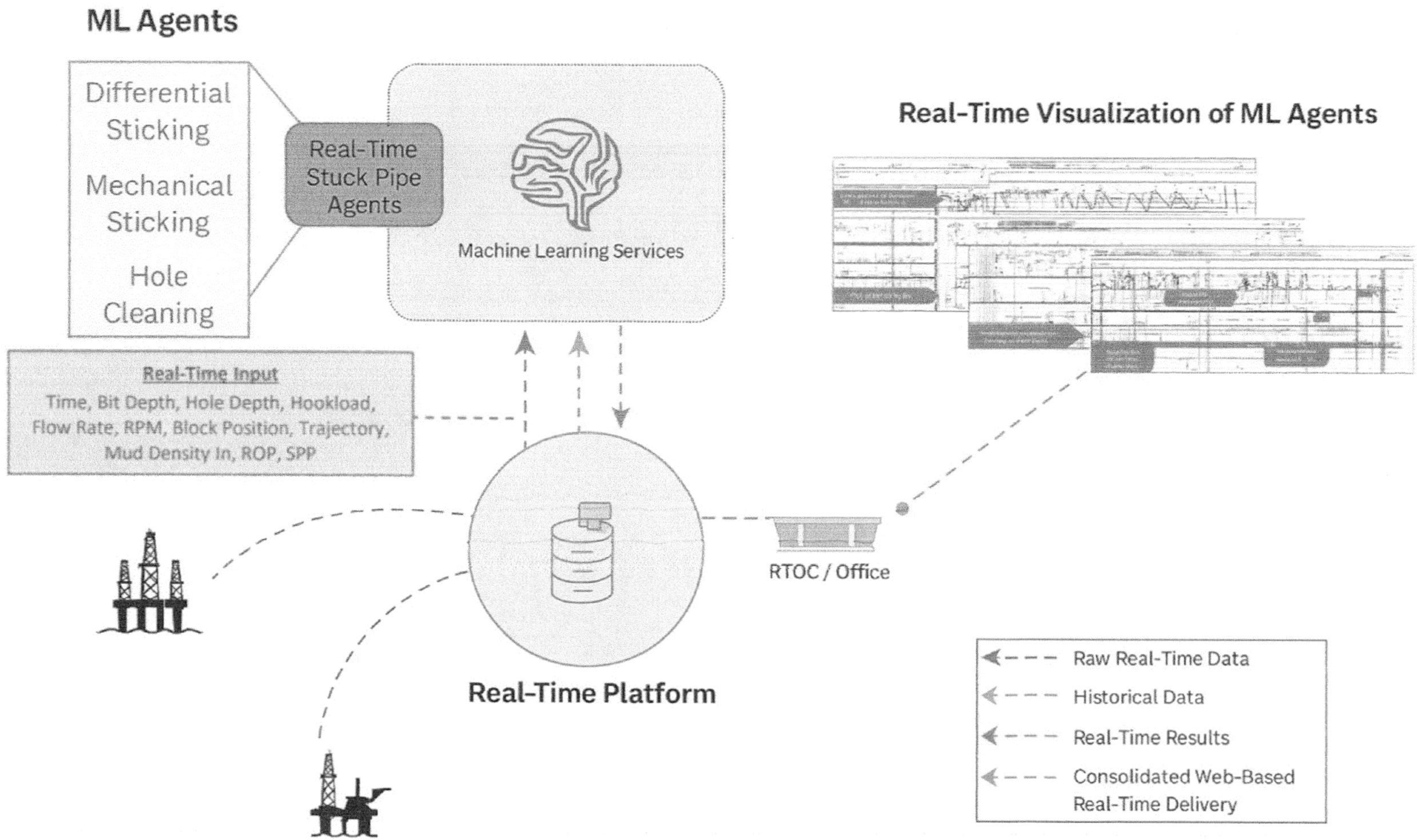

FIGURE 9.10 ML real-time stuck pipe agent.

Source: Readapted from Payrazyan and Robinson (2023).

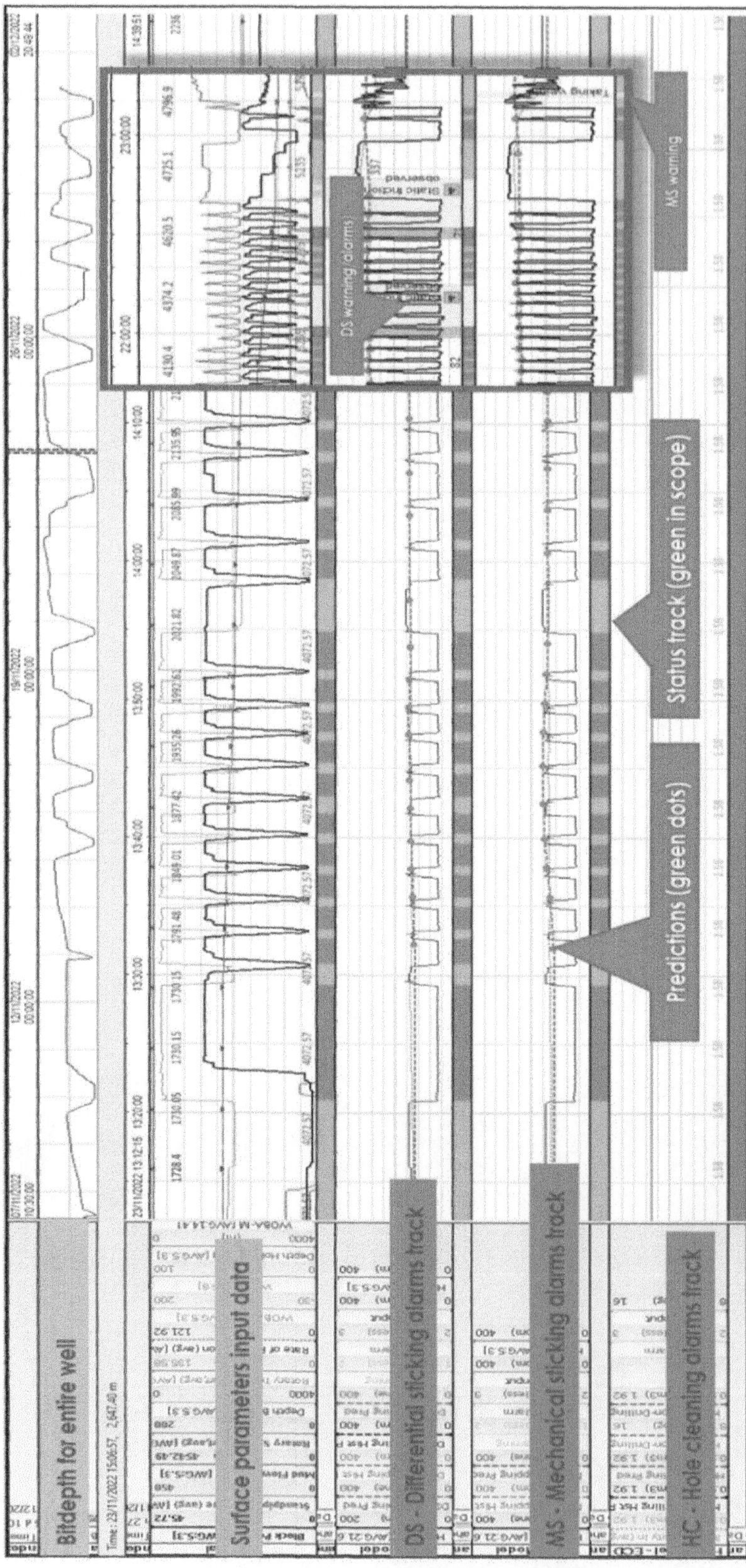

FIGURE 9.11 Real-time predictions ahead base offset analysis can be used for planning phase.

TABLE 9.3 Performance analysis of ML agent models on wells

WELL	*STUCK PIPE EVENT*	*DEPTH*	*ESTIMATED NPT*	*SUMMARY OF RESULTS*
SP06	One stuck pipe event	~446 m	~24 hours	Initial symptoms observed ~2 hours in advance. Multiple torque risk warnings provided by the MS model. Hole-cleaning risk warning and DS static friction warning prior to the event. Occurred while drilling.
SP03	One stuck pipe event	~523 m	~32 hours	Initial symptoms observed ~3 hours in advance. Hole-cleaning risk warnings and multiple drag anomaly risk warnings provided by the HC and MS model prior to the event. Occurred while drilling.

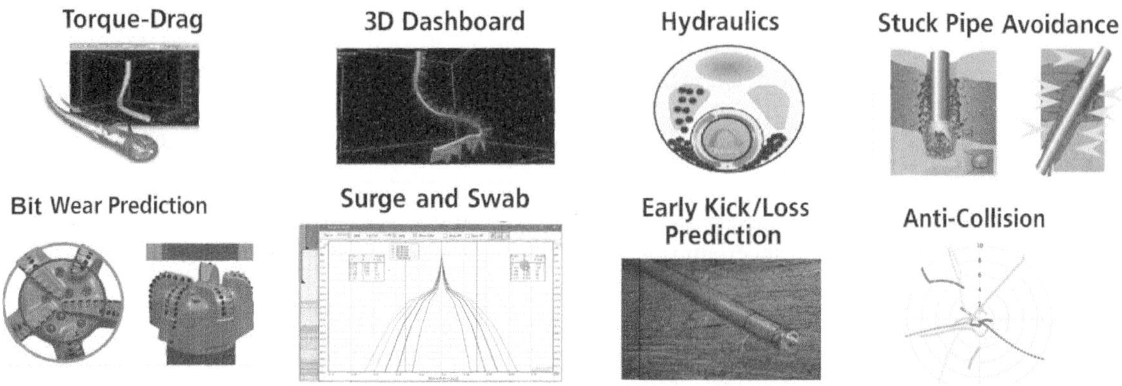

FIGURE 9.12 Visualisations for replaying offset well data.

and opportunities for improvement. This module helps us identify patterns and trends from nearby wells, which informs better decision-making and enhances our overall drilling performance.

This figure shows different visualisations for replaying offset well data, highlighting key events and trends that aid in our analysis.

Torque & Drag Module

The Torque & Drag module is our early warning system for mechanical issues. It helps us to understand the forces acting on the drill string and predict any problems with torque and drag. By doing this, we can prevent stuck pipe incidents and keep our drilling operations running smoothly.

Hydraulics Module

The Hydraulics module is all about managing the fluids in the wellbore. It helps us optimise mud properties and flow rates, which is crucial for efficient hole cleaning and preventing damage to the formation. This module ensures that the well stays in good shape and drilling performance is top-notch.

Swab and Surge Module

Swab and surge analysis is essential for avoiding well control issues. The Swab and Surge module predicts pressure changes during tripping operations and helps us manage the risks associated with swabbing and surging. This keeps everything safe and efficient while we're drilling.

3D Visualisation Module

The 3D Visualisation module gives us a detailed view of the wellbore and the surrounding formations. It helps us visualise complex well paths and spot potential issues, making it easier to make informed decisions in real time. This module is key for understanding the spatial relationships within the wellbore.

Collision Avoidance Monitoring Module

The Collision Avoidance Monitoring module is our safeguard against wellbore collisions. It continuously monitors the position of the drill bit and nearby wells to prevent any dangerous and costly collisions. This ensures that our drilling operations are safe and efficient.

Early Kick Loss Prediction

Early Kick Loss Prediction is all about being proactive. This module analyses real-time data to predict early signs of kicks and losses, allowing us to intervene before any well control incidents occur. It's a crucial tool for preventing blowouts and keeping everything under control.

Stuck Avoidance Module

The Stuck Avoidance module focuses on preventing stuck pipe incidents. It analyses real-time data and historical trends to provide recommendations that help us avoid conditions that could lead to stuck pipe. This ensures that our drilling operations continue smoothly and efficiently.

9.6 DRILLING WELL ON SIMULATOR (DWOS)

From planning to execution, advanced simulations play a crucial role in improving our well delivery process by minimising risk and uncertainty. The DWOS room is equipped with advanced components to enhance the training experience, as shown in Figure 9.13.

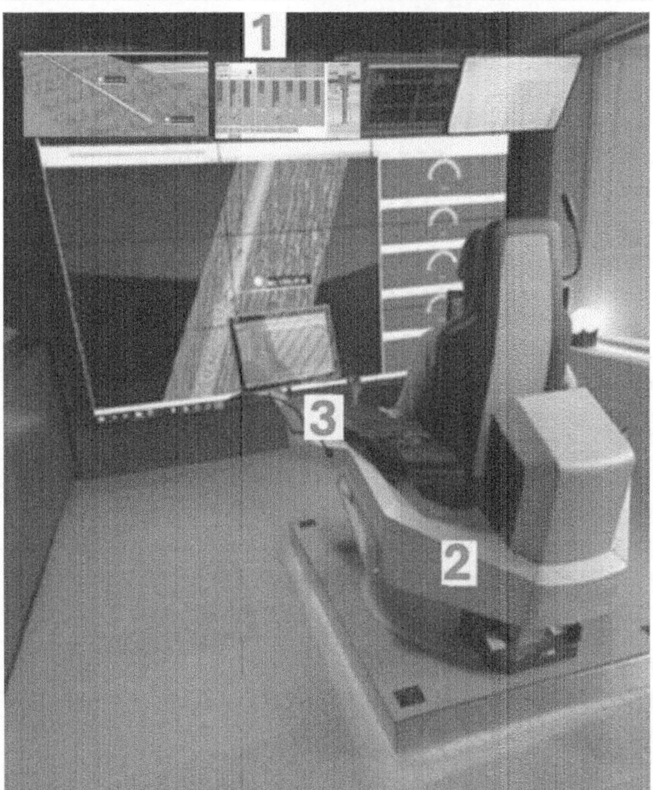

1-Monitors: High-resolution screens show digital twins of drilling operations, like the bottom hole and subsurface conditions. (Bimastianto et al., 2020).
2-Advanced Chair: An ergonomic chair with various features for comfort and extended use, helping operators stay focused.
3-Joystick: A joystick for precise control and interaction with the simulation, allowing effective navigation and manipulation of the virtual drilling environment.

FIGURE 9.13 Key components of the DWOS room setup.

This figure shows the key components of the DWOS room, including a 3D digital twin that provides a perfect view of the bottom hole situation, highlighting possible faults, washout areas in the section, and potential stringers.

Drill the Well on Simulator (DWOS) offers team-based scenario training for conventional drilling, Extended Reach Drilling (ERD), Managed Pressure Drilling (MPD), deepwater, High-Pressure High-Temperature (HPHT), and other complex well conditions. DWOS provides unique functionality for drilling training with an innovative and realistic, dynamic drilling simulation experience.

One of the important aspects of DWOS, as highlighted in this book, is minimising risk during the planning phase by considering and testing different risk scenarios before execution. This includes scenarios for ERD, MPD, deepwater, HPHT, and other complex well conditions.

In this book, we focus on various risk scenarios for the planning phase to minimise risk and uncertainty, as follows:

High-Speed Drilling Challenges Risk Scenario

Compared to a decade ago, some operators are now drilling wells at speeds exceeding 300 meters per hour, whereas previously, speeds ranged from 30 to 120 meters per hour. High-speed drilling, such as 350 meters per hour, presents unique challenges. The simulator helps us address and overcome these challenges more effectively in Extended Reach Drilling (ERD) or multilateral wells, ensuring efficiency without compromising safety. We have successfully demonstrated that this strategy works perfectly in some wells.

Stuck Pipe Risk Scenario

One of the biggest risks in drilling is getting the drill pipe stuck. Our simulator helps us quality check our stuck pipe procedures, ensuring we have effective protocols in place to address and mitigate these incidents. This reduces downtime and associated costs.

Digital RAPS Testing Risk Scenario

The simulator allows us to validate the reliability and accuracy of the Rig Action Plan (RAP), enhancing our decision-making during drilling operations and minimising the risk of errors.

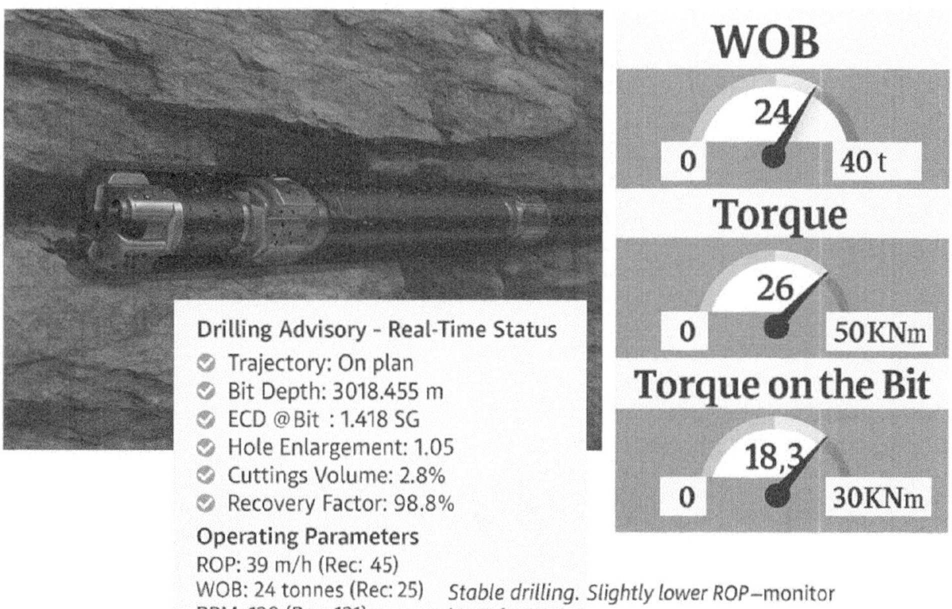

FIGURE 9.14 DWOS real-time advisory and 3D bottom-hole visualization during stable drilling conditions.

Re-Playing Drilled Wells Risk Scenario

By re-playing drilled wells on the simulator, we can analyse past operations to identify areas for improvement. This helps us apply lessons learned to future projects, reducing the risk of repeating mistakes.

Hole Cleaning Risk Scenario

Effective hole cleaning is essential to prevent blockages and ensure smooth drilling operations. The simulator helps us optimise our hole cleaning procedures, minimising the risk of operational disruptions.

Sense Check Other Simulations Risk Scenario

Cross-verifying the accuracy and effectiveness of various simulation models is important. The simulator allows us to sense check other simulations, ensuring consistency and reliability in our operations.

Losses Risk Scenario

Fluid losses during drilling can lead to significant operational challenges and increased costs. The simulator helps us develop strategies to minimise these losses, reducing the risk of complications.

Optimum Drilling Parameters Scenario

Finding the best drilling parameters is key to maximising efficiency and minimising risks. The simulator allows us to test and determine the optimum parameters, enhancing overall drilling performance.

Well Control Risk Scenario

Well control incidents, such as blowouts, are serious risks in drilling operations. The simulator helps us enhance our well control procedures, ensuring we can prevent and manage these incidents effectively, keeping our personnel and equipment safe.

Emerging Technologies 10

10.1 DOWNHOLE SENSORS

10.1.1 Introduction

Real-time data is essential for improving drilling efficiency and optimising performance in modern drilling operations. By continuously monitoring both surface data (e.g., mud pit volume, flow rate, hook load, surface RPM, and cutting analysis) and downhole data (e.g., bottom hole pressure, torque on bit, and vibration), engineers can quickly identify potential issues, such as lost circulation, drill string sticking, or equipment malfunctions. Early detection enables timely corrective actions, preventing problems from worsening.

Additionally, real-time data is crucial for calibrating physics-based models like hydraulics and torque-and-drag in digital twin-based drilling operations. By utilising data such as hook load and torque, engineers can extract the appropriate coefficient of friction, refining torque-and-drag model predictions in real time to ensure safe drill string mechanics. Accurately calibrated hydraulic models also ensure wellbore stability and prevent kick-related issues. These models' predictions describe the downhole condition used if data transmission is missing due to faulty sensors.

Downhole sensors are specialised downhole tools/instruments that are installed in the oil and gas wells. Downhole sensors measure a wide range of data in real time during drilling operations, enabling operators to monitor well conditions, optimise drilling performance, and ensure safety. These sensors are typically integrated into Measurement While Drilling (MWD)/Logging While Drilling (LWD) systems and are positioned close to the drill bit to provide timely insights into the formation and drilling dynamics. The next subsection presents a brief description of the downhole sensors and applications:

- MWD/LWD Sensors
  - for Directional Drilling
  - for Formation Evaluation
- Fibre optics sensor for Smart Well Completion

10.1.2 MWD/LWD Sensors

- Directional Drilling and Drilling Dynamics Sensors

Mud motors and Rotary Steerable Systems (RSS) are essential for controlling wellbore direction during drilling operations. These systems optimise drilling performance, help with accurate placement, and

DOI: 10.1201/9781003618171-10

improve safety by providing real-time data for adjusting drilling parameters. The typical MWD sensors measure:

- Flow Rate Sensors: Measure the flow of drilling fluid.
- Pressure Sensors: Measure downhole pressure.
- Temperature Sensors: Measure downhole temperature.
- Inclination and Azimuth Sensors: Measure the angle and direction of the wellbore.
- Weight on Bit (WOB) Sensors: Measure the force applied to the drill bit.
- Torque on Bit (TOB) Sensors: Measure the torque on the drill bit.
- Vibration Sensors: Detect vibrations.
- Formation Evaluation Sensors.

Formation Evaluation with LWD tools measures the geological formation properties being drilled in real-time data. LWD technology allows for *continuous monitoring and evaluation* of the formation's properties during the drilling operation. The typical LWD tools are equipped with several sensors to measure various formation properties such as:

- Gamma Ray Sensor:
  - Measures natural gamma radiation emitted by the formation.
  - Helps differentiate between shale and non-shale (e.g., sandstone or limestone) layers.
- Resistivity Sensor:
  - Measures the electrical resistivity of the formation.
  - Indicates the presence of hydrocarbons. Low resistivity often suggests the presence of water, while higher resistivity may indicate the presence of hydrocarbons (oil or gas).
- Density Sensor:
  - Measures the bulk density of the formation.
- Porosity Sensor:
  - Measures hydrogen content in the formation, providing additional information about the fluid content and porosity.
  - Neutron porosity tools are used in combination with density tools to calculate porosity more accurately. They also help distinguish between water, oil, and gas.
- Formation Pressure Sensor:
  - Measures the pressure of the formation in real time.
- Acoustic/Sonic Sensor:
  - Measures the velocity of sound waves through the formation.
  - Used to estimate formation's mechanical, elastic, in-situ stress, and petrophysical properties (e.g., porosity, permeability).
  - Used for the estimation of formation pressure.

Real-time data is essential for improving drilling performance, safety, and efficiency. With the integration of MWD sensors for *directional drilling and formation evaluation*, engineers can continuously monitor and adjust drilling parameters, identify potential issues, and calibrate models to predict downhole conditions accurately. These advancements help in enhancing wellbore placement, preventing equipment malfunctions, ensuring safe, optimised drilling operations, and optimising reservoir development.

Example 1: Application of Formation Sensor Data
During the well planning phase, it is common practice to develop a well pressure prognosis constrained by the fracture pressure and collapse pressure gradient. These gradients define the safe operational mud weight window and are critical for ensuring wellbore stability.

The gradients are derived from formation properties such as formation strength, elasticity, in-situ stresses, and pore pressure. As noted earlier, density (Figure 10.1) and sonic logs (Figure 10.2) are used

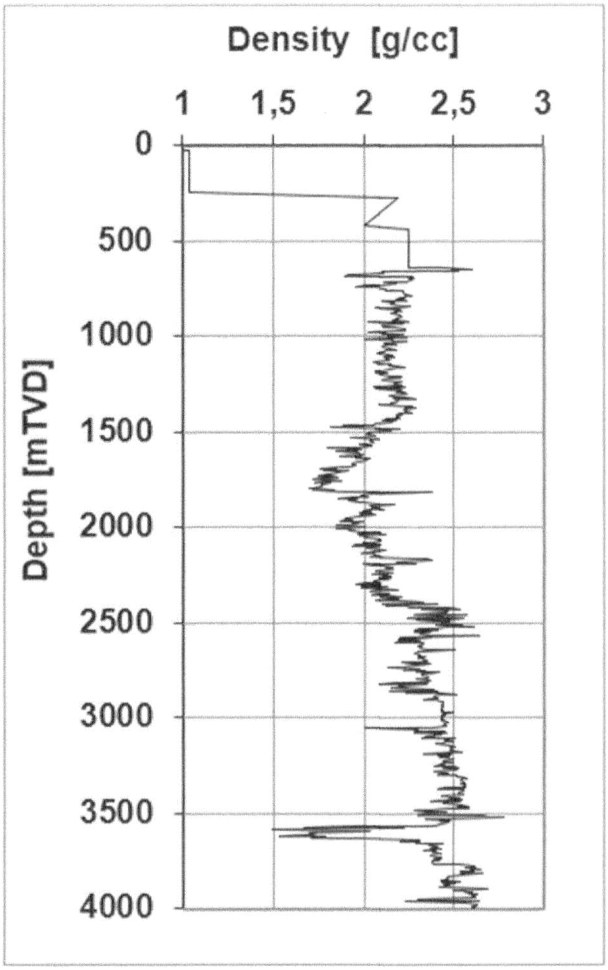

FIGURE 10.1 Density log.

to calculate rock properties such as Young's modulus, Poisson's ratio, and uniaxial compressive strength (UCS) through established empirical correlations. In this example, using the downhole recorded logs shown below, the following formation properties will be quantified:

- Overburden stress
- Pore pressure
- Uniaxial compressive strength
- Young's modulus

10.1.2.1 Overburden Stress (Vertical Stress)

The vertical stress (σ_v) is also known as overburden stress. It acts vertically and is orthogonal to the maximum horizontal stress (σ_{hmax}) and the minimum horizontal stress (σ_{hin}). This stress results from the weight of the overlying rock layers and the fluid column above the point of interest, and it plays a significant role in determining wellbore stability. It is estimated as:

$$\sigma_v = \rho_w g Z_w + g \int_{Z_w}^{Z} \rho_{b,z} dZ \tag{10.1}$$

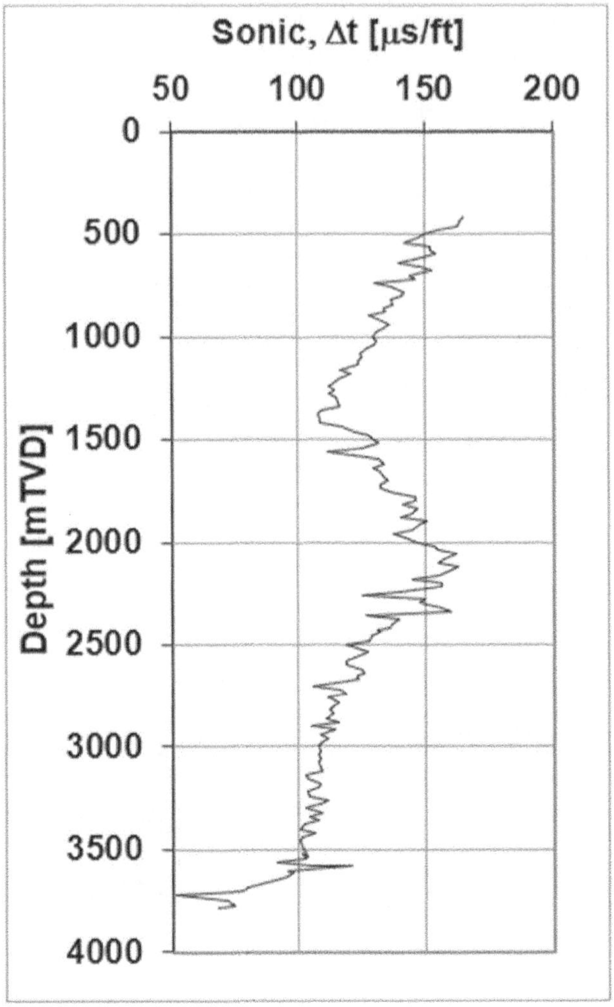

FIGURE 10.2 Sonic log.

where ρ_w is the water density, Z_w is the water depth, and ρb is the density of the formation being drilled, measured from the density log tool.

Figure 10.3 shows the overburden pressure, which is computed by integrating the density profile.

10.1.2.2 Pore Pressure Determination

Pore pressure can be determined through both *pre-drilling and post-drilling* methods.

- *Pre-drilling* techniques depend on seismic data, including P-wave (Vp) and S-wave (Vs) velocities, resistivity, and sonic logs.
- *Post-drilling* techniques employ direct measurements such as drill stem tests (DST) and production and repeat formation tests (RFT).

For this example, we will use the Eaton equation, which is a commonly used empirical model that uses resistivity and sonic log data (Eaton, 1975). The Eaton model reads:

$$P_p = \sigma_v - (\sigma_v - P_n)\left(\frac{\Delta t_n}{\Delta t}\right)^3 \tag{10.2}$$

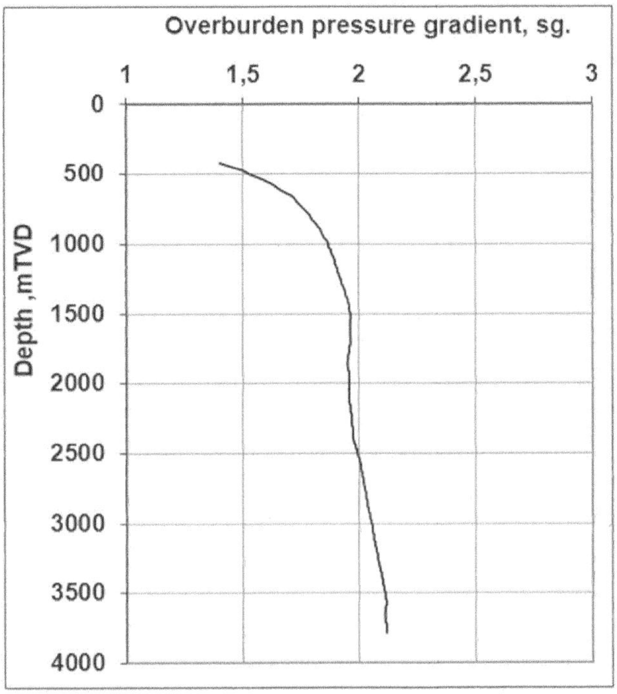

FIGURE 10.3 Overburden pressure gradient computed by using the density log.

where,

- Δt_n is the normal trendline at the normal pressure, P_n (1.03 sg),
- Δt is the sonic transit time log-response,
- Pp is formation fluid hydrostatic pressure, and s_v is overburden gradient,
- n=3 is the Eaton exponent, which varies from region to region.

Figure 10.4 illustrates the decreasing trend of normal compaction with increasing depth. When applying Eaton's method, the first step is to determine the normal transit time (Δt_n), which is derived from data above the compaction or shale zone. Selecting the correct Normal Compaction Trend (NCT) is critical. If the NCT is chosen from a zone affected by elevated pore pressure, it will lead to inaccurate pore pressure predictions. Therefore, the NCT must be identified from a properly compacted, normally pressured interval.

Following the abovementioned procedure, the sonic dataset up to depth 1200 m (above the broken line) was used to generate the normal trendline (Figure 10.2). Before constructing the trendline, the data was smoothed using a moving average filter to reduce high-frequency noise and highlight the overall signal trend. As illustrated in Figure 10.5, the sonic log displays a good fit with an exponential function. By expressing the normal sonic as a function of depth, the trendline was generated through extrapolation into the deeper section of the formation.

10.1.2.3 Pore Pressure Prediction Using Eaton's Method

Figure 10.6 presents the sonic log alongside the normal compaction trendline. These datasets, combined with the overburden pressure data shown in Figure 10.3, were used as inputs for Eaton's empirical model to estimate pore pressure. The calculations were carried out using three different values of Eaton's exponent: n = 1.2, 1.5, and 2.0. The resulting pore pressure profiles are illustrated in Figure 10.7. Since the Eaton exponent can vary depending on regional geology, the generated pore pressure profiles should be calibrated against known formation pressure data, such as from a post-drill formation pressure test.

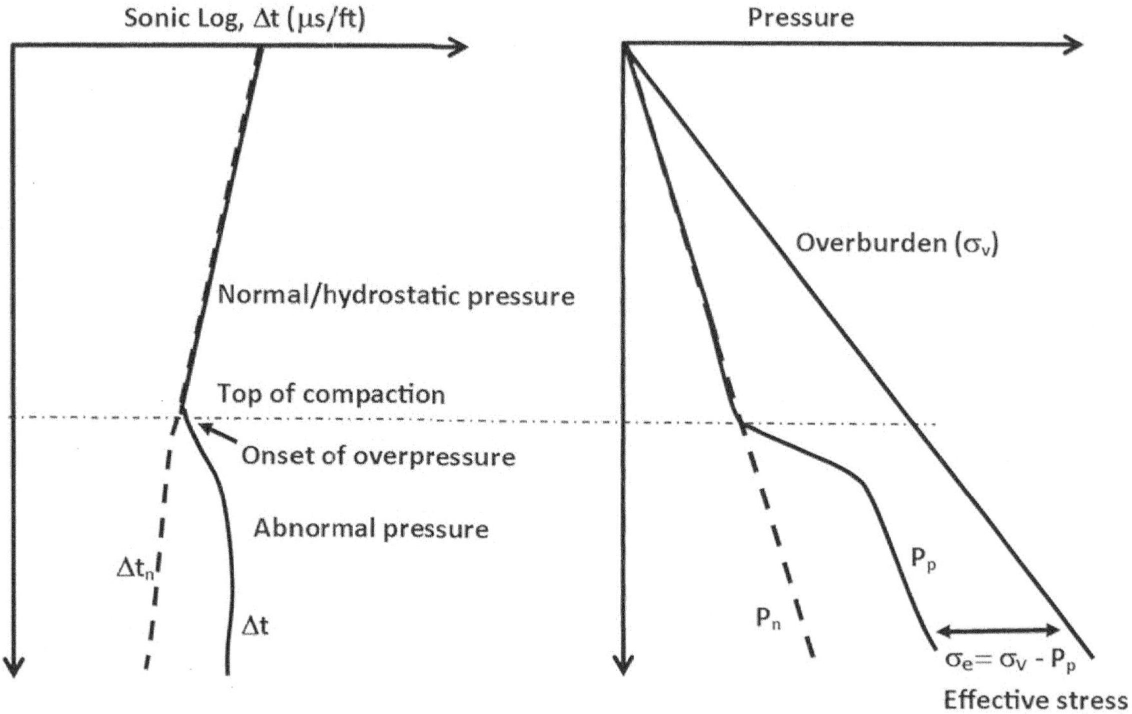

FIGURE 10.4 Illustration of pore pressure determination.

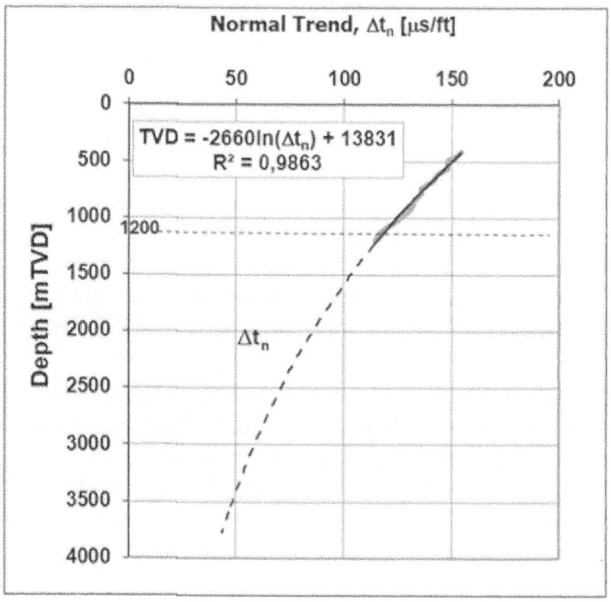

FIGURE 10.5 Normal Trend line generation using the normal pressures region dataset.

10.1.2.4 Uniaxial Compressive Strength

Uniaxial Compressive Strength (UCS) is a rock property used in wellbore collapse modelling, representing the rock's resistance to compressive stress. The UCS is determined through laboratory testing. However, continuous core extraction during drilling is costly and often impractical. As an alternative, empirical

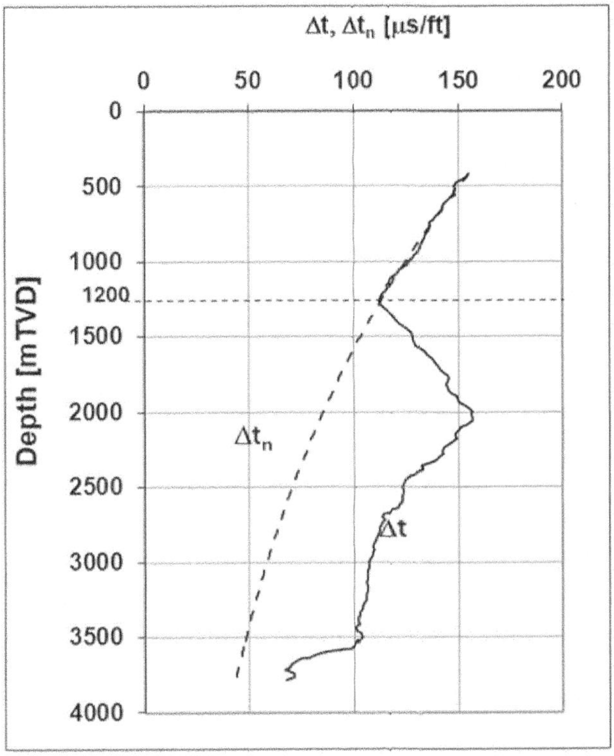

FIGURE 10.6 Normal and sonic logs vs depth.

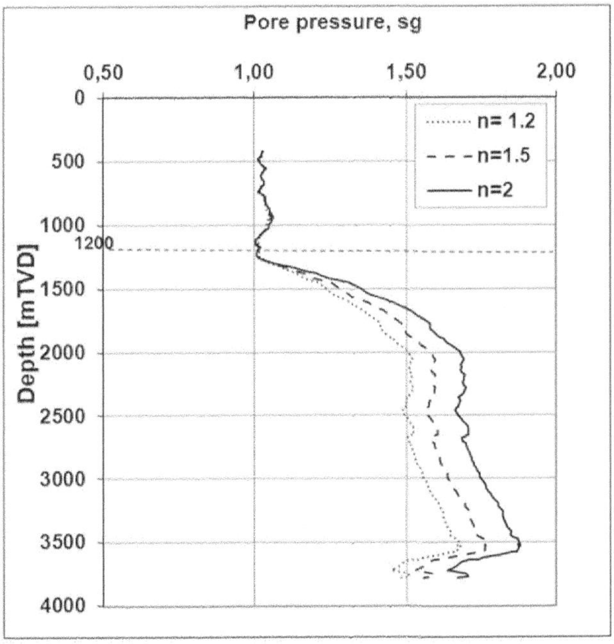

FIGURE 10.7 Pore pressure prediction for an exponent n =1.2, 1.5, and 2.

models provide a practical solution. Horsrud (2001) developed a widely used model for North Sea formations that estimates UCS from compressional wave velocity (Vp) obtained from sonic logs:

$$C_0 = 0,77 * V_p^{2.93} \tag{10.3}$$

where C_0 is uniaxial compressive strength (UCS) in MPa and V_p is compressional wave velocity in km/s.

Young's Modulus
Several empirical models have been developed that relate the compressional wave velocity to Young's modulus. Among others, Horsrud's (Horsrud, 2001) model reads:

$$E = 0.076V_p^{3.223} \tag{10.4}$$

where E is Young's modulus (GPa) and Vp is compressional wave velocity (km/s).

10.1.2.5 Formation Strength and Elastic Properties

Figures 10.8 and 10.9 present the uniaxial compressive strength (UCS) and Young's modulus, respectively. These formation strength and elastic properties were derived from the sonic log (Figure 10.2) using established empirical correlations (Equations (10.3) and (10.4)). As previously discussed, these mechanical parameters are critical for wellbore stability design.

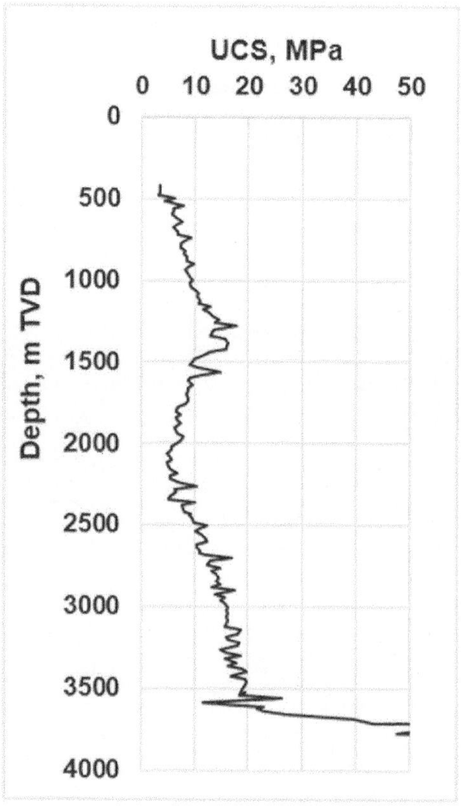

FIGURE 10.8 Uniaxial compressive strength.

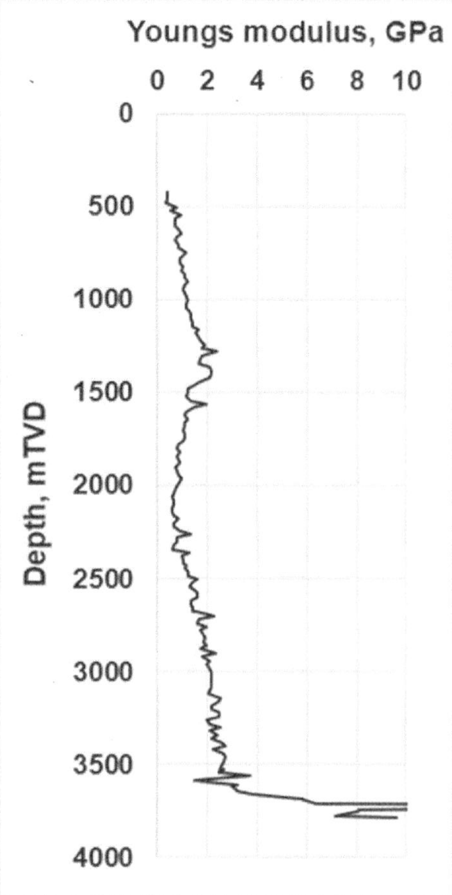

FIGURE 10.9 Young's modulus of the formation.

Example 2: Hydraulic Model Prediction and Calibration With Downhole Sensor Data

Hydraulic models are typically developed based on a set of simplifying assumptions, which often limit their ability to fully represent complex, real-world operational conditions. In a study by Lohne et al. (2008), field-measured hydraulic data from wells in the North Sea were compared with predictions from the hydraulic well model. The comparison revealed significant discrepancies between the measured data and the model outputs, highlighting the limitations of the model under actual field conditions, as illustrated in Figure 10.10 by using different field dataset.

Several factors contribute to the weakness of the model. To address the missing physics, the authors have included a calibration factor, c, in the pressure loss model as

$$\Delta P = c\frac{2f\rho V^2}{D_H}L \tag{10.5}$$

where ΔP is pressure loss, f is the friction factor, ρ is the density of the fluid, V is the flow velocity, D_H is the flow path, and L is the length of the flow line

Based on the measured data, the generated calibration factor is shown in Figure 10.11.

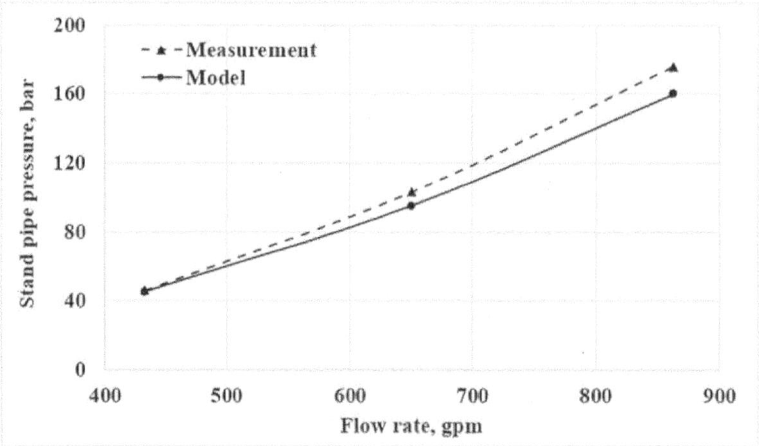

FIGURE 10.10 Comparison between measurement and model prediction.

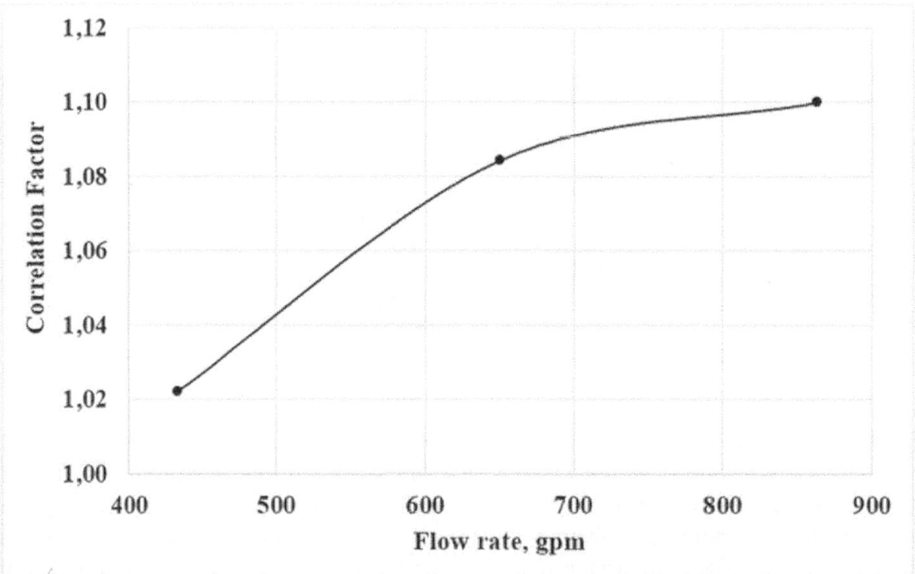

FIGURE 10.11 Calibration.

The hydraulic model is calibrated using a calibration factor, resulting in an excellent match with the measured data, as illustrated in Figure 10.12. These results underscore the critical role of real-time downhole measurements in achieving accurate model calibration.

10.1.3 Fibre Optics Sensor

Conventional wells are typically drilled in vertical, horizontal, or multilateral trajectories to maximise reservoir contact. However, they often rely on periodic well interventions, such as wireline or coiled tubing operations, to acquire essential downhole data like pressure, temperature, and flow rates. These interventions can be costly, time-consuming, and operationally risky, particularly in complex or deepwater environments.

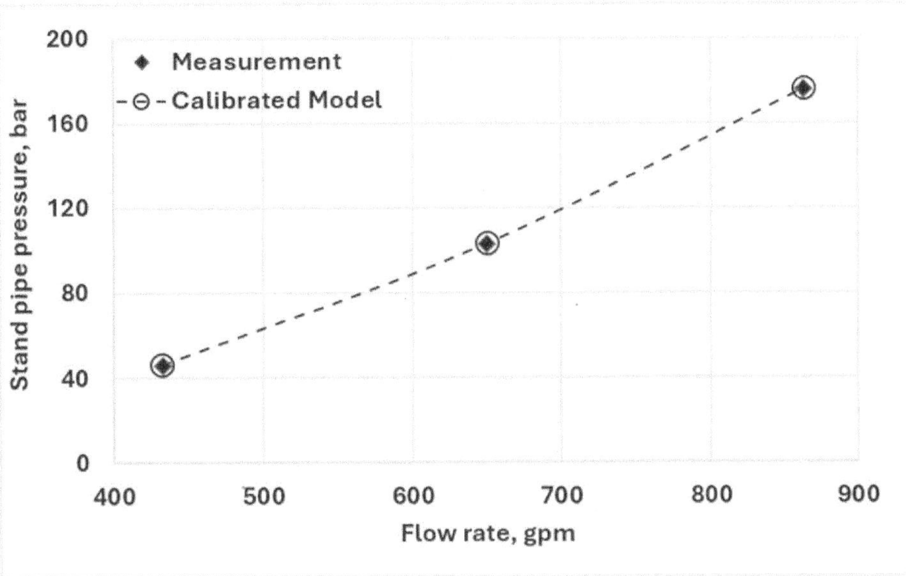

FIGURE 10.12 Model calibrated with measurement.

Source: Lohne et al. (2008).

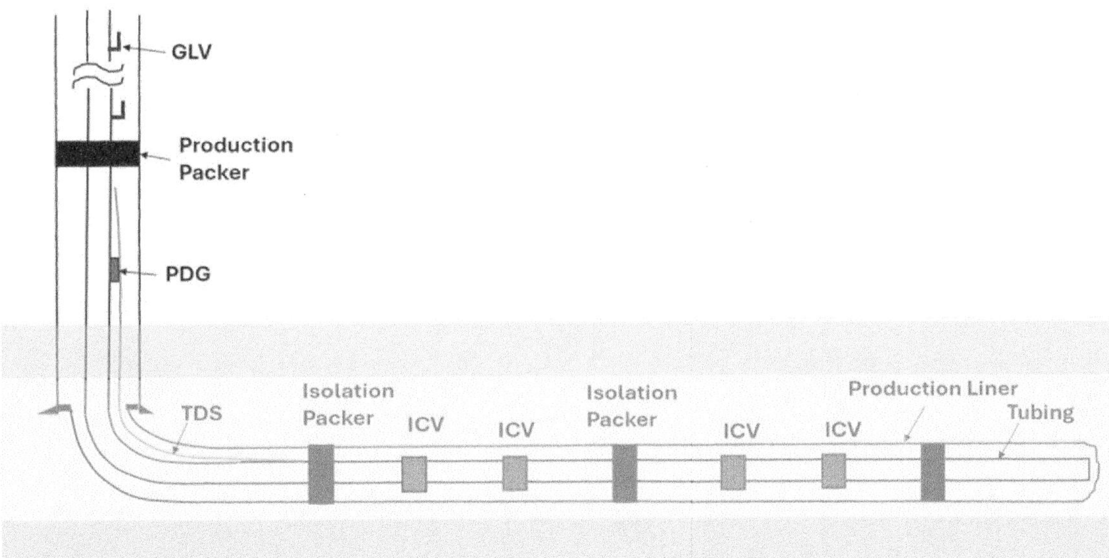

FIGURE 10.13 Illustration of a smart well.

As illustrated in Figure 10.13, *Smart wells* enhance *conventional completions* by integrating mechanical and electronic downhole equipment, including permanent sensors, packers, and interval control valves (ICVs). These systems enable continuous monitoring and real-time data transmission from the wellbore, giving operators detailed insight into downhole conditions.

The advantages of smart wells:

- Remote zone production control without physical intervention.
- Remote real-time production optimisation based on live data.
- Cost saving: By avoiding workovers and reducing intervention frequency.

- Reservoir management: By managing pressure and flow rate per zone.
- Enhanced recovery.
- Extend the field life.

Early smart well systems depended on conventional *electronic sensors and electrical control* systems, but these technologies often experience increased failure rates at elevated temperatures. As a result, the industry shifted toward more robust solutions. Modern smart wells now commonly employ *fibre optic sensors* for improved reliability and performance in high-temperature, high-pressure (HTHP) environments.

10.1.3.1 How Does Fibre Optics Work?

Fibre optics operate by transmitting light signals through thin strands of glass or plastic fibres. These light signals are generated and modulated to carry data, then converted back into electrical signals at the receiving end. Figure 10.14 illustrates this. One of the main advantages of fibre optic communication is its ability to transmit data over long distances with minimal signal loss and electromagnetic interference.

The core mechanism that enables light to travel efficiently through the fibre is called *total internal reflection*. This occurs when light travelling within the fibre core hits the core-cladding boundary at an angle greater than the critical angle, causing it to reflect entirely within the core rather than escaping. This allows the light to "bounce" its way down the fibre, preserving signal strength and integrity over long distances. Figure 10.15 illustrates the total internal reflection of light in a section of optical fibre.

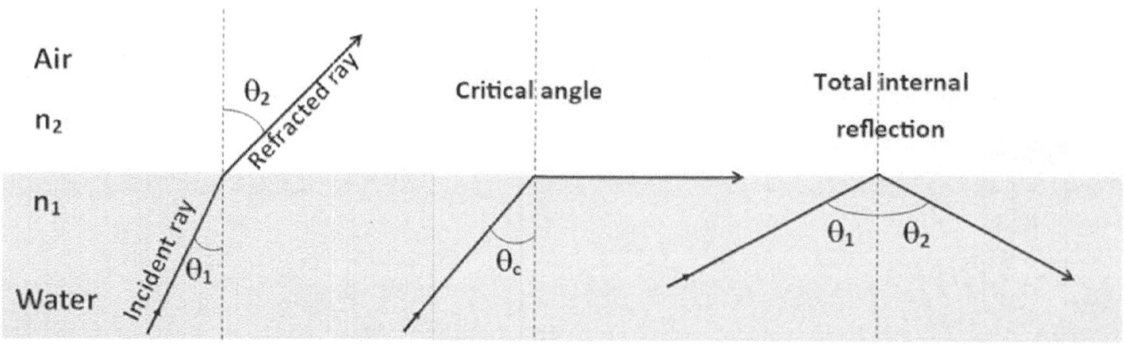

FIGURE 10.14 Illustrates refraction, critical angle, and total internal reflection between two media (air and water).

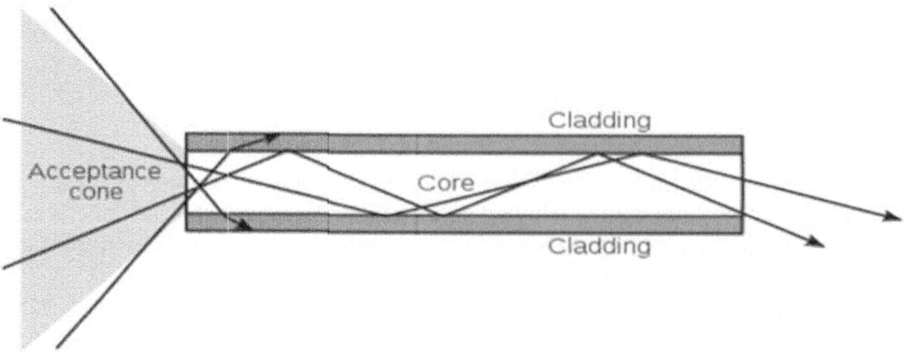

FIGURE 10.15 Illustration of the light's total internal reflection in a section of optical fibre.

The typical fibre optic sensor technologies, among others, are:

- DTS (Distributed Temperature Sensing)
- DAS (Distributed Acoustic Sensing)
- DSS (Distributed Strain Sensing)
- DAS (Distributed Vibration Sensing)
- Pressure and temperature (PT) gauges
- Flow meters

Example 3: Fibre Optics Application
Among others, Silva Junior et al. (2012) reviewed the fibre optics sensor technologies and applications in petroleum wells. These sensors are utilised across various domains, including *well condition monitoring, well performance evaluation, well stimulation, flow assurance, advanced well completions*, and *reservoir characterisation.* Fibre optic sensors enable the measurement and detection of critical parameters, among others, such as:

- strain
- pressure
- velocity
- flow rate
- vibrations
- acoustics
- electric current
- temperature
- chemicals
- corrosion
- tubular leaks.

10.1.4 4D Caliper: Wellbore Geometry in 4D

The wellbore shape significantly impacts many well operations, such as cement volume, casing running, and wellbore friction. Traditionally, wellbore geometry is determined using 2- and 4-arm calipers. Recently, nanoscale impulse radar technology has been implemented in a 9½" 4D Caliper Logging-While-Drilling tool. This technology is still under development and is expected to evolve further in the future.

The 4D Caliper technology allows for the determination of the borehole's shape and size while tripping in and/or out without downhole rotation. This capability helps us understand the 4D (time-lapse) development of the borehole and provides the latest borehole shape and volume before subsequent operations, such as running liners or casing and cementing. It also addresses wellbore and drill string integrity (Figure 10.16).

A novel ultra-wide-band impulse radar (electromagnetic)-based technology is available in the industry, implemented in a 9½" O.D 4D Caliper™ Logging-While-Drilling product. This technology is utilised in 4D Borehole Imaging Services to acquire and derive borehole shape and size over time, allowing for time-lapse imaging. This system uses nanoscale impulse radar sensors sampled at 400 times per second and combines data from multiple passes (trip-in, drilling, and trip-out) to deliver precise borehole shape and size measurements along with true 4D time-lapse imaging.

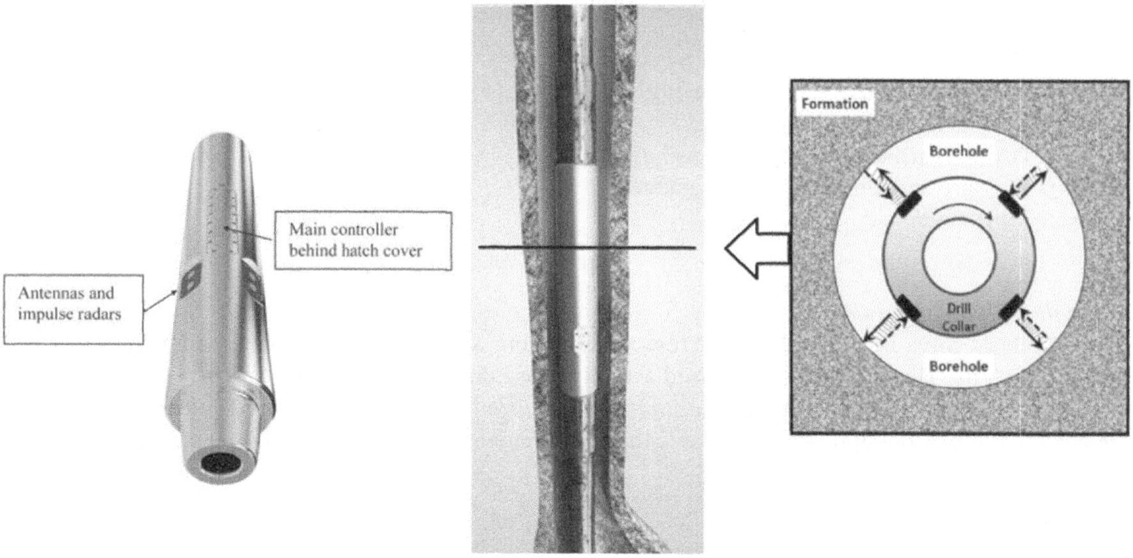

FIGURE 10.16 The 4D caliper, as part of the drilling BHA, images and maps the borehole for shape and size purposes while rotating.

Source: Adapted and revised from Øy et al. (2023).

10.1.4.1 What is the 4D Caliper?

The 4D Caliper is a cutting-edge tool for drilling operations that provides detailed 3D images of the borehole. It helps drillers avoid sidetracks and reduce non-productive time, enhancing the efficiency and effectiveness of drilling processes.

10.1.4.2 Why Use the 4D Caliper?

The 4D Caliper offers several benefits:

- **Optimises drilling and tubular run efficiency**: By providing a 3D image of the borehole's exact shape, it helps drillers avoid sidetracks and reduce non-productive time.
- **Improves drilling performance**: It identifies critical issues like key seatings, cutting beds in horizontal or highly deviated parts of the well, spiral patterns, and ledges inside the hole. These insights ensure better drilling results and more precise hole cleaning.
- **Enhances well integrity and reduces cement operation costs**: Accurate measurements of cement volumes lead to more efficient and cost-effective cementing operations.
- **Facilitates better slot recovery and P&A decisions**: With precise measurements of casing wear and eccentricity, it helps make informed decisions for slot recovery and plug and abandonment (P&A) operations.

10.1.4.3 Examples from Field Measurements

The 4D Caliper reveals important features like key seatings, spiral patterns, washout zones, and cutting beds, which are crucial for understanding the borehole's condition. Here are some examples:

- **Good borehole shape**: A well-maintained borehole has a consistent diameter with very few deviations, indicating a precise and stable drilling process. The detailed images from the 4D Caliper help ensure the borehole stays uniform, reducing the risk of structural problems and improving overall drilling performance. (Figure 10.17)

FIGURE 10.17 Good borehole shape.

Source: Adapted and edited from WELL ID AS (2025).

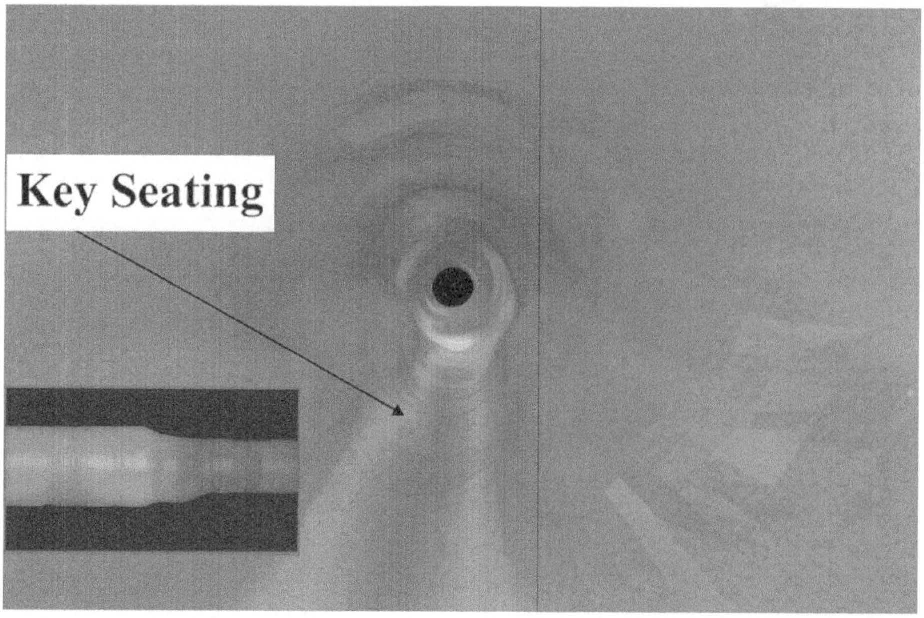

FIGURE 10.18 Key seatings.

Source: Adapted and edited from WELL ID AS (2025).

- **Key seatings**: These are areas where the drill string might get stuck, causing delays and operational issues. The 4D Caliper helps detect these areas early, allowing for timely corrective actions to ensure smooth drilling operations. (Figure 10.18)
- **Spiral patterns**: Spirals indicate minor deviations in the hole shape, which can affect the stability and integrity of the well. The 4D Caliper helps monitor and minimise these deviations, maintaining a more consistent borehole shape. (Figure 10.19)

- **Cutting beds**: Found in horizontal or highly deviated parts of the well, cutting beds can lead to poor hole cleaning and an increased risk of stuck pipe. The 4D Caliper provides detailed images that help identify and manage these cutting beds effectively, ensuring better drilling performance. (Figure 10.20)
- **Washout zones**: These are areas where the borehole diameter has increased due to erosion, compromising the well's stability. The 4D Caliper's precise imaging helps detect and address these washout zones, ensuring a more stable and secure borehole. (Figure 10.21)

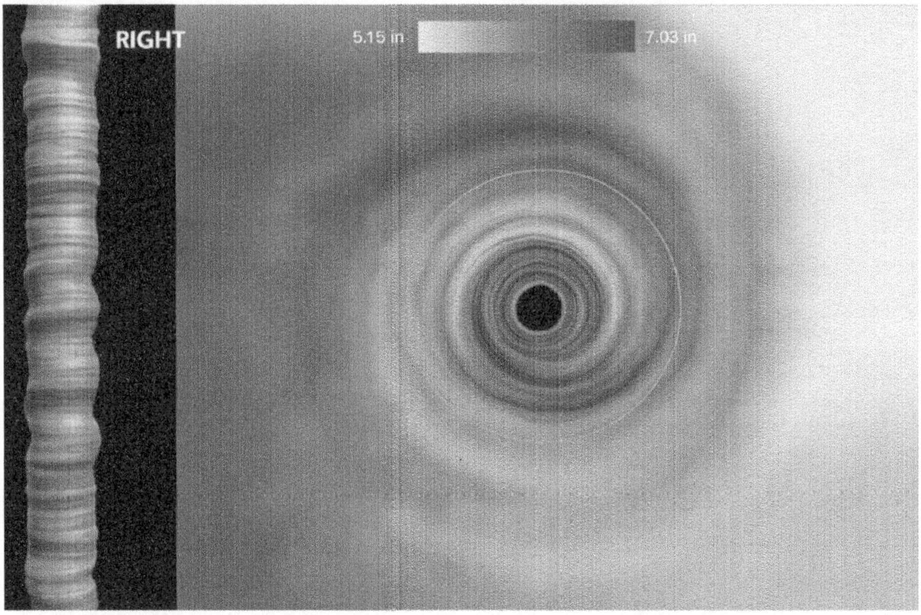

FIGURE 10.19 Spiral patterns.

Source: Adapted and edited from WELL ID AS (2025).

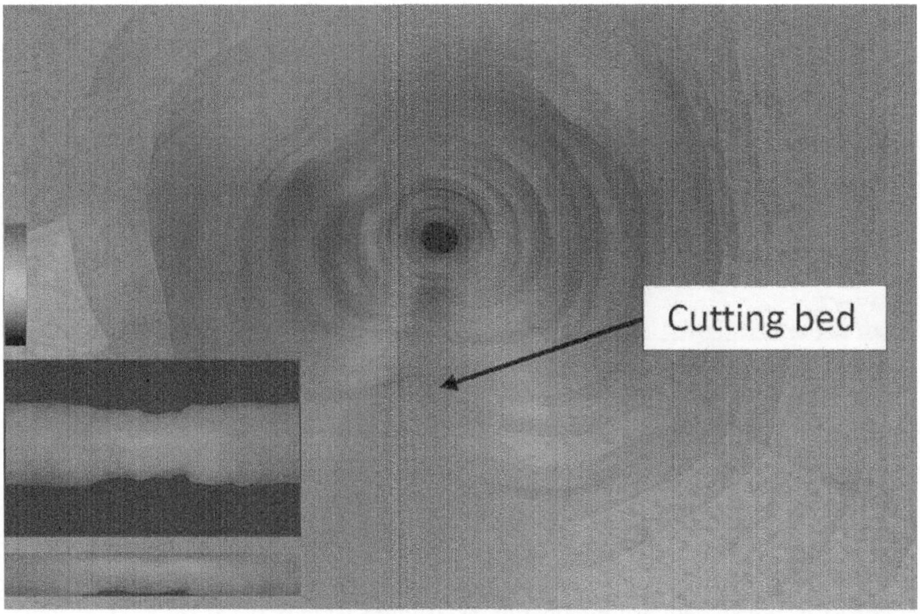

FIGURE 10.20 Cutting beds.

Source: Adapted and edited from WELL ID AS (2025).

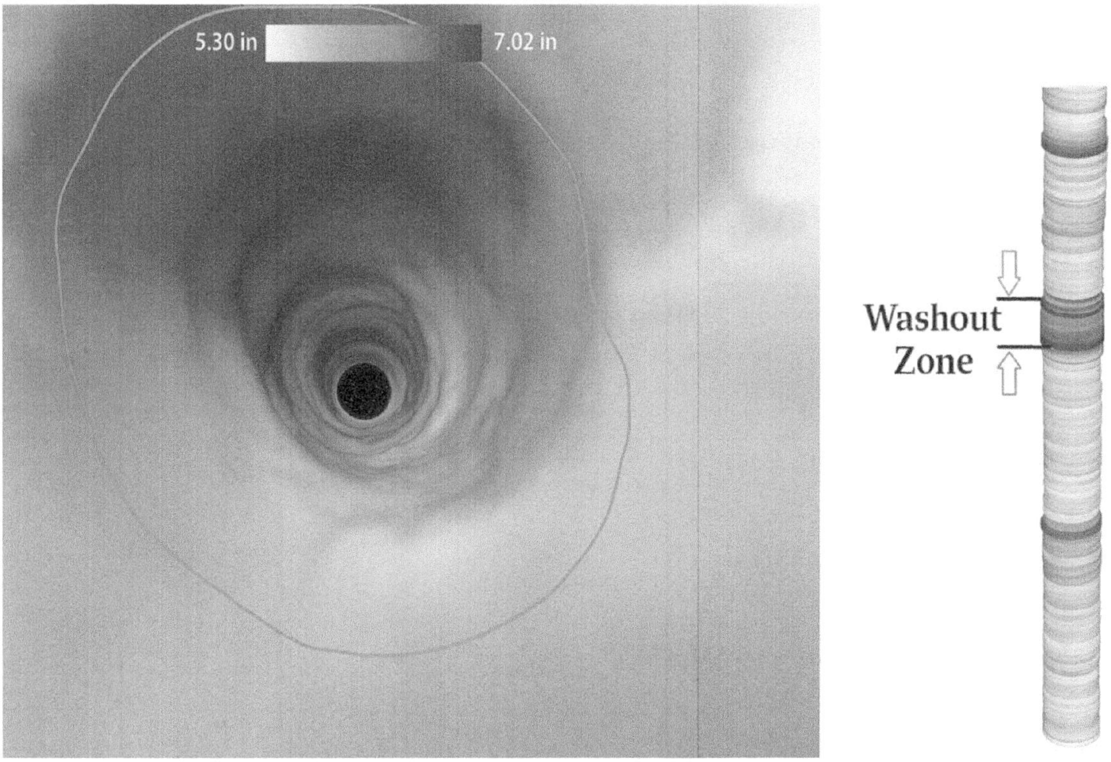

FIGURE 10.21 Washout zones.

Source: Adapted, combined, and modified for illustrative purposes from WELL ID AS (2025).

10.1.4.4 4D Caliper Placement Scenarios

The 4D Caliper tool can be placed in various positions within the drill string to optimise borehole imaging and data collection. Refer to Figure 10.22 for an illustration of different placements of the 6-¾" O.D Tool and 9½" O.D Tool in the drill string (along-string) and top of BHA placement.

As technology advances, various companies are developing new tools and techniques to enhance drilling operations. Accurate data from these tools can be integrated into software to create precise advisory systems for drilling operations. Based on this advisory system, we can update well designs using actual hole enlargement ratios for different models, such as hydraulics and hole cleaning models. This is crucial during the planning phase, as it allows us to simulate wells with real data from 4D Caliper technology in offset wells, rather than relying on theoretical hole sizes.

10.2 AUTOMATED MUD SYSTEMS

Recently, a major international operator rolled out a remotely controlled, real-time drilling fluids monitoring system for both onshore and offshore operations. This automated mud system, which includes several components, marks a big step forward in the drilling industry.

We won't get into the technical details, but real-time fluid analysis and visualisation during operations are key features. It is important to note that this technology is still evolving, with various companies working on improvements. One of the commercialised technologies in the market is introduced as follows:

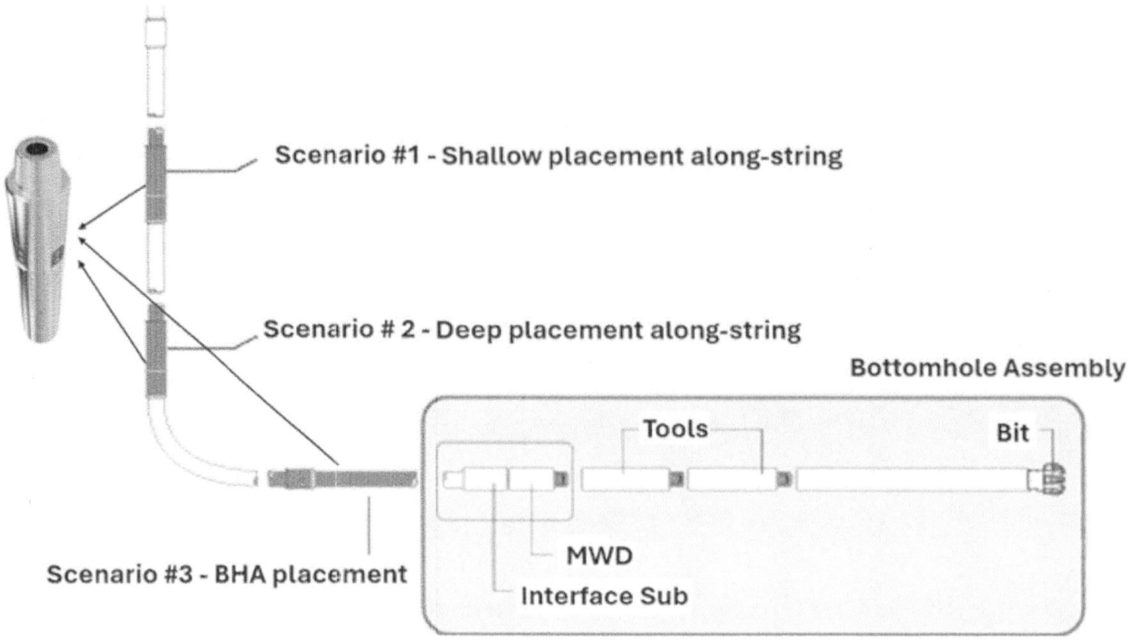

FIGURE 10.22 4D Caliper placement scenarios in drilling string.

Source: Adapted and revised from WELL ID AS.

- Real-time fluid analysis for optimal well operations.
- Adds more instruments for comprehensive fluid analysis.
- Monitors fluid processes in real-time via IoT.
- Sends automated fluid reports via email.
- Uses AI to provide real-time advice based on fluid data.
- Optimises fluid management using integrated data.

The real-time fluid unit continuously measures drilling fluid rheology and density, providing real-time data to drilling operations. The unit performs automated real-time mud analysis of the following:

- Rheology (Acc. to API Fann 35 readings)
- Density
- Gel Strength
- Temperature to provide standard conditions

Each real-time fluid unit skid requires a mud supply and a mud return. The mud supply flows via a mud strainer basket hole. This strainer is supplied as part of the installation.

Real-Time Fluid Analysis Construction Options:
Real-time fluid analysis consists of piping and instrumentation combined into an integrated frame (skid), semi-integrated frame, or custom installation for online, real-time, and continuous automatic measurement.

1. Framed skid: A fully assembled unit with all instrumentation, valves, pipework, and local cabling. Requires good material handling access and floor area due to its size. Needs to be lifted into place and hooked up.
2. Semi-integrated skid: Comes in partially assembled parts, easier to move due to smaller size. May require custom design engineering and on-site assembly.

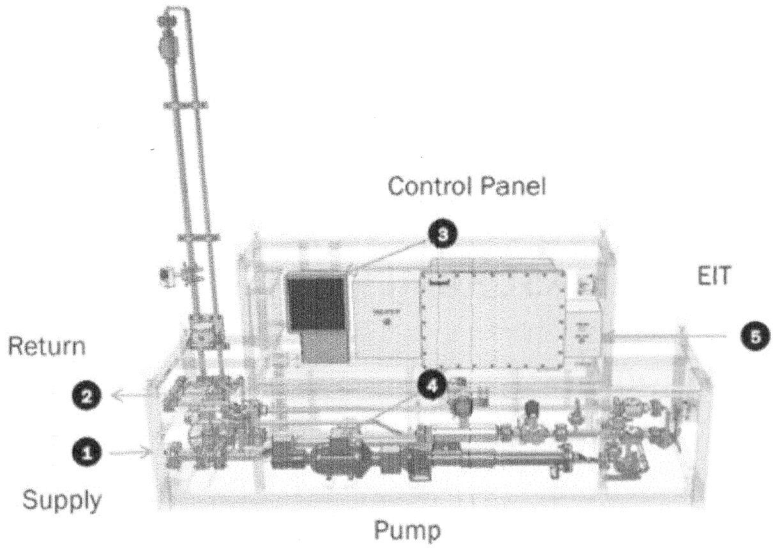

FIGURE 10.23 Automated mud systems technology: framed skid.

Source: Revised figure from Intelligent Mud Solutions (IMS Technology).

3. Integrated skid: Used where material handling access is restricted, or floor space is limited. Not pre-assembled; parts must be installed on site, typically against a bulkhead for support. Requires custom design engineering.

Hook-up connections to the fluid process are typically via 3-inch supply hoses for inflow and 1-inch hoses for return flow. Permanent installations use hard piping. There is also a 1-inch drain and temporary 1-inch purge and flush lines with quick connections.

Control System:

A control system is provided alongside the real-time drilling fluids monitoring system, consisting of:

- Control cabinet
- Human Machine Interface (HMI)

This can be an Ex-certified control system with explosion-proof and increased safety cabinets, or a control system for installation in a non-hazardous location, which is proposed for these installations.

The real-time product line offers several advantages over the conventional FAN 35 system. These benefits include easy installation, efficient operation, and compatibility with various fluids. Here are some of the key benefits:

- **Plug-and-Play installation:** Easy setup with 3" supply and 1" return.
- **Sample volume:** Less than 60 L/min.
- **Autonomous operation:** Fully self-operating system.
- **Fluid compatibility:** Works with OBM, WBM, and alternative fluids.
- **Particle tolerance:** Handles particles up to 6 mm.
- **Closed loop system:** Ensures a fully enclosed process.
- **Remote capabilities:** Includes remote support, monitoring, troubleshooting, and software updates.

Online real-time software provides an easy-to-use digital fluid analysis platform Accessible worldwide on any device. This intuitive visualisation of fluid measurements helps users make valuable decisions.

The digital platform offers:

- Real-time trending and logging of data from any fluid analysis sensor.
- Autonomous fluid analysis reports with automatic distribution to pre-identified receivers.
- Advisory AI helps fluid specialists optimize fluid performance.

10.3 WIRED PIPE AND ALONG-STRING MEASUREMENT SYSTEM

10.3.1 Introduction

Wired pipe telemetry (WDT) provides a significant advantage over conventional mud pulse telemetry (MPT) by enabling high-speed, continuous data transmission through the drill string exceeding 57,600 bits per second (Foster et al., 2018). In contrast, MPT transmits data at a much slower rate of 20 bits per second (Saleh et al., 2019) and experiences interruptions when rig pumps are turned off.

Figure 10.24 illustrates the WDT tool and the data transfer setup. The WDT system typically consists of five main components (Schils et al., 2016).

- *Surface network controller:*
  - The network controller securely transmits tool commands and real-time data to multiple users from a local interface. Surface cabling, specifically designed for rig operations, includes all required cables and junction boxes for installation.
  - Installed throughout the rig structure, this cabling ensures data transmission from the data swivels to the surface network.

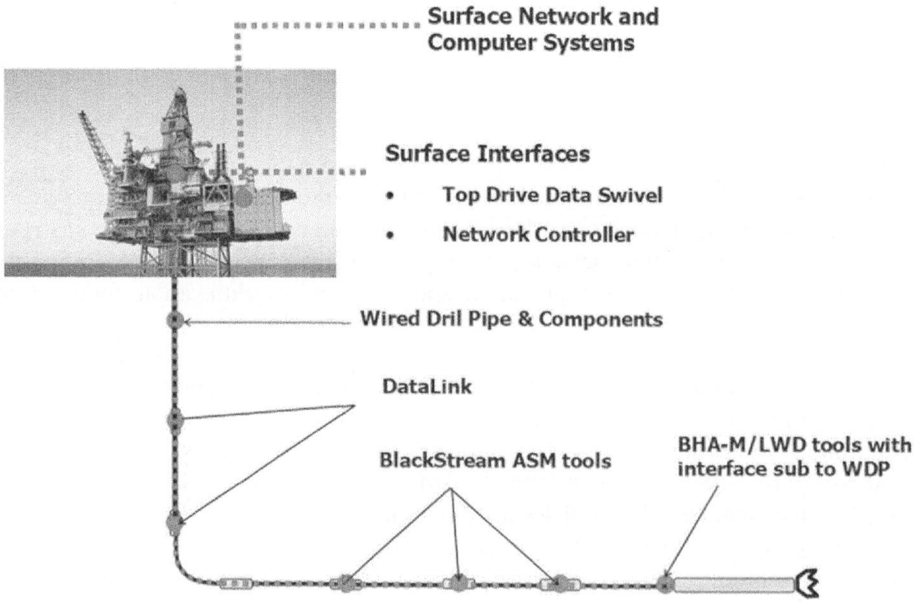

FIGURE 10.24 An illustration of the typical WDT system including BlackStream ASM tools and data transfer.
Source: Revised from Schils et al. (2016).

- *Top drive swivel:*
    - The top drive swivel serves as the interface between the stationary system at the surface and the rotating system at the bottom of the top drive assembly. It consists of a sub, which facilitates the transmission of a network signal into the swivel, allowing for continuous communication between the stationary and rotating components of the system.
- *Wired drill string components (including all pipe and auxiliary string components):*
    - While wired pipes resemble standard drill pipes, they incorporate additional components that enable data transmission between the surface and downhole tools. Each component of the drill string network includes two electromagnetic coils connected by DataCable.
- *DataLink (data signal "booster"):*
    - DataLinks are downhole network electronics subs that repeat, filter, and amplify the signal along the string, ensuring a consistent and reliable data transmission to the surface.
- *Network interface sub (situated on top of the BHA):*
    - Interface between the MWD/LWD tools and the wired pipe network.

The NOV BlackStream™ Along-String Measurement (ASM) tool is a collar-based, downhole drilling dynamics measurement system. Its compact and versatile design allows for flexible placement along the drill string, enabling the collection of high-frequency data such as temperature, annular pressure, rotational speed, and three-axis vibration at regular intervals.

Designed to integrate with the IntelliServ™ networked drill string, the ASM transmits data to the surface in real time via the IntelliServ™ high-speed wired drill pipe telemetry network. Powered by batteries, ASM tools operate independently of fluid and flow conditions, enabling continuous and reliable data acquisition throughout the drilling process.

Wired pipe telemetry enables real-time data extraction from advanced systems such as MWD, LWD, and rotary steerable systems. This advancement enhances Bottom Hole Assembly (BHA) control, geosteering, and drilling dynamics with greater precision. The continuous and high-speed data transmission via the WDP allows operators to make more informed decisions, optimising drilling performance and safety throughout the drilling process.

10.3.2 Wired Pipe Telemetry Application

Field case studies have demonstrated that the successful implementation of WDT technology can significantly enhance drilling operations by providing a range of operational and performance benefits. The main advantages observed in the field include (Edwards et al., 2013; Schils et al., 2016; Teelken et al., 2016; Sehsah et al., 2017; Israel et al., 2018):

- Improved wellbore placement
- Increased drilling rates by ROP optimisation
- Decreased downhole tool failures
- Optimised hole cleaning
- Detect formation ahead of the bit
- Improved UBO and MPD operations
- Controlled downhole WOB
- Improved real-time downhole monitoring
- Early kick detection and wellbore stability controls
- Faster detection of swab and surge pressures
- Fluid/mud level detection
- Reduction in wireline operational requirements
- Increased reservoir exposure
- Loss event detection

WDPT reduces non-productive time (NPT), which saves the rig rate by the following time-saving categories:

- Data transmission
- Drilling performance
- Bit run reduction (decrease in tripping jobs)
- Hole cleaning
- Trip speed optimisation

10.4 DEVELOPMENT TOWARDS AUTONOMOUS DRILLING

Drilling automation has seen significant progress over the past few years, marking a substantial shift from traditional drilling methods. This shift focuses on the dynamic adjustment of parameters in response to downhole conditions in real time, utilising sensors, and automation. While a comprehensive discussion of all advancements would require extensive chapters, which is beyond the scope of this section, we will highlight key advancements in robotisation, auto drilling, and computer vision, focusing on their core concepts.

10.4.1 Robotisation, Automation, and Autonomous Control

In most cases, robotics is a branch within the broader field of automation, focusing specifically on physical robots. It incorporates multiple disciplines to design, build, program, and use robotic machines. These robots can transform the way we work. Let's look at three types of robots: industrial robots, service robots, and humanoid robots.

- **Industrial robots** automate tasks in factories, are precise and stationary, and are used in welding, assembly, and material handling.
- **Service robots** perform tasks in public spaces and homes, interact with humans, like robot vacuums and greeters.
- **Humanoid robots** resemble humans and mimic emotions, like Sophia, with AI to recognise and respond to gestures.

Generally, the goal of robotics is to automate tasks or processes. Some companies in the drilling industry offer automated systems for tubular and tool handling, suitable for both onshore and offshore drilling rigs, including new builds and retrofits. By integrating robotic components with system automation, operators can enhance efficiency and performance. The full robotic system automates repetitive tasks on the drill floor, improving safety and consistency.

Key components include (Figure 10.25):

- **Drill Floor Robot**: Automates pipe and tool operations.
- **Robotic Roughneck**: Fully electric, adjustable torque wrench, and backup-tong.
- **Robotic Pipe Deck Handler**: Hands-free pipe handling from deck to well centre.
- **Robotic Pipe Handler**: Flexible robot for tubular handling.
- **Robotic Pipe Feeder**: Fast, precise tubular handling from storage to well centre.

10.4.2 Next-Generation Auto Driller

The primary purpose of the Auto Driller, developed in the late 20th century and enhanced over time, is to maintain set points for weight on bit (WOB), differential pressure, torque, and rate of penetration (ROP). It

FIGURE 10.25 Overview of robotics automated drilling components and systems.

Source: Adapted and compiled from Canrig Robotic Technologies AS.

continuously monitors and adjusts these parameters, providing operators with enhanced visibility through its high-resolution display.

The main goal of the Next-Generation Auto Driller is to maintain stable control of drilling parameters, reducing time per stand and maximising single-run sections. Variations in ROP, WOB, differential pressure, torque, and rotary speed can damage the bit, motor, and BHA, diminishing their performance and lifespan. Initial efforts focused on identifying the optimal set point combination to enhance stability and net ROP. Although this approach yielded some improvements, the system remained unstable with formation changes. Consequently, the auto driller's tuning parameters were adjusted to enhance system stability across various formations.

10.4.3 Establishing the Safe Operating Envelope for Digital Drilling

The Safe Operating Envelope defines the limits for key drilling parameters to avoid dysfunction and achieve desired outcomes. It is essential for maintaining drilling efficiency and safety by ensuring operations remain within safe and optimal parameters.

Digital transformation in drilling has led to the development of various applications and algorithms to enhance drilling efficiencies, detect and interpret drilling events, and manage dysfunctions. However, different solutions often target similar problems using diverse methodologies, which can lead to conflicting recommendations. For example, a vibration advisor might recommend increasing RPM and reducing WOB to mitigate stick/slip vibration. Normally, these RPM and WOB adjustments are temporary measures to resolve vibration issues and do not pose significant risks to ECD or hole cleaning. Higher RPMs can actually improve cutting transport. Managing vibration is a conventional method with varying approaches across the industry, and it is not yet standardised.

The Safe Operating Envelope helps harmonise these conflicting recommendations by defining operational limits that ensure safe and efficient drilling. It provides a framework for integrating multiple advisory systems, ensuring consistent and optimal drilling performance. In summary, the Safe Operating

FIGURE 10.26 Overview of automated computer vision system.

Source: Revised from Svendsen et al. (2025).

Envelope is crucial for maintaining drilling efficiency and safety. It helps in defining the operational limits and ensuring that the drilling process remains within safe and optimal parameters.

10.4.4 Automated Computer Vision System

The Automated Computer Vision System is a technology that enhances drilling operations through digital shaker surveillance. It leverages advanced computer vision to provide real-time insights into shaker performance, cuttings recovery, and wellbore stability (Svendsen et al., 2025).

Key components include (Figure 10.26):

- **Explosion-Proof Cameras**: These cameras capture real-time images in hazardous environments, ensuring reliable performance and safety compliance.
- **Edge Computer Vision Technology**: This technology processes images locally at the rig site, providing quick and reliable information about shaker load and cuttings volume.
- **Driller's Smart Screen**: A hybrid dashboard that displays real-time data on shaker load, hole cleaning metrics, and other critical parameters, enhancing situational awareness.
- **Real-Time Alerts**: The system offers 24/7 monitoring and alerts, reducing the need for crew visits to the shakers and improving safety.

The system captures images of cuttings from the shakers using explosion-proof cameras. These images are processed locally using edge computer vision technology to measure shaker load and detect anomalies. The processed information is then displayed on a smart screen, providing the drilling team with real-time insights. This helps optimise drilling performance, ensure wellbore stability, and support safety practices.

Benefits include:

- **Enhances Situational Awareness**: Provides real-time insights into shaker performance and wellbore stability.
- **Optimises Drilling Performance**: Monitors and improves cuttings recovery and shaker load distribution.
- **Ensures Safety Compliance**: Reduces exposure to hazardous zones and supports process safety best practices.
- **Reduces Environmental Impact**: Improves drilling efficiency, reduces emissions, and minimises non-productive time.

Enhancing ROP Strategies

11

11.1 INTRODUCTION

When deviated offshore drilling started in the 1980s a number of critical elements were identified. Some of these are wellbore stability, wellbore friction, and hydraulics. To handle wellbore cleaning a number of experimental flowlines were built in several countries. Around 1990 experimental data were used to design wellbore cleaning models. API RP 13D used in Section 2.4 gives a good summary. Mainly the outcome was how much drilled cuttings the wellbore could handle before the annulus packed off. This was coupled to a model that coupled cuttings production with ROP. API RP 13D provides a number of graphs where pump rate and ROP are selected based on drilling fluid properties, wellbore inclination, and wellbore dimensions. In addition, the standard also summarises the elementary physics around drilled cuttings. Although the standard is old, it has been updated a number of times and is giving a correct picture of the process.

A typical drilling rate in the North Sea averaged 50–60 metres per hour in the 12¼" section for decades.

Instantaneous ROP is measured over a short period or distance while drilling is in progress, providing a snapshot of the drilling performance at a specific moment. Average ROP is calculated over a longer interval, typically the entire section drilled, giving a broader view of the overall drilling efficiency.

This low ROP was used because of concerns about plugging the annulus with stuck drill string as a result. This ROP is also reflected in the API RP 13D standard. We can say that the ROP was held back to avoid wellbore problems such as wellbore fill and stuck pipe.

Over the years much more experience has been gained, and the older drilling practice was challenged. Drilling practices and routines were improved, drilling muds have been improved, and modern drilling rigs have more mud pumps and improved mud handling systems to mention a few. Basically, the operators gradually increased ROP and observed that with better control they did not get the problems they expected. Today the operators can drill at 100–200 metres per hour in 17½–6½" and 200–300 metres per hour in the 12¼" section without major problems. This improvement in ROP is basically obtained because technical and operational experience obtained built confidence by gradually challenging the limits and established truths. The earlier API RP 13D was used assuming a high risk, which is now reduced by practical experience.

Achieving high ROP (Rate of Penetration) is influenced by a combination of factors, including drilling practices, formation type and compressive strength, mud properties, hole cleaning efficiency, bit design, shaker capacity, surface cuttings capacity, directional plan, and connection speeds and others. Each of these elements plays a crucial role in optimizing drilling performance and efficiency.

While each element plays a crucial role, for an operation to achieve high ROP successfully it also means that the following operations achieve the required objectives; meaning that one enables successful running and cementing of casing/liner. To enable this, all of the "key elements" must be designed for.

The second objective for increased ROP is to minimize open hole exposure time to ensure that operations are successful by minimizing associated/consequential risks when running and cementing casing.

In the following, we will give a short review of the evolution in drilling practice and equipment that has allowed us to drill at high ROP.

DOI: 10.1201/9781003618171-11

The primary objective of maximizing the ROP in drilling operations is to achieve fast and efficient drilling while maintaining control over risks and ensuring the acquisition of high-quality data. While high ROP reduces daily/rental cost it increases the rate at which equipment is used and/or consumed, typically increasing the average daily cost for an operation, although total cost for a well is reduced. Subsequently, by delivering a well for less cost consistently can enable the sanctioning of new projects and wells targeting lesser volumes. This becomes more and more relevant in the age of more mature fields and smaller discoveries.

Figures 11.1 and 11.2 illustrate the potential time savings on a typical ERD and MLW project involving several wells by increasing the drilling speed to 75 metres per hour in the 12¼" section and optimizing the ROP in the reservoir section. Assuming a total of 97,000 metres to be drilled in the reservoir and 200,000

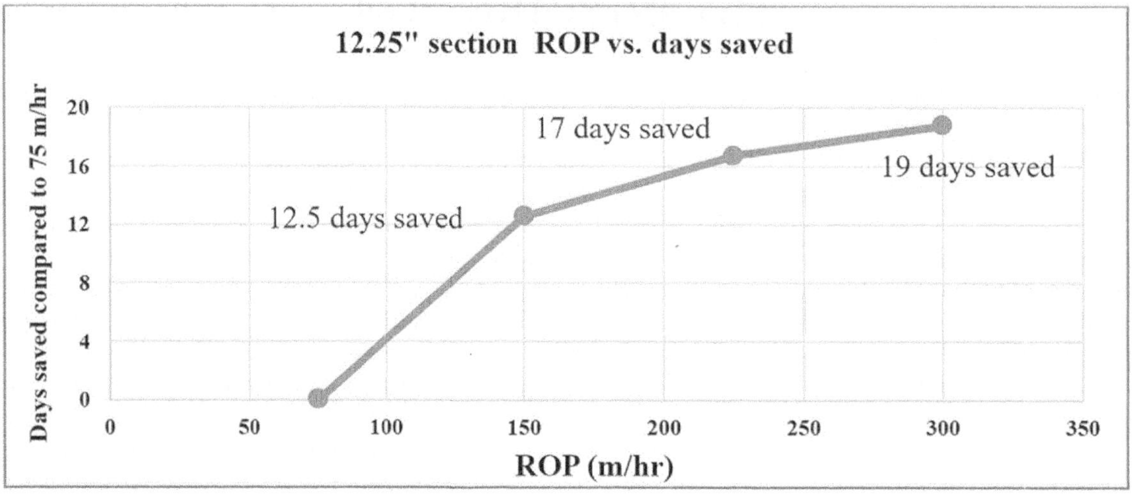

FIGURE 11.1 Impact of increasing drilling speed to 75 metres per hour on project duration in the 12.25" section.

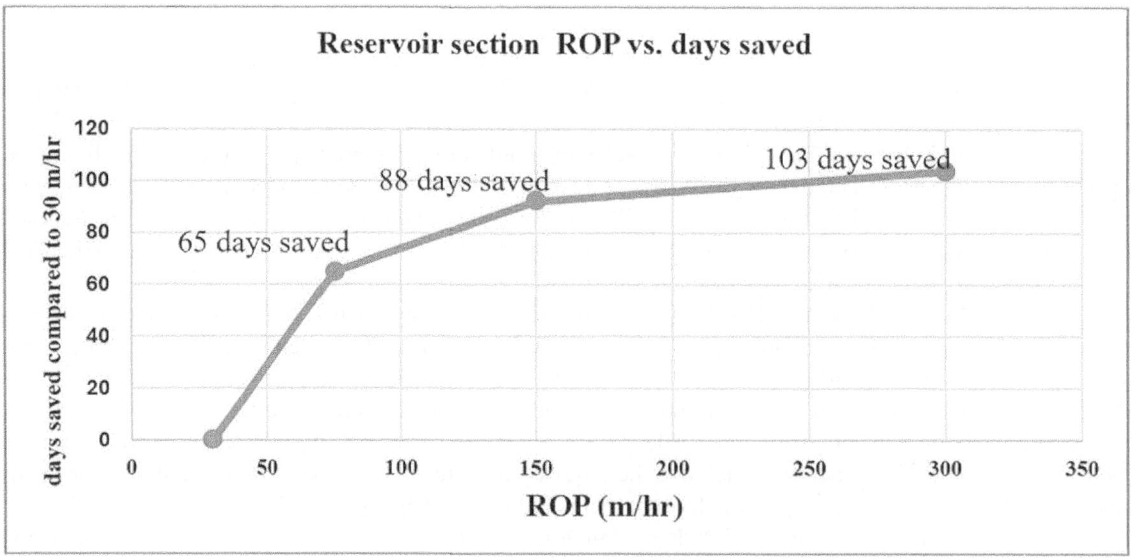

FIGURE 11.2 Impact of increasing drilling speed to 30 metres per hour on project duration in the reservoir section.

metres in the overburden, drilling at 300 metres per hour instead of the limited speeds of 75 metres per hour in the overburden and 30 metres per hour in the reservoir can save approximately 122 days.

However, it is important to note that while higher absolute ROP numbers can typically be achieved in the 12¼" section, the incremental gain may be more significant in both reservoir sections and larger hole sizes. For example, on a typical project, the 16.5" section is about 2–3 km long, while the 12¼" section is 1–1.5 km. Despite this, the section where the longest drilling time is spent is the reservoir, even though it is the shortest. Therefore, an incremental improvement in reservoir drilling can have a bigger impact on financial aspects rather than just maximizing the ROP value itself.

For instance, increasing the ROP from 150 m/hr to 200 m/hr in the 12¼" hole section has a much smaller impact compared to increasing the ROP for example from 10 m/hr to 60 m/hr in the 8½" hole section. This demonstrates that focusing on incremental improvements in reservoir sections can be more beneficial overall.

11.2 TIME VERSUS DEPTH ANALYSIS AND RATE OF PENETRATION (ROP) CALCULATIONS

A time versus depth analysis is essential for organizing and executing well construction activities efficiently. It breaks down the drilling process into structured phases, repeatable actions, and context-dependent tasks, ensuring everything runs smoothly and safely (see Table 11.1).

Figure 11.3 illustrates the sequential phases and runs involved in a drilling operation. Each phase represents a significant stage in the well-construction process, with specific drilling and casing runs.

The green line in Figure 11.3 shows the bit depth during the drilling operation. This line is key for understanding how the drill bit progresses through the subsurface layers. By tracking bit depth, operators can monitor drilling efficiency and effectiveness in real time.

Why the time versus depth graph matters:

- Monitoring Progress: It visually shows how fast the drill bit is moving, helping spot any delays or issues.
- Optimizing Drilling: By analysing bit depth with other parameters like ROP, operators can make better decisions to speed up drilling.
- Spotting Issues: Any deviation in the green line can signal potential problems that need fixing to stay on track.

TABLE 11.1 Time versus depth analysis levels

LEVEL	*DESCRIPTION*
Well Level	The analysis covers the entire drilling project, from initial planning to well completion. This includes site analysis, obtaining permits, and selecting the right equipment.
Phase Level	Each phase represents a major stage in the drilling process, from when equipment enters the borehole to when it's retrieved. Phases are goal-oriented and may span days or weeks, depending on depth, geology, and technical challenges.
Run Level	Runs are specific sequences of activities within each phase, forming the building blocks of daily operations. They are repeated cyclically and are essential for progressing through each phase efficiently.
Tasks	Tasks are specific activities performed during actions, adapting to real-time conditions. They ensure that each action is executed effectively and safely.

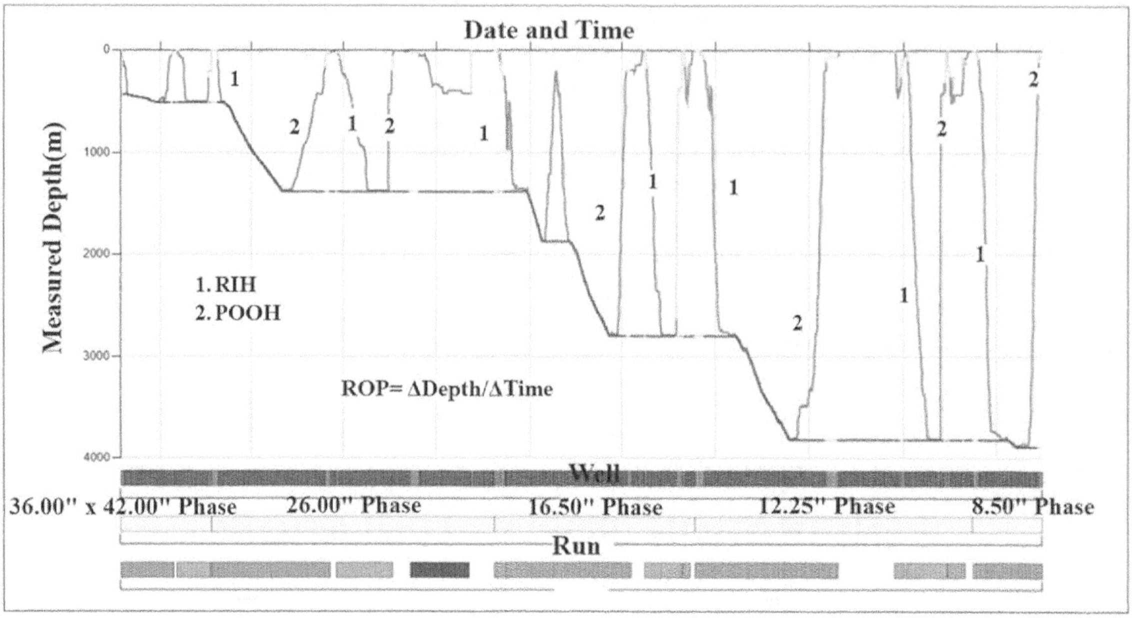

FIGURE 11.3 Time versus depth for various phases and runs in the drilling operation.

- Reducing Uncertainty: Accurate bit depth tracking helps ensure the drilling follows the planned schedule and avoids unexpected challenges or manages them appropriately.
- Improving Efficiency and Safety: Using this graph in analysis helps operators understand drilling performance better and make necessary adjustments for improved efficiency and safety.

In a comprehensive study by Dupriest et al. (2011), 40 factors that limit the ROP in drilling operations were identified. These factors are crucial for maximizing drilling performance and include:

- Cuttings Handling: Efficient removal of cuttings to prevent regrinding and bit wear.
- Logging While Drilling (LWD): Ensuring accurate data collection without hindering drilling speed.
- Hole Cleaning: Maintaining a clean borehole to avoid blockages and reduce drag.
- Vibrations: Minimizing vibrations to protect equipment and maintain steady drilling.
- Bit Balling: Preventing the accumulation of cuttings on the bit, which can reduce cutting efficiency.
- Inadequate Depth of Cut: Ensuring the bit cuts deep enough to maximize penetration without excessive wear.
- Bottomhole Balling/Chip Hold Down Effect: Managing the accumulation of cuttings at the bottom of the hole to maintain effective drilling.
- Tool Steering: Accurate steering to maintain the desired well path and avoid unnecessary corrections.
- Weight on Bit (WOB)/Hard Formation: Applying the right amount of weight to the bit, especially in hard formations, to maximize drilling speed without damaging the bit.

Following the identification of crucial factors limiting the ROP such as efficient cuttings handling, maintaining a clean borehole for effective hole cleaning, minimizing vibrations, preventing bit balling and bottomhole balling, ensuring adequate depth of cut, accurate tool steering, and appropriately applying WOB in various formations, the implementation of real-time drilling optimization tools becomes paramount. Figure 11.4, titled the Modified stability chart (Mirani–Samuel stability plot) or ROP optimization chart, offers a dynamic and comprehensive approach to enhance ROP by visually addressing many of these limitations within a RPM–WOB domain.

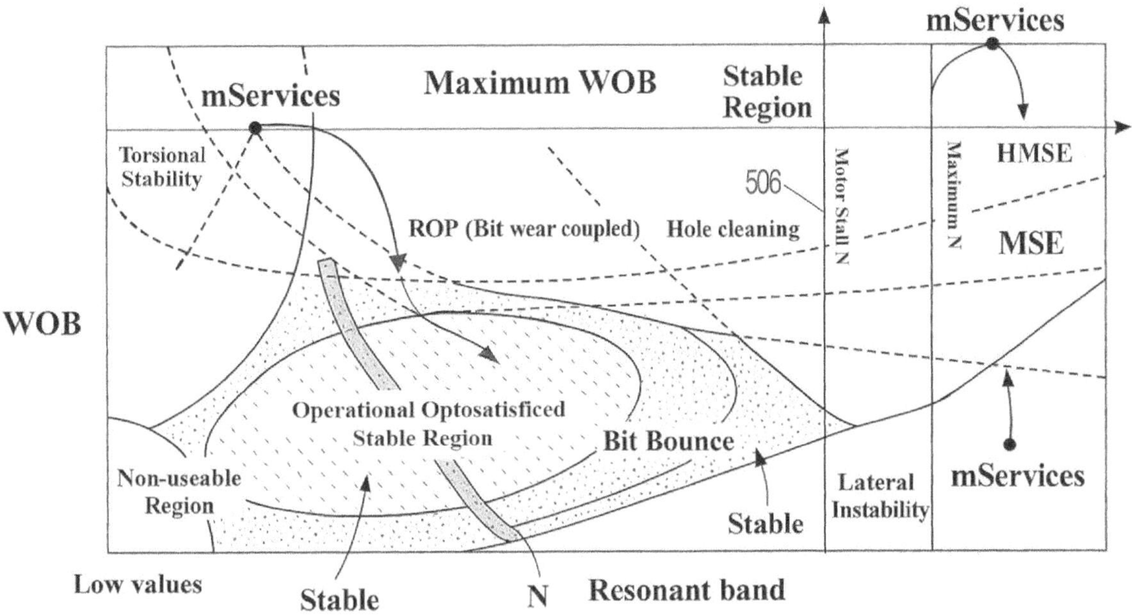

FIGURE 11.4 Revised stability chart (Mirani–Samuel stability plot) or ROP optimization chart.

This chart, driven by interconnected microservices, provides a real-time representation of the dynamically stable operating region. By operating within this stability envelope, drillers can actively mitigate vibrations, including torsional (stick-slip) and lateral (whirl), which are significant hindrances to efficient drilling and can cause equipment damage. The torsional and lateral lines on the chart clearly demarcate unstable zones, guiding operations towards smoother and more effective drilling. Similarly, avoiding the critical RPM bands for resonant frequency and bit bounce minimizes structural instability and axial vibration, contributing to a more consistent ROP. Furthermore, Figure 11.4 aids in optimizing the application of WOB in relation to formation hardness. The chart, particularly the ROP line, helps identify effective WOB and RPM combinations for achieving a desirable ROP while considering bit wear. Operating below the ROP line might offer a higher initial ROP but at the cost of excessive bit wear, whereas operating significantly above it leads to diminishing ROP gains and the "flounder point". The goal is to find the optimal balance within the stable region.

The holecleaning line on Figure 11.4 directly addresses the crucial aspect of maintaining a clean borehole. Operating above this line indicates insufficient flow rate for effective cuttings handling, potentially leading to bit balling and reduced penetration. Staying below this line ensures that cuttings are adequately removed, allowing the bit to continuously engage with fresh formation.

The concept of the optosatisficed space (a "good enough" optimal range that balances multiple objectives under real-world conditions) within Figure 11.4 is central to this enhanced approach. It represents the operating parameters that are "good enough" optimal, balancing the competing factors of ROP maximization, dynamic stability, bit wear, and hole cleaning. By aiming to operate within this optosatisficed window, drillers can achieve a significantly improved ROP while proactively managing the various limitations identified in comprehensive studies, leading to more efficient and cost-effective drilling operations.

11.3 RIG ACTION PLAN (RAP)

A Rig Action Plan (RAP) is an essential and critical link between planning and execution phase of an operation. Today, these procedures, which include instructions and recommendations to a rig crew for the upcoming activities, are typically provided to the driller in a PDF, or Word or through a digital RAP systems document. Every company has a different name for it: drilling procedure, standing instructions for driller, detailed operation procedure, plan of action, etc.

The RAP is a comprehensive document designed to outline the strategic approach and detailed steps necessary for the successful execution of drilling operations. This plan ensures that all team members are aligned and prepared to achieve high performance while maintaining safety and operational efficiency.

Key components of a rig action plan:

Objectives:

- Clearly defined goals for the operation, such as drilling specific sections of the well, performing formation integrity tests (FIT), and reaching target depths.

Well status:

- Current status of the well, including preparations, installations, and pressure tests performed.

Well control:

- Primary and secondary well control measures, including the use of specific drilling fluids and equipment.

Major operational risks:

- Identification of potential risks, such as losses below the casing shoe, unexpected formation integrity test values, and hole cleaning challenges.

Performance goals:

- Success factors, quality goals, and key performance indicators (KPIs) to measure the efficiency and effectiveness of the operation.

Preparations and equipment checklist:

- Detailed preparations required, including ensuring critical items are on board, priming and testing systems, and preparing contingency plans.

Operational steps:

- Detailed description of how to execute the operations.
- Clear operational limits on how to execute different operations to avoid unwanted events, such as BHA failure, losses, well control situation, etc.
- Additional notes, if relevant.

Attachments:

- Supporting documents such as section segment categorization, BHA tally, bit specifications, and various procedures and decision trees.

By following the RAP, the rig crew can execute the operation safely and efficiently, minimizing risks and ensuring the well objectives are met. The RAP also facilitates communication and coordination among different team members, ensuring everyone is aware of their roles and responsibilities.

Figure 11.5 (shown on the next page) presents a typical RAP template. This figure provides a visual overview of the wellbore structure, key operational parameters, and critical information required for the drilling operation.

Focusing on one part of the RAP for ROP through multidisciplinary teams from the planning phase to the execution phase, instead of purely working on ROP itself or focusing on a mathematical model of ROP, can be an effective approach. The primary issue in our planning and operations often stems from an over-reliance on mathematical modelling, which is based on limited low ROP data to achieve a good R-squared value without challenging the model and approach. Some research calculates R-squared values to demonstrate good predictions, which is useful for showing model performance. However, these models often do not address how much faster we can drill in certain formations.

Instead of focusing solely on mathematical models, we should consider how much we can improve drilling speed while accounting for all operational limitations. RAPs provide a systematic approach that transforms operational parameters such as ROP from luck to science by integrating practical, real-time adjustments, and continuous improvement.

While RAPs are a comprehensive framework designed to enhance drilling performance by systematically planning, executing, and monitoring drilling activities, they offer a structured method to enhance operational parameters in drilling. ROP is one of the important aspects of RAP that we can focus on effectively.

Here's why:

1. Pre-Drilling Preparation:
    - Formation Analysis: Review offset well data and identify "sweet spots" for ROP gains.
    - Tool Selection: Match bit type to formation hardness and confirm motor/turbodrill compatibility.
    - Hydraulics Design: Calculate optimal flow rates and nozzle sizes for hole cleaning.
2. Practical Application: RAP provides actionable steps that drilling teams can follow on-site, making it easier to implement and see immediate results.
3. Real-Time Adjustments: RAP allows for dynamic adjustments based on real-time data, which is crucial for maintaining optimal ROP.
4. Comprehensive Strategy: RAP integrates various aspects like bit selection, hydraulics, and vibration mitigation, offering a holistic approach to ROP enhancement.
5. Continuous Improvement: By reviewing and updating RAPs based on actual performance, you can ensure ongoing optimization and learning.
6. Operational Parameters Optimization: The RAP includes various attachments for optimizing operational parameters, such as hydraulic simulations, vibration analysis, torque and drag simulations, all of which are directly related to enhancing ROP.
7. Collaboration and Training: RAP fosters collaboration among the drilling team and ensures everyone is aligned with the optimization goals. This collective effort can be more impactful than individual calculations.

Table 11.2 outlines the ideal parameters for recommended flow rates and RPM for different hole sizes. These parameters are designed to optimize drilling performance, but it's important to note that they may not be achievable in every condition due to varying geological and operational factors.

While the table provides a good foundation for drilling operations, before discussing the RAP, it's important to emphasize that, compared to the past, we now have two key capabilities. Firstly, we have more advanced models for engineering simulations, and secondly, we have improved tools to visualize different perspectives of well engineering based on offset wells. The industry has made significant advancements that go beyond these ideal parameters. Modern tools such as hydraulic and hole cleaning models, along with advanced drilling practices and technologies, have enabled much higher ROP than those listed in the table.

Rig Action Plan No. XX				Alliance logo
RIG/UNIT		<Activity name>	DATE	DD.MM.20
WELL	[Well name]		REV	*0/1/2/3*
PAGE	1 of 4		STATUS	Set status...

Approved by:	Approved by:	Approved by:
...lDrilling Supervisor	Rig Contractor Drilling Section Leader	Offshore Service Lead

SIMOPS Overview

Area	Ongoing work / potential clashes
Main	
Aux	
Moonpool	
Subsea	
Deck	

Distribution	
Deviations	•
Objectives	•
Well status	• Ref. attached status schematic
Well control	• Primary: • Secondary: • Shearable item: • Last BOP test:
Main operational risks	•
References	•

Technical data and operational limits

Min. ID in system
Max. OD in system
Capacities
Relevant equipment specifications

Toolbox Talks

The inclusion of TBT within the RAP as an operational line have been removed. The intention of this is:

- *Avoid prescriptive instruction*
- *Empower those involved in the operation to take TBT at the most appropriate time and with correct content. Whenever possible, TBT held offline.*

Responsible	Preparations and equipment checklist
AD Sign:	BHA handling (optional for prep categories: use "Category" style) ☐ Elevator inserts ☐ Slips ☐ Etc... Other things ☐ Etc. ☐ Etc.2

Operations

1. Operation 1
   a. Step 1
      - Sub-step
        - Sub-sup-step (ask yourself: do you really need to use this?)
      - Value to be noted down: ________ kNm (torque, weight ...)

   b. Step 2
   c. Step 3

Note: *add notes relevant to this operational step*

2. Operation 2
   a. Step 1

Note:

Next operation

(Main:) RAP Y – <Activity name>

(Aux: Optional)

Attachments

1.	Tally
2.	BHA schematic
3.	Motor spec
4.	

FIGURE 11.5 Typical template for RAP.

TABLE 11.2 Ideal parameters for flow rates and RPM by hole size based on good old drilling practices

HOLE SIZE	*MAX GPM*	*MIN GPM*	*BEST RPM*	*MIN RPM*
16"–17½"	900–1200	800 gpm, with ROP at 20 m/hr (≈65'/ hr)	120–180	120
12¼"	800–1100	650–700 gpm, with ROP at 10–15m/hr (≈30–50'/ hr) 800 gpm, with ROP at 20–30 m/hr (≈65–100' / hr)	150–180	120
9⅞"	700–900	500 gpm, with ROP at 10–20 m/hr (≈33–65' / hr)	120–150	100
8½"	450–600	350–400 gpm, with ROP at 10–20 m/hr (≈33–65' / hr)	70–100	60

We now have access to a wealth of offset well data, which we can analyse to achieve more accurate and precise engineering simulations. By examining these offset wells, we can identify relevant wellbores based on distance and other filters, presenting key facts and challenges. This process allows us to explore detailed information for each offset wellbore in both 2D and 3D formats, summarise findings as an "experience" or "summary," and store the analysis for future reference and use in engineering simulations.

For example, for each engineering simulation, we need accurate data on formation properties, cutting size, cutting density, hole shape and size, zones prone to washout, and many other factors. If we have data on hole enlargement ratios in a specific area, we can incorporate this into our hydraulic model simulations. This kind of detailed information is incredibly useful for improving the accuracy and effectiveness of our engineering simulations.

11.3.1 RAP Framework and Process

The RAP framework and process are designed to streamline the selection and implementation of activities for drilling operations. While the RAP template includes a comprehensive list of relevant activities (Figure 11.6), it's important to note that this approach is specific to certain operators. There are systems that enable this way of working, but it is not universally adopted by all operators. Engineers can choose activities that align with their planned operations from these templates, ensuring efficiency and safety in their drilling processes.

The RAP framework optimizes operations by tailoring activities to the specific rig, well, and geological formations, ensuring efficiency and effectiveness.

Once finalized by the offshore leadership, the RAP is published and made accessible to the driller and other personnel on the rig.

The high level of readiness for RAPs sent offshore is crucial for several reasons:

- Most of the content can and should be completed by the onshore team.
- This approach frees up time for offshore personnel to focus on following up operations rather than writing RAPs.

Typically, the only changes made offshore should be updates such as:

- Updating from planned to actual depths, lengths, etc.
- Updating displacement volumes for cement, etc.
- Any other changes should be discussed with the onshore team.

Achieving a high ROP in drilling operations requires more than just having the right technology, equipment, and skilled personnel. When all these elements are in place, they create the potential for high ROP. However, to consistently reach and maintain this high performance, an accurate and detailed RAP is crucial.

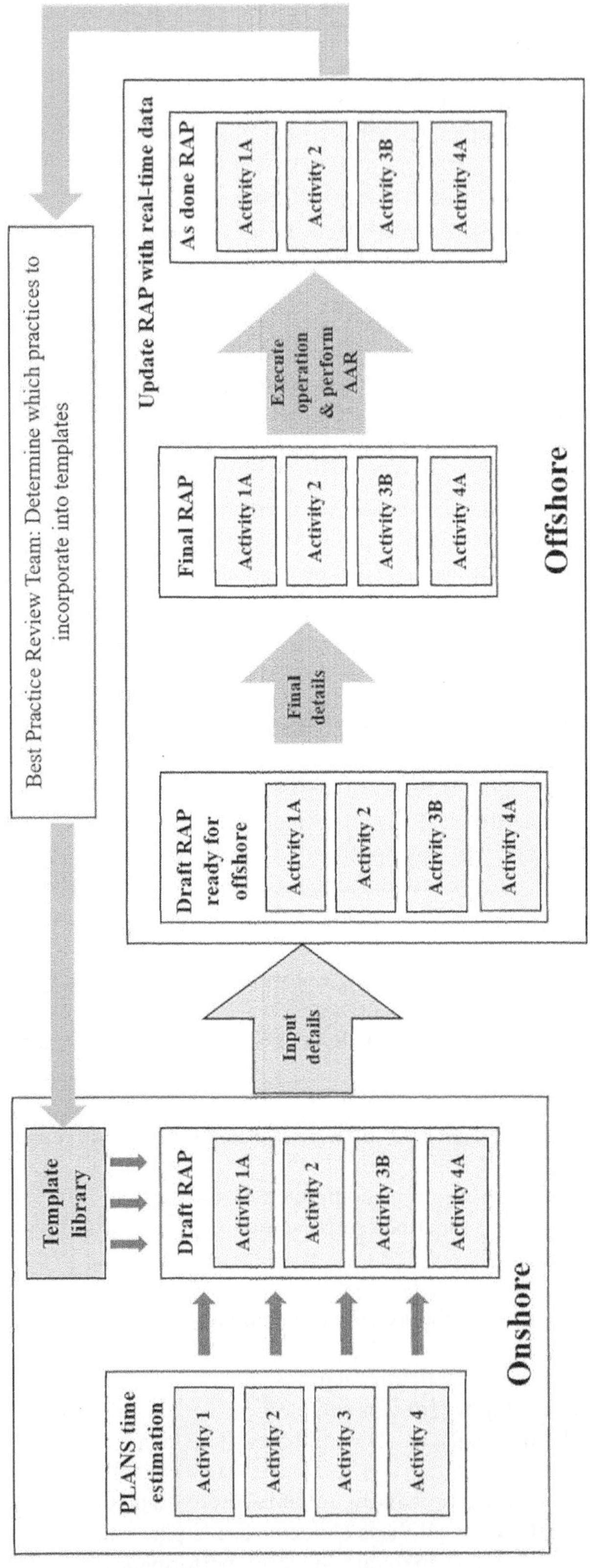

FIGURE 11.6 Example of RAP framework and process overview.

Cuttings Load % (total)				
Max ROP	3800 lpm	4000 lpm	4200 lpm	4400 lpm
50	2.011%	1.903%	1.735%	1.664%
70	2.761%	2.502%	2.352%	2.275%
90	3.497%	3.359%	3.157%	2.998%
110	4.256%	4.157%	3.988%	3.740%
130	5.088%	4.860%	4.623%	4.243%
150	5.919%	5.646%	5.345%	4.977%
165	6.569%	6.164%	5.747%	5.474%
180	7.249%	6.693%	6.348%	5.912%
210	8.505%	8.104%	7.479%	6.855%

ECD (sg) @ TD				
Max ROP	3800 lpm	4000 lpm	4200 lpm	4400 lpm
50	1.546	1.546	1.546	1.546
70	1.553	1.551	1.551	1.552
90	1.559	1.559	1.559	1.558
110	1.566	1.567	1.566	1.565
130	1.574	1.573	1.572	1.569
150	1.581	1.580	1.578	1.576
165	1.587	1.585	1.582	1.581
180	1.594	1.590	1.588	1.585
210	1.605	1.602	1.598	1.593

Minimum Transport efficiency (%) in Open Hole				
Max ROP	3800 lpm	4000 lpm	4200 lpm	4400 lpm
50	52%	52%	52%	52%
70	48%	48%	48%	48%
90	45%	45%	45%	45%
110	43%	43%	42%	42%
130	40%	40%	40%	40%
150	38%	38%	38%	38%
165	37%	37%	37%	37%
180	36%	36%	36%	36%
210	34%	34%	34%	34%

FIGURE 11.7 Advisory system for hydraulics and hole cleaning in the 12¼" hole section.

A well-prepared RAP provides clear instructions, operational parameters, and risk mitigation strategies, ensuring that the drilling process is optimized for maximum efficiency. This structured approach supports high ROP by minimizing downtime, preventing issues, and enabling real-time adjustments based on changing conditions. Therefore, achieving high ROP is not a matter of chance but the result of meticulous planning and execution, guided by a precise RAP or eRAP.

In the RAP, various considerations and limitations must be addressed. Advanced software applications can significantly enhance this RAP. For instance, we can utilize advanced software for hydraulics and hole cleaning, as well as for drilling string vibration analysis. Consequently, the role of these advisory systems in different aspects of the RAP can be greatly improved.

In Figure 11.7, we have an advisory system focused on hydraulics and hole cleaning scenarios, such as Equivalent Circulating Density (ECD) and cuttings accumulation, with different ROP strategies as a typical example for the 12¼" hole section.

11.4 CASE STUDIES OF HIGH-EFFICIENCY DRILLING OF ROP ON THE NCS

In the Norwegian Continental Shelf (NCS), data from the top 25 wells that were each drilled in just 24 hours showcase the efficiency and capability of modern drilling technology with one of the pioneering operators (Figure 11.8). These wells, mostly featuring 12¼" hole sections, were drilled using a 6th-generation rig. Each well was drilled with varying ROP, considering the hardness of the formations encountered. In some of these wells, operators drilled at rates exceeding 300 metres per hour in the Skade and Utsira formations. However, the average ROP across these wells was less than the mentioned ROP due to the varying geological conditions and formation properties.

In addition to the above figures, one should be aware that when drilling at higher rates and considering average ROP rather than instantaneous ROP, connection times become more important. The impact of connection times can be seen in Figure 11.9. It is apparent that when exceeding 200 metres per hour in an instantaneous ROP, if one has long connection times, the time spent on a stand and average ROP is more impacted by connections. As an example, if drilling with 200 metres per hour over a stand, and a connection takes 5 minutes, then this connection equates to 30% of the time spent on the stand. So, to achieve a high average ROP one must both drill fast and make quick connections.

Reducing connection time is crucial because it directly impacts the ROP, which measures how quickly the drill bit advances through the rock. Longer connection times result in more downtime, leading to lower ROP and extended drilling durations. Conversely, minimizing connection time can enhance ROP, allowing faster and more efficient drilling. For example, reducing connection times from 7 minutes to just 1.5 to 2.5 minutes can significantly improve drilling performance.

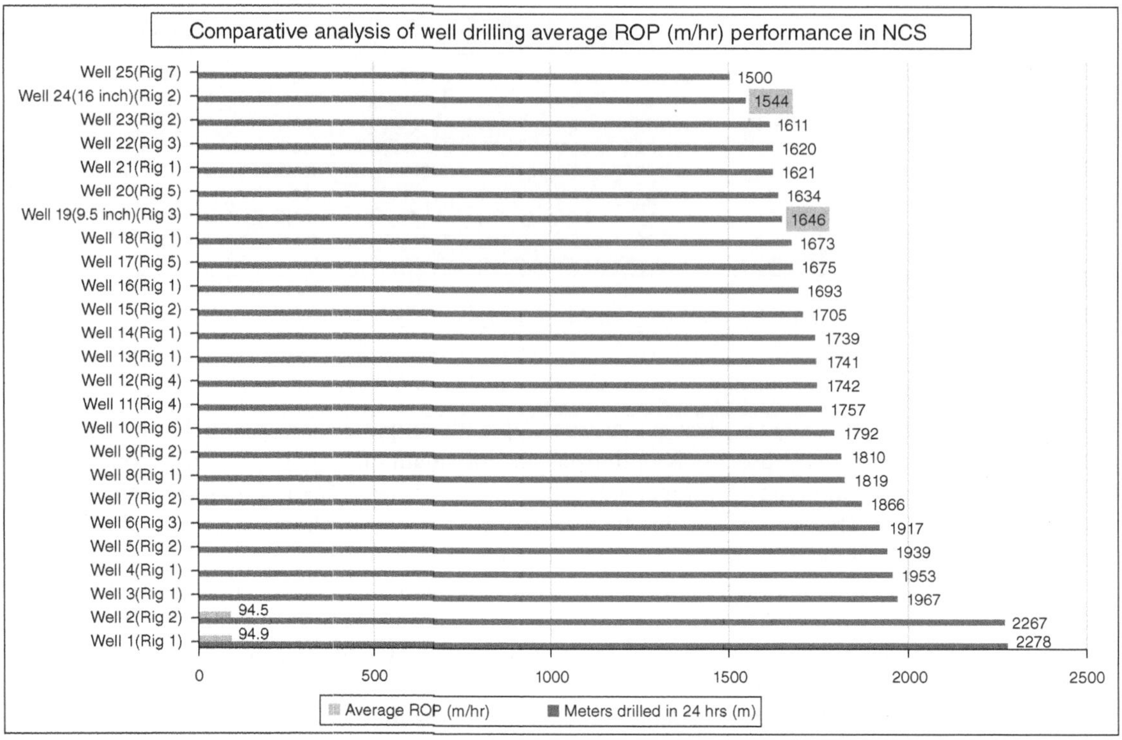

FIGURE 11.8 Visual representation of the 25 wells, highlighting the total meterage drilled in 24 hours.

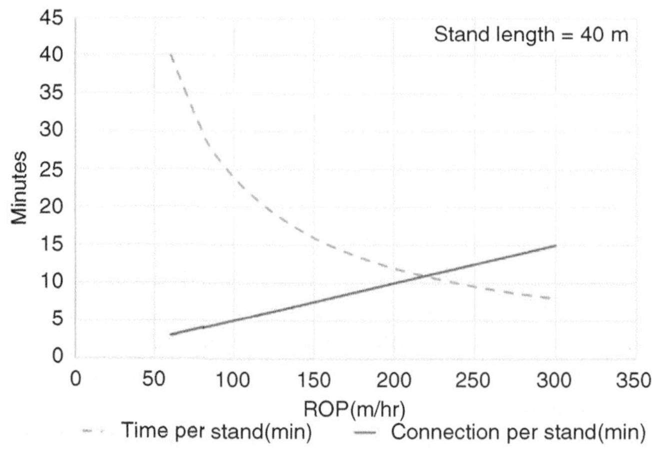

FIGURE 11.9 Impact of connection times on average ROP.

First, let's focus on the schematic of the wellbore we want to drill, including formation types and information as shown in Figure 11.10. After that, we will examine the operational trends for one of the very high ROP wells from Figure 11.8.

Figure 11.11 shows a detailed operational depth trend for ROP, Flow rate, RPM, WOB, SPP, ECD, and bit depth. This comprehensive visualization helps in understanding the interplay between

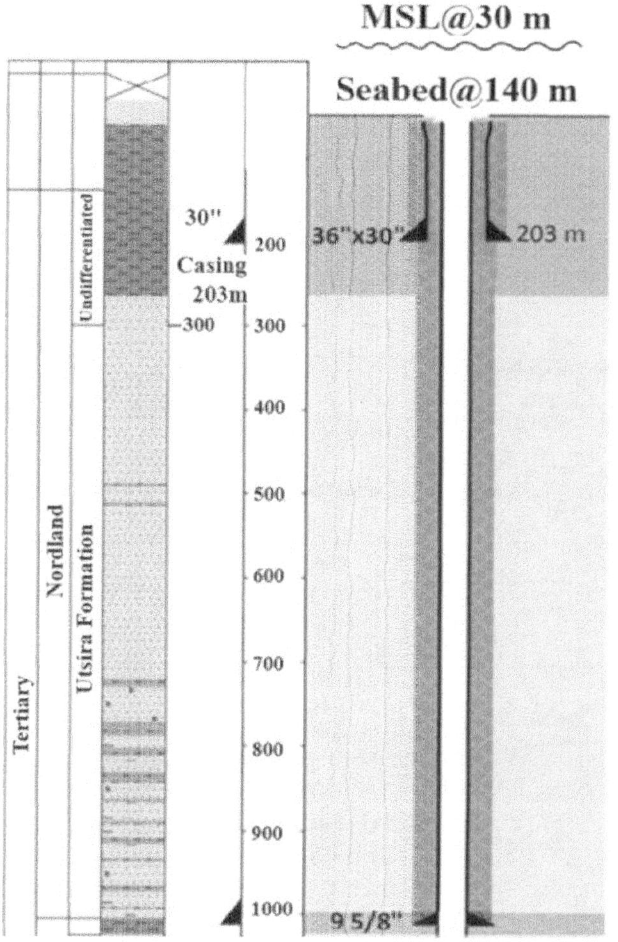

FIGURE 11.10 Wellbore schematic with formation information.

operational parameters and geological formations, providing valuable insights for optimizing drilling performance.

A simplified BHA was used, excluding porosity and density tools, and limiting flow rates to 2500–2800 lpm to avoid washing out loose sand sections. High-viscosity sweeps were frequently pumped, achieving a high ROP of 250 m/hr. However, low flow rates and fast ROP led to poor MWD decoding, preventing logging or ECD data collection. Towards the end, ROP was reduced to 60 m/hr and then 15 m/hr to identify shale and call TD at 986m MD. The hole was displaced to 1.3 sg WBM before pulling out, with no tight hole observed.

Drilling continued with controlled parameters, increasing ROP to 350 m/hr through the Utsira Sand. This high speed was maintained by setting a maximum flow rate of 2800 lpm to avoid damaging the hole further. The efficient drilling through the Utsira formation demonstrated the capability to achieve high ROPs under optimal conditions. From 485m to 530m, drilling continued at 300 m/hr with 2500 lpm, maintaining high ECD up to 1.26 sg. The last 6 m were drilled at 15 m/hr while displacing to KCl mud 1.2 sg.

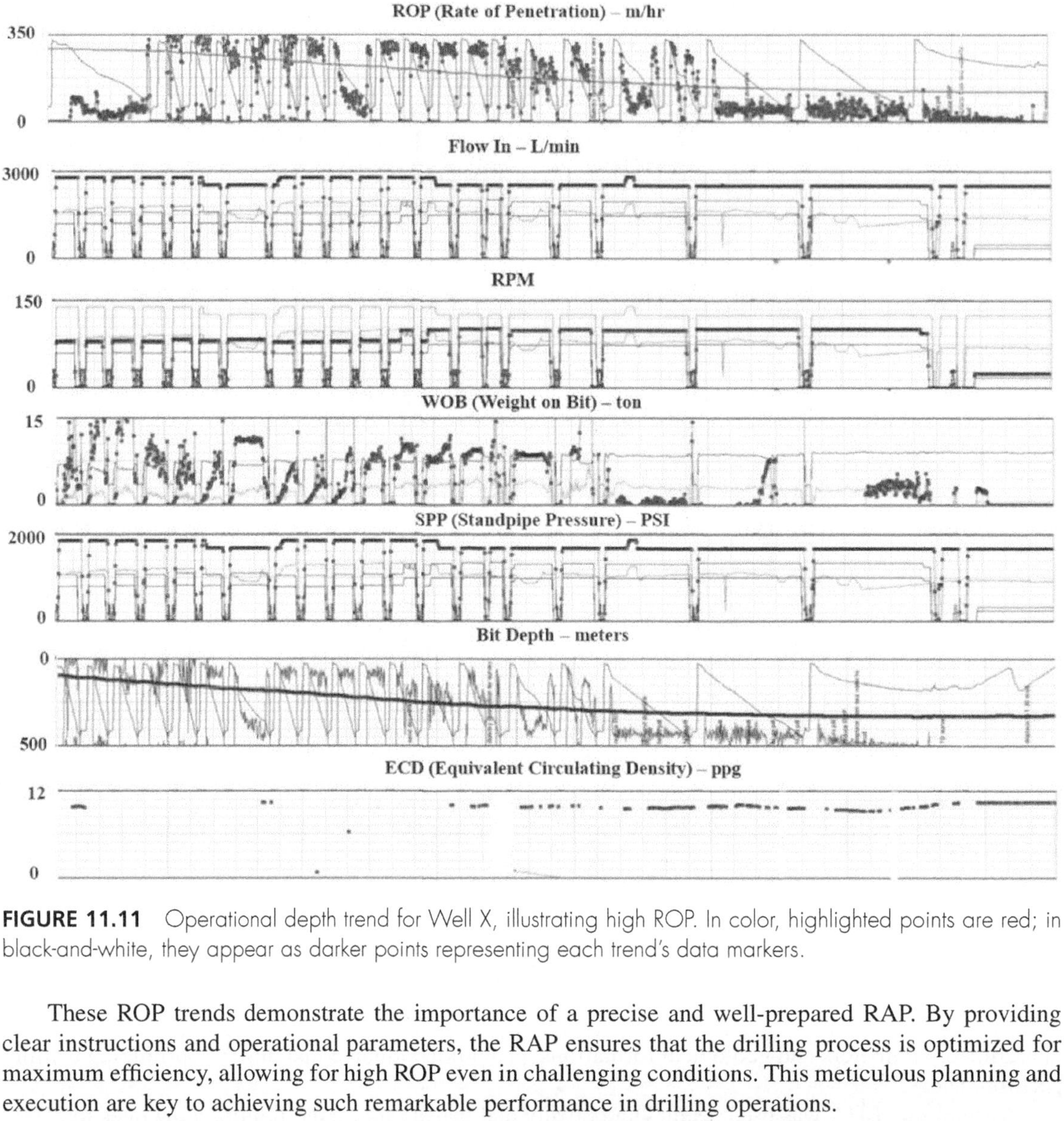

FIGURE 11.11 Operational depth trend for Well X, illustrating high ROP. In color, highlighted points are red; in black-and-white, they appear as darker points representing each trend's data markers.

These ROP trends demonstrate the importance of a precise and well-prepared RAP. By providing clear instructions and operational parameters, the RAP ensures that the drilling process is optimized for maximum efficiency, allowing for high ROP even in challenging conditions. This meticulous planning and execution are key to achieving such remarkable performance in drilling operations.

To achieve high performance in drilling operations, such as maximum ROP, it is essential to have a well-prepared and precise RAP. A comprehensive RAP ensures that all aspects of the drilling process are optimized, from equipment readiness to operational parameters. By leveraging the latest advancements in technology and best practices, we can significantly enhance ROP and overall efficiency.

Here are some key developments:

1. Automation and Real-Time Data Analytics:
   - Dynamic Parameter Adjustment: Modern rigs use automated systems to continuously optimize drilling parameters such as weight on bit, RPM, and mud flow in real time. This allows the system to adapt to changes in the formation, maintaining peak ROP.
   - Predictive Algorithms: Advanced AI and machine learning algorithms predict formation characteristics, enabling pre-emptive adjustments to avoid slowdowns and optimize drilling performance.

2. Advanced Mechanical Systems:
    - Top Drive Systems: These systems enable continuous rotation and precise control of the drill string, improving torque management and reducing the time needed for connections.
    - High-Efficiency Drill Bits: The use of Polycrystalline Diamond Compact (PDC) bits and hybrid technologies, combined with rigs designed to maximize their performance, significantly enhances cutting efficiency.
3. Vibration and Stability Control:
    - Sensors are used to detect harmful vibrations. Automated systems then make corrections to mitigate bit wear and maintain a consistent ROP, ensuring smoother and more efficient drilling operations.
    - Implement vibration avoidance, when historical data suggest that vibration-induced damage leads to tool failures, to drill section in one run with optimum ROP rather than max ROP. Overall, the time saving will be greater by avoiding trips compared to the incremental saving one has from increasing ROP with subsequent tool failure.
4. Improved Hydraulics and Hole Cleaning:
    - Enhanced mud systems and hydraulic efficiency prevent the accumulation of cuttings, reducing drag and maintaining a high ROP. This ensures that the wellbore remains clean and drilling operations proceed smoothly.
5. Energy and Power Management:
    - Optimized power distribution systems ensure that maximum energy is directed to the drill bit, even in complex well profiles. This maximizes drilling efficiency and helps achieve higher ROP.
6. Optimized Drilling Fluids:
    - High-Performance Mud Systems: Using drilling fluids with optimized rheological properties can enhance hole cleaning and reduce friction, leading to higher ROP.
    - Real-Time Monitoring: Continuous monitoring and adjustment of drilling fluid properties help maintain optimal conditions for high ROP.
    - Mud systems with "low fracture gels" that minimizes pressure spikes when bringing pumps back up after a connection. This helps reduce the risk of inducing losses.
7. Optimized Connection Procedure:
    - Scrutinize connection procedures such that these are optimized for as quick downtime as possible.
    - Focus on how quickly the pumps are brought down (negative pressure spikes) and how quickly pumps are turned on again (positive pressure spikes) to minimize pressure fluctuations on the formation.
    - Avoidance of "drilling of weight" to save time on drilling a full stand.
    - Be aware of potential induced vibrations if rotations is increased after a connection if bit is still on bottom. Evaluate to pull of bottom before setting slips.
8. Data Analytics and Machine Learning:
    - Predictive Maintenance: Using data analytics to predict equipment failures can minimize downtime and maintain high ROP.
    - Optimization Algorithms: Machine learning algorithms analyse historical drilling data to optimize drilling parameters for maximum ROP.

By leveraging these advancements, the industry has been able to achieve and even exceed ROPs of 300 metres per hour, showcasing the continuous innovation and adaptation to new technologies that drive efficiency and productivity in modern drilling operations.

These advancements in 6th-generation rigs are crucial for improving the ROP and overall drilling efficiency. By leveraging automation, advanced mechanical systems, and digital integration, the industry can achieve higher ROPs and more efficient drilling operations. Continuous innovation and adaptation to new technologies are key to maintaining high efficiency and productivity in modern drilling.

11.5 OVERCOMING ROP LIMITERS

Understanding the factors that limit the ROP in drilling operations is essential for optimizing performance and efficiency. Different hole sizes and the corresponding Logging While Drilling (LWD) and Measurement While Drilling (MWD) tools can play a role in determining the maximum achievable ROP. In this discussion, we delve into the primary limiters affecting ROP for various hole sizes and present Table 11.3 summarising these factors. By addressing these bottlenecks, operators can enhance their drilling performance and achieve better results.

Table 11.4 illustrates the limitations of tools used in specific wells in the NCS as a primary example (Figure 11.12). Table 11.3 is a typical example. These limitations may vary on a case-by-case basis, and the table is provided only as a representative reference.

TABLE 11.3 Example table of hole sizes, LWD/MWD tools, and ROP limiters

HOLE	*LWD/MWD*	*LIMITER*	*LWD MAX. ROP (M/HR)*
32″	Dir (Directional control,)	Verticality	N/A
24″	GR, Res	Hole cleaning	N/A
17.5″	GR, Res	Hole cleaning	N/A
12.25″	GR, Res, Den/Neu, Sonic, Deep Res (geosteering)	Den/Neu	150 m/hr
		Sonic	120–225 m/hr
8.5″	GR, NBGR, Res, Den (Image)/Neu, Sonic, Deep Res, FPWD	Deep Res	75 m/hr (WDP)
			30 m/hr (RDP)

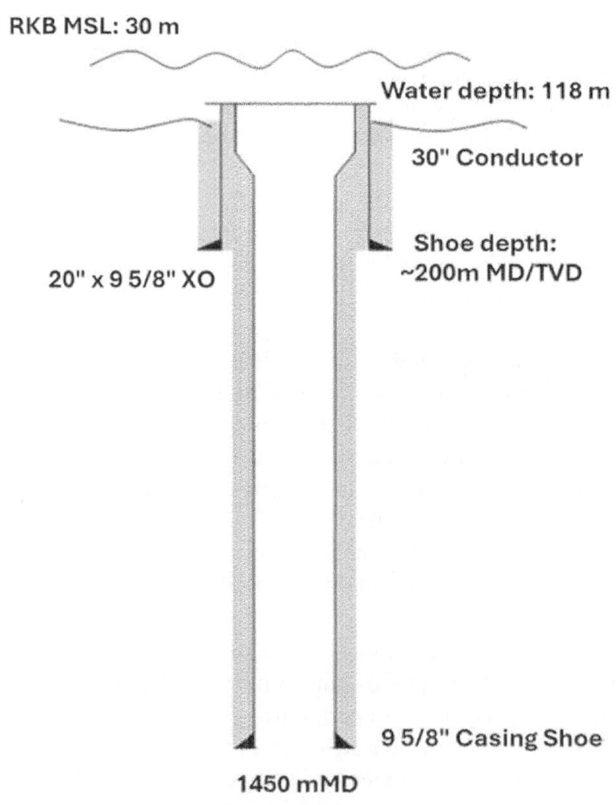

FIGURE 11.12 ROP evaluation in the 12¼" hole section.

The specifications sheet indicates that drilling faster than 220 metres per hour in the 12¼" hole section is possible based on the offset well analysis and mechanical properties of the formation, such as rock hardness, porosity, and fracture density. Achieving such high ROPs consistently requires careful consideration of various factors, including the efficiency of the drilling fluid and real-time monitoring of drilling parameters.

Table 11.4 represents the limitations related to Data Acquisition (Data Aq), Bottom Hole Assembly (BHA), minimum and maximum flow rates (lpm), hole cleaning (hydraulics), and cuttings handling ($26m^3$/hr) used in specific wells in the NCS.

To overcome the ROP limiters, it is essential to:

- Optimize Drilling Parameters: Adjust WOB, rotary speed (RPM), and mud properties to suit the specific formation characteristics. This optimization helps achieve the best possible penetration rates while maintaining the integrity of the wellbore.
- Use Advanced Drilling Tools: Implement the latest Logging While Drilling (LWD) and Measurement While Drilling (MWD) technologies to provide real-time data and enhance decision-making. These tools offer valuable insights into subsurface conditions, allowing for more precise adjustments during drilling.
- Improve Hole Cleaning: Ensure efficient removal of cuttings to prevent re-drilling and maintain high ROP. Effective hole cleaning reduces the risk of stuck pipe incidents and enhances overall drilling efficiency.
- Enhance Bit Design: Utilize bits designed for specific formations to improve penetration rates and tool longevity. Advanced bit designs can handle varying rock hardness and abrasiveness, leading to better performance and reduced wear.
- Monitor Data Acquisition (Data Aq): Ensure that data acquisition systems are functioning optimally to provide accurate and timely information. This helps in making informed decisions and adjusting drilling parameters effectively.
- Manage Bottom Hole Assembly (BHA): Optimize the configuration of the BHA to suit the specific drilling conditions. This includes selecting the appropriate tools and components to maximize ROP and minimize downtime.
- Control Flow Rates (Min/max flow): Adjust the flow rates to ensure efficient hole cleaning and cuttings removal. Proper flow rate management is crucial for maintaining high ROP and preventing issues related to cuttings handling.
- Handle Cuttings Efficiently: Ensure that the cuttings handling system is capable of managing the expected volume of cuttings. Efficient cuttings handling prevents re-drilling and maintains drilling efficiency.

By addressing these key points, operators can enhance their drilling performance and achieve better results, ultimately pushing the boundaries of modern drilling operations. Continuous innovation and adaptation to new technologies are crucial for maintaining high efficiency and productivity in the industry.

Table 11.5 provides an overview of drilling performance across various sections of the wellbore depicted in Figure 11.8. As the second example, the 12.25-inch hole size demonstrates the best ROP performance. Table 11.5 compares Gross ROP, Net ROP, and time metrics such as Weight to Slips, Slips to Slips, and Slips to Weight, highlighting the efficiency and effectiveness of drilling operations in each section (Figure 11.13).

The 12¼" hole section strategy aimed to maximize drilling speed by dividing the trajectory into three distinct parts, each with specific constraints. The fastest interval drilled was 2278 metres in 24 hours (Figure 11.14).

- Part 1 – The Tricky Zone (1789 mMD – 2450 mMD): This section was constrained by the DPZ#7 fracture gradient and the Seal 8 stringer. The target ROP was over 200 metres per hour, with a maximum Equivalent Circulating Density (ECD) of 1.82 s.g. The ROP was reduced to 60 metres per hour near the stringer to manage risks.

TABLE 11.4 Example of limitations to maximizing rate of penetration (ROP)

SECTION	*MW*	*INTERVAL & LENGTH*	*MAX ROP & FLOWRATE*				*COMMENTS*	*MAX ECD (PRIMARILY SHMIN PPFG)*	*TARGET ROP (M/HR)*
			DATA AQ (M/HR)	*BHA MIN/MAX FLOW (LPM)*	*HOLE CLEANING (HYDRAULICS) (M/HR)*	*CUTTINGS HANDLING (26 M³/H)*			
36″ section	1.03sg	148–200m Length: 52m	N/A	2600–3600	N/A	N/A		N/A	10
12 ¼″ section	1.03sg	200–1164m Length: 964m	Unrestricted	4000–4600	Unrestricted	N/A	Monitor drilling parameters	1.34sg EMW	200
12 ¼″ section	1.20sg	1164–1380m Length: 216m	200	4000–4600	180	N/A	Hydraulics limiting factor 7% cuttings load w/4000 lpm	1.34sg EMW	180
12 ¼″ section	1.20sg	1380–1450m Length: 70m	50	4000–4600	180	N/A	Data acquisition limiting factor	1.34sg EMW	50
8 ½″ section	1.30sg	1450–1966m Length: 516m	115	1500–1900	>115	Unrestricted	Data acquisition limiting factor 3.8% cuttings load w/ 1900 lpm. ECD: At TD = 1.50sg, at shoe: 1.48sg	1.53sg EMW	115
8 ½″ section	1.30sg	1966–2091m Length: 125m	44	1500–1900	>44	Unrestricted	Data acquisition limiting factor 1.4% cuttings load w/ 1900 lpm. ECD: At TD = 1.48sg, at shoe: 1.45sg	1.50sg EMW	44
8 ½″ section	1.30sg	2091–2310m Length: 219m	30	1500–1900	>30	Unrestricted	Data acquisition limiting factor 1.0% cuttings load w/ 1900 lpm. ECD: At TD = 1.47sg, at shoe: 1.45sg	1.50sg EMW	30

TABLE 11.5 Drilling performance overview with focus on 12.25" hole section

SECTION	*GROSS ROP (M/HR)*	*NET ROP (M/HR)*	*WEIGHT TO SLIPS (MIN)*	*SLIPS TO SLIPS (MIN)*	*SLIPS TO WEIGHT (MIN)*
16 ½"	74.2	96.9	1.2	3.6	1.3
12 ¼"	79.5	108.5	-	3.2	-
8 ½" Y1	23.2	33.3	-	3.4	2.0
8 ½" x 10" Y2	16.4	23.7	2.2	3.4	2.4
6 ½" Y2	17.8	22.9	1.6	4.3	2.5

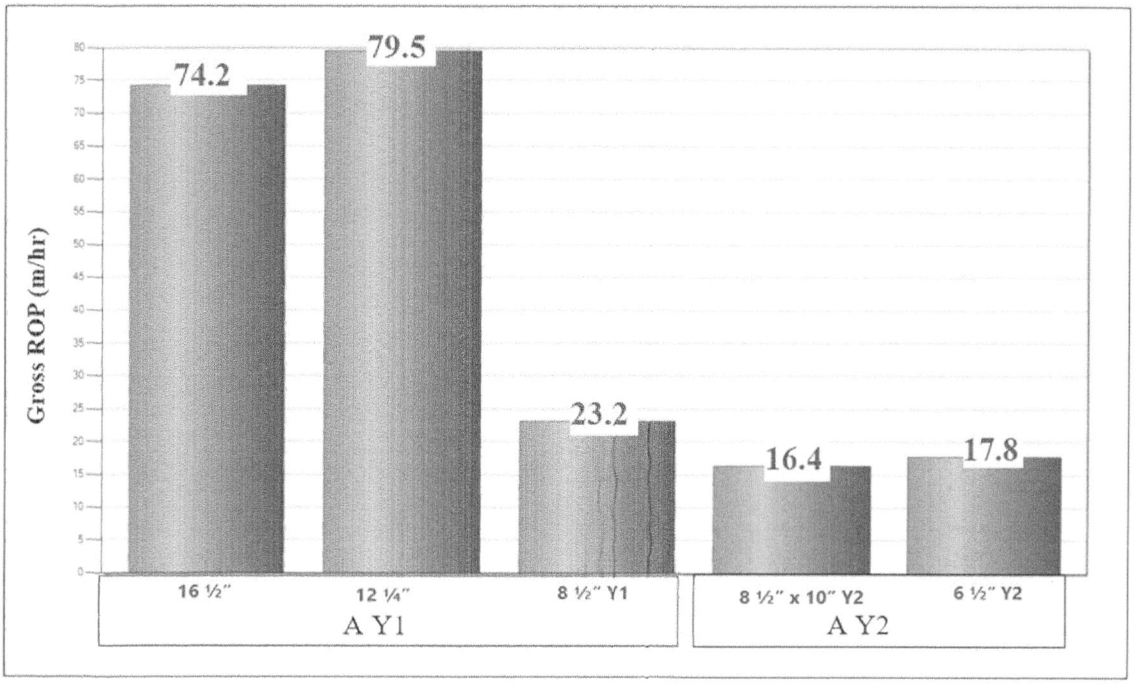

FIGURE 11.13 ROP performance highlighting best results in 12¼" hole section.

- Part 2 – The Highway (2450 mMD – 3920 mMD): This section had minimal constraints, allowing for a target ROP of over 240 metres per hour and a maximum ECD of 1.84 s.g.
- Part 3 – The Landing (3920 mMD – 4445 mMD): This section involved directional work for proper reservoir entry and addressed anti-collision issues. The target ROP was 175 metres per hour up to 4250 mMD, then reduced to 40 metres per hour to ensure thorough well cleaning.

Table 11.6 provides detailed drilling parameters for the 12¼" hole section, highlighting the operational metrics and performance across different depths.

Figure 11.15 includes the procedures and attachments that support the recommended drilling parameters for this section, which are directly related to ROP enhancement, based on various simulations such as hydraulics simulations.

Technical Data and Operational Limits:

- Specifications and constraints for drilling.

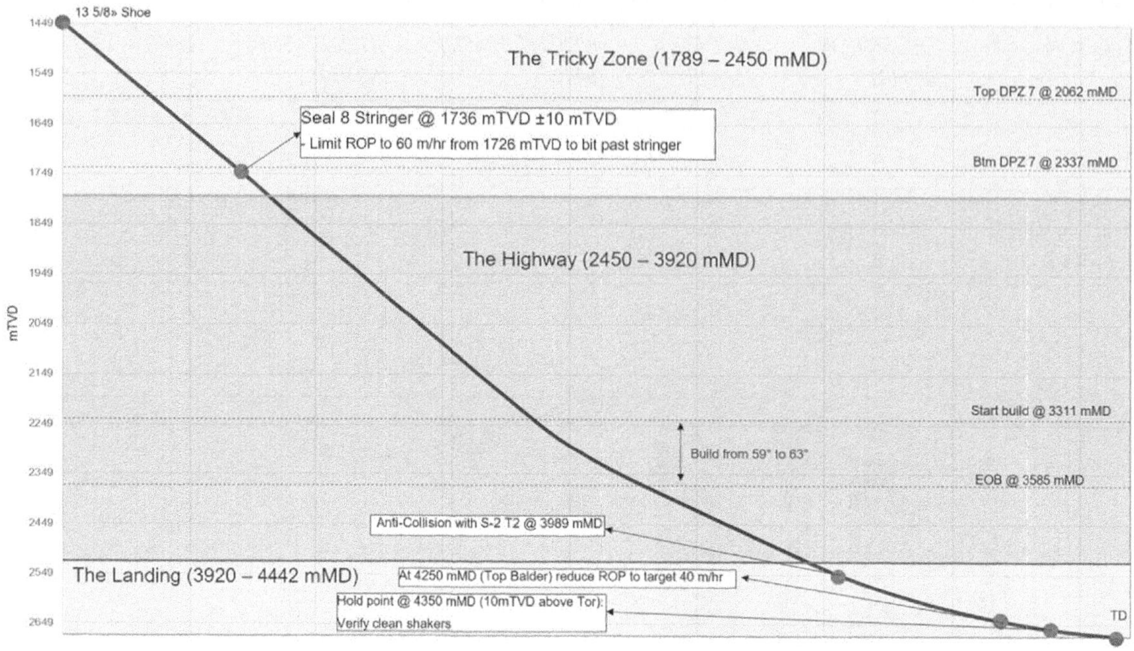

FIGURE 11.14 Drilling strategy for the 12¼" hole section: Maximizing ROP through targeted intervals.

TABLE 11.6 Drilling parameters 12.25" section

DEPTH	*FLOW*	*SPP*	*ROTATION*	*TORQUE*	*WOB*	*ESD*	*ECD*	*ROP INST.*	*ROP AVG.*	*COMMENT*
mMD	**LPM**	**Bar**	**RPM**	**kNm**	**MT**	**s.g**	**s.g**	**m/hr**	**m/hr**	
1782–1924	4100	257	120	13–17	8–12	1.74	1.80	50–128	40.6	
1924–2315	4100–4500	257	120	13–17	10–15	1.75	1.80	125–225	86.9	
2315–2379	4500	330	160	20–23	5–7	1.75	1.79–1.80	60	58	Reduced ROP to 60 m/hr when drilling into DPZ8
2379–2973	4600	344	160	20–23	5–7	1.75	1.81	90–225	108	
2973–3628	4600	388	160	20–23	16–21	1.75	1.82	250–275	113.9	
3628–4152	4600	388	160	20–23	16–21	1.75	1.83	250–275	104.8	
4152–4446	4530	430	170	30–36	16–21	1.75	1.81–1.80	100–20	32.6	Reduced ROP to align for reservoir landing and reduce cuttings load

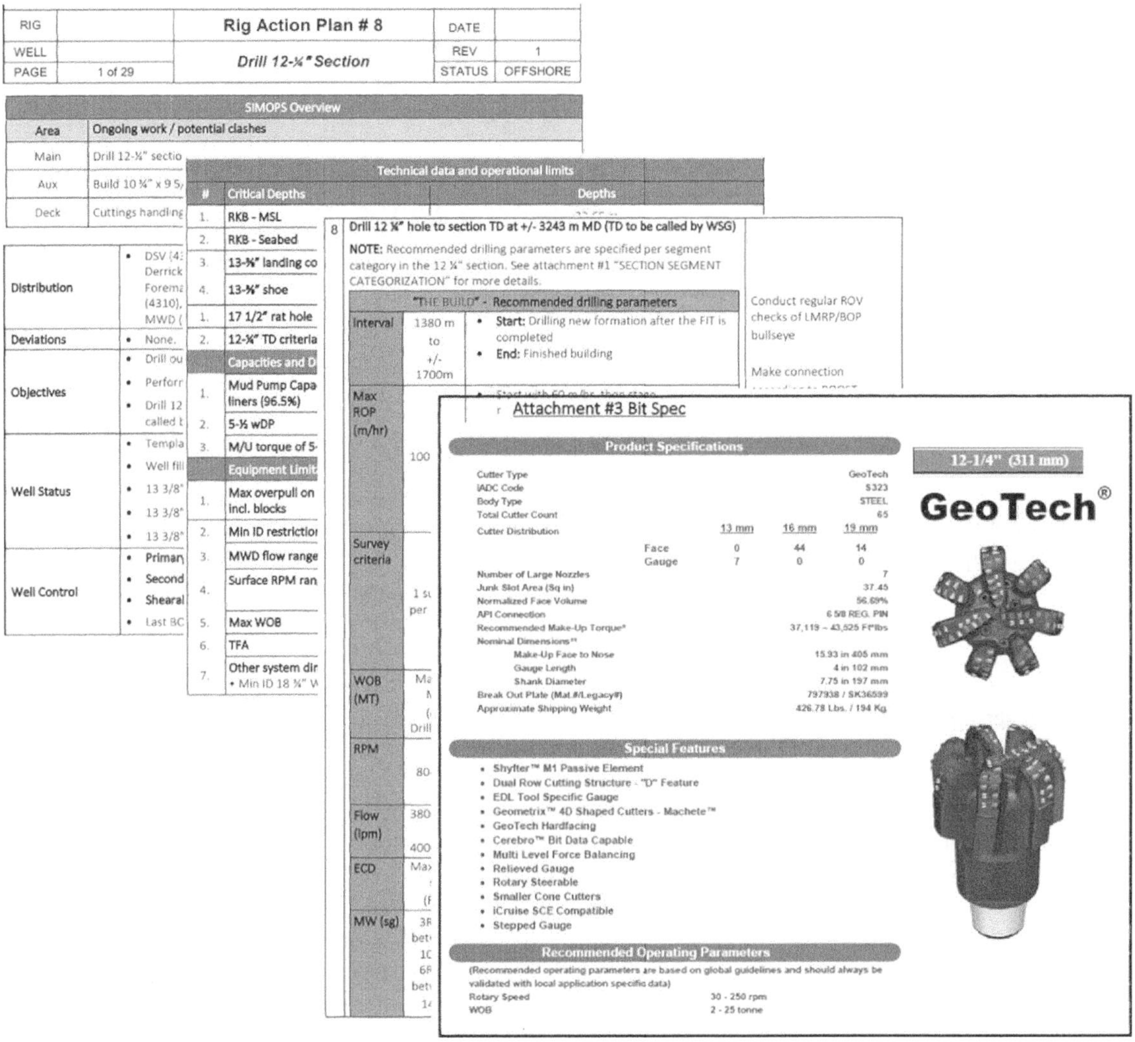

FIGURE 11.15 RAP documentation for the 12¼" hole section.

Preparations and Equipment Checklist:

- List of necessary preparations and equipment.

Operations:

- Step-by-step drilling procedures.

Attachments:

1. Best practice drilling information
2. Formation tops geological prognosis
3. Bit specifications
4. Preliminary 12¼" BHA
5. Lost circulation material (LCM) tree
6. Well barrier schematic
7. 12¼" Lower formation schematic

This structured approach ensures that all aspects of the drilling operation are thoroughly planned and executed, enhancing overall performance and efficiency, including the ROP.

11.6 HYDRAULIC ENGINEERING SIMULATIONS WITH DIFFERENT ROP STRATEGIES

In this section, we address several important questions:

1. What is the acceptable amount of cuttings in the sections for the operation?
2. Should we use torque and drag as indicators of wellbore cleanliness?
3. Should the maximum ROP check be stopped based on these indicators?
4. What should be the limit for the unlimited ROP check?

During the planning phase, hydraulic simulations are crucial for optimizing ROP strategies. These simulations help determine the best fluid properties, well trajectory, and tubular configurations. By analysing offset well information, we can make adjustments to drilling parameters, ensuring maximum ROP and efficient drilling operations. Therefore, we can achieve accurate simulations to be considered in the RAP.

When setting up well configurations for engineering simulations, it's important to consider several key aspects of the well (Figure 11.16):

- Fluid Properties: Include details such as rheology, gelling, PVT (Pressure-Volume-Temperature) characteristics, and more.
- Trajectory: Map out the well's path, including inclination, azimuth, true vertical depth (TVD), total measured depth (TMD), dogleg severity (DLS), and other relevant parameters.
- Tubular Specifications: Specify components like outer diameter (OD), inner diameter (ID), weight per length, and type of the bottom hole assembly (BHA).
- Bit Characteristics: Describe the bit type, nozzle size, and other specifications.
- Wellbore Geometry: Detail the geometry, including open hole sections.
- Temperature Profile: Map out temperature changes with depth.
- Pressure Profile: Detail the pressure changes within the well, including pore pressure, fracture pressure, and collapse pressure.
- Formation Characteristics: Describe the geological formations encountered, including thickness, top formation, permeability, conductivity, porosity, and more.
- Reservoir Details: Include information about the reservoir's size, shape, and fluid properties.
- Rig Specifications: Describe the rig's capabilities and limitations, such as blowout preventer (BOP), flowline, and other relevant equipment.

We conducted hydraulic simulations to ensure effective hole cleaning within the current drilling parameters outlined in the rig action plan. These simulations tested different rates of penetration (ROP) and pump rates, while keeping the rotary speed constant at 160 RPM, as shown in Table 11.7. The simulations indicate that an annular velocity of around 45 to 52 metres per minute is needed for good hole cleaning in this section. The results from the second and third simulations show that there is no significant risk of cutting accumulation when drilling at rates of 80 metres per hour and 100 metres per hour with pump rates of 5250 and 5500 litres per minute in Table 11.8, respectively.

One important factor to consider is the cutting load. If we increase the threshold for cutting load to, for example, 14%, we can significantly boost the ROP while staying within the safe mud window. This

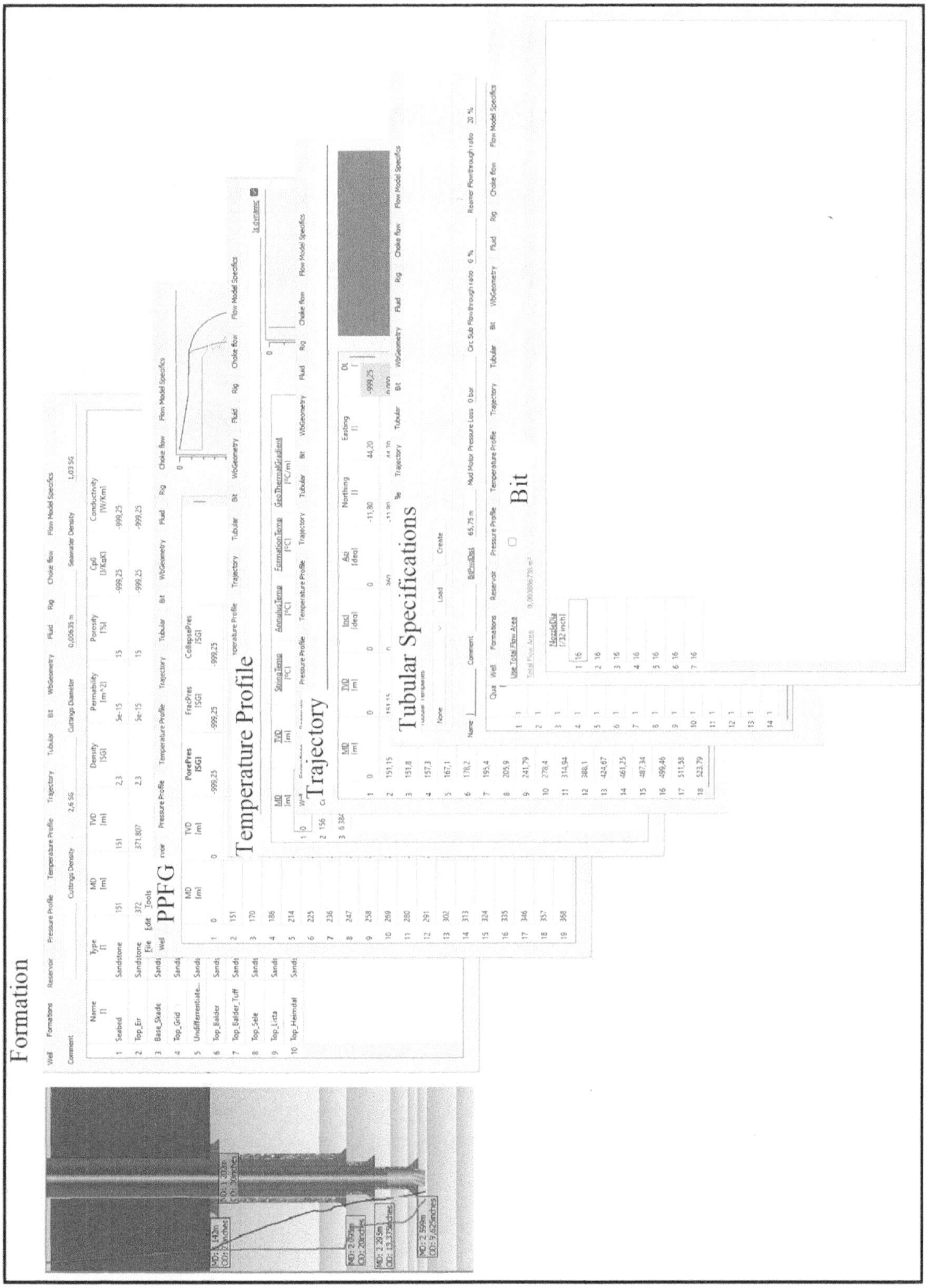

FIGURE 11.16 Different screenshots for data entry in well configuration.

TABLE 11.7 Simulation input drilling 16-½″ section

SIMULATION	*DEPTH (M)*	*ROP STRATEGIES (M/HR)*	*RPM*	*FLOWRATE (LPM)*	*MWIN (SG)*
1	845–2055	60	160	5000	1.40
2	845–2055	80	160	5250	1.40
3	845–2055	100	160	5500	1.40

TABLE 11.8 Simulation output drilling 16-½″ section

SIMULATION	*ECD AT 16″ CASING SHOE (SG)*	*MAX ECD AT PWD (SG)*	*MAX SPP (BAR)*	*ANNULAR VELOCITY AT 16″ CASING (M/MIN)*	*ANNULAR VELOCITY AT OH (M/MIN)*	*CTR PROFILE (%)*	*CUTTING LOAD (%)*
1	1.444–1.451	1.454	259	32.29–33.91	43.68–51.55	93–95	2.67–2.81
2	1.450–1.456	1.460	283	33.94–35.65	45.91–54.22	93–96	3.33–3.75
3	1.456–1.464	1.468	307	35.77–37.51	48.29–56.84	93–96	3.89–4.6

adjustment depends on various guidelines, operational experience in the field, and the threshold levels defined by their software for simulations and monitoring the well during real-time operations.

The pursuit of a high ROP in drilling operations requires a combination of advanced technology, meticulous planning, and continuous innovation. Over the years, the industry has evolved significantly, improving drilling practices and technologies to achieve higher ROPs safely and efficiently.

A well-prepared RAP is essential for optimizing drilling performance. By incorporating advanced engineering simulations and analysing offset well data, the RAP ensures that drilling operations are efficient and effective. Addressing ROP limiters such as cuttings handling, hole cleaning, vibrations, and bit design is crucial for enhancing performance. Modern advancements in automation, real-time data analytics, and advanced mechanical systems play a significant role in overcoming these challenges.

Hydraulic engineering simulations are vital for optimizing ROP strategies. These simulations help determine the best fluid properties, well trajectory, and tubular configurations, ensuring effective hole cleaning and efficient drilling operations. By leveraging these insights, operators can maintain high ROP while managing critical factors like cuttings load.

In summary, achieving high ROP is the result of meticulous planning, execution, and continuous improvement. By utilizing advanced technologies and comprehensive engineering simulations, the industry can achieve remarkable performance and efficiency in drilling operations.

Appendix A

A System for Experience Transfer

A.1 SELECTION OF ELEMENTS TO EVALUATE

Drilling of a well incorporates many elements. Of major concern is the expense of the well. Some of the cost elements are listed below:

- Rental cost for drilling rig
- Cost of auxiliary transportation, helicopter, boats
- Cost of drill bits and casing
- Cost of services like mud logger and directional driller
- Cost of drilling crew, and operator's personnel

This list can be expanded. However, we will in this section focus on the cost of the drilling operation itself, and not on the auxiliary services, also because time reduction in drilling directly reduces the auxiliary expenditures. A drilling operation is different from many industrial processes, as its cost depends on one single element; whether the drilling process is ongoing or not. If the operation stops for some reason, the complete set-up is idle, and the cost is a loss function. The efficiency of the whole operation depends on the state of the drilling process.

Another key element is the rig cost. In offshore drilling, the daily rig rate is a dominating cost element. Usually, the contracts are written on a daily rate, rather than as a lump sum. It should here be emphasised that the forms of future contracts can become important elements in cost optimisation, not only for the daily cost/lump sum, but also for the drilling contractor/operator split of responsibilities. The contract itself is therefore an important element in cost optimisation.

We will in this appendix take a technical approach. We will study the time consumption of various operations, with the understanding that it is a direct cost factor because of the high rig rate. This analysis can be expanded, but we believe that a technical approach is an absolutely necessary step in minimising the total expenditures of drilling a well, which form the basis for further work on contract forms.

A.2 TIME ANALYSIS OF FIELD DATA

A Norwegian operator pre-drilled six wells in the period 1990–1993 before installing the production platform. These operations were later extensively analysed because a significant saving was visualised when drilling another 50 production wells in the field, by systematically improving on a learning curve. The present section is a short summary of the results from this study.

The casing strings of the six wells were set at approximately the same depths, but the borehole inclinations were quite different, with reservoir sections drilled from 20 to nearly 60 degrees inclination. Obviously, the lengths of each section were different because of this. To make the data comparable, we therefore length-normalised all data. That is, we divided each time element with the hole length it represented. In this way we obtained data that were representative for all inclinations.

A.2.1 Breakdown of Time Used for Each Hole Section

In the following time consumption for each of the hole sections will be compared. All time data are length-normalised. The minimum cost/length factor gives the best result. Therefore, for each operation listed, the best result is shown in ***bold*** numbers. The data are taken from the operating company's experience transfer data base.

The 24/26 in. Hole Sections

Table A.1 is a summary of the specific time consumption for the surface hole sections during various operations. We observe that all wells showed some optima. However, some overall trends are:

- Time used on wellhead installations shows a definite improvement from the first to the last well.
- Most other operations show improvement from the first three to the last three wells.
- Landing and cementing of casing was performed with a minimum time in well E.
- Well D performed best in general, both under the total drilling time, and under the total time which include other associated operations.

TABLE A.1 Specific time consumption during drilling of the 24/26 in. hole sections

	SPECIFIC TIME CONSUMPTION (MIN/METRE)					
OPERATION	*WELL A*	*WELL B*	*WELL C*	*WELL D*	*WELL E*	*WELL F*
BOP activities	***0.00***	0.27	0.80	0.14	0.05	***0.00***
BOP/wellhead equip.	10.11	5.62	7.30	4.67	4.27	3.84
Other	0.09	***0.03***	***0.00***	***0.00***	***0.00***	***0.00***
Casing	4.09	3.22	3.06	2.91	***2.36***	2.59
Survey	–	0.15	–	***0.10***	0.27	***0.10***
Press detection	0.23	***0.09***	0.19	0.14	***0.09***	0.10
Circ./cond.	0.92	0.49	0.85	***0.38***	0.50	0.48
Drill	***0.92***	1.58	3.06	2.34	2.77	2.40
Open hole	4.41	***0.00***	0.57	***0.00***	***0.00***	***0.00***
Ream	0.46	0.94	***0.00***	***0.00***	***0.00***	***0.00***
Trip	5.01	2.79	3.11	***1.24***	1.82	2.21
Underream	***0.00***	1.61	2.59	2.24	2.64	2.74
Total drilling	11.72	7.41	10.17	***6.20***	7.73	7.82
Total	26.23	16.79	21.52	***14.17***	14.77	14.45

TABLE A.2 Specific drilling time for the 16 in. section

	SPECIFIC TIME CONSUMPTION (MIN/METRE)					
OPERATION	*WELL A*	*WELL B*	*WELL C*	*WELL D*	*WELL E*	*WELL F*
BOP activities	0.16	***0.15***	0.49	0.58	0.84	0.72
BOP/wellhead equip.	0.56	0.56	0.27	***0.18***	0.88	0.84
Other	0.03	0.11	***0.00***	***0.00***	***0.00***	***0.00***
Casing	3.26	2.95	1.96	***1.75***	2.78	5.15
Survey	–	–	–	–	–	–
Press detection	–	–	–	–	–	–
Circ./cond.	0.63	0.67	1.22	0.40	***0.32***	0.38
Drill	4.22	3.36	3.27	3.43	3.54	***3.16***
Open hole	0.00	0.00	0.00	0.00	0.00	0.00
Ream	0.30	0.11	1.22	0.21	0.02	***0.00***
Trip	3.59	2.47	2.69	1.72	***1.37***	3.16
Underream	0.00	0.00	0.00	0.00	0.00	0.00
Total drilling	8.74	6.61	8.40	5.76	***5.24***	6.71
Total	12.76	10.39	11.12	***8.27***	9.75	13.42

The 16 in. Hole Sections

Table A.2 shows the specific time used during drilling of the 16 in. hole sections. A quick comparison results in the following conclusions:

- There are no improvements regarding wellhead handling.
- The casing operation and the total time was lowest in well D, which was overall considered the best well.
- Wells E and F performed best from a drilling operations point of view.

The 12-1/4 in. Hole Sections

Evaluation of the specific time consumption for the 12-1/4 in. hole sections shown in Table A.3, results in the following observations:

- The wellhead operations were most efficient in well F.
- The casing operation was most efficient in well F.
- Well D seems best from a drilling operation point of view.
- An overall evaluation shows that well D performed best.

The 8-1/2 in. Hole Sections

In wells A and F, a 9-5/8 in. casing were set through the reservoir. In these wells there were therefore no drilled 8-1/2 in. sections. The remaining fours wells had this section drilled, as shown in Table A.4. An overall evaluation of Table A.4 results in:

- Well D performed best when installing the wellhead equipment.
- Well E performed best under the casing operations.
- Well C has best total evaluation, followed by well B.

TABLE A.3 Specific time consumption for the 12-1/4 in. hole sections

	SPECIFIC TIME CONSUMPTION (MIN/METRE)					
OPERATION	*WELL A*	*WELL B*	*WELL C*	*WELL D*	*WELL E*	*WELL F*
BOP activities	1.58	***0.10***	0.81	0.69	0.38	0.30
BOP/wellhead equip.	0.43	0.54	0.35	0.27	0.44	***0.22***
Other	***0.00***	***0.00***	***0.00***	***0.00***	0.03	0.03
Casing	3.03	3.04	2.53	3.12	3.40	***2.45***
Survey	0.74	0.10	–	–	***0.03***	***0.03***
Press detection	0.03	–	–	–	–	–
Circ./cond.	1.12	0.54	0.54	***0.30***	0.76	0.87
Drill	7.39	0.51	4.83	***2.82***	4.75	6.28
Open hole	0.00	0.00	0.00	0.00	0.00	0.00
Ream	0.79	0.00	0.35	***0.00***	0.21	0.14
Trip	7.61	4.61	2.69	***2.16***	3.37	4.05
Underream	0.00	0.00	0.00	0.00	0.00	0.00
Total drilling	16.91	14.66	8.40	***5.28***	9.08	
Total	22.72	18.43	12.08	***9.36***	13.36	

TABLE A.4 Specific drilling time for the 8-1/2 in. hole sections

	SPECIFIC TIME CONSUMPTION (MIN/METRE)					
OPERATION	*WELL A*	*WELL B*	*WELL C*	*WELL D*	*WELL E*	*WELL F*
BOP activities	–	1.56	2.23	***1.06***	1.46	–
BOP/wellhead equip.	–	0.42	***0.00***	***0.00***	***0.00***	–
Other	–	0.00	0.00	0.00	0.00	–
Casing	–	6.12	5.16	9.60	***4.79***	–
Survey	–	–	–	–	0.14	–
Press detection	–	0.06	–	–	***0.05***	–
Circ./cond.	–	1.02	***0.27***	0.86	1.05	–
Drill	–	***3.24***	4.95	10.93	9.13	–
Open hole	–	0.00	0.00	0.00	0.00	–
Ream	–	1.14	0.76	***0.13***	0.18	–
Trip	–	5.34	***4.02***	7.98	7.53	–
Underream	–	0.00	0.00	0.00	0.00	–
Total drilling	–	10.74	***10.00***	19.60	17.90	–
Total	–	18.90	***17.39***	30.26	24.34	–

A.2.2 Total Time and Interruptions

We will now briefly repeat the time consumption from the previous tables, and expand the discussion a little. In Table A.5 the total time per metre drilled is shown for the wells. We observe that there is some spread. However, well D is clearly the best well, with minimum time spent for both the 24/26 in. section and the 12-1/4 in. section.

Table A.6 includes drilling, and the other operations listed in Tables A.1 to A.4. For this case well D was best in three sections, and well C in one section. We will come back to the specific time spent when deriving a prognosis later.

TABLE A.5 Comparison with specific drilling time for the six wells

	SPECIFIC TIME CONSUMPTION (MIN/METRE)					
	WELL A	*WELL B*	*WELL C*	*WELL D*	*WELL E*	*WELL F*
24"/26"	11.72	7.41	10.17	***6.20***	7.73	7.82
16"	8.94	6.61	8.40	5.76	***5.24***	6.71
12¼"	(16.91)	14.66	8.40	***5.28***	9.08	(11.33)
8½"	–	10.74	***10.00***	19.60	17.90	–

TABLE A.6 Comparison with total specific time for the six wells. In addition to drilling time, other operations like casing setting and cementing are included

	SPECIFIC TIME CONSUMPTION (MIN/METRE)					
	WELL A	*WELL B*	*WELL C*	*WELL D*	*WELL E*	*WELL F*
24"/26"	26.23	16.79	21.52	***14.17***	14.77	14.45
16"	12.76	10.39	11.12	***8.27***	9.75	13.42
12¼"	(22.72)	18.43	12.08	***9.36***	13.36	(14.35)
8½"	–	18.90	***17.39***	30.26	24.34	–

TABLE A.7 Time consumption due to interruptions in hours

	SPECIFIC TIME CONSUMPTION (MIN/METRE)					
	WELL A	*WELL B*	*WELL C*	*WELL D*	*WELL E*	*WELL F*
24"/26"	204.5	330.5	106.5	169	25	13
16"	45	1.5	25	230	477	66
12¼"	488.5	6.5	37	29.5	56.5	33
8½"	–	61	181	57.5	207	–
	738	399.5	349.5	486	765.5	112

The experience data base also contains information about interruptions. These are unforeseen events. One common example is the storms off the coast of Norway during the winter season. Sometimes drilling must halt, and the marine riser disconnected.

Table A.7 lists total time spent on interruptions. Well E destroys the trend of improvements. On an average basis, 475 hours were spent on interruptions for each well. Waiting of weather accounts for only 89 of these hours on an average basis.

Tables A.8 to A.11 show more details about the interruption time. A lot of the time is spent on equipment repair. We will not discuss the details of these tables further, only point to the observation that this type of comparative analysis is important in order to determine the correcting activities. To reduce down time, compensating activities should be planned before the well is drilled.

A.2.3 Time Consumption Related to Hole Stability

A borehole stability analysis has also been performed outside the scope of the experience transfer data base. In the following a short review will be given, and the time lost due to borehole stability problems will be assessed.

TABLE A.8 Interruption time in the 26/24 in. hole sections

	INTERRUPTION					
PHASE	*WELL A*	*WELL B*	*WELL C*	*WELL D*	*WELL E*	*WELL F*
Fish		138.00	50.50	4.00		
Lost circ.						
Maintain						
Maintain/rep.	2.50	1.50	2.00	0.50	1.50	
Other	9.50		2.00			
Sidetrack		123.00				
Strike						
Subsea equip. f.	113.00	8.00	13.00	99.50	5.00	13.00
Subsurface equip. f.		7.00		37.00	17.00	
Surface equip. f.	17.00	7.00			1.50	
Wait	62.50	46.00	39.00	28.00		
Well control						
Well control metres, section	653.00	988.00	637.00	629.00	660.00	625.00
Total (hrs)	204.50	330.50	106.50	169.00	25.00	13.00

TABLE A.9 Interruption time in the 16 in. hole sections

	INTERRUPTION					
PHASE	*WELL A*	*WELL B*	*WELL C*	*WELL D*	*WELL E*	*WELL F*
Fish				139.00	65.50	
Lost circ.				16.50		
Maintain						
Maintain/rep.			1.50	41.50	2.50	2.00
Other						
Sidetrack					15.50	
Strike						
Subsea equip. f.	38.00		1.00		205.00	5.50
Subsurface equip. f.			4.00	24.00		
Surface equip. f.	7.00	1.50	18.50	9.00	45.50	1.00
Wait					143.00	57.50
Well control						
Well control metres, section	910.00	803.00	982.00	979.20	1425.00	787.00
Total (hrs)	45.00	1.50	25.00	230.00	477.00	66.00

The field is located in a highly faulted area. The tectonics give rise to anisotropic stresses. In the southern part of the field, the maximum horizontal stress exceeds the overburden load. Some practical consequences are:

- The tectonic stresses are probably worsening the borehole stability problems.
- The in-situ stresses control the fracture propagation directions. If the field will later be stimulated through fracturing, the efficiency of production or injection is directly tied to the in-situ stresses.
- Knowledge of the stress levels and orientations may aid the well path selection process by planning to minimise borehole stability-related problems.

In the following several borehole stability-related elements will be discussed.

TABLE A.10 Interruption time in the 12-1/4 in. hole sections

	INTERRUPTION					
PHASE	*WELL A*	*WELL B*	*WELL C*	*WELL D*	*WELL E*	*WELL F*
Fish	44.00			2.00		
Lost circ.	281.00			13.00	20.00	
Maintain						
Maintain/rep.	0.50	3.50	2.50	10.00	1.50	1.50
Other						
Sidetrack						
Strike						
Subsea equip. f.	121.00				33.50	7.50
Subsurface equip. f.			2.00		1.50	
Surface equip. f.	30.50	3.00	32.50	4.50		1.50
Wait	7.50					22.50
Well control	4.00					
Well control metres, section	1178.00	612.00	782.00	1000.00	1024.00	1104.00
Total (hrs)	488.50	6.50	37.00	29.50	56.50	33.00

TABLE A.11 Interruption time in the 8-1/2 in. hole sections

	INTERRUPTION					
PHASE	*WELL A*	*WELL B*	*WELL C*	*WELL D*	*WELL E*	*WELL F*
Fish		31.00	47.00	13.00		
Lost circ.			37.50			
Maintain						
Maintain/rep.		3.50	1.50	4.50	3.00	
Other			1.50			
Sidetrack						
Strike						
Subsea equip. f.				7.50	152.00	
Subsurface equip. f.		16.50	3.00		18.50	
Surface equip. f.		10.00	23.50	8.00	0.50	
Wait			67.00	24.50	33.00	
Well control						
Well control metres, section		500.00	552.00	453.00	657.00	
Total (hrs)		61.00	181.00	57.50	207.00	

Borehole Elongation

During the last 10–15 years a number of exploration wells and production wells have been drilled in this field. Moderate borehole collapse is a common feature in all these wells. In the reservoir sections the hole increase in the shale layers is typically 2–4 in., while the sand stringers in between are typically in-gauge. This is not an ideal situation with respect to cement placement, and is believed to have resulted in poor zonal isolation in several instances.

The collapse problem is clearly time dependent. If a hole section can be cased off within 4 days after drilling, little collapse results. However, logging and coring and interruptions often prohibit this.

The mud weights used today are higher than those used during earlier exploration drilling. Although this may have had a positive effect on hole collapse, the wells are now highly inclined which make them even more prone to enlargement. Circulation losses often set an upper limit on the mud weight.

The operating company have also used several mud systems in an attempt to reduce the collapse problems. In the earlier exploration wells gypsum/polymer mud was used, while now KCl/polymer mud of various salinities are used. Again, there are no significant improvements with respect to borehole collapse.

Circulation Losses

Circulation losses were experienced in several wells. This is believed to be governed by the minimum in-situ stress. Also, under several of the loss incidents, there were signs of pack-off due to insufficient hole cleaning. Under these circumstances, the formation may have sufficient integrity under normal circumstances, but an excessive pressure builds up in the borehole due to plugging.

Table A.12 summarises the reported circulation loss incidents. Most of these occurred during the cementing operation, but before the cement was displaced behind the casing. It is therefore probably tied to the hole cleaning problem mentioned above.

One particular experience observed during this pre-drilling phase was the connection between hole strength and mud quality. When using new mud during leak-off testing, a lower leak-off gradient apparently occurred compared to when using used mud with a higher solids content. The procedure during leak-off testing was therefore altered and now requires used drilling mud during testing.

Table A.12 below summarises the reported losses during the pre drilling phase.

Differential Sticking

In five of the six wells differential sticking was reported. The basic mechanism is that the pressure over balance inside the borehole forces the bottom hole assembly toward a permeable formation. The three key elements are the differential pressure, the existence of permeable formations, and a large projected area which the pressure can act on. The following number of incidents were reported:

1 in well A
1 in well B
5 in well C
1 in well D
2 in well E
0 in well F

TABLE A.12 Summary of circulation losses reported in the six wells

	DEPTH INTERVAL (MTVD)	*LOSS GRADIENT (S.G.)*
Well E	1741–1872	1.75
	2281–2547	1.81–1.84
Well C	2705–2759	1.81
Well A	1138–1146	–
	2327–2552	–
	2826	–
Well B	1167–1231	–
Well D	1143–1187	1.55
	1876–1918	1.64
	2407–2701	–
Well F	–	–

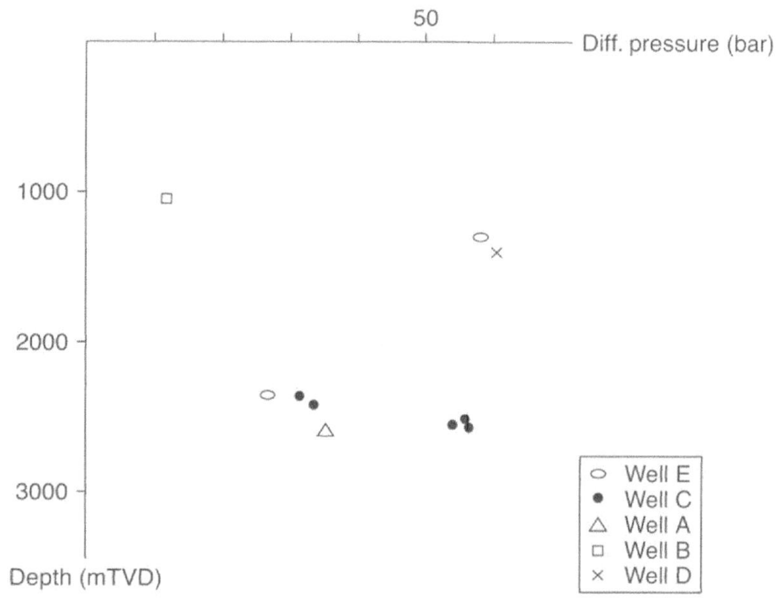

FIGURE A.1 The differential pressure recorded during the pipe sticking events.

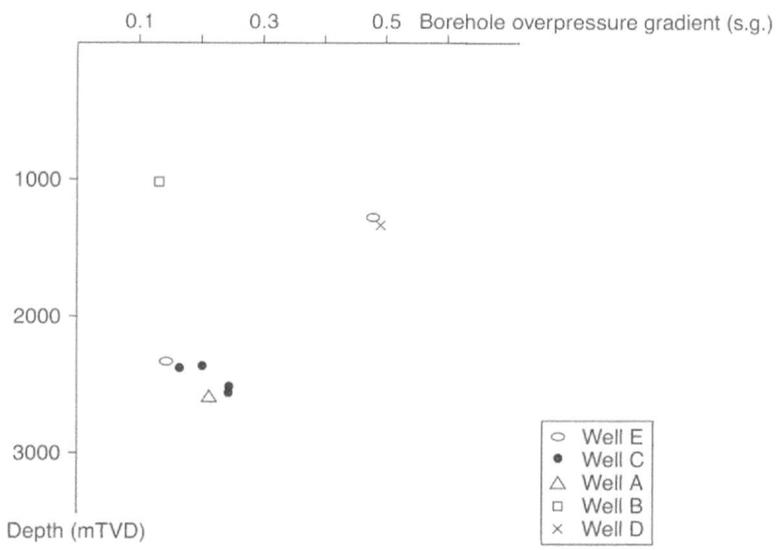

FIGURE A.2 The overbalance gradient during the pipe sticking events.

Figures A.1 and A.2 show the pressures and the gradients recorded during the reported differential sticking incidents. However, inspection of the lowest recordings question the identified mechanism. It is likely that some of these incidents are mechanical sticking rather than differential sticking, and probably tied to the issue of hole cleaning.

Tight Hole

Tight hole conditions can be taken as a measure of the state of the hole. If significant tight hole exists, it may be an indicator of a hole that may be subjected to a later collapse. Section 2.1 in this book covers the optimal mud weight selection to minimise collapse and tight hole conditions.

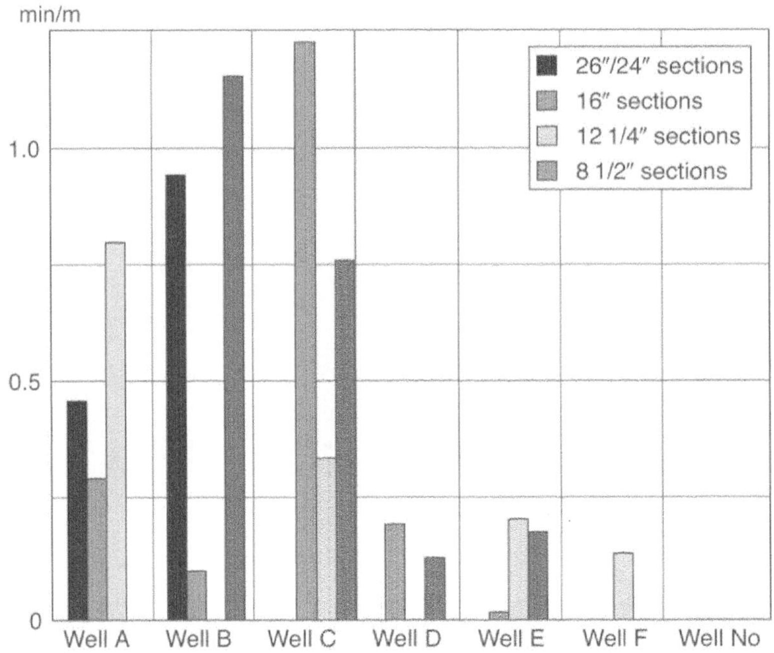

FIGURE A.3 Specific time consumption during reaming.

When tight hole arises, the hole has to be reamed and washed, which is a time-consuming process. Figure A.3 shows the specific reaming times for the six wells. After the third well was drilled, a new mud weight design principle was introduced, called the "median line principle". This principle is derived in Section 2.1. We observe that the reaming time was significantly reduced for the last three wells, with the last well having only negligible reaming. It should also be emphasised that this improvement shown in Figure A.3 also contains an element of a learning curve for the drilling personnel.

Summary

Table A.13 lists the various types of borehole stability problems recorded during pre-drilling of the six wells. Although well F had only minor problems, it is not representative for future wells as no coring took place, and only a limited formation evaluation program was performed.

Finally, Table A.14 gives an overview over the time consumption related to borehole stability.

Before closing this discussion on borehole stability, it should be observed that if the average time to drill each well is 60 days, unforeseen borehole stability problems account for 15% of the time consumption. This percentage has not seen a dramatic improvement the last decade, and for future wells, borehole stability is going to become an even more important issue.

A.2.4 Generating an Ideal Time Prognosis

The previous discussion comprises many elements, which have to be addressed separately. We will focus on the main elements from a well design point of view, and demonstrate how to utilise some of the information available.

During well planning it is difficult to estimate time consumption due to interruptions. In the time plan one must also incorporate contingency planning. Therefore, the time curves have often no common reference, and are therefore not directly comparable.

TABLE A.13 Summary of borehole stability problems for the six wells

HOLE SECTION:	*WELL A*	*WELL B*	*WELL C*	*WELL D*	*WELL E*	*WELL F*
24″		FS				
16″	TS		TH, FS	TH, FS, TS	TH, DF, TS	
12¼″	TH, FS DF, TS	TS		FS, TS	TH, FS	TH, FS
8½″		TH, FS	TH, FS DF, TS	TS	TH, FS, TS	

TH =Tight hole; FS = Stuck drillstring/casing; DF = Differential sticking; TS = Loss of circulation.

TABLE A.14 Time consumption related to borehole stability

EVENT	*TIME USED (DAYS)*
Circulation losses	15
Tight hole	2
Squeeze cementing	15
Stuck casings	20
Fishing	2
Total	52 days
Per well	52/6 = 8.7 days

Earlier in this appendix we have analysed the time consumption for various operations. This included the drilling time and the total time to drill a well including tripping, casing setting, formation evaluation, and interruptions. We will now use the best times from each section, and generate an ideal time consumption. This ideal time is achievable if no problems arise. We know, however, that some time loss will always occur.

From Tables A.1 to A.4 we can define the best achievable time consumption as follows:

In the 24/26 in. hole section well D performed best
In the 16 in. hole section wells D and E performed best
In the 12-1/4 in. section well D performed best
In the 8-1/2 in. section well C performed best.

These best data are listed in Table A.15 below.

An example demonstrates the application of the time prognosis.

Figure A.4 shows the time curve for well D. Also shown is the minimum prognosis derived from the best data of this evaluation. We observe that even if this well probably had the best performance of all six wells, considerable time was spent on unplanned events such as interruptions and hole stability problems. In fact, as the actual drilling time was 57 days compared to the minimum prognosis of 33 days, time consumption was 73% higher.

This example demonstrates the nature of a real drilling operation, where many factors influence the progress. The ideal minimum time is not realisable, but by careful planning the above percentage should be kept at a minimum.

TABLE A.15 Lowest time consumption achieved for the drilling progress

HOLE SECTION (INCH)	*DRILLING RATE (MIN/M)*
24/26	6.20
16	5.24
12¼	5.28
8½	10

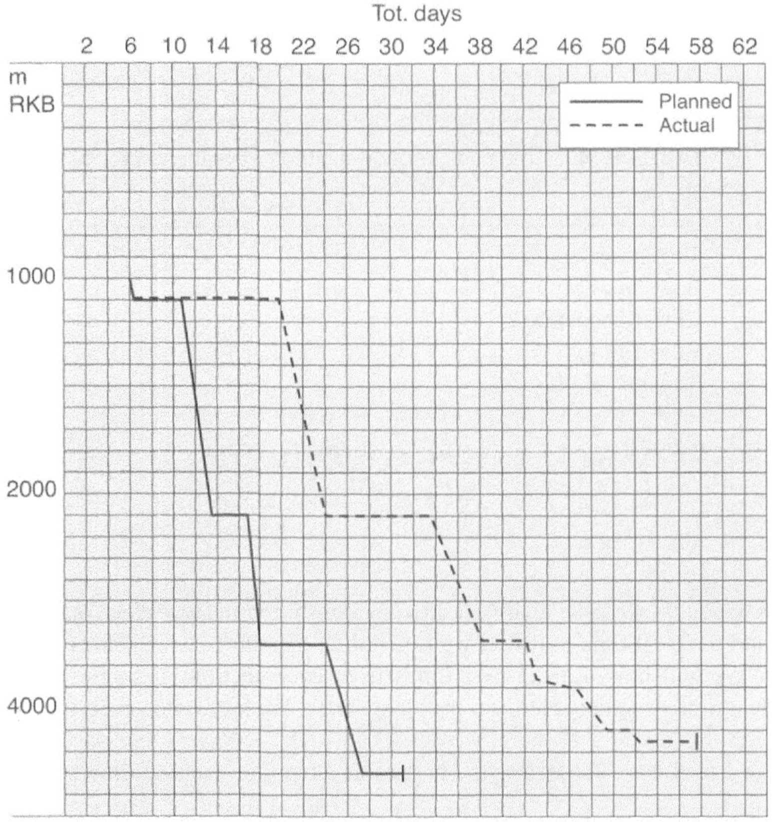

FIGURE A.4 Minimum time and actual time for well D.

A.3 A SIMPLE SYSTEM FOR EXPERIENCE TRANSFER

Studying the statistics, one finds that 10–20% of the rig time is spent handling unforeseen problems. This statistic has not improved significantly in the last 10 years. We have of course made some progress, but the wells have to some extent changed character. During the last 10 years long reach and horizontal wells have evolved into lengths never achieved before.

The industry is continuously changing. Currently there are trends towards deeper sub-sea wells, and new technology like slim-hole drilling is under development. Therefore, even if we make progress in one area, we will still be faced with new challenges, because the limits are always extended. An important way to improve is to learn from failures, therefore experience transfer is important.

Date: Time:	Report no:	No of pages enclosed:
Well:	Hole/section:	Depth interval:
Rig/platform:	Time interval/section:	

Instructions:

- Drilling engineer: Mark out key words below that closest describes the reported events
- Library (SAD): Store copies in well file

Casing type:

- ☐ Surface casing
- ☐ Conductor casing
- ☐ Intermediate casing
- ☐ Production casing
- ☐ Production liner

Well type:

- ☐ Exploration
- ☐ Appraisal
- ☐ Production

Key words:

Time delay:

- ☐ Waiting on weather
- ☐ Other time delay (specify)______________

Operations:

- ☐ Drilling
- ☐ Coring
- ☐ Logging
- ☐ Casing
- ☐ Completion
- ☐ Other operation (specify)______________

Hole stability:

- ☐ Loss/gain
- ☐ Differential sticking
- ☐ Tight hole
- ☐ Cavings
- ☐ Well control
- ☐ Other hole stability (specify)______________

System, equipment:

- ☐ Mud
- ☐ Mudlogging
- ☐ Cementing
- ☐ Logging equipment
- ☐ Casing equipment
- ☐ Directional drilling
- ☐ Hydraulics
- ☐ Drillstring
- ☐ Wireline
- ☐ Topside system (specify)______________
- ☐ Other syst. equipment (specify)______________
- ☐ Ref. ERF report no.

Safety:

- ☐ Los time accident
- ☐ Near miss
- ☐ Other safety (specify)______________

Remarks:

FIGURE A.5 Experience summary form.

The aforementioned problems are frequently of borehole stability type. Casing landing problems, stuck pipe, hole cleaning problems, and cementing problems are often a result of borehole stability problems. These are difficult to handle, and it is fair to say that we do not yet possess a full understanding of them. My approach is to try to analyse all problems and to evaluate each well design in light of these.

There is considerable potential in evaluating past failures. Yet, many operators have built large databases with various degree of improvement. The data quantity is of little help unless we perform quality evaluations. The drilling process is very complex and contains many elements, but in this appendix we have focused on the well itself.

In the following a simple experience transfer system will be proposed. The main function is as follows. Figure A.5 shows a form which identifies certain problems causing rig downtime. This form is filled out for every hole section drilled. The forms are stored and used to identify certain problem types. For example, during planning of production well number 11 a check in the Experience Summary Form book revealed that there had been lost circulation problems in wells 1, 3, 4, and 7. The reports from these wells were then checked out for further analysis.

Although a very simple system, it provides information very quickly, and can improve the efficiency of the well design process significantly.

Appendix B

Evaluation of Ballooning in Deep Wells

B.1 INTRODUCTION

During drilling of wells it is sometimes difficult to analyse observations about the drilling operation. One example is the excess gas often observed in high pressure wells. The decision-making process is tedious and difficult since we do not possess a full understanding of the underlying mechanisms. In this appendix we will shed light on some of these odd observations, but we realise that we do not yet have all the answers.

During drilling it is often observed that the return mud volume varies to some extent, giving either a too low or a too high return rate. Field personnel know this effect, which is considered normal. This effect is often called ballooning. Under critical circumstances, however, it may be difficult to separate this ballooning effect from responses such as having a mud loss or taking a kick. Therefore, if we manage to understand the physics of ballooning, we may separate this effect out, and evaluate the residue as a potential well control problem.

Gill (1987; 1989) has published a number of papers about ballooning. Since Gill describes the field observations rather well, we will quote some of his observations:

Symptoms of ballooning include:

- Loss of mud (30 to 100 bbl), at less-than-measured fracture strength, while circulating.
- Gaining mud back (30 to 100 bbl), if it is allowed, with pumps off.
- Reading about 300 psi shut-in-drill-pipe-pressure, if the well is shut in immediately after pumps are stopped and the pressure differential induced around the borehole is not allowed to relax and flow the lost mud back.
- Unnatural amounts of "reflex gas" milked temporarily from newly drilled sands after connections and trips, leading to still further mud weight increases.

In simple terms, a total volume change of 100–200 bbls may take place during a drilling operation due to ballooning. This volume is of the same order as a kick potentially leading to a blowout.

Although Gill describes the behaviour well, physical interpretation will be expanded in this section. He mainly explains the phenomenon as being caused by ballooning shales, that is, he assumes that the borehole wall can flex in either direction due to pressure variations. In the following we will propose an alternative explanation.

B.2 RESULTS FROM SIMULATION STUDIES

A Norwegian operator was planning a deep high-pressure well. To reduce the element of uncertainty in the critical phases, it was decided to undertake a study of ballooning. The study was composed of the following elements:

- pressure drop versus flow rate modelling
- temperature-dependent mud behaviour
- compressibility modelling of the drilling mud
- pressure-related volume expansion of casing and open hole

Each of the elements above is built into a simulator, which will not be described here. We will, rather, study the results. Two scenarios were investigated.

Figures B.1 and B.2 show the well design for the two cases. Ballooning case 1 is designed to address the critical phase of positioning the production casing without penetrating into the reservoir, and ballooning case 2 is designed to investigate effects during drilling of the critical reservoir section.

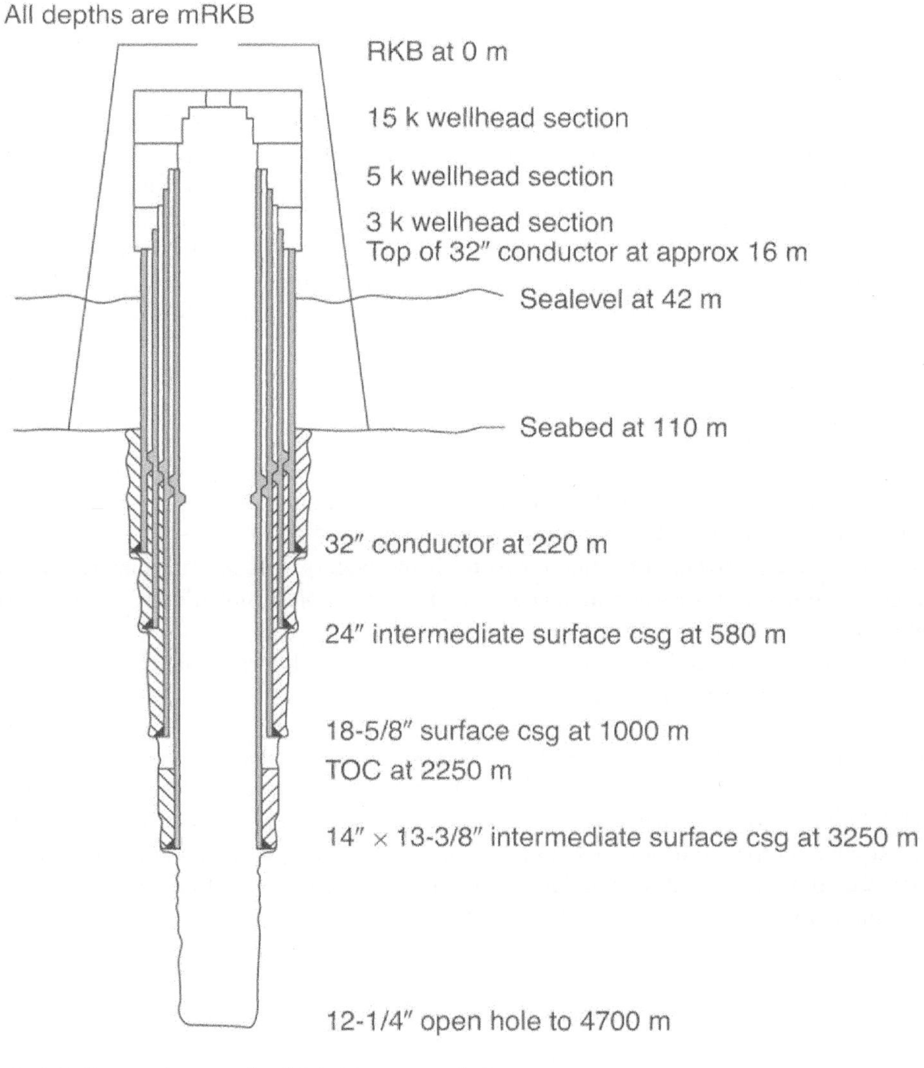

FIGURE B.1 Wellbore schematic for 12-1/4 in. open hole.

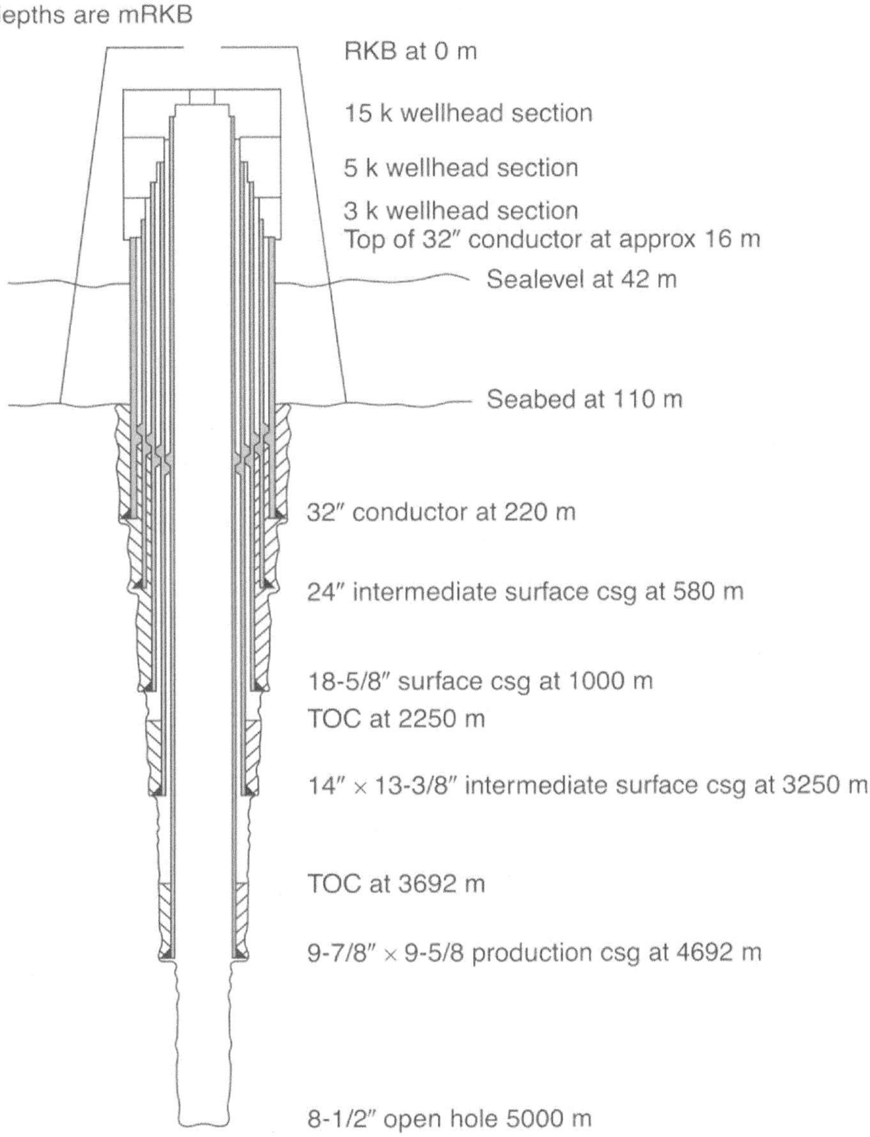

FIGURE B.2 Wellbore schematic for 8-1/2 in. open hole.

B.2.1 Pressure Effects

Using the well designs shown in Figures B.1 and B.2 it was first investigated what effect the mud flow has on the mud expansion. When the mud pumps are stopped, the frictional losses inside the hydraulic system cease, resulting in a slightly smaller mud pressure, which again allows for volume change within the mud, casing and the rock of the open hole. The results are displayed in Table B.1.

Table B.1 shows that the volume expansion taking place when the mud pumps are stopped is mainly due to the compressibility of the drilling mud. The volume increase due to rebound of casing and open hole amounts for only 5 and 14% for the two cases considered. The result is reasonable, since the casing and the rock are relatively stiff bodies, but the mud has a certain compressibility.

From the above example we will conclude that the volume increase seen during drill pipe connection operations is mainly due to expansion of the drilling mud.

The pressure effects discussed above are relatively simple. In the following we will look into a much more complex mechanism, namely the effects of temperature.

B.2.2 Temperature Effects

Before looking into the actual results from the simulations, we will discuss the temperature process. In Figure B.3a a static temperature profile in the well is shown. If the well is left static for a longer period of time, this is the expected profile. However, during drilling we are pumping mud into the drillpipe with return up the annulus. Since the mud is exposed to the temperature profile of the formation, it will become heated while going down, and cooled on its way up the annulus. The drilling fluid coming out of the annulus is usually warmer than the fluid pumped in.

We realise that the drilling fluid transports heat from the bottom of the hole and up the hole. During continuous flow we will therefore obtain a profile as illustrated in Figure B.3a. Since the drilling mud has a cooling effect, it may actually lower the bottom hole temperature as well. The hydraulic system is actually acting as a counter current heat exchanger, as illustrated in Figure B.3c. Figure B.3b illustrates that the outside and the inside of the drillpipe may have a slightly different temperature profile as well. Corre et al. (1984) gave an early attempt to model the system thermodynamically. The solution is very complex.

The well as a thermodynamic system has certain properties. The formation temperature is acting as an infinite source, and the convective heat loss is rather small. Therefore, the change in temperature is rather slow, and the process is nearly always transient or pseudo steady state. This is an important property of the system, which we will address in the following.

Several scenarios were constructed around the same two cases of Figures B.1 and B.2, to investigate the temperature effects. The temperature effects behave as slowly varying, or transient, in time. That is, the volume change is mainly a direct result of the heating or cooling. To simulate a realistic case, we will assume the scenario of Figure B.1. Furthermore, we assume that mud has been circulated at a constant rate for 10 hours. Then the pumps are stopped, and we observe the changes for the next 10 hours.

The input flow rate is shown in Figure B.4a., showing first 10 hours pumping, then 10 hours of no flow. The temperature of the return mud is shown in Figure B.4b. We observe that continuous pumping leads to a non-linear heating of the active mud system. This is well known from practical well operations. Furthermore, we see also that even at the end of the cycle (20 hrs), the surface annulus temperature is still 43°C, which implies that the mud system is not identical to the initial temperature profile.

In Figure B.3a we indicated that the bottom hole temperature is actually lowered when circulating. For case 1, the simulated profile is shown in Figure B.4c. The predicted minimum of 60°C is reached after about one hour. After that the bottom hole temperature increases because the temperature of the whole active mud system is increasing. However, after 20 hours, the bottom hole temperature is still not equal to the static temperature of 170°C. The bottom hole temperature is an important calibration point for this type

TABLE B.1 Short-time volume increase in the total hydraulic system if the mud pump is stopped

	12¼" HOLE (CASE 1)	*8½" HOLE (CASE 2)*
Flowrate (l/min)	2500	1500
Volume increase, mud (l)	220	160
Volume increase, open hole (l)	5	9
Volume increase, casing (l)	7	18

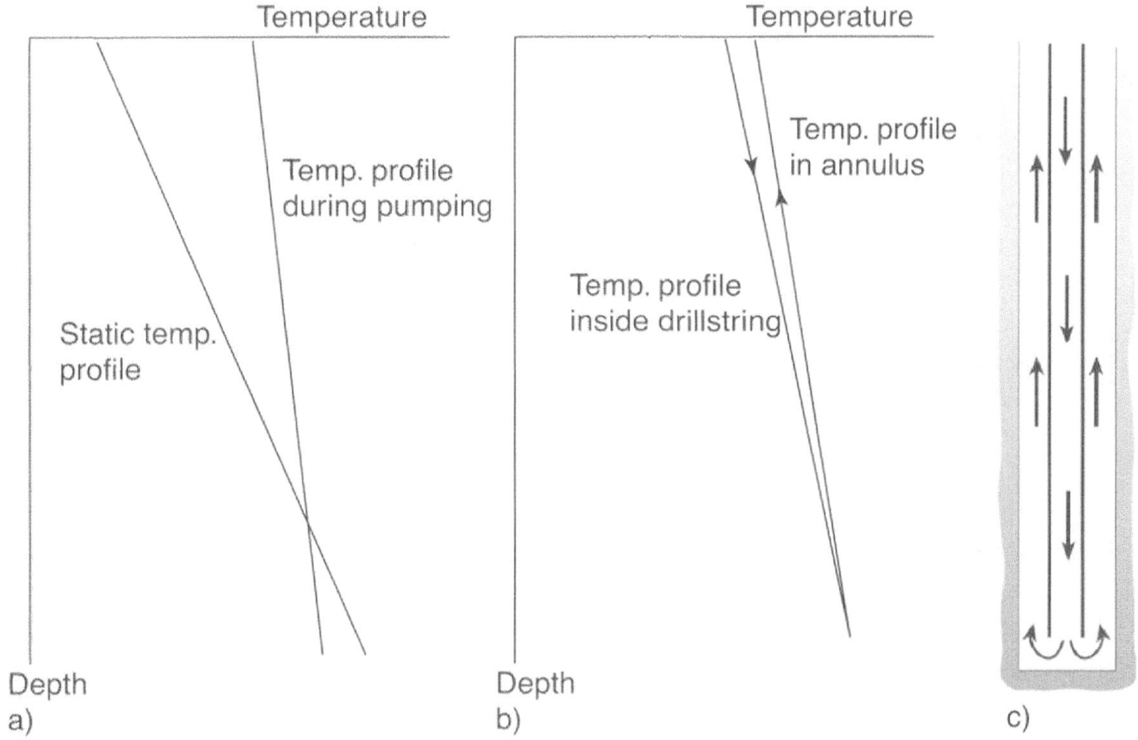

FIGURE B.3 Static and dynamic temperature profiles.

of simulation. At present we have little data, but we do know of a well where a bottom hole temperature drop from 180°C to 110°C was recorded. We therefore believe that this is a realistic case.

Figure B.4d shows the "ballooning" effect of the mud due to these temperature changes. After about two hours there is nearly 600 litres less mud in return compared to the input. After about seven hours, the mud volume is in balance, but at 10 hours there is 400 litres in excess. The additional 150 litres at 10 hrs is due to expansion when the frictional pressure drop is removed, as the mud pumps are stopped.

Finally, Figure B.4e shows the equivalent density below the drill bit. The nonlinear behaviour of the previous curves is reflected here. Also note that the static density gradient has decreased from 1.80 s.g. to 1.77 s.g.

The complex and time-dependent nature of the temperature effects on ballooning are clearly observed in the previous example. In the following we will make a brief comparison of the two cases of Figure B.1 and B.2, assuming first 10 hrs pumping followed by 10 hrs with no pumping.

Table B.2 illustrates the somewhat confusing picture. The temperature effects on the drilling fluid show in this case a variation in the order of –550 to +660 litres, a total variation in the order of 8 barrels.

However, when comparing Gill's observations (Gill, 1989) we observe that Table B.2 demonstrates a behaviour very similar to the observations. However, the field observations from the reference show a much larger variation in volume than the examples shown in Table B.2. The simulations were therefore repeated with a flow rate of 4000 l/min for case 1, and now the volume increase was 12,000 litres or 75 bbls.

It is believed that the physics of the ballooning process is described with the previous examples. However, as the heat transfer process is both significant and complex, more work must be performed to establish more general criteria.

In the following, we will look into a simple way to estimate some of these effects in the field.

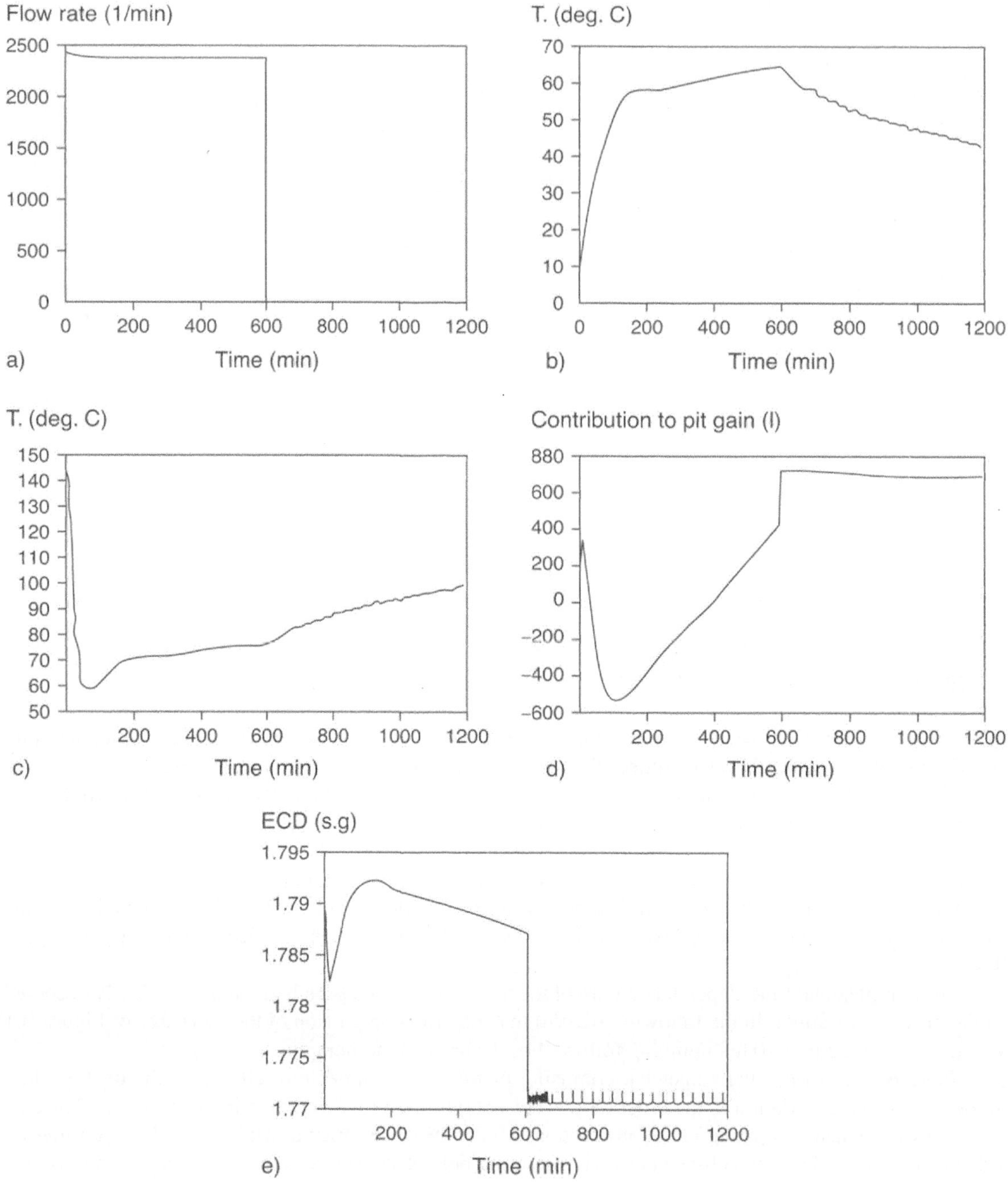

FIGURE B.4 Long-time behaviour of the mud system due to temperature effects. a) Inlet flow rate, b) mud temperature at outlet, c) mud compression and expansion, and d) equivalent circulating density.

TABLE B.2 Volume changes at various times

	12 ¼" HOLE (CASE 1)	*8 ½" HOLE (CASE 2)*
Flowrate (l/min)	2400	1500
Volume change after 2 hrs (l)	-550	-460
Volume change after 5 hrs	-200	-520
Volume change after 10 hrs	+380	-500
After the pumps are stopped:		
Volume change after 10 hrs	+660	-320
Volume change after 20 hrs	+620	-70

B.3 COUPLING BETWEEN PIT GAIN AND MUD DENSITY

On a floating rig the volumetric measurement accuracy is in the order of cubic metres. This is mainly due to rig motions, and level height sensor location in the mud pit. On fixed platforms such as production platforms and jack-up drilling rigs, the volumetric accuracy is higher. At present, rig personnel claim to measure the return mud volume with an accuracy down to 100 litres. If this is the case, we can use this information to calculate the effective density of the drilling mud.

Figure B.5 shows a conceptual picture of the ballooning concept. To the left is a well shown with a given volume. In the middle a too low return case is illustrated. If we had only one well volume of drilling mud at our disposal, the annulus volume would have dropped a height h. Likewise, the excess mud case is shown to the right. Receiving more mud in return can be visualised as a hypothetical annulus extended to a height h.

First of all, a given mud volume can be converted into an equivalent annulus height as follows:

$$h = \text{Mud gain or loss/Annular capacity of the borehole.} \quad \text{(B.1)}$$

Performing a mass balance on the mud in the well, we will first consider the too low return case. We pumped down a mud measured at surface conditions with a density of ρ. After some time we would have obtained the situation shown in the middle figure of Figure B.5, if we had only one annulus volume at our disposal. The mud in the well is now more dense than initially. However, the well is full, such that the additional pressure exerted by the height h can be approximated as:

$$0.098\rho h \quad \text{(B.2)}$$

and the effective pressure at the bottom of the well as:

$$0.098\rho D + 0.098\rho h \quad \text{(B.3)}$$

Expressed with the effective density at the bottom of the well:

$$0.098\rho_1 D \quad \text{(B.4)}$$

Equating Equations B.3 and B.4 yields the approximate expression for the effective mud density as a function of too low mud return:

$$\rho_1 = \rho\left(1 + \frac{h}{D}\right) \tag{B.5}$$

Performing a similar evaluation for the excess mud return of Figure B.5, we obtain a resulting effective mud density which is lighter than initially. The resulting equation becomes:

$$\rho_1 = \rho\left(1 - \frac{h}{D}\right) \tag{B.6}$$

We will in the following use the above simplified equations to estimate the effective mud densities during the previous cases 1 and 2, and to compare these to the results from the more complex simulators.

Examples

Example 1. For the 12-1/4 in. hole evaluated in case 1, we obtain the following data:

Flow rate:	4000 l/min
Maximum pit gain:	12m^3
Mud density at surface:	1.8 s.g.
Capacity of well:	1/15.78 m^3/m
Depth of well:	4700 m

The excess mud return is equivalent to the following height in the well:

$$h = 12 \text{ m}^3 \times 15.78 \text{ m/m}^3 = 189 \text{ m}$$

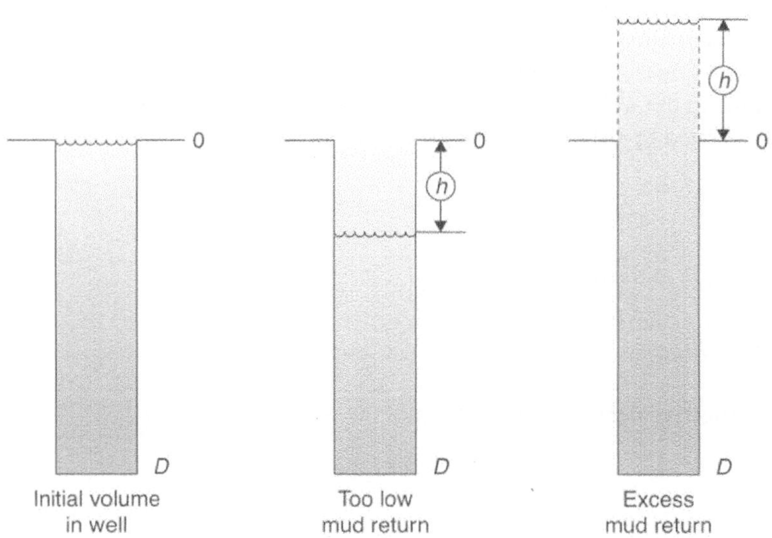

FIGURE B.5 Conceptual model to estimate the effective mud density.

This is equivalent to a reduced effective mud density equal to:

$$\rho_1 = 1.80\left(1 - \frac{189 \text{ m}}{4700 \text{ m}}\right) = 1.73 \text{s.g.}$$

The simulator predicted an effective mud density of 1.72 s.g. for the same case, so the two methods are in reasonable agreement.

Example 2. Here we want to investigate the mud density during drilling of the 8-1/2 in. hole section as defined in case 2. The following input data are given:

Flow rate:	2500 l/min
Maximum pit gain:	1.4 m^3
Mud density at surface:	2.12 s.g.
Capacity of well:	1/41.77m^3/m
Depth of well:	5000 m

The excess mud return is equivalent to the following height in the well:

$$h = 1.4 \text{ m}^3 \times 41.77 \text{ m} / \text{m}^3 = 58.5 \text{ m}$$

This is equivalent to a reduced effective mud density equal to:

$$\rho_1 = 2.12\left(1 - \frac{58.5 \text{ m}}{5000 \text{ m}}\right) = 2.095 \text{ s.g.}$$

The simulator predicted an effective mud density of 2.08 s.g. for the same case, so the two methods are again in reasonable agreement.

The above density correction can be valuable to adjust well data towards more realistic values. Also, we observe that considerable errors can be introduced if we are not correcting our density data.

Finally, the method shown directly couples the mud return rate to the effective density in the well. By introducing the element of ballooning we may reduce the element of unknowns in critical wells. We also have a tool to redesign the operation if difficult situations arise.

B.4 THE GAS PROBLEM

In the introduction, the problem of excess gas was mentioned. During drilling of critical wells, often more gas is brought to the surface than expected. Sometimes abnormal events take place, for example the excess gas may increase after the mud weight is increased. Gas reading is probably the single most important pore pressure indicator in high-pressure wells, and we understand the difficulties of interpretation when such events take place.

We cannot provide a complete picture of mechanisms associated with gas readings, but will in the following address a few.

- The effective mud density is lower due to the effects described in this appendix.
- Surge and swab pressures arising during tripping may "milk" gas from the formation into the wellbore.

- An abnormally high mud weight will compress the borehole wall in an attempt to increase its diameter. This may give rise to a higher pore pressure in the vicinity of the borehole.
- Creep or compaction effects coupled with fluid flow may locally alter the pore pressure.
- The gas readings used today are only qualitative. Future work should quantify these to establish the correct amount of drilled out gas.

B.5 SUMMARY

In this section we have looked into volumetric changes within the mud system, usually called ballooning. The extent of ballooning is large in certain instances, with a magnitude large enough to mask pore pressure responses.

A simulator was built to investigate the ballooning phenomenon. It was concluded that most of the ballooning was due to compression and expansion of the drilling mud, and that the volume change associated with the borehole and casing is relatively small.

Short-time volume change, as for example back-flow during connections, can be explained as expansion of the mud when the frictional pressure drop of the hydraulic system vanishes. This is a relatively small effect, up to a few hundred litres.

Long-time volume gain or losses are due to thermal effects within the complete hydraulic system. This can in certain instances amount to many cubic metres. This is a transient or a pseudo-steady state effect with a time factor of hours. In other words, short-time effects are mainly pressure relief, while long-time effects are mainly due to a slowly changing temperature profile.

Finally, a simple field method was proposed to estimate the effective density at the bottom of the well.

Other aspects and discussions of practical applications of mud cooling are discussed by Maury & Guenot (1993) and Corre & Guenot (1984).

Problem

A kick was taken during drilling of an HPHT well. After successfully bullheading the kick, the well was shut in and observed. The wellhead pressure increased slowly over a period of a few weeks, and when opened some mud returned to the surface. At that time the conclusion was that there was a very high pore pressure in the well. However, we will investigate if temperature expansion is another explanation.

The 8-1/2 in. hole section is drilled to 4500 m. Mud density at standard conditions is 2.10 s.g. A volume imbalance is seen of 10 m^3. Calculate:

a) The effective mud density if there is excess mud in the active pit.
b) The effective mud density if there is less mud in the active pit.
c) Discuss if this is a realistic explanation of the well behaviour. Propose alternative explanations.

References

Aadnøy, B.S. (1990). Inversion technique to determine the in-situ stress field from fracturing data. *Journal of Petroleum Science and Engineering*, Vol. 4, 127–141.

Aadnoy, B.S. (1991). Effects of reservoir depletion on borehole stability. *Journal of Petroleum Science and Engineering*, Vol. 6, 57–61.

Aadnoy, B.S. (1998). Geomechanical analysis for deep-water drilling. Paper SPE 39339 presented at the IADC/SPE Drilling Conference, Dallas, Texas, 3–6 March.

Aadnoy, B.S. and Andersen, K. (2001). Design of oil wells using analytical friction models. *Journal of Petroleum Science and Engineering*, Vol. 32, 53–71.

Aadnoy, B.S. and Andersen, K. (1998). Friction Analysis for Long-Reach Wells. Paper IADC/SPE 39391 presented at the IADC/SPE Drilling Conference held in Dallas, Texas 3–6 March 1998.

Aadnoy, B.S. and Bakoy, P. (1992). Relief well breakthrough in a North Sea problem well. *Journal of Petroleum Science and Engineering*, Vol. 8, 133–152.

Aadnoy, B.S. and Belayneh, M. (2004). Elasto-plastic fracturing model for Wellbore stability using non-penetrating fluids. *Journal of Petroleum Science and Engineering*, Vol. 45, 179–192.

Aadnoy, B.S. and Chenevert, M.E. (1987). Stability of highly inclined boreholes. *SPE Drilling Engineering*, Vol. 2, No. 4, 364–374.

Aadnoy, B.S. and Kaarstad, E. (2006). Theory and application of buoyancy in wells. IADC/SPE 101795 presented at the 2006 IADC/SPE Asia Pacific Drilling Tech. Conf. and Exhibition, Bangkok, Thailand, November 13–15.

Aadnoy, B.S. and Larsen, K. (1989). Method for fracture gradient prediction for vertical and inclined boreholes. *SPE Drilling Engineering*, Vol. 4, No. 2, 99–103.

Aadnøy, B.S. and Looyeh, R. (2019). *Petroleum Rock Mechanics: Drilling Operations and Well Design –* Second Edition. Elsevier Publishing. Paperback ISBN 978-0-12-815903.

Aadnoy, B.S., Bratli, R.K. and Lindholm, C. (1994). In-situ stress modelling of the Snorre field. Paper presented at Eurock 94, Delft, the Netherlands. Proceedings: Rock Mechanics in Petroleum Engineering, A.A. Balkema, 1994, ISBN 90 5410 502X, 871–878.

Aadnoy, B.S., Fazaelizadeh, M. and Hareland, G. (2010). A 3-dimensional analytical model for Wellbore friction. *Journal of Canadian Petroleum Technolology*, Vol. 49, No. 10, 25–36. Paper Number: SPE-141515-PA.

Aadnoy, B.S., Larsen, K. and Berg, P.C. (1999). Analysis of stuck pipe in deviated boreholes. Paper SPE 56628 presented at the 1999 SPE Annual Technical Conf. and Exhibition, Houston, Texas, October 3–6.

Aadnoy, B.S., Larsen, K. and Berg, P.C. (2003). Analysis of stuck pipe in deviated boreholes. *Journal of Petroleum Science and Engineering*, Vol. 37, 195–212.

Aadnoy, B.S., Mostafavi, V. and Hareland, G. (2009). Fracture mechanics interpretation of leak-off tests. Paper SPE 126452 presented at the Kuwait International Petroleum Conference and Exhibition, December 14–16, 2009.

Aadnoy, B.S., Soteland, T. and Ellingsen, B. (1991). Casing point selection at shallow depth. *Journal of Petroleum Science and Engineering*, Vol. 6, 45–55.

Aadnoy, B.S., Belayneh, M., Arriado, M. and Flatebo, R. (2008). Design of well barriers to combat circulation losses. *SPE Drilling and Completion*, Vol. 23, 295–300.

Aarrestad, T.V. and Blikra, H., (1994). Torque and drag-Two factors in extended reach drilling. *Journal of Petroleum Technology*, Vol. 46, 800–803.

Aasen, J.A. and Aadnoy, B.S.(2007). Three-dimensional well tubular design improves margins in critical wells. *Journal of Petroleum Science and Engineering*, Vol. 56, No. 4, 232–240.

Abdo, J. and Haneef, M.D. (2013). Clay nanoparticles modified drilling fluids for drilling of deep hydrocarbon wells. *Applied Clay Science*, Vol. 86, 76–82.

Aftab, A., Ismail, A.R. and Ibupoto, Z.H. (2017). Enhancing the rheological properties and shale inhibition behavior of water-based mud using nano silica, multi-walled carbon nanotubes, and graphene nanoplatelet. *Egyptian Journal of Petroleum*, Vol. 26, 291–299.

Aftab, A., Ismail, A.R., Khokhar, S. and Ibupoto Z.H. (2016). Novel zinc oxide nanoparticles deposited acrylamide composite used for enhancing the performance of water-based drilling fluids at elevated temperature conditions. *Journal of Petroleum Science and Engineering*, Vol. 146, 1142–1157.

Ahmed, N., Alam, M.S. and Salam, M.A. (2020). Experimental analysis of drilling fluid prepared by mixing iron (III) oxide nanoparticles with a KCl-Glycol-PHPA polymer-based mud used in drilling operation. *Journal of Petroleum Exploration and Production Technology*, Vol. 10, 3389–3397.

Aker, B.P. (2025). Next generation field development planning. 2025, from https://akerbp.com/en/next-generation-field-development-planning/

Al Tamimi, H., Moreno, C. and Dolle, N. (2024). Enabling digital transformation by automating the well trajectory design. SPE-222275-MS Paper presented at the ADIPEC, Abu Dhabi, UAE, November 2024.

Alfsenen, T.E., Heggen, S., Blikra, H. and Tjoetta, H. (1995). Pushing the limits for extended reach drilling: World record from platform Statfjord C, well C2. SPE *Drilling & Completion*, Vol. 10, No. 2, 71–76.

Al-saba, M.T., Al Fadhli, A., Marafi, A., Hussain, A., Bander, F. and Al Dushaishi, M.F. (2018). Application of nanoparticles in improving rheological properties of water-based drilling fluids. SPE-192239-MS paper presented at the SPE Kingdom of Saudi Arabia Annual Technical Symposium and Exhibition, Dammam, Saudi Arabia, April 2018.

Altindal, M.C., Nivlet, P., Tabib, M., Rasheed, A., Kristiansen, T.G. and Khosravanian, R. (2024). Anomaly detection in multivariate time series of drilling data. *Geoenergy Science and Engineering,* Vol. 237, Article 212778.

Alvi, M.A.A., Belayneh, M., Fjelde, K.K., Saasen, A. and Bandyopadhyay, S. (2021). Effect of Hydrophobic Iron Oxide nanoparticles on the properties of oil-based drilling fluid. *Journal of Energy Resource Technology*, Vol. 143, No. 4, 043001.

API Bulletin 5C3. (1989). *Bulletin on Formulas and Calculations for Casing, Drill Pipe and Line Pipe Properties.* Fifth edition, July 1989. American Petroleum Institute.

API Bulletin 5C3 (1994): Bulletin on formulas and calculations for casing, tubing, drill pipe and line pipe properties., Sixth edition, October 1994 Americal Petroleum Institute.

API RP 13D. (2009). *Recommended Practice on the Rheology and Hydraulics of Oil-Well Drilling Fluids*, API RP 13D, Fourth Edition. American Petroleum Institute.

API. (2006). *Rheology and Hydraulics of Oil-well Drilling Fluids.* API Recommended Practice 13D, Fifth edition, June 2006.

Bandyopadhyay, S., Singh, G., Sandvig, I., Sandvig, A., Mathieu, R., Anil Kumar, P. and Glomm, W.R. (2014). Synthesis and in vitro cellular interactions of superparamagnetic Iron nanoparticles with a crystalline gold shell. *Applied Surface Science*, Vol. 316, No. 10, 171–178.

Baram, M.S. (2011). Final Report on the Investigation of the Macondo Well Blowout. Available at: https://scholarship.law.bu.edu/faculty_scholarship/3227

Beharie, C., Francis, S. and Øvestad, K.-H. (2015). Resin: An alternative barrier solution material. SPE 173852-MS paper presented at the SPE Bergen One Day Seminar, Bergen, Norway, April 2015.

Belayneh, M. and Aadnoy, B.S. (2003). Fracture mechanics model of tensile borehole failure. Paper presented at the Sixth International Conference on Analysis of Discontinuous Deformation (ICADD), Trondheim, Norway, October 5–8, 2003.

Belayneh, M., Aadnøy, B.S. and Sharman, T. (2016). Dynamic and static sagging characterization and performance of four oil-based muds. OMAE2016-54457, Proceedings of the ASME 2016 35th International Conference on Ocean, Offshore and Arctic Engineering OMAE2016, June 19-24, 2016, Busan, South Korea.

Berland, S.A. (1993). Modelling and field evaluation of time-dependent borehole collapse. M.S. thesis in petroleum engineering, Rogaland University Centre, Stavanger. (In Norwegian).

Bimastianto, P., Khambete, S., AlSaadi, H., Couzigou, E., Al-Marzouqi, A., Chevallier, B., Qadir, A., Pausin, W. and Vallet, L. (2020). Digital twin implementation on current development drilling, benefits, and way forward. Paper presented at the Abu Dhabi International Petroleum Exhibition & Conference, Abu Dhabi, UAE, November 2020.

Bleler, R. (1990). Selecting a drilling fluid. *Journal of Petroleum Technology*, Vol. 42, 832–834.

Bourgoyne, A.T., Millheim, K.K., Chenevert, M.E. and Young, F.S. (1986). Applied drilling engineering. *SPE Textbook Series*, Vol. 2, ISBN 1-55563-001-4.

Bradley, W.B. (1979). Failure of inclined boreholes. *Journal of Energy Resources Technology*, Vol. 102, 232–239.

Bradley, W.B. (1990). A task force approach to reducing stuck pipe costs. SPE/IADC paper 21999 presented at the 1990 SPE/IADC Drilling Conference, Amsterdam, October 1990.

Breckels, I.M. and van Eeckelen, H.A.M. (1982). Relationship between horizontal stress and depth in sedimentary basins. *Journal of Petroleum Technology*, Vol. 34, 2191–2199.

Bui, B., Saasen, A., Maxey, J., Ozbayoglu, M.E., Miska, S.Z., Yu, M. and Takach. N.E. (2012). Viscoelastic properties of oil-based drilling fluids. *Annual Transactions of the Nordic Rheology Society*, Vol. 20, 33–47.

Caenn, R., Darley, H.C.H. and Gray, G.R. (2011). *Composition and Properties of Drilling and Completion Fluids*, 6th ed.; Gulf Professional Publishing, Houston, TX, USA; ISBN 978-0-12-383858-2.

Cao, J., Gocmen, E., Nabavi, J., Oedegaard, S. I., Kristiansen, T. G., Khosravanian, R. and Solem, K. (2025). Optimizing drilling operations in real-time: The Role of Digital Twins in reducing risks and enhancing performance. SPE-223819-MS Paper presented at the SPE/IADC International Drilling Conference and Exhibition, Stavanger, Norway, March 2025.

Cartalos, U. and Dupuis, D. (1993). An analysis accounting for the combined effect of drillstring rotation and eccentricity on pressure losses in slimhole drilling. Paper SPE/IADC 25769 presented at the 1993 SPE/IADC Drilling Conference in Amsterdam, February 23–25, 1993.

Casson, N. (1959). *Rheology of Disperse Systems*. Pergamon Press, London, 84.

Cayeux, E., Pelfrene, G., Mihai, R., Dvergsnes, E., Tjøstheim, B.P., Baume, A., Khosravanian, R. and Kristiansen, T.G. (2023). A new paradigm for automatic well path generation using multidisciplinary constraints. SPE-215021-MS Paper presented at the SPE Annual Technical Conference and Exhibition, San Antonio, Texas, USA, October 2023.

Charlez, P. and Heugas, O. (1991). Evaluation of optimal mud weight in soft shale levels. *Rock Mechanics as a Multidisiplinary Science*, J.C. Rogiers (ed.), Balkema, Rotterdam: CRC Press.

Clark, R.K. et al. (1976). Polyacrylamide-potassium-chloride mud for drilling water sensitive shales. *Journal of Petroleum Technology*, Vol. 261, 719–727.

Collins, R.E. and Graves, W.G. (1978). A new rheological model for non-Newtonian fluids. SPE 7654.

Corre, B., Eymard, R. and Guenot, A. (1984). Numerical computation of temperature distribution in a wellbore while drilling. Paper SPE 13208 presented at the 59th Annual Tech. Conference and Exhibition, Houston, Texas, September 16–19, 1984.

Crockett, A.R., Okusu, N.M. and Cleary, M.P. 1986. A complete integrated model for design and real-time analysis of hydraulic fracturing operations. Paper SPE 15069 presented at the 56th California Regional Meeting of the Society of Petroleum Engineers, Oakland, California, April 2–4, 1986.

Dahl, N. and Solli, T. (1993). The structural evolution of the Snorre field and surrounding areas. *Geological Society, London, Petroleum Geology Conference Series*, Vol. 4, 1159–1166.

Daines, S.R. (1982). Prediction of fracture pressures for wildcat wells. *Journal of Petroleum Technology*, Vol. 34, 863–872.

Davidovits, J. (1991). Geopolymers: inorganic polymeric new materials. *Journal of Thermal Analysis*, Vol. 37, 1633–1656.

De Bree, P. and Walters, J.V. (1989). Micro/minifrac test procedures and interpretation for in situ stress determination. *International Journal of Rock Mechanics Science & Geomechanics Abstract,* Vol. 26. No. 6, 515–521.

Dehghani, F., Kalantariasl, A. and Peyvandi. K. (2023). The effect of a novel nanocomposite in water-based drilling fluid on filtration and Rheology factors. *Transactions of the Indian National Academy of Engineering*, Vol. 8, 639–645.

Dehghani, F., Kalantariasl, A., Saboori, R. and Sabbaghi, S. (2019). Performance of carbonate calcium nanoparticles as filtration loss control agent of water-based drilling fluid. *SN Applied Sciences*, Vol. 1, 1466.

Dong, X., Sun, J., Huang, X., Lv, K., Zhou, Z. and Gao, C. (2022). Nano-laponite/polymer composite as a filtration reducer on water-based drilling fluid and mechanism study. *Royal Society Open Science*, Vol. 9, No. 10, 220385.

Doschek, M., Wamani, D., Sveinall, H., James, R. and Faschingbauer, R. (n.d.). Case history: A description of the technical experiences, benefits, and challenges of a real-time drilling fluids monitoring system when integrated into a drilling campaign. Presented by: Henrik Sveinall, Intelligent Mud Solutions AS, Auglendsmyra 8, 4016 Stavanger, Norway.

Dunbar, M.E., Warren, T.M. and Kadaster, A.G. (1986). Bit sticking caused by borehole deformation. *SPE Drilling Engineering*, Vol. 1, No. 4, 417–425.

Dupriest, F.E., Elks, W.C., Ottesen, S., Pastusek, P.E., Zook, J.R. and Aphale, C.R. (2011). Borehole-quality design and practices to maximize drill-rate performance. SPE *Drilling & Completion,* Vol. 26, No. 02, 303–316. Paper Number: SPE-134580-PA.

Duxson, P., Fernandez-Jimenez, A., Provis, J.L., Lukey, G.C., Palomo, A. and van Deventer, J.S.J. (2007). Geopolymer technology: The current state of the art. *Journal of Materials Science*, Vol. 42, 2917–2933.

Eaton, B. (1975). The equation for geopressure prediction from well logs. Society of Petroleum Engineers of AIME.

Edwards, S.T., Coley, C.J., Whitley, N.A., Keck, R.G., Ramnath, V., Foster, T., Coghill, K. and Honey, M.A. (2013). Summary of wired drill pipe (IntelliServ Networked Drillstring) field trials and deployment in BP. SPE/IADC 163560 Paper presented at the SPE/IADC Drilling Conference, Amsterdam, The Netherlands, March 2013.

Ershadi, V., Ebadi, T., Rabani, A.R., Ershadi, L. and Soltanian, H. (2011). The effect of Nano Silica on cement matrix permeability in oil well to decrease the pollution of receptive environment. *International Journal of Environmental Science and Development*, Vol. 2, No. 2, 128–132.

Fatemeh, M., Fjelde, K.K., Jostein, S., Saasen, A. and Vrålstad, T. (2015). A study of possible solutions for cost-efficient Subsea Well Abandonment. I: ASME 2015 34th International Conference on Ocean, Offshore and Arctic Engineering. The American Society of Mechanical Engineers (ASME). ISBN 978-0-7918-5647-5.

Fatemeh, M., Fjelde, K.K., Saasen, A. and Vrålstad, T. (2014). An investigation of different approaches for probabilistic cost and time estimation of Rigless P&A in Subsea multi-well campaign. I: SPE Bergen One Day Seminar, Bergen, Norway, 2 April 2014 - Conference Proceedings. Society of Petroleum Engineers. ISBN 978-1-61399-323-1.

Fatemeh, M., Fjelde, K.K., Saasen, A. and Vrålstad, T. (2015). Essential aspects in probabilistic cost and duration forecasting for Subsea Multi-Well abandonment: Simplicity, industrial applicability and accuracy. I: SPE-173850-MS- SPE Bergen One Day Seminar 2015 - Seminar digital proceedings. Society of Petroleum Engineers. ISBN 978-1-61399-384-2.

Fatemeh, M., Fjelde, K.K., Saasen, A. and Vrålstad, T. (2015). Probabilistic cost and time estimation of rigless plug and abandonment. *Journal of Petroleum Technology*, Vol. 67, 117–119. ISSN 0149-2136.

Fazaelizadeh, M, Hareland, G. and Aadnoy, B.S. (2010). Application of new 3-D analytical model for directional wellbore friction. Aadnoy. Canadian Journal of Petroleum Technology.

Fjaer, E., Holt, R., Raaen, A.M. and Risnes, R. (1992). Petroleum-related rock mechanics. *Developments in Petroleum Science*, Vol. 33, Elsevier Science Publishers B.V., Amsterdam, ISBN 0-444-88913-2.

Fjelde, K.K. (2012). Plug and Abandonment Lecture Materials, Lecture Note, University of Stavanger.

Foster, R. and Macmillan, R. (2018). High-speed telemetry on wired drill pipe, history, and impact on drilling process. OOTC-28735-MS Paper presented at the Offshore Technology Conference, Houston, Texas, USA. April 2018.

Fowler, J.R., Klementich, E.F. and Chappell, J.F. (1983). Analysis and testing of factors affecting collapse performance of casing. *Transactions of the ASME*, Vol. 105, 574–579.

French, F.R. and McLean, M.R. (1993). Development drilling problems in high-pressure reservoirs. *Journal of Petroleum Technology*, Vol. 45, 772–777.

Fulks, J. and Carragher, P. (2020). Alternative abandonment material: Bismuth BiSN. SPE-202354-MS paper presented at the SPE Asia Pacific Oil & Gas Conference and Exhibition, Australia, October 20–22, 2020.

Gasda, S.E., Bachu, S. and Celia, M.A. (2004). Spatial characterization of the location of potentially leaky wells penetrating a deep saline aquifer in a mature sedimentary basin. *Environmental Geology*, Vol. 46, 707–720.

Gill, J.A. (1987). Well logs reveal true pressures where drilling responses fail. *Oil and Gas Journal*, Vol. 85, 41–45.

Gill, J.A. (1989). How borehole ballooning alters drilling responses. *Oil and Gas Journal*, Vol. 87, 43–52.

Gray-Stephens, D., Cook, J.M. and Sheppard, M.C. (1994). Influence of pore pressure on drilling response in hard shales. *SPE Drilling Engineering*, Vol. 9, No. 4, 263–270.

Halal, A.S. and Michell, R.F. (1994). Casing design for trapped annular pressure buildup. *SPE Drilling Engineering*, Vol. 9, No. 2, 107–114.

Hale, A.H., Mody, F.K. and Salisbury, D.P. (1993). The influence of chemical potential on wellbore stability. *SPE Drilling & Completion*, Vol. 8, No. 3, 207–216.

Halliburton. (2025). Digital Well Construction – DecisionSpace® 365 Enterprise, 2025, from www.halliburton.com/en/software/decisionspace-365-enterprise/digital-well-construction.

Hareland, G. and Hoberock, L.L. (1993). Use of drilling parameters to predict in-situ stress bounds. Paper SPE/IADC 25727 presented at the 1993 SPE/IADC Drilling Conference, Amsterdam, February 23–25, 1993.

Hazov, V.A. and Hurshudov, V.A. (1993). Leak-off tests help determine wellbore compressibility. *Oil and Gas Journal*, Vol. 91, 71–73.

He, Z., Zhu, X., Wang, J., Mu, M. and Wang, Y. (2019). Comparison of CO_2 emissions from OPC and recycled cement production. *Construction and Building Materials*, Vol. 211, 965–973.

Hemphill, T. and Larsen, T.I. (1993). Hole-cleaning capabilities of oil-based and water-based drilling fluids: A comparative experimental study. Paper SPE 26328 presented at the 68th Annual Tecn. Conf. and Exhibition, Houston, Texas, October 3–6, 1993.

Hempkins, W.B., Kingsborough, R.H., Lohec, W.E. and Nini, C.J. (1987). Multivariate statistical analysis of stuck drillpipe situations. *SPE Drilling Engineering*, Vol. 2, No. 3, 237–244.

Herschel, W.H. and Bulkley, R. (1926). Measurement of consistency as applied to Rubber-Benzene solutions. *Proceedings of the ASTM*, Vol. 26, 621.

Hodgson, R. and Hassard, P. (2006). Reducing the learning curve through the use of an advanced drilling simulator. IADC/SPE 98107 Paper presented at the IADC/SPE Drilling Conference, Miami, Florida, USA, February 2006.

Holmquist, J.L. and Nadia, A. (1939). A theoretical and experimental approach to the problems of collapse of deep-well casing. Drilling and Production Practices, API, Dallas, 932.

Horsrud, P. (2001). Estimating mechanical properties of Shale from empirical correlations, SPE Drilling & Completion, June 2001.

Ibrahim, Z.Z., Hamidi, H., Afzal, W. and Huseyin, M. (2024). Investigating the effects of nano-Fe_3O_4 and MWCNTs on the filtration and rheological properties of water-based muds at elevated temperature and pressure. *Colloids and Surfaces A: Physicochemical and Engineering Aspects*, Vol. 701, 134912.

IFP. (1991). *Drilling Data Handbook*. Institut Francais du Petrole. ISBN 0-87201-206-9.

Ikeda, S. (1992). Advanced casing design against stress corrosion cracking of grade P-110 casing in lignosulfonate mud. *SPE Drilling Engineering*, Vol. 7, No. 2, 141–147.

Iskandar, F.F., Abiddin, M.S.Z., Nazzeri, N., Aziz, A.A. and Atemin, A. (2018). Integrated Real-Time Operation Centre: A Complete Solution towards Effective & Efficient Drilling Operation. OTC-28598-MS Paper presented at the Offshore Technology Conference Asia, Kuala Lumpur, Malaysia, March 2018.

Israel, R., McCrae, D., Sperry, N., Gorham, B., Thompson, J., Raese, K., Pink, S. and Coit, A. (2018). Delivering Drilling Automation II – Novel automation platform and wired drill pipe deployed on Arctic drilling operations. SPE-191574-MS Paper presented at the SPE Annual Technical Conference and Exhibition, Dallas, Texas, USA.

Jafariesfad, N., Gong, Y., Geiker, M.R. and Skalle, P. (2016). Nano-sized MgO with engineered expansive property for oil well cement systems. SPE-180038-MS in SPE Bergen One Day Seminar. Norway, April 20, 2016.

Johnson, R., Jellison, M.J. and Klementich, E.F. (1987). Triaxial-load-capacity diagrams provide a new approach to casing and tubing design analysis. *SPE Drilling Engineering*, Vol. 2, No. 3, 268–274.

Jonathan, F.N., Jørpeland, J-T., Schwartze, C. (2018). Case history: A new approach to section milling: Leaving the Swarf Behind! OTC-28757-MS paper presented at the Offshore Technology Conference held in Houston, Texas, USA, 30 April–3 May 2018.

Jøntvedt, E., Fjeldheim, M, Løchen, J, Howard, S., Leon, S., Busengdal, C. and Gyland, K.R. (2018). Deployment of Cesium Formate Drill-In and openhole completion fluid in the martin linge high pressure, high permeability gas reservoir enhances total's operational efficiency and radically improves well performance. SPE-189550-MS. International Conference and Exhibition on Formation Damage Control held in Lafayette, Louisiana, USA, February 7–9, 2018.

Kaarstad, E. and Aadnoy, B.S. (2008). Improved prediction of shallow sediment fracturing for offshore applications. *SPE Drilling and Completion*, Vol. 23, 88–92.

Kaarstad, E., Aadnoy, B.S. and Fjelde, T. (2009). A study of temperature-dependent friction in wellbore fluids. SPE/IADC 119768 paper presented at the SPE/IADC Drilling Conference and Exhibition held in Amsterdam, The Netherlands, March 17–19, 2009.

Kastor, R.L. (1986). Triaxial casing design for Burst. Paper IADC/SPE 14727 presented at the 1986 IADC/SPE Drilling Conference, Dallas, Texas, February 10–12, 1986.

Khosravanian, R. and Aadnoy, B.S. (2021). *Methods for Petroleum Well Optimization: Automation and Data Solutions*. Gulf Professional Publishing. Cambridge, Massachusetts ISBN: 978032390231.

Kirsch, G. (1898). "Die Theorie der Elastizitat und die Bedurfnisse der Festigkeitslehre," *Zantralblatt Verlin Deutscher Ingenieure*, Vol. 42, 797–807.

Knudsen, K., Leon, G.A., Sanabria, A.E., Ansari, A. and Pino, R.M. (2015). First application of thermal activated Resin as unconventional LCM in the Middle East. SPE-177430-MS paper presented at the Abu Dhabi International Petroleum Exhibition and Conference held in Abu Dhabi, UAE, November 9–12, 2015.

Kristiansen, T., Dyngeland, T., Kinn, S., Flatebø, R. and Aarseth, N.A. (2018). Activating Shale to form well barriers: Theory and field examples. SPE-191607-MS paper presented at the 2018 SPE Annual Technical Conference, Dallas, Texas, September 24–26, 2018.

Krus, H. and Prieur, J.M. (1991). High-pressure well design. *SPE Drilling Engineering,* Vol. 6, No. 4, 240–244.

Kyllingstad, A., Horpestad, J.L., Kristiansen, A. and Aadnoy, B.S. (1993). Factors limiting the quantitative use of mud-logging data. Paper SPE 25319 presented at the SPE Asia Pacific Oil and Gas Conference, Singapore, February 8–10, 1993.

Lafarge, E.G. (2004). Industrially interesting approaches to "low-CO_2" cements. *Cement and Concrete Research*, Vol. 34, 1489–1498.

Larsen, A.G. (2016). HydraWell intervention. Meeting 19th February 2016. Handout. HydraWell Intervention AS. HydraSystem Technologies.

Lermo, L.K. (1993). Hydraulic optimisation in deviated and horizontal wells. M.S. thesis in petroleum engineering, Rogaland University Centre. (In Norwegian).

Li, H., Xiao, H-g. and Ou, J-p. (2004). A study on mechanical and pressure-sensitive properties of cement mortar with nanophase materials. *Cement and Concrete Research*, Vol. 34, No. 3, 435–438.

Li, Z., Wang, H., He, S., Lu, Y. and Wang, M. (2006). Investigations on the preparation and mechanical properties of the nano-alumina reinforced cement composite. *Materials Letters*, Vol. 60, No. 3, 356–359.

Liu, X., Chen, L., Liu, A. and Wang, X. (2012). Effect of Nano-$CaCO_3$ on properties of cement paste. *International Conference on Future Energy, Environment, and Materials. Energy Procedia*, Vol. 16, 991–996.

Lockett, T.J., Richardson, S.M. and Worraker, W.J. (1993). The importance of rotation effects for efficient cuttings removal during drilling. Paper SPE/IADC 25768, presented at the 1993 SPE/IADC Drilling Conference, Amsterdam, February 23–26, 1993.

Lohne, H.P., Gravdal, J.E., Dvergsnes, E.W., Nygaard, G. and Vefring, E.H. (2008). Automatic calibration of real-time computer models in intelligent drilling control systems - results from a North Sea Field Trial. IPTC 12707, Proceddings, International Petroleum Technology Conference, Kuala Lumpur, Malaysia, December 3–5.

Luo,Y., Bern, P.A. and Chambers, B.D. (1992). Flow-rate predictions for cleaning deviated wells. Paper IADC/SPE 23884 presented at the 1992 IADC/SPE Drilling Conference, New Orleans, February 18–21.

Luo,Y., Bern, P.A., Chambers, B.D. and Kellingray, D.S. (1994). Simple charts to determine hole cleaning requirements in deviated wells. Paper IADC/SPE 27486 presented at the 1994 IADC/SPE Drilling Conference, Dallas, February 15–18.

Marken, C.D. and Saasen, A. (1992). The influence of drilling conditions on annular pressure losses. Paper SPE 24598 presented at the 67th Annual Techn. Conf. and Exhibition, Washington, DC, October 4–7, 1992.

Marshall, D.W., Asahi, H. and Ueno, M. (1994). Revised casing-design criteria for exploration wells containing H_2S. *SPE Drilling & Completion*, Vol. 9, No. 2, 115–123.

Maury, V. (1993). An overview of tunnel, underground excavations and borehole collapse mechanisms. Edited by J. A. Hudson *Comprehensive Rock Engineering*. Pergamon Press, Vol, 4, 369–411.

Maury, V. and Guenot, A. (1993). Practical advantages of mud cooling systems for drilling. Paper SPE/IADC 25732 presented at the 1993 SPE/IADC Drilling Conference, Amsterdam, February 23–25, 1993.

Maxey, J. (2007). Rheological analysis of static and dynamic Sag in drilling fluids. *Annual Transactions of the Nordic Rheology Society*, Vol. 15, 2007.

McCann, R.C, Quigley, M.S., Zamora, M. and Slater, K.S. (1993). Effects of high-speed pipe rotation on pressures in narrow annuli. Paper SPE 26343 presented at the 68th Annual Tecn. Conf. and Exhibition, Houston, Texas, October 3–6, 1993.

McLean, M.R. and Addis, M.A. (1990). Wellbore stability analysis: A review of current methods of analysis and their field application. Paper IADC/SPE 19941 presented at the 1990 IADC/SPE Drilling Conference, Houston, Texad, February 27-March 2, 1990.

Meek, J.W. and Harris, K. (1993). Repairing casing leaks with small-particle-size cement. *SPE 21972-PA Production & Facilities*, Vol. 8, No. 1, 45–50.

Mezger, T.G. (2011). *The Rheology Handbook - For Users of Rotational and Oscillatory Rheometers*. Vincentz Network, Hanover, Germany, 3rd revised edition. ISBN 978-3-86630-890-9.

Monrose, H. and Boyer, S. (1992). Casing corrosion: Origin and detection. *The Log Analyst*, 507–519.

Morita, N., Whitfill, D.L., Nygaard, O. and Bale, A. (1988). A quick method to determine subsidence, reservoir compaction, and in-situ stress induced by reservoir depletion. *Journal of Petroleum Technology*, Vol. 41, No. 1, 71–79.

Murtaza, M., Rahman, M.K. and Al-Majed, A.A. (2016). Mechanical and microstructural studies of nanoclay-based oil well cement mix under high pressure and temperature application. IPTC-18991-MS Paper presented at the International Petroleum Technology Conference, Bangkok, Thailand, November 2016.

Murtaza, M., Rana, A., Ahmad, H.M., Mahmoud, M., Patil, S., Kamal, M.S. and Al Sheri, D. (2023). Performance evaluation of Iron Oxide and Graphite Nanoparticles in water-based drilling muds at HPHT conditions. SPE-213963-MS Paper presented at the Middle East Oil, Gas and Geosciences Show, Manama, Bahrain, February 2023.

Nassar, N.N. and Husein, M.M. (2007). Study and modeling of Iron Hydroxide nanoparticle uptake by AOT (w/o) microemulsions. *Langmuir*, Vol. 23, 13093–13103.

NORSOK D-010. (2013). *Well Integrity in Drilling and Well Operations* (Rev. 4, June 2013).

NPD. (1991). Acts, regulations and provisions for the petroleum activity. The Norwegian Petroleum Directorate, December 1991.

Nwaoji, C.O. Hareland, G., Husein, M., Nygaard, R. and Zakaria, M.F. (2013). Wellbore strengthening-nano-particle drilling fluid experimental design using hydraulic fracture apparatus. SPE/IADC 163434 paper presented at the SPE/IADC Drilling Conference and Exhibition held in Amsterdam, The Netherlands, March 5–7, 2013.

O'Brien, D.E. and Chenevert, M.E. (1973). Stabilising sensitive shales with inhibited, potassium based drilling fluids. *Journal of Petroleum Technology*, Vol. 25, 1089–1100.

Oil & Gas UK. (2015). *Guidelines for the Abandonment of Wells*, Issue 5. London: Oil and Gas UK. ISBN 1 903 004 50 0.

Oil & Gas UK. (2015). *Guidelines on Qualification of Materials for the Abandonment of Wells*, Issue 2. London: Oil and Gas UK.

Oudeman, P. and Bacarreza, L.J. (1995). Field trial results of annular pressure behavior in a high-pressure/high-temperature well. *SPE Drilling & Completion,* Vol. 10, No. 2, 84–88.

Øy, L., Bye, I., Lauritzen, M., Herbert, M., Grunwald, K., Rød, A.M. and Kane, D. (2023). 4D radar imaging of wellbore geometry while drilling. SPE-212499-MS Paper presented at the SPE/IADC International Drilling Conference and Exhibition, Stavanger, Norway, March 7–9, 2023.

Parrott, A. and Warshaw, L. (2017). *Industry 4.0 and the Digital Twin: Manufacturing Meets its Match.* Deloitte University Press. www2.deloitte.com/us/en/insights/focus/industry-4-0/digital-twin-technology-smart-factory.html

Pastusek, P., Barajas, P., Hinojosa, A., Clayton, P., Whatley, M., Soni, S., Manocha, M., Saxena, S., Dixit, R. and Payette, G. (2022). Auto driller improvements and open-source modeling reduce drilling time and trips. SPE-210312-MS Paper presented at the SPE Annual Technical Conference and Exhibition, Houston, Texas, USA.

Pate'-Cornell, M.E. (1993). Learning from the Piper Alpha Accident: A postmortem analysis of technical and organizational factors. *Risk Analysis*, Vol. 13, No. 2, 215–232.

Patil, R.C. and Deshpande, A. (2012). Use of nanomaterials in cementing applications. SPE-155607-MS paper presented at the SPE International Oilfield Nanotechnology Conference and Exhibition, Noordwijk, The Netherlands, June 2012.

Payne, M.L. and Swanson, J.D. (1990). Application of probabilistic reliability methods to tubular design. *SPE Drilling Engineering*, Vol. 5, No. 4, 299–305.

Payrazyan, V.K. and Robinson, T.S. (2023). Leveraging targeted machine learning for early warning and prevention of stuck pipe, tight holes, pack offs, hole cleaning issues, and other potential drilling hazards. OTC-32169-MS Paper presented at the Offshore Technology Conference, Houston, Texas, USA, May 1–4, 2023.

Rabia, H. (1985). *Oilwell Drilling Engineering*. ISBN 0-86010-661-6. Graham & Trotman Ltd, London.

Rahimirad, M. and Baghbadorani, J.D. (2012). Properties of oil well cement reinforced by Carbon Nanotubes. SPE-156985-MS Paper presented at the SPE International Oilfield Nanotechnology Conference and Exhibition, June 12–14, 2012.

Rios, R. and Ars, F. (2021). Plug and abandonment materials – technology landscape. OTC-30938-MS paper presented at the Offshore Technology Conference held in Houston, Texas, USA, August 16–19, 2021, Offshore Technology.

Robertson, R.E. and Stiff, H.A., Jr. (1976). An improved mathematical model for relating shear stress to shear rate in drilling fluids and cement slurries. *Society of Petroleum Engineers Journal* Vol. *16*, 31–36.

Roegiers, J.C. and Detournay, E. (1988). Considerations on failure initiation in inclined boreholes. Paper presented at the 29th U.S. Symposium on Rock Mechanics (USRMS), Minneapolis, Minnesota, June 1988. Paper Number: ARMA-88-0461.

Roji, R., Egyed, C. and Lips, J.P. (2012). Nano-engineered oil well cement improves flexibility and increases compressive strength: A laboratory study. SPE-156501-MS Paper presented at the SPE

International Oilfield Nanotechnology Conference and Exhibition, Noordwijk, The Netherlands, June 2012.

Rygg, V., Grannes, I. and Svindland, A. (2010). Permanent abandonment of a North Sea Well using unconsolidated well plugging material. SPE-133446-MS paper presented at the SPE Deepwater Drilling and Completions Conference, Galveston, Texas, USA, October 2010.

Saasen, A., Fjelde, K.K., Vrålstad, T., Raksagati, S. and Moeinikia, F. (2013). Plug and abandonment of offshore exploration wells. I: Offshore Technology Conference. Society of Petroleum Engineers. ISBN 978-1-61399-241-8.

Saasen, A., Wold, S., Ribesen, B.T., Tran, T.N., Huse, A. Vidar, R., Grannes, I. and Svindland, A. (2010). Permanent abandonment of a North Sea Well using unconsolidated well plugging material. SPE-133446-MS paper presented at the SPE Deepwater Drilling and Completions Conference, Galveston, Texas, USA, October 2010.

Sadeghalvaad, M. and Sabbaghi, S. (2015). The effect of the TiO_2/polyacrylamide nanocomposite on water-based drilling fluid properties. *Powder Technology*, Vol. 272, 113–119.

Saleh, M.M., Wu, A. and Fu, Q. (2019). A review of mud pulse telemetry signal impairments, modeling and suppression methods. *Journal of Petroleum Exploration and Production Technology*. Vol. 9, 779–792.

Samuel, R. (2023). The traditional drilling optimization conceptually deadlocked for realtime well engineering: An alternate solution. SPE-214848-MS Paper presented at the SPE Annual Technical Conference and Exhibition, San Antonio, Texas, USA, October, 2023.

Santarelli, F.J. and Carminati, S. (1995). Do shales swell? A critical review of available evidence. Paper SPE/IADC 29421 presented at the 1995 SPE/IADC Drilling Conference, Amsterdam, February 28–March 2, 741–756.

Santarelli, F.J. and Dardeau, C. (1992). Drilling through highly fractured formations. Paper SPE 24592 presented at the 67th Ann. Tech. Conf. and Exhibition, Washington DC, October 4–7.

Santoro, D., Spelta, A., Ferrara, P., Bianchini, L. P., Bartucci, G., Zarra, A. F., Pregliasco, M., Fontana, A., Pinto Vieira, L., Lahou, K., Nabavi, J., Huseynov, P., Gocmen, E. B., Lye, J. and Loh, J. (2023). Well simulation and real time operations engineering overview and results for onshore, offshore and Subsea drilling operations. SPE-215594-MS SPE Offshore Europe Conference & Exhibition, Aberdeen, Scotland, UK.

Schils, S., Teelken, R., van Burkleo, B., Rossa, O.J. and Edwards, N. (2016). The use of wired drillpipe technology in a complex drilling environment increased drilling efficiency and reduced well times. SPE-178863-MS Paper presented at the IADC/SPE Drilling Conference and Exhibition, Fort Worth, Texas, USA.

Schoenmakers, J.M. (1987). Casing wear during drilling: Simulation, prediction, and control. *SPE Drilling Engineering*, Vol. 2, No. 4, 375–381.

Scott, P.D. and Zamora, M. (2004). Barite-Sag management: Challenges, strategies, opportunities. SPE-87136-MS IADC/SPE Drilling Conference, Dallas, Texas, March 2–4, 2004.

Sehsah, O., Ghazzawi, A., Vie, G.J., Al-Tajar, T., Ali, A., Al-Mohammed, A.H., Itani, M., Ullah, S., Escalera, H., & Balka, M. (2017). Intelligent drilling system: Expanding the envelope of wired drill pipe. SPE-188321-MS Paper presented at the Abu Dhabi International Petroleum Exhibition & Conference, Abu Dhabi, UAE.

Seymour, K.P. and MacAndrew, R. (1994). Design, drilling and testing of a deviated HPHT exploration well in the North Sea. *SPE Drilling Engineering*, Vol. 9, No. 4, 244–248.

Sharma, M.M, Zhang, R., Chenevert, M.E., Ji, L., Guo, Q. and Friedheim, J. (2012). A new family of nanoparticle-based drilling fluids. SPE-160045-MS paper presented at the SPE Annual Technical Conference and Exhibition, San Antonio, Texas, USA, October 2012.

Shenghua, L., Yujuan, M., Chaochao, Q., Sun, T., Liu, J. and Zhou, Q. (2013). Effect of graphene oxide nanosheets of microstructure and mechanical properties of cement composites. *Construction and Building Materials*, Vol. 49, 121–127.

Sheppard, M.C., Wick, C. and Burgess, T. (1987). Designing well path to reduce drag and torque. *SPE Drilling Engineering*, Vol. 2, No. 4, 344–350.

Shuaibi, F., Harthi, M., Large, S., Obilaja, J.-F., Senani, M., Moreno Gomez, C., Mahrazy, K., Hussain, M., Al Busaidi, M., Savels, T. and Dolle, N. (2021). Leveraging Game AI to transform integrated Brownfield well planning. SPE-207947-MS Paper presented at the Abu Dhabi International Petroleum Exhibition & Conference, Abu Dhabi, UAE, November 2021.

Sifferman, T.R. and Becker, T.E. (1992). Hole cleaning in full-scale inclined wellbores. *SPE Drilling Engineering*, Vol. 7, No. 2, 115–120.

Silva Junior, M.F., Muradov, K.M. and Davies, D.R. (2012). Review, analysis, and comparison of intelligent well monitoring systems. SPE 150195 paper prepared at the SPE Intelligent Energy International held in Utrecht, The Netherlands, March 27–29, 2012.

Simpson, J.P., Dearing, H.L. and Salisbury, D.P. (1989). Downhole simulation cell shows unexpected effects of shale hydration on borehole wall. *SPE Drilling Engineering*, Vol. 4, 24–30.

Solvi, L.J.R., Revheim, O., Schaefer, S. and Schutte, F.J. (2019). An electronic rig action plan - information carrier equally applicable to the driller and the automation platform. Society of Petroleum Engineers. SPE-195959-MS Presented at the SPE Annual Technical Conference and Exhibition, Calgary, Alberta, Canada, 30 September–2 October 2019.

Stair, M.A. and McInturff, T.L. (1986). Casing and tubing design considerations for deep sour gas wells. *SPE Drilling Engineering*, Vol. 1, No. 3, 221–232.

Steiger, R.P. (1982). Fundamentals and use of potassium/polymer drilling fluids to minimise drilling and completion problems associated with hydratable clays. *Journal of Petroleum Technology*, Vol. 34, 1661–1670.

Steinar, S. (2016). Statoil ASA. Meeting, 9th May 2016. Handout.

Svendsen, K.E., Kristiansen, T.G., Martin, J., Askø, A., Bjørlo, J., Khosravanian, R., Holt, C. and Ruel, F. (2025). Automated computer vision system for real-time detection of drilled cuttings and cavings. SPE-223785-MS Paper presented at the SPE/IADC International Drilling Conference and Exhibition, Stavanger, Norway, March 2025.

Taha, N.M. and Lee, S. (2015). Nano Graphene application improving drilling fluids performance. IPTC-18539-MS paper presented at the International Petroleum Technology Conference held in Doha, Qatar, December 6–9, 2015.

Teelken, R., Jaring, C., Craig, A., Allan, R., Roberts, H., Hatch, A. and Shepherd, N. (2016). Implementation of wired drill pipe saved multiple days per well, by addressing performance limiters, increasing drilling efficiency. SPE-178798-MS Paper presented at the IADC/SPE Drilling Conference and Exhibition, Fort Worth, Texas, USA.

Thorstensen, E., Vadset, K., Straume, M.K., Delabroy, L., Abdelal, G., Carragher, P., Fulks, J. and Mason, D.L. (2022). Bismuth plugs used to cap all wells during the Final Phase of the Valhall DP abandonment Campaign, Offshore Norway. SPE 31897-MS paper presented at the Offshore Technology Conference, April 25, 2022.

Torbergsen, H-E.B., Haga, H.B., Sangesland, S., Aadnøy, B.S., Sæby, J., Johnsen, S., Rausand, M. and Lundeteigen, M-A. (2012). An introduction to well integrity. Rev. 0, 04 December 2012.

van Oort, E., Juenger, M., Aldin, M., Thombare, A., McDonald, M., Lucas, A. and Ditlevsen, F. (2022). Simplifying well abandonments using Shale as a barrier. *SPE Drillinf & Completion,* Vol. 37, No. 04, 282–294. Paper Number: SPE-199654-PA.

Vignes, B. and Aadnoy, B.S. (2008). Well-integrity issues offshore Norway. Paper IADC/SPE 112535 presented at the 2008 SPE/IADC Drilling Conference, Orlando, Florida, March 4–6, 2008.

Vignes, B. and Aadnoy, B.S. (2010). Well-integrity issues offshore Norway. *SPE Production & Operations*, Vol. 25, No. 02, 145–150. SPE-112535-PA.

Vryzas, Z., Mahmoud, O., Nasr-El-Din, H.A. and Kelessidis, V.C. (2015). Development and testing of novel drilling fluids using Fe2O3 and SiO2 nanoparticles for enhanced drilling operations. IPTC-18381-MS Paper presented at the International Petroleum Technology Conference, Doha, Qatar, December 2015.

White, J.P. and Dawson, R. (1987). Casing wear: Laboratory measurements and field predictions. *SPE Drilling Engineering*, Vol. 2, No. 1, 56–62.

William, J.K.M., Ponmani, S., Samuel, R.R., Jitendra, N. and Sangwai, S. (2014). Effect of CuO and ZnO nanofluids in xanthan gum on thermal, electrical, and high-pressure rheology of water-based drilling fluids. *Journal of Petroleum Science and Engineering*, Vol. 117, 15–27.

Williams, S., Carlsen, T., Constable, K. and Guldahl, A. (2009). Identification and qualification of shale annular barriers using wireline logs during plug and abandonment operations. SPE/IADC 119321 paper presented at the SPE/IADC Drilling Conference and Exhibition held in Amsterdam, The Netherlands, March 17–19, 2009.

Zakaria, M.F., Husein, M. and Hareland, G. (2012). Novel nanoparticle-based drilling fluid with improved characteristics. SPE 156992 paper presented at the SPE International Oilfield Nanotechnology Conference held in Noordwijk, The Netherlands, June 12–14, 2012.

Zamora, M. and Bleier, R. (1977). Prediction of drilling mud Rheology using a simplified Herschel-Bulkley model. *Journal of Pressure Vessel Technology*, 485–490.

Zamora, M. and Power, D. (2002). Making a Case for AADE Hydraulics and the Unified Rheological Model, AADE-02-DFWM-HO-13. Houston, TX, USA, 23, 2002.

Zhang, R., Cheng, X., Hou, P. and Ye, Z. (2015). Influences of nano-TiO_2 on the properties of cement-based materials: Hydration and drying shrinkage. *Construction and Building Materials*, Vol. 81, 35–41, ISSN 0950-0618.

Index